Mathematics in Industry

Volume 40

Mathematics in Industry focuses on the research and educational aspects of mathematics used in industry and other business enterprises. Books for *Mathematics in Industry* are in the following categories: research monographs, problem-oriented multi-author collections, textbooks with a problem-oriented approach, and conference proceedings. All manuscripts undergo rigorous peer review before acceptance. Relevance to the actual practical use of mathematics in the industry is the distinguishing feature of the books in the *Mathematics in Industry* series.

Jin Keun Seo
Editor

Deep Learning and Medical Applications

Springer

Editor
Jin Keun Seo
School of Mathematics and Computing
(Computational Science and Engineering)
Yonsei University
Seoul, Korea (Republic of)

ISSN 1612-3956 ISSN 2198-3283 (electronic)
Mathematics in Industry
ISBN 978-981-99-1838-6 ISBN 978-981-99-1839-3 (eBook)
https://doi.org/10.1007/978-981-99-1839-3

Mathematics Subject Classification: 00A06, 00A69, 00B10, 15A29, 35R30, 49N45, 65F22, 68U10, 68T05, 78A70

This Springer imprint is published by the registered company Springer Nature Singapore Pte Ltd.
The registered company address is: 152 Beach Road, #21-01/04 Gateway East, Singapore 189721, Singapore

This book is dedicated to the great teacher, Eugene Fabes (1937–1997), who lived for others rather than himself.

Preface I

This book deals with AI-based medical image analysis, which is experiencing a paradigm shift due to a marked and rapid advance in deep learning techniques. Recently, there have been developed numerous innovative AI-based programs that are helping doctors dramatically reduce fatigue and improve clinical workflows. These AI technologies will continue to evolve to improve the satisfaction of both doctors and patients by reducing the tedious work of doctors and the time spent examining patients. This book is written for both professionals and graduate students who want to develop AI-based medical image analysis programs with practical meaning and value. This book is very practical and provides deep learning techniques and algorithm implementation details, as well as in-depth knowledge of the clinical environment, data acquisition systems, and computational mathematics.

Seoul, Korea (Republic of)
February 2023

Jin Keun Seo

Preface II

Artificial Intelligence (AI)-based medical image analysis is experiencing a paradigm shift due to a marked and rapid advance in Deep Learning (DL) techniques. In recent years, numerous innovative DL-based programs have been developed to dramatically reduce the work of doctors and improve clinical workflows. Such programs are superior in their ability to utilize prior anatomical knowledge compared to conventional regularized data fitting methods. They can properly configure the network architecture and training data to reflect the clinician's measurement procedures. A major current limitation of DL models is the difficulty in rigorously interpreting how DL models make decisions. Despite the lack of rigorous analysis in DL, recent rapid advances indicate that DL methodologies will see continued improvements in performance as training data and experience accumulate over time.

DL is a subset of Machine Learning (ML), which Arthur Lee Samuel (1901–1990) defined as "a field of study that gives computers the ability to learn without being explicitly programmed." The core of ML is representation learning that allows a machine to learn representation of features from a data set to perform a special task. DL in medical image analysis is used for learning about multiple levels of feature abstraction and representation of inputs (e.g., X-ray, computerized tomography, magnetic resonance imaging, and ultrasound images) to generate the requisite output (e.g., segmentation, detection, classification, and image enhancement).

In March 2016, ML took a big leap forward when AlphaGo, an ML program for the game Go made by DeepMind, defeated the world's best player, Lee Sedol, in Korea. This historic match has attracted significant attention in both the scientific and popular press because of the tough computational challenges associated with playing Go proficiently. Go has a huge number of cases, which makes its complexity vastly greater than that of chess; therefore, it was regarded to be almost impossible to handle by explicit mathematical means. However, AlphaGo seemingly handled this huge complexity without explicit programming. This historical success led to opportunities for DL's applications in the medical field.

Understanding the opportunities and limitations of DL is important for setting future research directions. It should be noted that the current level of DL has limitations in applying it in a complex medical environment where a lot of information is mixed. For the time being, it seems to be difficult for DL to reach a level that can replace medical doctors. Just as a human cannot beat a puma in a 100m race, deep learning in the medical field has limitations that are difficult to overcome. In 2017, Jupiter Medical Center (the first USA community hospital) announced the employment of cognitive computing platform IBM Watson for oncology to provide insights into physicians to help them deliver personalized, evidence-based cancer treatment. However, in February 2021, the *Wall Street Journal* reported that IBM was trying to sell Watson Health because it was unprofitable. At the current level of AI, it is difficult for AI to make decisions with subtle boundaries, taking into account individual patient circumstances and complex medical environments. Nirav R. Shah at Stanford University's Clinical Excellence Research Center explained well why it is difficult for AI to replace physicians. One of the great things about a physician is not only knowing the right thing to do in a complex set of circumstances but also the ability to act on that knowledge and to do what's right. Nirav mentioned: "*With regards to cancer, we are talking about a constellation of thousands of diseases, even if the focus is on one type of cancer. For example, what we call 'breast cancer' can be caused by many different underlying genetic mutations and should not really be lumped together under one heading. AI can work well when there is uniformity and large data sets around a simple correlation or association. By having many data points around a single question, neural networks can learn. With cancer, we are breaking several of these principles.*" *(Forbes Magazine, Feb 27 2021)* For DL to succeed in the medical field, it must be acknowledged that the DL approach has its limitations. Unfortunately, many recent papers seem to have no economic value, are too exaggerated in their claims, or too idealistic. DL is just an image feature extraction calculator, not magic. Developing DL models in the medical field requires an accurate understanding of the medical environment, medical devices and practical value, and acknowledging the limitations of where AI can and does not work well. The adoption of medical AI in conservative healthcare is not easy until we provide interpretable AI solutions.

Despite the limitations of current DL, DL methods have many potential applications in the medical field and have already achieved many successes. DL tools can support complex clinician workflows and automate routine tasks that are unnecessary, time-consuming, and tiring for clinicians, just as convenience stores are using intelligent self-checkout counters for faster checkout, reduced labor costs, and understanding customer behavior.

DL techniques appear to overcome the limitations of existing mathematical methods in handling various ill-posed problems. In the field of medical imaging, highly underdetermined problems arise from a desire to reduce the sampling requirement for medical image reconstruction as much as possible, while maintaining high resolution. Undersampled MRI, sparse view CT and interior tomography are typical examples of underdetermined problems (violating Nyquist sampling criteria), where the goal is to minimize the ratio of the number of equations (measured data) to the number of unknowns (pixels of the image). In the classical sense of Hadamard, a highly underdetermined problem (far fewer equations than unknowns) is severely ill-posed. Traditionally, regularization techniques have been widely used to impose specific prior distributions on the expected images. These include using a Banach space norm to enforce sparsity in expressions. However, in medical imaging, norm-based regularization might not be able to selectively preserve clinically useful features that are small in terms of regularization norm. DL seems to have strong capabilities to explore the prior information of the expected solutions through training data.

This book tries to strike a balance between theory, numerical practice, and clinical application, and ultimately aims to apply this to the healthcare industries. The book is structured for practical application and aims to provide deep learning techniques and algorithm implementation details, as well as in-depth knowledge of the clinical environment, data acquisition systems, and computational mathematics. Deep learning methods have a strong capability to explore the prior information of the expected images via training data, which allows them to deal with the uncertainty of solutions to highly ill-posed. These mathematical issues are discussed in this book. This book covers:

- deep learning applications in various medical imaging modalities including ultrasound, X-ray, MRI, and CT;
- challenging problems in automated medical image analysis and reconstruction;
- supporting mathematical theories;
- in-depth explanations on emerging issues on dental cone-beam CT, fetal ultrasound, and bioimpedance.

This book was written mostly by my doctoral students, with the help of medical doctors, dentists, and medical company CEOs. This book was supported by Samsung Science & Technology Foundation (No. SRFC-IT1902-09).

Seoul, Korea (Republic of)
February 2023

Jin Keun Seo

Acknowledgements

First, the editor would like to thank all the authors contributing to this book. Second, all authors would like to thank the Samsung Science and Technology Foundation for its support (No. SRFC-IT1902-09). Many thanks to HDXWIIL for providing CBCT instrumentation and ground truth data. And, we would like to express our deepest gratitude to several people who have helped make this book practical rather than pedantic.

Contents

Contributors

Taigyntuya Bayaraa Department of Mathematics, School of Applied Sciences, Mongolian University of Science and Technology, Ulaanbaatar, Mongolia

Hyun Cheol Cho AI Vision Group, Samsung Medison, Seoul, Korea

Chang Min Hyun School of Mathematics and Computing (Computational Science and Engineering), Yonsei University, Seoul, Korea

Tae Jun Jang School of Mathematics and Computing (Computational Science and Engineering), Yonsei University, Seoul, Korea

Ariungerel Jargal Department of Mathematics, School of Applied Sciences, Mongolian University of Science and Technology, Ulaanbaatar, Mongolia

Hong Jung HDXWILL, Seoul, Korea

Kang Cheol Kim School of Mathematics and Computing (Computational Science and Engineering), Yonsei University, Seoul, Korea

Hyeuknam Kwon Division of Software, Yonsei University, Wonju, Korea

Ja-Young Kwon Department of Obstetrics and Gynecology, Institute of Women's Life Medical Science, Yonsei University College of Medicine, Seoul, Korea

Sang-Hwy Lee Department of Oral and Maxillofacial Surgery, Oral Science Research Center, College of Dentistry, Yonsei University, Seoul, Korea

Sung Min Lee School of Mathematics and Computing (Computational Science and Engineering), Yonsei University, Seoul, Korea

Sung Wook Park AI Vision Group, Samsung Medison, Seoul, Korea

Jin Keun Seo School of Mathematics and Computing (Computational Science and Engineering), Yonsei University, Seoul, Korea

Siyu Sun School of Mathematics and Computing (Computational Science and Engineering), Yonsei University, Seoul, Korea

Hye Sun Yun School of Mathematics and Computing (Computational Science and Engineering), Yonsei University, Seoul, Korea

Acronyms

AE	"AutoEncoder" is an unsupervised learning technique for useful data representation in such a way that a feed-forward neural net is trained to reproduce its input at the output
AI	"Artificial Intelligence" is a set of algorithms that try to make machines have human-like cognitive abilities
CBCT	"Cone Beam Computerized Tomography" uses a cone-shaped X-ray beam to produce tomographic images of a human head
CNN	"Convolutional Neural Network" is a type of DL that is based on the shared weight architecture of the convolution kernels
CNS	"Central Nervous System" is the part of the nervous system consisting of the brain and spinal cord
DL	"Deep Learning" is a subset of ML that uses multiple layers to progressively extract higher level features from an input to draw human-like conclusions
DR	"Dimension Reduction" is the process of reducing the dimensionality of data to find a low-dimensional representation while maintaining the meaningful characteristics of the original data
EIT	"Electrical Impedance Tomography" aims to visualize the conductivity distribution within the body by applying electrical currents through an array of electrodes attached to its surface and measuring the resulting voltage
FCN	"Fully Convolutional Network" is a kind of CNN that is used for image segmentation
FOV	"Field Of View" is the area being scanned
GAN	"Generative Adversarial Network" consists of two NNs (generator and discriminator) that compete with each other to progressively improve their output towards an expected goal
IOS	"Intraoral Scanner" is a dental device that uses a light source to produce 3D images of a tooth surface and gingiva
KL-divergence	"Kullback–Leibler divergence" measures how much one probability distribution differs from another probability

ML	"Machine Learning" is a subfield of AI that provides systems the ability to automatically learn from data without being explicitly programmed
MRI	"Magnetic Resonance Imaging" is a medical imaging technique that uses the magnetic property of tissue to visualize structures in the body
NN	"Neural Network" is a type of ML with interconnected groups of nodes that mimic the simplification of neurons in the brain
PCA	"Principal Component Analysis" is a linear technique of DR that generates the principal coordinate axes in the order of importance in terms of the correlation between pixels for a particular set of images
R-CNN	"Region-based CNN" is a region-based object detection algorithm
RNN	"Recurrent Neural Network" is a type of NNs that contains loops, allowing information to be stored within the network
U-NET	"U-NET" is a U-shaped FCN that was developed for image segmentation
US	"Ultrasound" is a medical imaging technique that uses high-frequency sound waves and their echoes
VAE	"Variational AutoEncoder" is a special type of AE that encodes the input as a distribution over the latent space and the encoded latent variables are packed near normal distribution
YOLO	"YOLO (You Only Look Once)" is a real-time object detection algorithm

Chapter 1
Nonlinear Representation and Dimensionality Reduction

Hye Sun Yun, Ariungerel Jargal, Chang Min Hyun, and Jin Keun Seo

Abstract Digital medical images can be viewed as digital representations of real physical tissue properties that are convenient for handling, storing, transmitting, retrieving, and analyzing image information. In order to perform feature extraction/identification/classification from high-dimensional medical data, we need dimensionality reduction (DR), which can be achieved by identifying the statistical patterns in medical image data with highlighting their similarities and dissimilarities. Here, the dimension of medical images is the total number of pixels in the image. In DR, we try to find an useful representation of reduced dimensionality in high-dimensional data, which minimizes information loss by maximizing local data variance. Given medical image data, the key challenging issue is how can we efficiently extract a low-dimensional latent structure? Various dimension reduction techniques have been developed to process the high-dimensional data where the intrinsic dimensions are assumed to be much lower. In the very ideal case, data can be regressed linearly and DR can be performed by principal component analysis. This chapter explains the theories, principles, and practices of DR techniques.

1.1 Introduction

Historically, our scientists have always tried to find concise representations of observed scenes and various physical phenomena of interest. For example, the Fourier transform and wavelet transform are designed to represent signals and images

H. S. Yun · C. M. Hyun · J. K. Seo (✉)
School of Mathematics and Computing (Computational Science and Engineering), Yonsei University, Seoul, Korea
e-mail: seoj@yonsei.ac.kr

C. M. Hyun
e-mail: chammyhyun@yonsei.ac.kr

A. Jargal
Department of Mathematics, School of Applied Sciences, Mongolian University of Science and Technology, Ulaanbaatar, Mongolia

J. K. Seo (ed.), *Deep Learning and Medical Applications*, Mathematics in Industry 40,
https://doi.org/10.1007/978-981-99-1839-3_1

concisely using an orthonormal basis (avoiding redundant use). Partial differential equations have been used to concisely represent the key features of a variety of natural phenomena occurring in real-world applications from fluid flows to biology and medical imaging fields. For example, Maxwell's equations describe electromagnetic phenomena in a very concise way using mathematical tools such as vector fields (i.e. electric and magnetic fields), gradient, Curl, and divergence. Numerical computing tools such as the finite element method (FEM) are used to view the solutions of partial differential equations in an image form. Certainly, such concise expressions help to analyze physical phenomena and extract useful properties in detail.

Over the past few decades, the problem of low-dimensional representation has received considerable attention due to a growing need to effectively process and analyze high-dimensional data. The key here is to reduce the dimensionality of the data to a manifold with a much lower dimension than the original space without losing important features that can affect data analysis. Dimensionality reduction techniques are used in a variety of areas, including medical imaging, face recognition, data processing, pattern recognition, and others. Recently, deep expressions have played important roles in underdetermined problems [27–29, 61], facial recognition [26], anomaly detection [5], medical image classification [17], and others. In underdetermined problems (e.g., sparse-view CT, compressed sensing MRI), the low-dimensional representation problem involves significantly reducing the dimension of the solution space by using some form of prior information (or regularity) as a constraint of the solution space [29]. In the unconstrained face recognition example, the dimensionality reduction techniques such as autoencoders were used to learn pose-invariant representations and pose codes to disentangle unconstrained images into latent factors of variations. This chapter mainly focuses on low-dimensional representations arising in the field of medical imaging.

A medical image of 256×256 pixel size can be regarded as a point $\mathbf{x} = (x_1, \cdots, x_{256\times256})$ in the 256×256-dimensional Euclidean space $\mathbb{R}^{256\times256}$, where x_j (j-th axis coordinate) corresponds to the grayscale intensity at the j-th pixel (see Fig. 1.1). If the image $\mathbf{x}$ has 256×256 pixels, coded on 256 grayscale level, then $\mathbf{x}$ is regarded as a point in the discrete space $\{0, 1, \cdots, 255\}^{256\times256}$;

$$\mathbf{x} \in \{\underbrace{0, 1, \cdots, 255}_{\text{256 grayscale level}}\}^{256\times256}, \tag{1.1}$$

The number of all possible points in the discrete space $\{0, 1, \cdots, 255\}^{256\times256}$ is $256^{256\times256}$, which is much larger than the number of atoms in the universe. Imagine that $\mathscr{X}$ represents a distribution of all head tomographic images lying in the discrete space $\{0, 1, \cdots, 255\}^{256\times256}$. Even if tens of thousands of head tomographic images in $\mathscr{X}$ are collected every day for 100 years, they occupy a very small area in the entire space. Most of the discrete space $\{0, 1, \cdots, 255\}^{256\times256}$ is filled with images that look like noise. Hence, it may be possible to generate a low-dimensional representation $\{G(\mathbf{z}) : \mathbf{z} \in \{0, 1, \cdots, 255\}^k\}$ (with k being much smaller than 256×256) that is close to the distribution $\mathscr{X}$ in some metric. The key for the low-dimensional

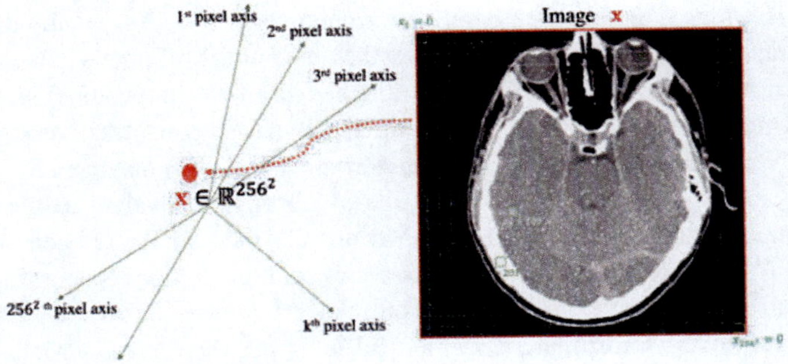

Fig. 1.1 A medical image can be regarded as a point $\mathbf{x} = (x_1, \cdots, x_{256^2})$ in pixel-dimensional Euclidean space, where x_j corresponds to the grayscale intensity at the j-th pixel

representation (or manifold learning) is to extract useful dependencies between pixels (or coordinates) from a set of sample images in $\mathscr{X}$.

Image perception exploits the spatial relationship between pixels by capturing local and global interconnections between pixels. To analyze the image $\mathbf{x}$, we need to encode the image in the order from global features to local features (or express $\mathbf{x}$ in a simple way). This is because the image $\mathbf{x} = (x_1, \cdots, x_{256\times256})$ expressed in the Cartesian coordinate system itself is difficult to analyze and quantitatively evaluate image characteristics without special training (e.g., training for radiologist). Feature encoding for medical image analysis aims to reduce the dimensionality of the input space by removing redundant, unrelated, noise, and artifact parts of the input $\mathbf{x}$.

For ease of explanation, we consider a set of 2D computerized tomography (CT) head images, $\mathscr{T}_{images} = \{\mathbf{x}^{(n)}\}_{n=1}^{N_{\text{data}}}$, with a total number of pixels of 256×256. Until the advent of deep learning (DL), linear techniques were mostly used to capture the features of CT images in $\mathscr{T}_{images}$. Principal component analysis (PCA) is the most widely used method as a data-driven DR, which is designed to reduce the number of random variables under consideration in a way that projected data onto a linear subspace spanned by the set of principal components maximizes the variance [31]. PCA generates the principal coordinate axes in the order of importance in terms of the correlation between pixels for a particular set of images. PCA can separate pairwise linear dependencies between pixels and the generated axes are orthogonal to each other. Linear DR methods (including PCA, truncated Fourier transform, and wavelet transform) have poor capability in low-dimensional representation because the pixel representation $\mathbf{x} = (x_1, \cdots, x_{256\times256})$ is sensitive to rotation and translation. Linear DR methods are inefficient to extract the unique characteristics of data components accurately while ignoring meaningless factors of the data.

Compressed sensing (CS) can be viewed as piecewise-linear DR. CS is based on the assumption that $\mathbf{x}$ has sparse representation under a basis $\{\mathbf{d}_j\}_{j=1}^{J}$, i.e.,

$$\mathbf{x} \approx \mathbf{D}\mathbf{h} \quad \text{subject to the constraint} \quad \|\mathbf{h}\|_{\ell_0} \le k \quad (k \ll 256 \times 256), \tag{1.2}$$

where $\mathbf{D}$ is a matrix whose j-th column corresponds to $\mathbf{d}_j$, $\|\mathbf{h}\|_{\ell_0}$ is the number of non-zero entries of $\mathbf{h}$, and k is much smaller than the number of pixels 256×256. For example, if the constraint on $\|\mathbf{h}\|_{\ell_0}$ is $k = 2$, x lies on a two-dimensional plane. Since the plane mentioned just before varies with $\mathbf{x}$, CS can be considered as semi-linear DR. The NP-hard problem of finding non-zero positions of $\mathbf{h}$ has been bypassed by the somewhat rigorous observation that the ℓ^1 norm is equivalent to the ℓ^0 norm under certain conditions [6, 7, 12, 13]. Various CS methods have been developed involving ℓ^1-norm regularization minimization, such as dictionary learning, sparse representation using wavelets, framelets, etc. These CS methods have been powerful for noise removal. Unfortunately, in medical imaging, there are a variety of small features, such that the difference in the data fidelity is very small compared with that in normalization, whether or not those small features are present. Hence, finding a more sophisticated normalization to keep small features remains a challenging problem. These CS approaches might not be fit for the tasks for selectively capturing small features while incorporating the global information effectively.

Historically, DR has been an important part of deep learning (DL) applications [25, 33]. DL methods are based on the assumption that head CT data are on or near a low-dimensional manifold embedded in high-dimensional ambient space. For efficient feature encoding, convolution neural networks (CNNs) encode multi-viewpoint data by convolving the input with filters (weights to be learned). The reason that DL is possible is that all the images in $\mathscr{T}_{images}$ possess a similar anatomical structure that consists of skull, gray matter, white matter, cerebellum, among others. In addition, every skull and tissue in the image has distinct features that can be represented nonlinearly by a relatively small number of latent variables, and so does the entire image. Moreover, the skull and tissues of the image are spatially interconnected, and even if a part of the image is missing, the missing part can be recovered with the help of the surrounding image information. Developing low-dimensional representations of input data (i.e., head CT images) can improve a DL network's generalization capabilities by enhancing out-of-distribution robustness. To achieve DL effectively, data normalization and standardization may be necessary (as a preprocessing step) to reduce diversity in images caused by variation among CT scanners or imaging protocols. We should note that DL models may produce incorrect results even if the input deviates even slightly from the training data manifold. The risk of adversarial attacks should be reduced to ensure the reliability of DL methods.

DL can be viewed as a nonlinear model approach where training data $\mathscr{T}_{images}$ is used for probing a lower-dimensional manifold. In the area of face recognition, face images can be expressed sparsely by the representative samples on a low dimensional manifold that is locally linear. Autoencoder(AE) techniques (as the natural evolution of PCA) are widely used to find a low-dimensional representation for the unknown manifold from the given training data $\mathscr{T}_{images}$. The AE consists of an encoder $\Phi : \mathbf{x} \mapsto \mathbf{h}$ for a compressed latent representation and a decoder $\Psi : \mathbf{h} \mapsto \mathbf{x}$ for providing $\Psi \circ \Phi(\mathbf{x}) \approx \mathbf{x}$ (i.e., an output image aims to be similar to the original input image). Recent papers have reported that the AE-based approaches show remarkable performances in several applications [11, 32, 61, 63]. An important issue in AE is whether the latent space is well organized so that the decoder can generate new content. the variational autoencoder (VAE) was devised to provide regularity in the latent space so that the decoder can act like a generator. The VAE model consists of the reconstruction loss as in the AE (i.e., the output should be similar to the input) and a regularization term (using the Kullback–Leibler divergence) for enforcing the latent probability distribution to be suitably distributed. This chapter explains these issues in detail. We also discuss the current limitations of AEs and VAEs. We also discuss some challenging issues to improve performance of AEs for high-dimensional medical image applications.

1.2 Mathematical Notations and Definitions

Throughout this chapter, a vector $\mathbf{x} \in \mathbb{R}^{N_{\text{pixel}}}$ represents a medical image (e.g., CT, MRI, ultrasound image), where N_{pixel} stands for the dimension of $\mathbf{x}$ (e.g., the total number of pixels). The notation $\mathbb{N}$ will be reserved for the set of nonnegative integers and $\mathbb{N}_L := \{0, 1, \cdots, L-1\}$. Let $\mathbb{Z}$ denote the set of integers and $\mathbb{Z}_L := \{-L+1, \cdots, 0, 1, \cdots, L-1\}$. For example, if $\mathbf{x} \in \mathbb{R}^{N_{\text{pixel}}}$ represents a two-dimensional image with $N_{\text{pixel}} = 256^2$, then the $x(\mathbf{m})$ can express the grayscale pixel value at the pixel position $\mathbf{m} = (m_1, m_2) \in \mathbb{N}_{256} \times \mathbb{N}_{256}$ and

$$\mathbf{x} = (\underbrace{x(0,0), \cdots, x(0,255)}_{x(0,:)}, \underbrace{x(1,0), \cdots, x(1,255)}_{x(1,:)}, \cdots\cdots, \underbrace{x(255,0), \cdots, x(255,255)}_{x(255,:)})^T, \tag{1.3}$$

where $x(m_1, :) = (x(m_1, 0), \cdots, x(m_1, 255))$ is reserved for the (m_1+1)-th row of the image $\mathbf{x}$ and the superscript T stands for the transpose. For simplicity, we also write

$$x(\mathbf{m}) = x_{256(m_1-1)+m_2} \text{for } \mathbf{m} = (m_1, m_2) \in \mathbb{N}_{256} \times \mathbb{N}_{256}. \tag{1.4}$$

Throughout this section, we assume that $\mathscr{X}$ is a subset of $\mathbb{R}^{N_{\text{pixel}}}$, which is the set of all possible CT head images that a particular CT machine can generate. For convenience of explanation only, we limited to CT head images. Readers are free to change the set of images they are currently studying.

The following symbols are used to explain deterministic AEs.

- $\mathscr{T}_{images} = \{\mathbf{x}^{(k)}\}_{k=1}^{N_{\text{data}}}$ denotes a set of training data and N_{data} is the total number of the training data. Note that $\mathscr{T}_{images} \subset \mathscr{X} \subset \mathbb{R}^{N_{\text{pixel}}}$.
- A deterministic encoder is denoted by a map $\Phi : \mathbf{x} \mapsto \mathbf{z} \in \mathbb{R}^{N_{\text{latent}}}$, where a fixed $\mathbf{x}$ always produces a fixed latent vector $\mathbf{z}$ and N_{latent} denotes the dimension of the latent space.
- A deterministic decoder is denoted by a map $\Psi : \mathbf{z} \mapsto \mathbf{x}$.
- $\sigma(\cdot)$ represents an element-wise activation function such as a sigmoid function or a rectified linear unit (ReLU).

A metric on the set $\mathscr{X}$, denoted by $d(\cdot, \cdot)$, will be used to give a distance between two images. The Euclidean distance $d(\mathbf{x}, \mathbf{x}')$ is calculated by the Pythagorean theorem:

$$d(\mathbf{x}, \mathbf{x}') = \sqrt{\sum_{\mathbf{m}=(0,0)}^{(255,255)} |x(\mathbf{m}) - x'(\mathbf{m})|^2}, \tag{1.5}$$

The following symbols are used to explain probabilistic AEs.

- A probabilistic encoder is denoted by a probability distribution $q(\mathbf{z}|\mathbf{x})$. A probabilistic decoder is denoted by $p(\mathbf{z}|\mathbf{x})$.
- $N(\mu, \Sigma)$ denotes a multivariate Gaussian distribution with mean $\mu = (\mu_1, \cdots, \mu_k) \in \mathbb{R}^k$ and covariance matrix $\Sigma \in \mathbb{R}^{k\times k}$. In this section, we only consider a diagonal covariance matrix $\Sigma = [\sigma_1^2, \cdots, \sigma_k^2]$, where σ denotes the standard deviation.
- Throughout this chapter, the same symbol σ will be used to represent two different concepts; the activation function and the standard deviation. The meaning of this symbol can be easily distinguished from the context.
- $\mathbf{z} \sim \mathcal{N}(\mu, \Sigma)$ if the probability density function of random variable $\mathbf{z}$ satisfies

$$\mathcal{N}(\mathbf{z}|\mu, \Sigma) = \frac{1}{(2\pi)^{k/2}\sqrt{\det(\Sigma)}} e^{-\frac{1}{2}(\mathbf{z}-\mu)^T\Sigma^{-1}(\mathbf{z}-\mu)}, \tag{1.6}$$

  Here, the coefficient $(2\pi)^{k/2}\sqrt{\det(\Sigma)}$ is a normalized constant so that its integral over z is 1.
- $D_{KL}[q(\mathbf{z}|\mathbf{x}) \parallel p(\mathbf{z}|\mathbf{x})]$ denotes Kullback–Leibler divergence, which is a measure of how different one probability distribution $q(\mathbf{z}|\mathbf{x})$ is from another $q(\mathbf{z}|\mathbf{x})$. It is defined by

$$D_{KL}[q(\mathbf{z}|\mathbf{x}) \parallel p(\mathbf{z}|\mathbf{x})] = \int q(\mathbf{z}|\mathbf{x}) \log\left(\frac{q(\mathbf{z}|\mathbf{x})}{p(\mathbf{x}|\mathbf{z})}\right) d\mathbf{z}. \tag{1.7}$$

1.3 Linear Dimensionality Reduction

Linear dimensionality reduction methods have been widely used to analyze high-dimensional data. All medical images can be represented as vectors in Hilbert space $\mathbb{H} = \mathbb{R}^{N_{\text{pixel}}}$. Linear dimensionality reduction can be viewed as a projection mapping an image from a high-dimensional space $\mathbb{H}$ to a subspace $\mathbb{V}$ of much lower dimensions.

Theorem 1.1 *Let $\mathbb{V}$ be a subspace of the Hilbert space $\mathbb{H} = \mathbb{R}^{N_{\text{pixel}}}$. There exists a projection map*

$$\mathscr{P}_{\mathbb{V}} : \mathbb{H} \to \mathbb{V} \tag{1.8}$$

such that

$$\mathscr{P}_{\mathbb{V}}\mathbf{x} = \underset{\mathbf{v}\in\mathbb{V}}{argmin}\ \|\mathbf{x} - \mathbf{v}\|. \tag{1.9}$$

Moreover, $\langle \mathbf{x} - \mathscr{P}_{\mathbb{V}}\mathbf{x}, \mathbf{v} \rangle = 0$ *for all* $\mathbf{v} \in \mathbb{V}$. *If* $\{\psi^{(1)}, \cdots, \psi^{(N)}\}$ *is an orthonormal basis of* $\mathbb{V}$, *then* $\mathscr{P}_{\mathbb{V}}\mathbf{x}$ *is expressed as*

$$\mathscr{P}_{\mathbb{V}}\mathbf{x} = \sum_{n=1}^{N} \langle \mathbf{x}, \psi^{(n)} \rangle \psi^{(n)}. \tag{1.10}$$

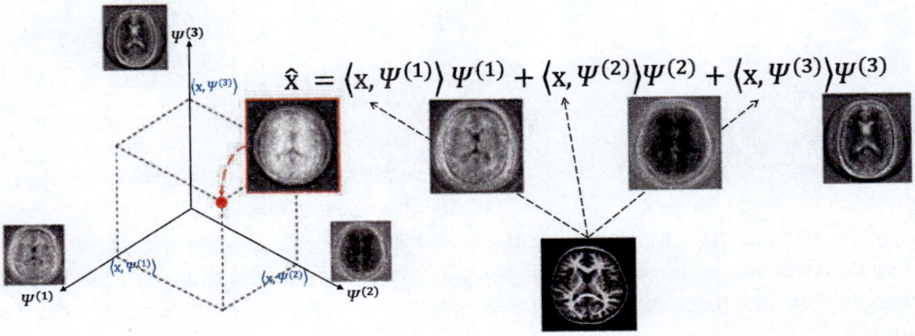

The projection map $\mathscr{P}_{\mathbb{V}}$ in Theorem 1.1 can be viewed as a feature projection that maps the data from the high-dimensional space to a low-dimensional space spanned by feature image vectors $\{\psi^{(1)}, \cdots, \psi^{(N)}\}$. These independent features can be extracted from a set of data $\{\mathbf{x}^{(1)}, \mathbf{x}^{(2)}, \cdots, \mathbf{x}^{(N_{\text{data}})}\}$.

Throughout this section, a vector $\mathbf{x} \in \mathbb{R}^{N_{\text{pixel}}}$ represents a two-dimensional image with pixel dimension $N_{\text{pixel}} = 256^2$. In this case, $x(\mathbf{m})$ is the grayscale intensity at pixel position $\mathbf{m} = (m_1, \mathbf{m}_2)$ in the Cartesian coordinate system.

1.3.1 Canonical Representation

Let $\psi_{\mathbf{m}}$ be the standard Canonical coordinate given by $\psi_{\mathbf{m}}(\mathbf{n}) = \delta(\mathbf{m} - \mathbf{n})$ with δ being the Dirac delta function, i.e.,

$$\psi_{\mathbf{m}} = (\underbrace{0, \cdots, 0}_{256m_1+m_2}, 1, 0, \cdots, 0) = \delta(\cdot - m_1, \cdot - m_2), \tag{1.11}$$

The set $\{\psi_{\mathbf{m}} \ : \ m_1, m_2 = 0, \cdots, 255\}$ constitutes the standard orthonormal basis for the Euclidian space $\mathbb{R}^{N_{\text{pixel}}}$. An image $\mathbf{x}$ can be viewed as a point in the high dimensional vector space $\mathbb{R}^{N_{\text{pixel}}}$ because it can be represented as

$$\mathbf{x} = \sum_{m_1=0}^{255} \sum_{m_2=0}^{255} \langle \mathbf{x}, \psi_{\mathbf{m}} \rangle \underbrace{\psi_{\mathbf{m}}}_{\mathbf{m}-\text{th coordinate}}, \tag{1.12}$$

where $\langle \mathbf{x}, \mathbf{x}' \rangle$ denotes the inner product of two vectors $\mathbf{x}, \mathbf{x}' \in \mathbb{R}^{N_{\text{pixel}}}$, that is defined by

$$\langle \mathbf{x}, \mathbf{x}' \rangle = \sum_{m_1=0}^{255} \sum_{m_2=0}^{255} x(\mathbf{m})\, x'(\mathbf{m}).$$

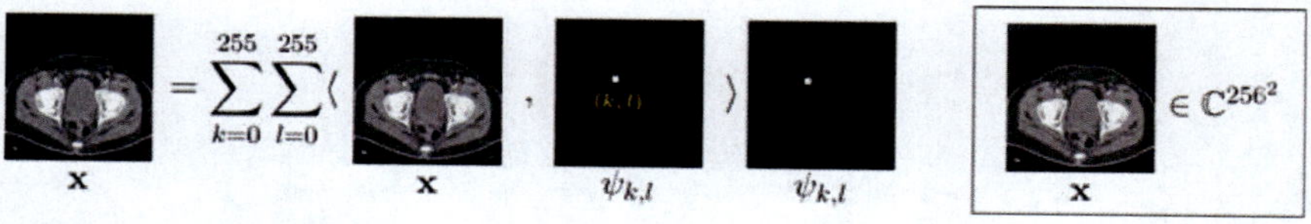

An important problem in medical image analysis is to establish the concept of distance between images. The Euclidean distance $d(\mathbf{x}, \mathbf{x}') = \|\mathbf{x} - \mathbf{x}'\|$ is commonly used because of its simplicity, but it doesn't seem to be appropriate as an image distance from the image recognition point of view. With this Euclidean distance, two images that are close are very different, while two images that are far away can be similar.

1.3.2 Fourier Encoding

The Fourier transform allows us to analyze the frequency component of $\mathbf{x}$. Let $N_{\text{pixel}} = 256^2$ and let $\{\psi_{\mathbf{m}} : \mathbf{m} = (0,0), \cdots, (255,255)\}$ be the standard Fourier basis, given by

$$\psi_{\mathbf{m}} = \frac{1}{256}\left(e^{i\frac{0m_1+0m_2}{256}}, \cdots, e^{i\frac{0m_1+(255)m_2}{256}}, e^{i\frac{1m_1+0m_2}{256}}, \cdots, e^{i\frac{1m_1+(255)m_2}{256}}, \cdots, \cdots\right) \in \mathbb{C}^{N_{\text{pixel}}}, \tag{1.13}$$

According to the 2D discrete Fourier transform, $\mathbf{x}$ can be expressed as

$$\mathbf{x} = \frac{1}{2\pi} \sum_{\mathbf{m}=(0,0)}^{(255,255)} \underbrace{\langle \mathbf{x}, \psi_{\mathbf{m}} \rangle}_{w(\mathbf{m})} \psi_{\mathbf{m}}, \tag{1.14}$$

where $\mathbf{w} \in \mathbb{C}^{L^2}$ is

$$w(\mathbf{m}) = \frac{1}{256} \sum_{\mathbf{n}=(0,0)}^{(255,255)} x(\mathbf{n}) \underbrace{\frac{1}{256} e^{-i\frac{\mathbf{m}\cdot\mathbf{n}}{256}}}_{\psi_{\mathbf{m}}(\mathbf{n})}. \tag{1.15}$$

The above identity (1.14) can be expressed in the following matrix multiplication form:

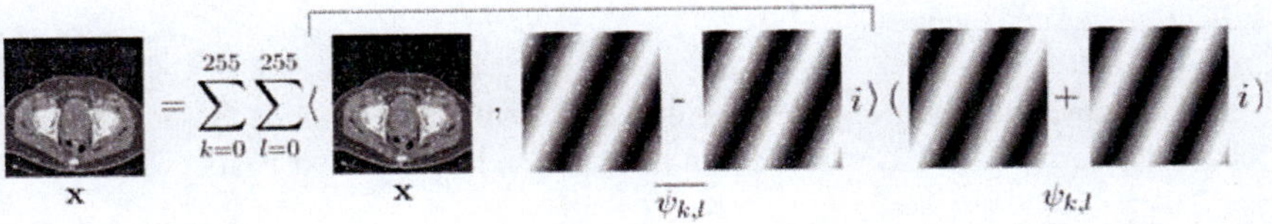

Fig. 1.2 Fourier transform

$$\mathbf{x} = \underbrace{\begin{pmatrix} \psi_{0,0}(0,0) & \cdots & \psi_{0,0}(0,255) & \psi_{0,0}(1,255) & \cdots\cdots & \psi_{0,0}(255,255) \\ \vdots & & \vdots & \vdots & & \vdots \\ \psi_{1,0}(0,0) & \cdots & \psi_{1,0}(0,255) & \psi_{1,0}(1,255) & \cdots\cdots & \psi_{1,0}(255,255) \\ \vdots & & \vdots & \vdots & & \vdots \\ \psi_{255,255}(0,0) & \cdots & \psi_{255,255}(0,255) & \psi_{255,255}(1,255) & \cdots\cdots & \psi_{255,255}(255,255) \end{pmatrix}}_{W^{-1}:\ \text{inverse of Discrete Fourier matrix}} \mathbf{h} \tag{1.16}$$

The linear system $W^{-1}\mathbf{h} = \mathbf{x}$ means that a linear combination of column vectors of W^{-1} is $\mathbf{x}$ (Fig. 1.2).

Let $\mathbb{V}$ be a subspace of $\mathbb{H}$ spanned by $\{\psi_{\mathbf{m}} : 0 \le m_1, m_2 < 63\}$. The dimension of $\mathbb{V}$ is 64^2, which is much smaller than $N_{\text{pixel}} = 256^2$. The projection $\mathscr{P}_{\mathbb{V}}\mathbf{x}$ in Theorem 1.1 can be considered as a dimensional reduction obtained by applying a low-pass filtering in the frequency domain (zeroing all frequency components above the cutoff frequency).

1.3.3 Wavelet Encoding

Wavelets have the advantage of providing simultaneous localization in the image and frequency domains, and are able to isolate the finer details of the image (Fig. 1.3).

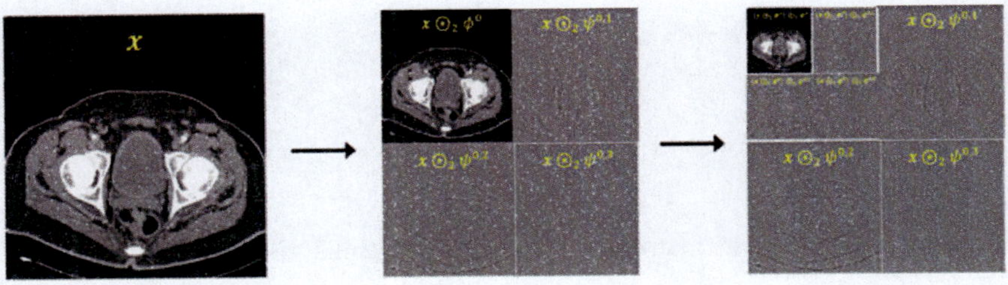

Fig. 1.3 Haar wavelet decomposition

1.3.3.1 Haar Wavelet

The Haar wavelet is the simplest wavelet. Let us explain the Haar decomposition and reconstruction process. The data $\mathbf{x}$ can be expressed by the Haar basis

$$\mathbf{x} = \sum_{l,k=0}^{127} \underbrace{\langle \mathbf{x}, \phi^0_{k,l} \rangle}_{\mathbf{x} \circledast_2 \overline{\phi^0}(k,l)} \phi^0_{k,l} + \sum_{q=1}^{3} \sum_{l,k=0}^{127} \underbrace{\langle \mathbf{x}, \psi^{0,q}_{k,l} \rangle}_{\mathbf{x} \circledast_2 \overline{\psi^{0,q}}(k,l)} \psi^{0,q}_{k,l}, \tag{1.17}$$

where

$$\phi^0_{k,l}(m,n) = \frac{1}{\|\phi^0_{k,l}\|} \left(\frac{1}{4} \sum_{s,t=0}^{1} \delta(m-2k-s, n-2l-t) \right) \sim 2 \begin{pmatrix} 1/4 & 1/4 \\ 1/4 & 1/4 \end{pmatrix},$$

$$\psi^{0,1}_{k,l}(m,n) = \frac{1}{\|\psi^{0,1}_{k,l}\|} \left(\frac{1}{4} \sum_{s,t=0}^{1} (-1)^s \delta(m-2k-s, n-2l-t) \right) \sim 2 \begin{pmatrix} 1/4 & -1/4 \\ 1/4 & -1/4 \end{pmatrix},$$

$$\psi^{0,2}_{k,l}(m,n) = \frac{1}{\|\psi^{0,2}_{k,l}\|} \left(\frac{1}{4} \sum_{s,t=0}^{1} (-1)^t \delta(m-2k-s, n-2l-t) \right) \sim 2 \begin{pmatrix} 1/4 & 1/4 \\ -1/4 & -1/4 \end{pmatrix},$$

$$\psi^{0,3}_{k,l}(m,n) = \frac{1}{\|\psi^{0,3}_{k,l}\|} \left(\frac{1}{4} \sum_{s,t=0}^{1} (-1)^{t+s} \delta(m-2k-s, n-2l-t) \right) \sim 2 \begin{pmatrix} 1/4 & -1/4 \\ -1/4 & 1/4 \end{pmatrix},$$

for $k, l = 0, \ldots, 127,\ q = 1, 2, 3$. Here, $\mathbf{x} \circledast_n \phi$ denotes the convolution of $\mathbf{x}$ with the filter ϕ and n-stride. The convolution of an image $\mathbf{x}$ and a filter $W = \begin{pmatrix} w_{00} & w_{01} \\ w_{10} & w_{11} \end{pmatrix}$ with stride 2, denoted by $\mathbf{x} \circledast_2 W$, is defined by

$$\mathbf{x} \circledast_2 W = \begin{pmatrix} \sum_{m,n=0}^{1} x(m,n) w_{m,n} & \sum_{m,n=0}^{1} x(m,n+2) w_{m,n} & \cdots & \sum_{m,n=0}^{1} x(m,n+254) w_{m,n} \\ \sum_{m,n=0}^{1} x(m+2,n) w_{m,n} & \sum_{m,n=0}^{1} x(m+2,n+2) w_{m,n} & \cdots & \sum_{m,n=0}^{1} x(m+2,n+254) w_{m,n} \\ \sum_{m,n=0}^{1} x(m+4,n) w_{m,n} & \sum_{m,n=0}^{1} x(m+4,n+2) w_{m,n} & \cdots & \sum_{m,n=0}^{1} x(m+4,n+254) w_{m,n} \\ \vdots & \vdots & & \vdots \\ \sum_{m,n=0}^{1} x(m+254,n) w_{m,n} & \sum_{m,n=0}^{1} x(m+254,n+2) w_{m,n} & \cdots & \sum_{m,n=0}^{1} x(m+254,n+254) w_{m,n} \end{pmatrix}. \tag{1.18}$$

Since ϕ^0 is related to the matrix $\frac{1}{2}\begin{pmatrix} 1 & 1 \\ 1 & 1 \end{pmatrix}$, it is convenient to write

$$\mathbf{x} \circledast_2 \phi^0 = \mathbf{x} \circledast_2 \begin{pmatrix} \frac{1}{2} & \frac{1}{2} \\ \frac{1}{2} & \frac{1}{2} \end{pmatrix}. \tag{1.19}$$

Similarly, $\psi^{0,1}$, $\psi^{0,2}$, $\psi^{0,3}$ are related to

$$\mathbf{x} \circledast_2 \psi^{0,1} = \mathbf{x} \circledast_2 \begin{pmatrix} \frac{1}{2} & -\frac{1}{2} \\ \frac{1}{2} & -\frac{1}{2} \end{pmatrix}, \qquad \mathbf{x} \circledast_2 \psi^{0,2} = \mathbf{x} \circledast_2 \begin{pmatrix} \frac{1}{2} & \frac{1}{2} \\ -\frac{1}{2} & -\frac{1}{2} \end{pmatrix},$$

$$\mathbf{x} \circledast_2 \psi^{0,3} = \mathbf{x} \circledast_2 \begin{pmatrix} \frac{1}{2} & -\frac{1}{2} \\ -\frac{1}{2} & \frac{1}{2} \end{pmatrix}.$$

According to the decomposition, $\mathbf{x}$ can be expressed by four different images $\mathbf{x} \circledast_2 \overline{\phi^0}$, $\mathbf{x} \circledast_2 \overline{\psi^{0,1}}$, $\mathbf{x} \circledast_2 \overline{\psi^{0,2}}$, $\mathbf{x} \circledast_2 \overline{\psi^{0,3}} \in \mathbb{C}^{128^2}$, which are defined by

$$\begin{aligned} \mathbf{x} \circledast_2 \overline{\phi^0}(k,l) &= \langle \mathbf{x}, \phi^0_{k,l} \rangle \quad \text{for } k,l = 0,\ldots,127, \\ \mathbf{x} \circledast_2 \overline{\psi^{0,q}}(k,l) &= \langle \mathbf{x}, \psi^{0,q}_{k,l} \rangle \quad \text{for } k,l = 0,\ldots,127,\ q = 1,2,3. \end{aligned} \tag{1.20}$$

Similarly, $\mathbf{x} \circledast_2 \overline{\phi^0}$ can be decomposed into

$$\mathbf{x} \circledast_2 \overline{\phi^0} = \sum_{l,k=0}^{63} \underbrace{\langle \mathbf{x} \circledast_2 \overline{\phi^0}, \phi^0_{k,l} \rangle}_{(\mathbf{x}\circledast_2\overline{\phi^0})\circledast_2\overline{\phi^0}(k,l)} \phi^0_{k,l} + \sum_{q=1}^{3} \sum_{l,k=0}^{63} \underbrace{\langle \mathbf{x} \circledast_2 \overline{\phi^0}, \psi^{0,q}_{k,l} \rangle}_{(\mathbf{x}\circledast_2\overline{\phi^0})\circledast_2\overline{\psi^{0,q}}(k,l)} \psi^{0,q}_{k,l}, \tag{1.21}$$

Let $\phi = \phi^0_{0,0}$ and $\psi^q = \psi^{0,q}_{0,0}$. Then, $\phi^0_{k,l}$ and $\psi^{0,q}_{k,l}$ can be expressed in terms of ϕ and ψ^q:

$$\begin{aligned} \phi^0_{k,l}(m,n) &= \phi(m-k, n-l) \quad \text{for } k,l = 0,\ldots,63, \\ \psi^{0,q}_{k,l}(m,n) &= \psi^q(m-k, n-l) \quad \text{for } k,l = 0,\ldots,63,\ q = 1,2,3. \end{aligned} \tag{1.22}$$

Then, the following sequence forms a multi-resolution analysis of $\mathbb{H}$ in the discrete sense:

$$V_7 = \left\{\mathbf{x} \circledast_2 \phi^0 : \mathbf{x} \in \mathbb{H}\right\} \supset V_6 = \left\{\mathbf{h} \circledast_2 \phi^0 : \mathbf{h} \in V_7\right\} \supset \cdots \supset V_1 = \left\{\mathbf{h} \circledast_2 \phi^0 : \mathbf{h} \in V_2\right\} \tag{1.23}$$

The convolution $\mathbf{x} \circledast_2 \phi^0$ is called the average pooling. Hence, the projection $\mathscr{P}_{V_7}\mathbf{x}$ can be regarded as a dimensional reduction obtained by the average pooling.

1.3.4 General Wavelet Basis

The data $\mathbf{x}$ can be expressed by a general wavelet basis.

$$\mathbf{x} = \sum_{l,k=0}^{127} \underbrace{\langle \mathbf{x}, \phi^0_{k,l} \rangle}_{\mathbf{x}\circledast_2\phi^0(k,l)} \phi^0_{k,l} + \sum_{q=1}^{3} \sum_{l,k=0}^{127} \underbrace{\langle \mathbf{x}, \psi^{0,q}_{k,l} \rangle}_{\mathbf{x}\circledast_2\psi^{0,q}(k,l)} \psi^{0,q}_{k,l}, \tag{1.24}$$

where

$$\phi^0_{k,l}(m,n) = \phi(m-k, n-l), \quad \psi^{0,1}_{k,l}(m,n) = \psi^1(m-k, n-l), \tag{1.25}$$
$$\psi^{0,2}_{k,l}(m,n) = \psi^2(m-k, n-l), \quad \psi^{0,3}_{k,l}(m,n) = \psi^3(m-k, n-l).$$

Similarly, $\mathbf{x} \circledast_2 \phi$ can be expressed by

$$\mathbf{x} \circledast_2 \phi = \sum_{l,k=0}^{63} \underbrace{\langle \mathbf{x}, \phi^1_{k,l}\rangle}_{\mathbf{x}\circledast_4 \phi^1(k,l)} \phi^1_{k,l} + \sum_{q=1}^{3}\sum_{l,k=0}^{63} \underbrace{\langle \mathbf{x}, \psi^{1,q}_{k,l}\rangle}_{\mathbf{x}\circledast_4 \psi^{1,q}(k,l)} \psi^{1,q}_{k,l}, \tag{1.26}$$

where

$$\phi^1_{k,l}(m,n) = \frac{1}{2^2}\phi([\frac{1}{2}m]-k, [\frac{1}{2}n]-l), \psi^{1,1}_{k,l}(m,n) = \frac{1}{2^2}\psi^1([\frac{1}{2}m]-k, [\frac{1}{2}n]-l),$$
$$\psi^{1,2}_{k,l}(m,n) = \frac{1}{2^2}\psi^2([\frac{1}{2}m]-k, [\frac{1}{2}n]-l), \psi^{1,3}_{k,l}(m,n) = \frac{1}{2^2}\psi^3([\frac{1}{2}m]-k, [\frac{1}{2}n]-l).$$

1.3.4.1 Framelet

The data $\mathbf{x}$ also can be represented by the following piecewise linear B-spline tight frame:

$$\mathbf{x} = \sum_{l,k=0}^{127} \underbrace{\langle \mathbf{x}, \phi^0_{k,l}\rangle}_{\mathbf{x}\circledast_2 \phi^0(k,l)} \phi^0_{k,l} + \sum_{q=1}^{8}\sum_{l,k=0}^{127} \underbrace{\langle \mathbf{x}, \psi^{0,q}_{k,l}\rangle}_{\mathbf{x}\circledast_2 \psi^{0,q}(k,l)} \psi^{0,q}_{k,l}, \tag{1.27}$$

where $\phi^0, \psi^{0,1}, \cdots, \psi^{0,8}$ are given by

$$\phi^0 \sim \tfrac{1}{16}\begin{pmatrix}1&2&1\\2&4&2\\1&2&1\end{pmatrix}, \quad \psi^{0,1} \sim \tfrac{\sqrt{2}}{16}\begin{pmatrix}1&0&-1\\2&0&-2\\1&0&-1\end{pmatrix}, \quad \psi^{0,2} \sim \tfrac{1}{16}\begin{pmatrix}-1&2&-1\\-2&4&-2\\-1&2&-1\end{pmatrix}$$
$$\psi^{0,3} \sim \tfrac{\sqrt{2}}{16}\begin{pmatrix}1&2&1\\0&0&0\\-1&-2&-1\end{pmatrix}, \quad \psi^{0,4} \sim \tfrac{1}{8}\begin{pmatrix}1&0&-1\\0&0&0\\-1&0&1\end{pmatrix}, \quad \psi^{0,5} \sim \tfrac{\sqrt{2}}{16}\begin{pmatrix}-1&2&-1\\0&0&0\\1&-2&1\end{pmatrix},$$
$$\phi^{0,6} \sim \tfrac{1}{16}\begin{pmatrix}-1&-2&-1\\2&4&2\\-1&-2&-1\end{pmatrix}, \quad \psi^{0,7} \sim \tfrac{\sqrt{2}}{16}\begin{pmatrix}-1&0&1\\2&0&-2\\-1&0&1\end{pmatrix}, \quad \psi^{0,8} \sim \tfrac{1}{16}\begin{pmatrix}1&-2&1\\-2&4&-2\\1&-2&1\end{pmatrix}. \tag{1.28}$$

Here, $\mathbf{x} \circledast_1 W$ with $W = \begin{pmatrix}w_{00}&w_{01}&w_{02}\\w_{10}&w_{11}&w_{12}\\w_{20}&w_{21}&w_{22}\end{pmatrix}$ is defined by

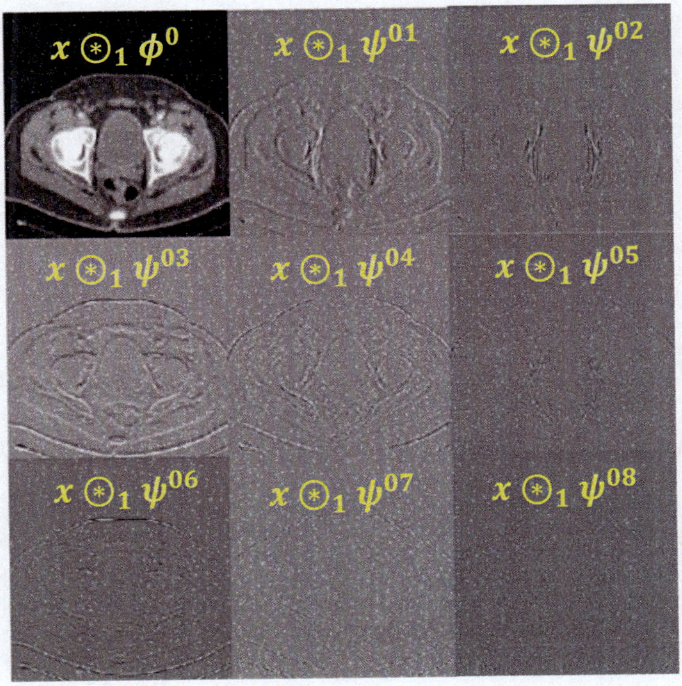

Fig. 1.4 Data representation for framelet decomposition

$$\mathbf{x} \circledast_1 W = \begin{pmatrix} \sum_{m,n=0}^{2} x(m-1, n-1)w_{m,n} & \sum_{m,n=0}^{2} x(m-1, n)w_{m,n} & \cdots & \sum_{m,n=0}^{2} x(m-1, n+254)w_{m,n} \\ \sum_{m,n=0}^{2} x(m, n-1)w_{m,n} & \sum_{m,n=0}^{2} x(m, n)w_{m,n} & \cdots & \sum_{m,n=0}^{2} x(m, n+254)w_{m,n} \\ \sum_{m,n=0}^{2} x(m+1, n-1)w_{m,n} & \sum_{m,n=0}^{2} x(m+1, n)w_{m,n} & \cdots & \sum_{m,n=0}^{2} x(m+1, n+254)w_{m,n} \\ \vdots & \vdots & & \vdots \\ \sum_{m,n=0}^{2} x(m+254, n-1)w_{m,n} & \sum_{m,n=0}^{2} x(m+254, n)w_{m,n} & \cdots & \sum_{m,n=0}^{2} x(m+254, n+254)w_{m,n} \end{pmatrix}. \tag{1.29}$$

Here, $x(m, n) = 0$ for $m, n = -1, 256$. The convolution $\mathbf{x} \circledast_1 W$ can be represented as the following linear form using a block Hankel matrix (Fig. 1.4):

$$\mathbf{x} \circledast_1 W = \underbrace{\begin{pmatrix} \mathscr{H}_{\mathbf{x}}^{0,0} & \mathscr{H}_{\mathbf{x}}^{0,1} & \mathscr{H}_{\mathbf{x}}^{0,2} \\ \mathscr{H}_{\mathbf{x}}^{1,0} & \mathscr{H}_{\mathbf{x}}^{1,1} & \mathscr{H}_{\mathbf{x}}^{1,2} \\ \mathscr{H}_{\mathbf{x}}^{2,0} & \mathscr{H}_{\mathbf{x}}^{2,1} & \mathscr{H}_{\mathbf{x}}^{2,2} \\ \vdots & \vdots & \vdots \\ \mathscr{H}_{\mathbf{x}}^{255,0} & \mathscr{H}_{\mathbf{x}}^{255,1} & \mathscr{H}_{\mathbf{x}}^{255,2} \end{pmatrix}}_{\mathscr{H}_{\mathbf{x}} \in \mathbb{C}^{256^2 \times 9}} \underbrace{\begin{pmatrix} w_{00} \\ w_{01} \\ w_{02} \\ w_{10} \\ \vdots \\ w_{22} \end{pmatrix}}_{\mathbf{w} \in \mathbb{C}^9}, \tag{1.30}$$

where

$$\mathscr{H}_{\mathbf{x}}^{0,0} = \begin{pmatrix} x(-1,-1) & x(-1,0) & x(-1,1) \\ x(-1,0) & x(-1,1) & x(-1,2) \\ \vdots & \vdots & \vdots \\ x(-1,254) & x(-1,255) & x(-1,256) \end{pmatrix}, \quad \mathscr{H}_{\mathbf{x}}^{0,1} = \begin{pmatrix} x(0,-1) & x(0,0) & x(0,1) \\ x(0,0) & x(0,1) & x(0,2) \\ \vdots & \vdots & \vdots \\ x(0,254) & x(0,255) & x(0,256) \end{pmatrix}, \cdots$$

$$\mathscr{H}_{\mathbf{x}}^{m,k} = \begin{pmatrix} x(m-1+k,-1) & x(m-1+k,0) & x(m-1+k,1) \\ x(m-1+k,0) & x(m-1+k,1) & x(m-1+k,2) \\ \vdots & \vdots & \vdots \\ x(m-1+k,254) & x(m-1+k,255) & x(m-1+k,256) \end{pmatrix}, \cdots \tag{1.31}$$

1.3.5 *Principal Component Analysis (PCA)*

1.3.5.1 Data, Covariance and Similarity Matrix

For ease of explanation, we assume that X is a datatable (or matrix) of images, lying in a high-dimensional space $R^{N_{\text{pixel}}}$ (e.g. $N_{\text{pixel}} = 93 \times 70$):

$$X = \begin{pmatrix} | & | & & | \\ \mathbf{x}^{(1)} & \mathbf{x}^{(2)} & \cdots & \mathbf{x}^{(N_{\text{data}})} \\ | & | & & | \end{pmatrix}^T = \begin{pmatrix} [\text{image}]^{(1)} \\ \vdots \\ [\text{image}]^{(N_{\text{data}})} \end{pmatrix}, \tag{1.32}$$

where the superscript T stands for the transpose. Each component $\mathbf{x}^{(n)}(\mathbf{m})$ of the image $\mathbf{x}^{(n)}$ represents the brightness of the $\mathbf{m}$th pixel.

The mean for the datatable X is given by

$$\bar{\mathbf{x}} := E(X) = \frac{1}{N_{\text{data}}} \sum_{n=1}^{N_{\text{data}}} \mathbf{x}^{(n)}, \tag{1.33}$$

where $E(X)$ denotes the expectation of X under the assumption that all $\mathbf{x}_n$ are equally likely. The standard deviation of datatable X (i.e. lowwise std(X)) is a measure of how spread out the data is:

$$\mathrm{std}(X) = \sqrt{\frac{1}{N_{\mathrm{data}}} \sum_{n=1}^{N_{\mathrm{data}}} \|\mathbf{x}^{(n)} - \bar{\mathbf{x}}\|^2}, \tag{1.34}$$

The variance is simply σ^2, the standard deviation squared.

The covariance matrix of the datatable X is $N_{\mathrm{pixel}} \times N_{\mathrm{pixel}}$ matrix which expresses the spread of data:

$$\mathrm{cov}(X, X) = \frac{1}{N_{\mathrm{data}} - 1} X^T X - \bar{\mathbf{x}}\bar{\mathbf{x}}^T, \tag{1.35}$$

It is easy to see that the covariance matrix $\mathrm{cov}(X, X)$ is symmetric and semi-definite positive.

The Gram matrix of the datatable X is an $N_{\mathrm{data}} \times N_{\mathrm{data}}$ matrix which represents the similarity (or dissimilarity) between data. Its ijth component is the inner product $G_{ij} = \langle \mathbf{x}_i - \bar{\mathbf{x}}, \mathbf{x}_j - \bar{\mathbf{x}} \rangle$:

$$G_X = (X - \mathbf{1}\bar{\mathbf{x}}^T)(X - \mathbf{1}\bar{\mathbf{x}}^T)^T = (I_{N_{\mathrm{data}}} - \frac{1}{N_{\mathrm{data}}}\mathbf{1}\mathbf{1}^T) X X^T (I_{N_{\mathrm{data}}} - \frac{1}{N_{\mathrm{data}}}\mathbf{1}\mathbf{1}^T), \tag{1.36}$$

where $\mathbf{1} = (1, \cdots, 1)^T \in \mathbb{R}^{N_{\mathrm{data}}}$ and $I_{N_{\mathrm{data}}}$ is the identity matrix of size N_{data}.

1.3.5.2 Singular Value Decomposition (SVD)

The singular value decomposition (SVD) of X allows us to split X into rank-one pieces ordered by their sizes:

$$\underbrace{X}_{N_{\mathrm{data}} \times N_{\mathrm{pixel}} \text{ matrix}} = \underbrace{\begin{pmatrix} \mathbf{u}_1 & \cdots & \mathbf{u}_{N_{\mathrm{data}}} \end{pmatrix}}_{U} \underbrace{\begin{pmatrix} \lambda_1 & & & 0 \cdots \\ & \ddots & & \\ & & \lambda_r & 0 \cdots \\ \mathbf{0} & \vdots & & \mathbf{0} \cdots \end{pmatrix}}_{\Lambda} \underbrace{\begin{pmatrix} \mathbf{w}_1^T \\ \vdots \\ \mathbf{w}_{N_{\mathrm{pixel}}}^T \end{pmatrix}}_{W} = \lambda_1 \mathbf{u}_1 \mathbf{w}_1^T + \cdots + \lambda_r \mathbf{u}_r \mathbf{w}_r^T, \tag{1.37}$$

where $\mathbf{u}_j, \mathbf{w}_j, \lambda_j$ satisfy the following:

- $\lambda_1 \geq \lambda_2 \geq \cdots \geq \lambda_r > 0 = \lambda_{r+1} = \lambda_{N_{\mathrm{pixel}}}$ are eigenvalues of X.
- $\mathbf{w}_1, \cdots, \mathbf{w}_{N_{\mathrm{pixel}}}$ are the corresponding unit eigenvectors of X.
- $\mathbf{u}_i = \frac{1}{\lambda_i} X \mathbf{w}_i$ for $i = 1, 2, \cdots, r$. Note that $\|\mathbf{u}_i\|^2 = \frac{1}{\lambda_i^2} \mathbf{w}_i^T X^T X \mathbf{w}_i = 1$.
- $UU^T = I_{N_{\mathrm{data}}}$ and $W^T W = I_{N_{\mathrm{pixel}}}$, where $I_{N_{\mathrm{data}}}$ is the identity matrix of order N_{data}.

The pseudo-inverse of X is

$$X^\dagger = W\Lambda^{-1}U^T = \frac{1}{\lambda_1}\mathbf{w}_1\mathbf{u}_1^T + \cdots + \frac{1}{\lambda_r}\mathbf{w}_r\mathbf{u}_r^T. \tag{1.38}$$

Note that the products $XX^\dagger$ and $X^\dagger X$ can be viewed as projection matrices:

- $XX^\dagger$ is the projection matrix onto the column space of X because

$$XX^\dagger\mathbf{u}_j = X\left(\frac{1}{\lambda_1}\mathbf{w}_1\mathbf{u}_1^T + \cdots + \frac{1}{\lambda_r}\mathbf{w}_r\mathbf{u}_r^T\right)\mathbf{u}_j = X(\frac{1}{\lambda_j}\mathbf{w}_j) = \mathbf{u}_j.$$

- $X^\dagger X$ is the projection matrix onto the row space of X because

$$X^\dagger X\mathbf{w}_j = X^\dagger(\lambda_j\mathbf{u}_j) = \mathbf{w}_j.$$

Hence, $\mathbf{w}^* = X^\dagger\mathbf{b}$ can be viewed as the least squares solution of $X\mathbf{w} = \mathbf{b}$ in the sense that

$$X^T X \underbrace{\mathbf{w}^*}_{X^\dagger\mathbf{b}} = X^T\mathbf{b} \quad \text{and} \quad \underbrace{\mathbf{w}^*}_{X^\dagger\mathbf{b}} = \text{argmin}_{\mathbf{w}}\|X\mathbf{w} - \mathbf{b}\|. \tag{1.39}$$

1.3.5.3 PCA Encoding

PCA is a method of expressing the datatable X to highlight both its similarities and dissimilarities, under the assumption that the data $\{\mathbf{x}_1, \cdots, \mathbf{x}_{N_{\text{data}}}\}$ lies approximately on a linear manifold (through linear regression), which is globally homeomorphic to $\mathbb{R}^k$. Assuming $k \ll \min(N_{\text{pixel}}, N)$, PCA performs dimensionality reduction, since the data can be projected onto the span of k principal components, which can be viewed as the best-fitting k-dimensional subspace.

To understand the principal components, let us consider the variance along a unit vector $\mathbf{d}$

$$\frac{1}{N_{\text{data}}}\sum_{n=1}^{N_{\text{data}}}(\mathbf{d}_1^T(\mathbf{x}_n - \bar{\mathbf{x}}))^2 = \mathbf{d}_1^T\left(\frac{1}{N_{\text{data}}}\sum_{n=1}^{N_{\text{data}}}(\mathbf{x}_n - \bar{\mathbf{x}})(\mathbf{x}_n - \bar{\mathbf{x}})^T\right)\mathbf{d}_1. \tag{1.40}$$

For notational simplicity, assume that the mean of the data is $\bar{\mathbf{x}} = \mathbf{0}$. The first principal component is the vector $\mathbf{d}_1 \in \mathbb{R}^{N_{\text{pixel}}}$ which maximizes the variance in datatable X

$$\mathbf{d}_1 = \underset{\|\mathbf{d}\|=1}{\text{argmax}}[\mathbf{d}^T X^T X\mathbf{d}]. \tag{1.41}$$

By using the Lagrange multiplier, (1.41) becomes the following unconstrainted maximization:

$$\mathbf{d}_1 = \underset{\mathbf{d}}{\text{argmax}}\left[\mathbf{d}^T X^T X\mathbf{d} + \lambda(1 - \|\mathbf{d}\|^2)\right]. \tag{1.42}$$

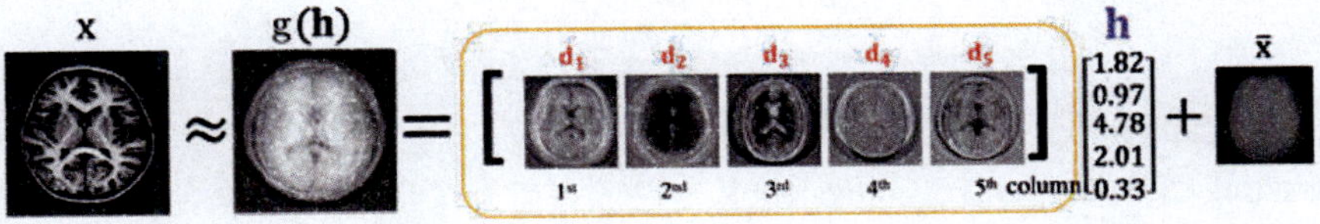

Fig. 1.5 Data representation using PCA

By differentiating $\left[\mathbf{d}^T X^T X\mathbf{d} + \lambda(1 - \|\mathbf{d}\|^2)\right]$ with respect to $\mathbf{d}$, we have

$$X^T X\mathbf{d} = \lambda\mathbf{d}. \tag{1.43}$$

Hence, $\mathbf{d}_1$ should be the unit eigenvector corresponding to the biggest eigenvalue (λ_1) of $X^T X$, because it maximize the variance (Fig. 1.5).

$$\mathbf{d}_1^T X^T X\mathbf{d}_1 = \lambda_1. \tag{1.44}$$

Let us explain PCA in a slightly different way. The decoder g can be expressed as a $N_{\text{pixel}} \times k$ matrix Ψ or W^T:

$$g(\mathbf{h}) = \Psi\mathbf{h} = \underbrace{\begin{pmatrix} | & | & & | \\ \mathbf{d}_1 & \mathbf{d}_2 & \cdots & \mathbf{d}_k \\ | & | & & | \end{pmatrix}}_{\Psi} \underbrace{\begin{pmatrix} h(1) \\ \vdots \\ h(k) \end{pmatrix}}_{\mathbf{h}} = \sum_{j=1}^{k} h(j)\,\mathbf{d}_j, \tag{1.45}$$

PCA constrains the columns of Ψ (i.e. $\mathbf{d}_1, \cdots, \mathbf{d}_k$) to be orthonormal vectors in $\mathbb{R}^{N_{\text{pixel}}}$:

$$\underbrace{\begin{pmatrix} | & | & & | \\ \mathbf{d}_1 & \mathbf{d}_2 & \cdots & \mathbf{d}_k \\ | & | & & | \end{pmatrix}^T}_{\mathscr{D}^T} \underbrace{\begin{pmatrix} | & | & & | \\ \mathbf{d}_1 & \mathbf{d}_2 & \cdots & \mathbf{d}_k \\ | & | & & | \end{pmatrix}}_{\mathscr{D}} = \underbrace{\begin{pmatrix} 1 & 0 & \cdots & 0 \\ 0 & 1 & \cdots & 0 \\ \vdots & \vdots & \ddots & \vdots \\ 0 & 0 & \cdots & 1 \end{pmatrix}}_{I_k} \tag{1.46}$$

These orthonormal vectors $\mathbf{d}_1, \cdots, \mathbf{d}_k$ are said to be principal components.

Given a data matrix $X = [\mathbf{x}_1, \cdots, \mathbf{x}_N]^T$, the decoding matrix Ψ can be chosen by

$$\Psi = \underset{\Psi \in \mathbb{R}^{256^2 \times k}}{\text{argmin}} \sum_{n=1}^{N} \|\mathbf{x}_i - \Psi\Psi^T\mathbf{x}_i\|^2 \text{subject to } \Psi^T\Psi = I_k. \tag{1.47}$$

Here, k is the number of principal components.

As the PCA reconstruction process, the first principal component $\mathbf{d}_1$ can be obtained by

$$\mathbf{d}_1 = \underset{\mathbf{d}\in\mathbb{R}^{256^2}}{\operatorname{argmin}} \sum_{n=1}^{N} \|\mathbf{x}_i - \mathbf{d}\mathbf{d}^T\mathbf{x}_i\|^2 \text{ subject to } \|\mathbf{d}\| = 1. \tag{1.48}$$

Finding $\mathbf{d}_1$ from (1.48) is equivalent to

$$\mathbf{d}_1 = \underset{\mathbf{d}\in\mathbb{R}^{256^2}}{\operatorname{argmin}} \|X - \mathbf{d}\mathbf{d}^T X\|_F^2 \text{ subject to } \|\mathbf{d}\| = 1, \tag{1.49}$$

where $\|\cdot\|_F$ represents the Frobenius norm (i.e. $\|X\|_F = \sqrt{\sum_{n=1}^N \|\mathbf{x}_i\|^2}$). Noting that $\|X - \mathbf{d}\mathbf{d}^T X\|_F^2 = (X - \mathbf{d}\mathbf{d}^T X)^T(X - \mathbf{d}\mathbf{d}^T X)$ with X being regraded as a constant, a simple computation shows that

$$\mathbf{d}_1 = \underset{\mathbf{d}\in\mathbb{R}^{256^2}}{\operatorname{argmin}} \left(-\operatorname{Tr}(X^T X \mathbf{d}\mathbf{d}^T)\right) = \underset{\mathbf{d}\in\mathbb{R}^{256^2}}{\operatorname{argmin}} \left(-\mathbf{d}^T X^T X \mathbf{d}\right) \text{ subject to } \|\mathbf{d}\| = 1. \tag{1.50}$$

Hence, $\mathbf{d}_1$ satisfies

$$\mathbf{d}_1 = \underset{\|\mathbf{d}\|=1}{\operatorname{argmax}} \left(\mathbf{d}^T X^T X \mathbf{d}\right). \tag{1.51}$$

To find the jth principal component $\mathbf{d}_j$, we subtract the first $j-1$ principal components from X:

$$X_j = X - \sum_{n=1}^{j} X\mathbf{d}_n\mathbf{d}_n^T, \tag{1.52}$$

Similarly, we can find $\mathbf{d}_j$ by minimizing

$$\mathbf{d}_j = \underset{\|\mathbf{d}\|=1}{\operatorname{argmax}} \left(\mathbf{d}^T X_j^T X_j \mathbf{d}\right). \tag{1.53}$$

Hence, the decoding matrix Ψ is a minimizer:

$$\Psi = \underset{\Psi^T\Psi = I_k}{\operatorname{argmax}} \operatorname{Tr}(\Psi^T X^T X \Psi). \tag{1.54}$$

Given decoding Ψ and data $\mathbf{x}$, we generate $\mathbf{c}$ in such a way that

$$\mathbf{h} = \underset{\mathbf{h}\in\mathbb{R}^k}{\operatorname{argmin}} \|\mathbf{x} - \Psi\mathbf{h}\|^2 = \underset{\mathbf{h}\in\mathbb{R}^k}{\operatorname{argmin}} (\mathbf{x} - \Psi\mathbf{h})^T(\mathbf{x} - \Psi\mathbf{h}). \tag{1.55}$$

Since $\mathbf{x}$ and D are regraded as constants, (1.55) can be simplified into

$$\mathbf{h} = \underset{\mathbf{h}\in\mathbb{R}^k}{\operatorname{argmin}} (-2\mathbf{x}^T\Psi\mathbf{h} + \mathbf{h}^T \underbrace{\Psi^T\Psi}_{I_k} \mathbf{h}) = \underset{\mathbf{h}\in\mathbb{R}^k}{\operatorname{argmin}} (-2\mathbf{x}^T\Psi\mathbf{h} + \mathbf{c}^T\mathbf{h}). \tag{1.56}$$

The minimizer **h** must satisfy

$$0 = \nabla_{\mathbf{h}}(-2\mathbf{x}^T \Psi \mathbf{h} + \mathbf{c}^T \mathbf{h}) = -2(\Psi^T \mathbf{x} - \mathbf{h}). \tag{1.57}$$

1.3.6 Regularization and Compressed Sensing

For $\mathbf{z} = (z_1, \cdots, z_{N_z})$, define ℓ_0-norm

$$\|\mathbf{z}\|_0 = \sharp\{j \ : \ z_j \neq 0\} \quad \text{(the number of non-zero entries of } \mathbf{z}) \tag{1.58}$$

and ℓ_1-norm

$$\|\mathbf{z}\| = \sum_{j}^{N_z} |z_j|. \tag{1.59}$$

Imagine that **x** is an approximate solution of the constraint linear problem:

$$\underbrace{\begin{pmatrix} \psi_{00} & \psi_{01} & \cdots\cdots & \psi_{0,N_z} \\ \psi_{10} & \psi_{11} & \cdots\cdots & \psi_{1,N_z} \\ \vdots & \vdots & & \vdots \\ \vdots & \vdots & & \vdots \\ \psi_{N_x,0} & \psi_{N_x 1} & \cdots\cdots & \psi_{N_x,N_z} \end{pmatrix}}_{\Psi} \underbrace{\begin{pmatrix} z_0 \\ z_1 \\ \vdots \\ \vdots \\ z_{N_z} \end{pmatrix}}_{\mathbf{z}} \approx \underbrace{\begin{pmatrix} x_0 \\ x_1 \\ \vdots \\ \vdots \\ x_{N_x} \end{pmatrix}}_{\mathbf{x}} \tag{1.60}$$

subject to the constraint

$$\|\mathbf{z}\|_{\ell_0} \leq k \quad (k \ll N_z). \tag{1.61}$$

This can be viewed as the problem of generating **x** by a sparse latent representation. The sparse representation problem can be NP-hard because the number of all cases of choosing k from N_z is huge. Instead of solving the NP-hard problem, one can consider the relaxed ℓ_1-minimization problem which is the closest convex minimization problem to the ℓ_0-minimization problem:

$$\mathbf{z} = \text{argmin}_{\mathbf{z}\in\mathbb{R}^n} \underbrace{\frac{1}{2}\|\mathbf{x} - \Psi\mathbf{z}\|_2^2}_{\Gamma(\mathbf{z})} + \lambda \underbrace{\|\mathbf{z}\|_{\ell_1}}_{\Upsilon(\mathbf{z})}, \tag{1.62}$$

where $\|\mathbf{z}\|_{\ell_1} = \sum_j |z(j)|$. Here, the first term enforces the residual $\Gamma(\mathbf{z}) = \frac{1}{2}\|\mathbf{x} - \Psi\mathbf{z}\|_2^2$ to be small and the regularization parameter $\lambda \geq 0$ controls the tradeoff between the approximation error $\Gamma(\mathbf{z})$ and the vector norm $\|\mathbf{z}\|_{\ell_1}$.

We should note that $\|\mathbf{z}\|_{\ell_1}$ is not differentiable at any point $\mathbf{z}$ being $\sharp\{j : u_j \neq 0\} > 0$. However, $\Upsilon(\mathbf{z})$ is still convex, and hence there exists the one-side directional derivative $\nabla\Upsilon(\mathbf{z}; \mathbf{d}) = \lim_{t\to 0^+} \frac{1}{t}(\Upsilon(\mathbf{z} - t\mathbf{d}) - \Upsilon(\mathbf{z}))$:

$$\nabla\Upsilon(\mathbf{z}, \mathbf{d}) = -\mathbf{w}^T\mathbf{A}^T(\mathbf{x} - \mathbf{A}\mathbf{z}) + \lambda \sum_{\{j:\, z_j \neq 0\}} d_j \mathrm{sign}(z_j) + \lambda \sum_{\{j:\, z_j = 0\}} |d_j|.$$

If $\mathbf{z}$ is a minimizer, then $\Upsilon(\mathbf{z} + t\mathbf{d}) \geq \Upsilon(\mathbf{z})$ for all $\mathbf{d}$ and $t > 0$ (goes up in all directions), and therefore $\nabla\Upsilon(\mathbf{z}; \mathbf{d}) \geq 0$ for all direction $\mathbf{d}$. To be precise, $\mathbf{z}$ is a minimizer $\iff$ $\nabla\Upsilon(\mathbf{z}; \mathbf{d}) \geq 0$ for all $\mathbf{d} \in R^n$ $\iff$

For all j in the set $J = \{1, \cdots, n\}$,

1. $\mathrm{sign}(z_j) \neq 0 \;\Rightarrow\; -\mathbf{e}_j^T \Psi^T(\mathbf{x} - \Psi\mathbf{z}) + \lambda\mathrm{sign}(z_j) = 0$
2. $z_j = 0 \;\Rightarrow\; |\mathbf{e}_j^T \Psi^T(\mathbf{x} - \Psi\mathbf{z})| \leq \lambda.$

Here, $\mathbf{e}_j$ is the unit vector with its j-th component being 1, and $\mathbf{e}_j\Psi$ is the j-th column of Ψ. The first statement comes from the existence of the directional derivative $\nabla\Upsilon(\mathbf{z}, \mathbf{e}_j) = -\nabla\Upsilon(\mathbf{z}, -\mathbf{e}_j) = 0$ for j being $\mathrm{sign}(z_j) \neq 0$. The second statement comes from the positivity of the one-side derivative $\nabla\Upsilon(\mathbf{z}, \pm\mathbf{e}_j) \geq 0$ for j being $\mathrm{sign}(z_j) = 0$.

1.4 Autoencoder and Manifold Learning

The Autoencoder (AE) is a representation learning technique that a feed-forward neural net is trained to reproduce its input at the output, where the network contains an internal "bottleneck" layer, which forces the network to develop a compact representation of the input data [25, 33]. More precisely, AEs are unsupervised neural networks that maintain the maximum information in the encoding phase ($\Phi : \mathbf{x} \mapsto \mathbf{h}$) so that a decoded image $\hat{\mathbf{x}} = \Psi(\mathbf{h}) = \Psi \circ \Phi(\mathbf{x})$ is close to the original input $\mathbf{x}$. For dimensionality reduction, AEs force the input $\mathbf{x}$ to pass through the network's bottleneck (i.e. the latent variable space) whose dimension (denoted by N_{latent}) is much less than the dimension (N_{pixel}) of the input $\mathbf{x}$ (i.e., $N_{\text{latent}} \ll N_{\text{pixel}}$).

The encoder Φ and the decoder Ψ are obtained using training data $\mathcal{T}_{images} = \{\mathbf{x}^{(k)}\}_{k=1}^{N_{\text{data}}}$:

$$(\Phi, \Psi) = \underset{(\Phi,\Psi)\in\mathbb{AE}}{\mathrm{argmin}} \frac{1}{N_{\text{data}}} \sum_{n=1}^{N_{\text{data}}} \|\Psi \circ \Phi(\mathbf{x}^{(n)}) - \mathbf{x}^{(n)}\|^2, \tag{1.63}$$

where $\mathbb{AE}$ is a learning network described in Fig. 1.6. The mappings Ψ and Φ will only be "approximately" inverse to each other. The encoder's output $\mathbf{h} = \Phi(\mathbf{x})$ can be viewed as a compressed latent representation, and the decoder Ψ converts $\mathbf{h}$ to an image similar to the original input. In summary, AEs learn an approximation of the

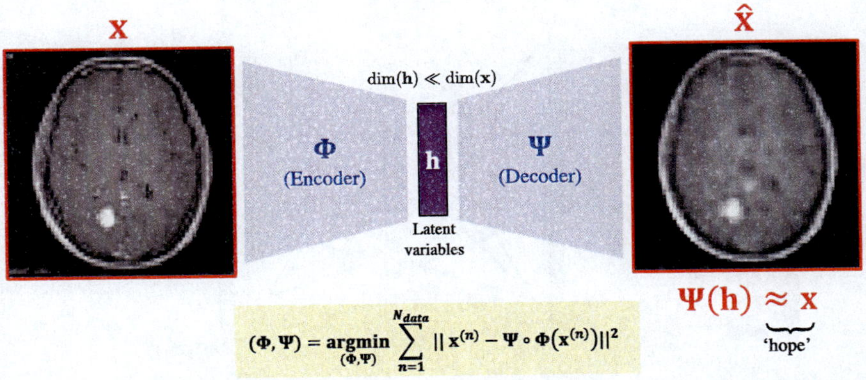

Fig. 1.6 The architecture of the autoencoder (AE). The AE is a low-dimensional representation learning technique in which a feed-forward neural net $\Psi \circ \Phi$ is trained to reproduce its input $\mathbf{x}$ at the output $\Psi \circ \Phi(\mathbf{x})$

identity function (i.e., lossless encoding $\Psi \circ \Phi(\mathbf{x}) \approx \mathbf{x}$) subject to the constraint of $N_{\text{latent}} \ll N_{\text{pixel}}$ (i.e., compressed latent representation or dimensionality reduction).

1.4.1 *Linear and Semi-linear Autoencoder*

The following linear AE with minimizing L^2-divergence is nearly equivalent to PCA:

$$\Phi = \underset{\Phi}{\text{argmin}} \frac{1}{N_{\text{data}}} \sum_{n=1}^{N_{\text{data}}} \|\Psi \Phi \mathbf{x}^{(n)} - \mathbf{x}^{(n)}\|^2, \tag{1.64}$$

where Φ is a $N_{\text{latent}} \times N_{\text{pixel}}$ encoding matrix and Ψ is a $N_{\text{pixel}} \times N_{\text{latent}}$ decoding matrix. The decoding Ψ can be set as the transpose of Φ. Unlike PCA, the column vectors of Φ may not be mutually orthogonal. Figure 1.9 shows a linear AE with $N_{\text{latent}} = 2$, and hence the output $\Psi \circ \Phi(\mathbf{x})$ lies on a two-dimensional space spanned by two column vectors of Ψ. Linear AEs with a lot of dimensionality reduction usually tend to lose a lot of image detail during the encoding process, an effect like Gibbs' phenomenon. Moreover, a linear interpolation of two images (i.e., an average of two images $t\Psi(\mathbf{h}^{(i)}) + (1-t)\Psi(\mathbf{h}^{(j)})$ produces mostly meaningless images.

An important goal of AE is for the decoder Ψ to generate various meaningful images by varying the latent vector $\mathbf{h}$. The range $\{\Psi(\mathbf{h}) : \mathbf{h} \in \mathbb{R}^{N_{\text{latent}}}\}$, being embedded into the high-dimensional space $\mathbb{R}^{N_{\text{pixel}}}$, is obtained by the training data $\mathscr{T}_{images} = \{\mathbf{x}^{(n)}\}_{n=1}^{N_{\text{data}}}$. Hence, we may assume that the data $\mathscr{T}_{images}$ lie nearly on a learned manifold $\mathscr{M} = \{\Psi(\mathbf{h}) : \mathbf{h} \in \mathbb{R}^{N_{\text{latent}}}\}$ whose Hausdorff dimension is N_{latent}. We hope that the interpolation of the latent space $(\Psi(t\mathbf{h}^{(i)} + (1-t)\mathbf{h}^{(j)}), 0 < t < 1)$

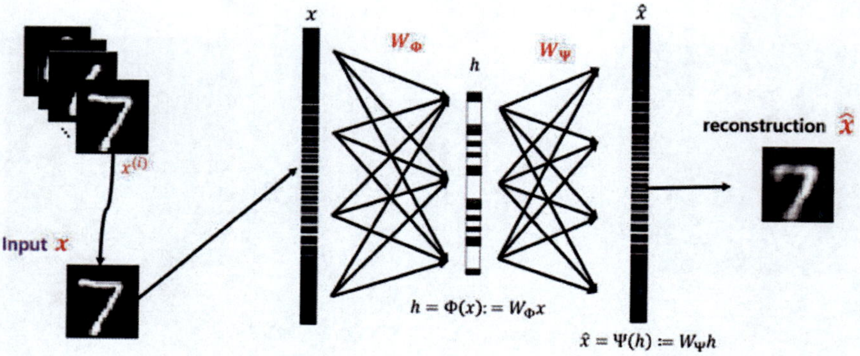

Fig. 1.7 This linear AE aims to find a $N_{\text{latent}} \times N_{\text{pixel}}$ matrix W_Φ (encoder) and $N_{\text{pixel}} \times N_{\text{latent}}$ matrix W_Ψ (decoder) such that the information loss $\|W_\Psi W_\Phi \mathbf{x} - \mathbf{x}\|$ for all data $\mathbf{x}$ is as small as possible

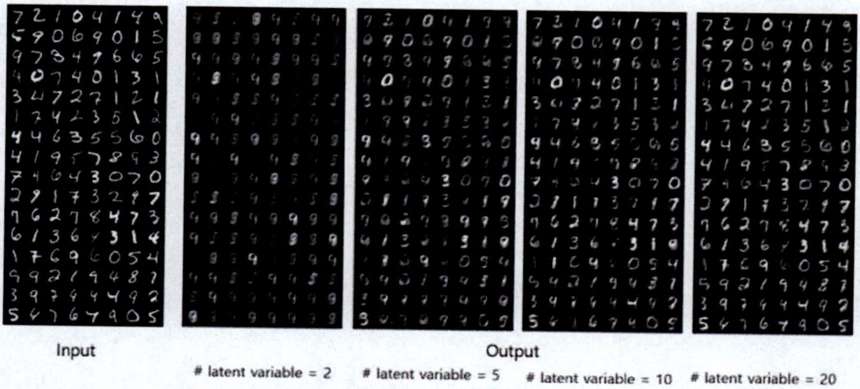

Fig. 1.8 Performance of linear AE with different $N_{\text{latent}} = 2, 5, 10, 20$

provides a medically meaningful interpolation between two images (a nonlinear interpolation between $\Psi(\mathbf{h}^{(i)}) = \mathbf{x}^{(i)}$ and $\Psi(\mathbf{h}^{(j)}) = \mathbf{x}^{(j)}$).

AEs can be used for image denoising [22]. Denoising AEs find the encoder Φ and decoder Ψ such that $\Psi \circ \Phi$ maps from noisy input to a clean image by filtering out noise while preserving important features. Its underlying idea is somewhat similar to existing methods such as low-pass filtering and compressed sensing (Figs. 1.7 and 1.8).

For the ease of explanation, let $\mathscr{T}_{images} = \{\mathbf{x}^{(n)}\}_{n=1}^{N_{\text{data}}}$ be a set of brain CT images. We hypothesize that most extant brain CT images lie near a low-dimensional manifold $\mathscr{M}$ with its Hausdorff dimension less than or equal to N_{latent}. We hope that the trained decoder Ψ will generate $\mathscr{M}$ (Figs. 1.9 and 1.10).

Restricted Boltzmann machines (RBMs) have been used as generative models that learn a probability distribution over the given data [26]. RBMs are two-layer NNs that have binary visible (input) and hidden units (latent variables). The term "restricted"

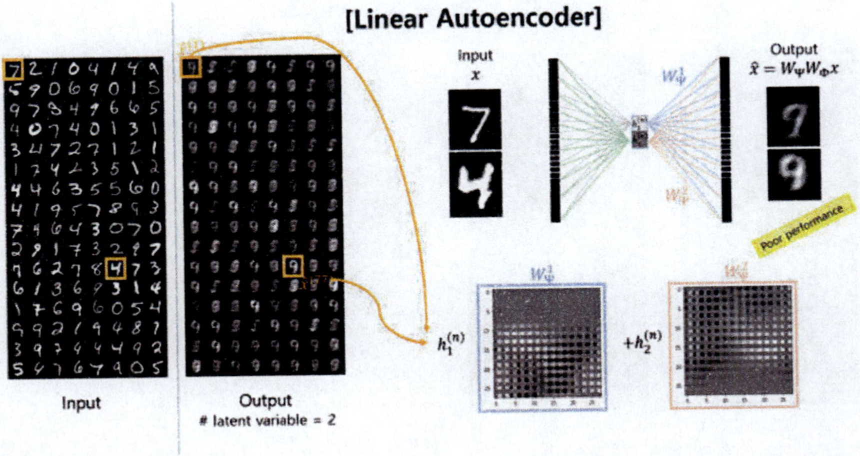

Fig. 1.9 This simple linear AE aims to find a $2 \times N_{\text{pixel}}$ matrix W_Φ and $N_{\text{pixel}} \times 2$ matrix W_Ψ which minimize $\sum_{\mathbf{x} \in \text{MNIST data}} \|W_\Psi W_\Phi \mathbf{x} - \mathbf{x}\|$. In the figure, W_Ψ^1 and W_Ψ^2 are the first and second row of W_Ψ

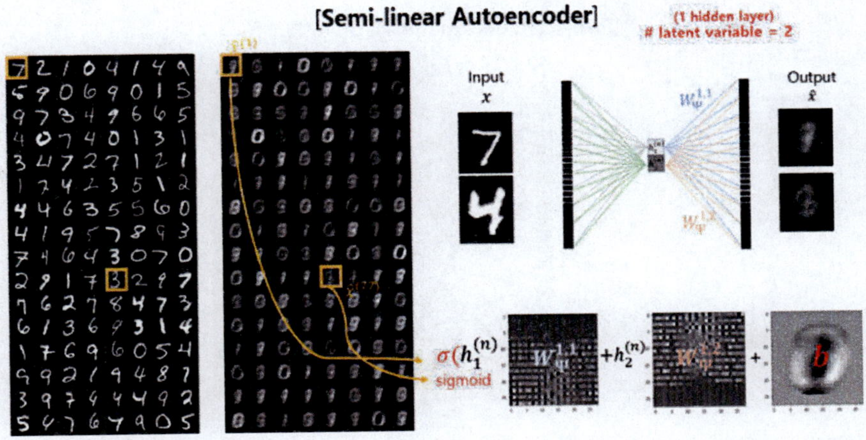

Fig. 1.10 Semi-linear AE by adding a sigmoid function in the decoder Ψ in Fig. 1.9. Here, the output is given by $\sigma(W_\Psi W_\Phi \mathbf{x} + \mathbf{b})$, where $\mathbf{b}$ is a bias vector

in RBM comes from the restriction that there is no communication between hidden units. At a hidden node, the latent vector is represented by $\mathbf{h} = \Phi(\mathbf{x}) := \sigma(W\mathbf{x} + \mathbf{b})$, where σ is the logistic sigmoid and W is a $N_{\text{latent}} \times N_{\text{pixel}}$ matrix. In the reconstruction phase, it uses the same matrix W to reproduce $\Psi(\mathbf{h}) = \sigma(W^T\mathbf{h} + \mathbf{b}')$, where W^T is the transpose of W. Hence, RBMs are symmetrical (Figs. 1.11, 1.12 and 1.13).

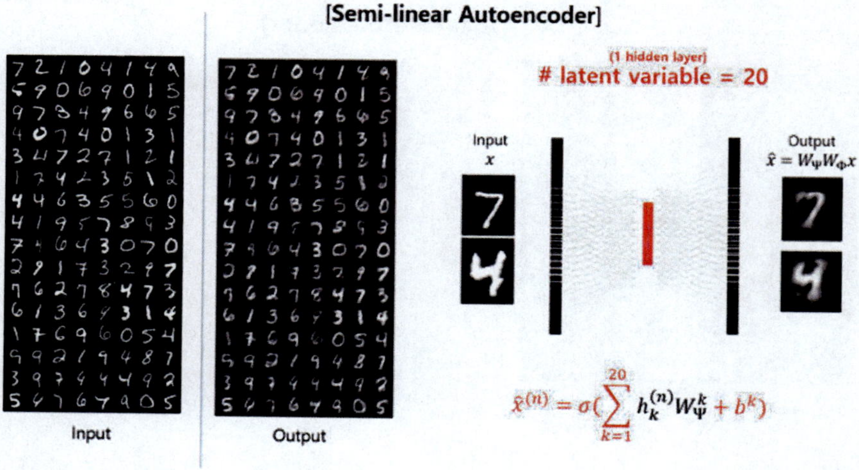

Fig. 1.11 Semi-linear AE with $N_{\text{latent}} = 20$. The number of unknown parameters is $N_{\text{pixel}} \times N_{\text{latent}} \times 2 + N_{\text{pixel}} = 784 \times 20 \times 2 + 784$. This approach is difficult to use when the input dimension is high

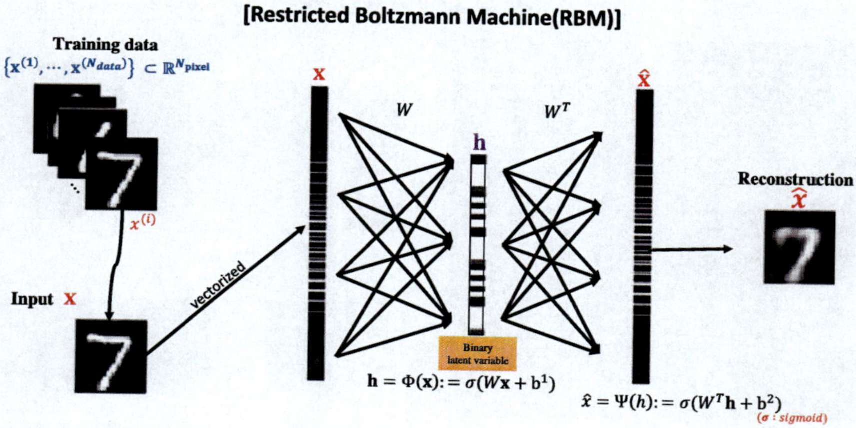

Fig. 1.12 This RBM is somewhat similar to the semi-linear AE in Fig. 1.11. The difference is that the decoder matrix W_Ψ is the transpose of the encoder matrix W_Ψ and the input unit and the hidden unit $\mathbf{h}$ are binary valued

In an RBM, we try to raise the probability $p_\Theta(\mathbf{x})$ (that will be defined later) by adjusting $\Theta = (W, \mathbf{b}, \mathbf{b}')$ to decrease the energy defined by

$$E(\mathbf{x}, \mathbf{h}) = -\mathbf{x} \cdot (W\mathbf{h}) - \mathbf{x} \cdot \mathbf{b}' - \mathbf{h} \cdot \mathbf{b}. \tag{1.65}$$

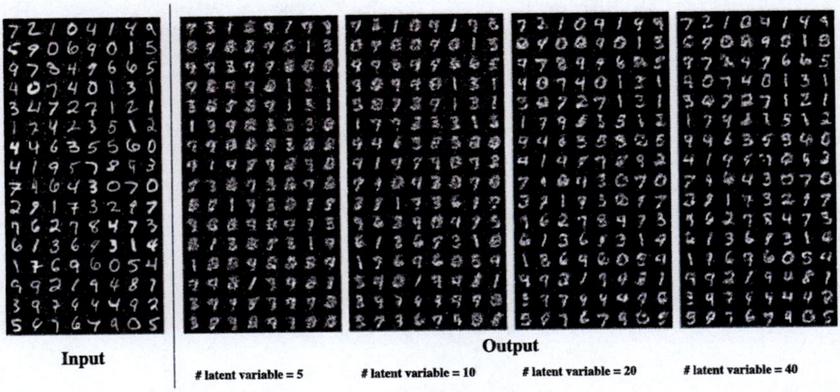

Fig. 1.13 Performance of RBM with different $N_{\text{latent}} = 5, 10, 20, 40$

The joint probability distribution for the pair $(\mathbf{x}, \mathbf{h})$ is defined by

$$p(\mathbf{x}, \mathbf{h}) = \frac{1}{Z} e^{-E(\mathbf{x},h)}, \tag{1.66}$$

where Z is a normalizing constant that is the sum of $e^{-E(\mathbf{x},h)}$ over all possible configurations. The aforementioned probability is the following marginal probability of $\mathbf{x}$:

$$p_\Theta(\mathbf{x}) = \sum_{\mathbf{h}} p(\mathbf{x}, \mathbf{h}) = \frac{1}{Z} \sum_{\mathbf{h}} e^{\mathbf{x}\cdot(W\mathbf{h}) + \mathbf{x}\cdot\mathbf{b}' + \mathbf{h}\cdot\mathbf{b}}, \tag{1.67}$$

Given the training data $\mathscr{T}_{images} = \{\mathbf{x}^{(k)}\}_{k=1}^{N_{\text{data}}}$, the loss function is

$$\text{Loss}(\Theta) = \frac{1}{N_{\text{data}}} \sum_{k}^{N_{\text{data}}} p_\Theta(\mathbf{x}^{(k)}), \tag{1.68}$$

Note that $p_\Theta(h_j = 1|\mathbf{x})$ (the conditional probability of $\mathbf{h}$ given $\mathbf{x}$) is the j-th component of $\sigma(W\mathbf{x} + \mathbf{b})$, and $p_\Theta(x_j = 1|\mathbf{h})$ (the conditional probability of $\mathbf{x}$ given $\mathbf{h}$) is the j-th component of $\sigma(W^T\mathbf{x} + \mathbf{b}')$.

1.4.2 Convolutional AE (CAE)

The simplest nonlinear encoder (compression) and decoder (decompression) are (Figs. 1.14, 1.15 and 1.16)

$$\Phi(\mathbf{x}) = \sigma(W \circledast \mathbf{x} + \mathbf{b}) \quad \text{and} \quad \Psi(\mathbf{h}) = \sigma(W' \circledast^\dagger \mathbf{h} + \mathbf{b}'), \tag{1.69}$$

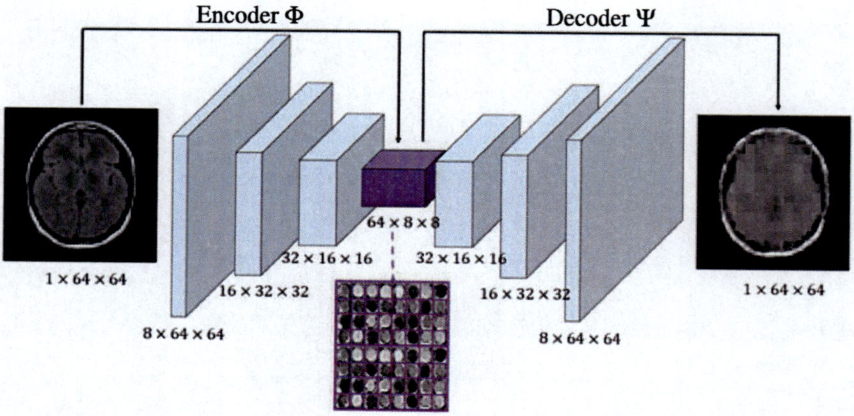

Fig. 1.14 Structure of CAE. CAE is a deep learning-based image compression method using a convolutional neural network (CNN) structure. Here, this network tries to represent the input image $\mathbf{x}$ with 64×64 pixels by $\Psi(\mathbf{h})$ with the number of latent variables $\mathbf{h}$ being $32 \times 8 \times 8$

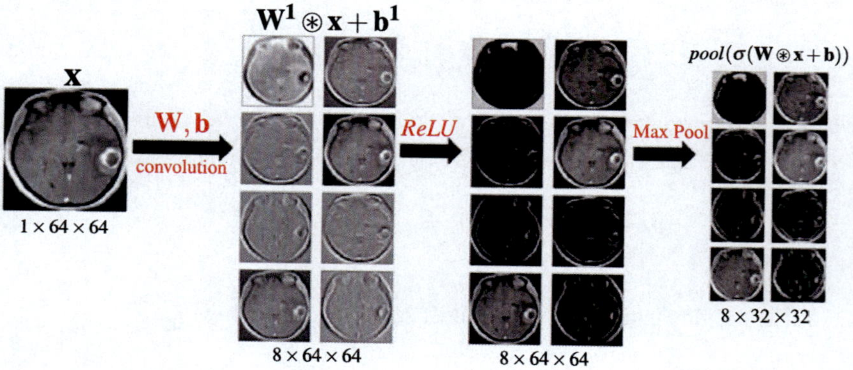

Fig. 1.15 The process of convolution operation. Here, 8 different convolution filters are applied to the input $\mathbf{x}$ to generate a feature map $\mathbf{W}^1 \circledast \mathbf{x} + \mathbf{b}^1$

where $W \circledast \mathbf{x}$ denotes the convolution [68] of $\mathbf{x}$ with weight W, $W' \circledast^{\dagger} \mathbf{h}$ denotes the up-convolution [68] of $\mathbf{h}$ with weight W', $\sigma(\cdot)$ represents an element-wise activation function, $\mathbf{h} \in \mathbb{R}^{N_{\text{latent}}}$, and $\mathbf{b}, \mathbf{b}' \in \mathbb{R}^{N_{\text{pixel}}}$. For clarity, let's take an example. Let $\mathbf{x}$ be an image of size 256×256. Let W be a weight W of size 5×5:

$$W = \begin{bmatrix} w(1,1) & \cdots & w(1,5) \\ \vdots & \ddots & \vdots \\ w(5,1) & \cdots & w(5,5) \end{bmatrix} \tag{1.70}$$

Then, the convolution $W \circledast_1 \mathbf{x}$ with stride 1 produces an image of dimension 252×252 that is given by

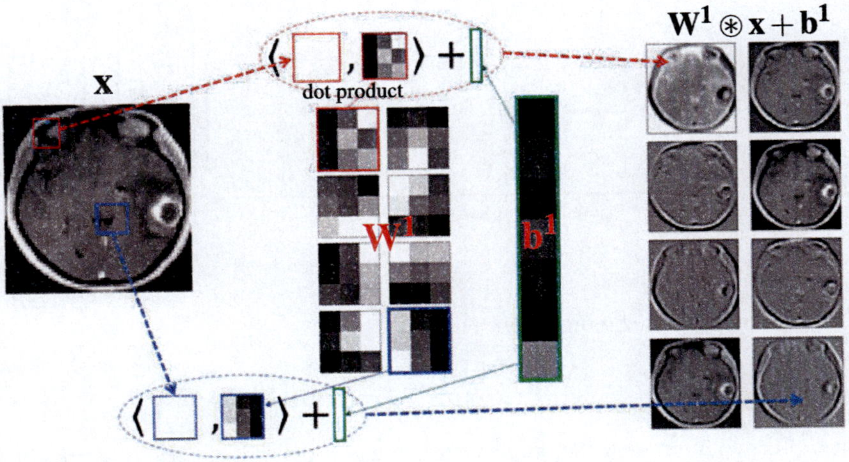

Fig. 1.16 Key structure of CNN in the first layer of encoding path. CNNs repeatedly use the operation $pool\ (\sigma(\mathbf{W} \circledast \mathbf{x} + \mathbf{b}))$, where σ is $ReLU$ and $pool$ is the max pooling operation

$$\mathbf{w} \circledast_1 \mathbf{x} = \underbrace{\begin{bmatrix} \sum_{i,j=1}^{5} w(i,j)x(i,j) & \sum_{i,j=1}^{5} w(i,j)x(i,j+1) & \cdots & \sum_{i,j=1}^{5} w(i,j)x(i,j+251) \\ \sum_{i,j=1}^{5} w(i,j)x(i+1,j) & \sum_{i,j=1}^{5} w(i,j)x(i+1,j+1) & \cdots & \sum_{i,j=1}^{5} w(i,j)x(i+1,j+251) \\ \vdots & \vdots & \ddots & \vdots \\ \sum_{i,j=1}^{5} w(i,j)x(i+251,j) & \sum_{i,j=1}^{5} w(i,j)x(i+251,j+1) & \cdots & \sum_{i,j=1}^{5} w(i,j)x(i+251,j+251) \end{bmatrix}}_{\text{This is a } 252\times252 \text{ matrix, where 252 comes from } 252=256-5+1.} \tag{1.71}$$

where $x(i, j)$ denotes the gray-scale value at pixel position (i, j). To understand the up-convolution $W' \circledast^{\dagger} \mathbf{h}$, we consider the following simple example:

$$\underbrace{\begin{pmatrix} w(1,1) & w(1,2) \\ w(2,1) & w(2,2) \end{pmatrix}}_{W'} \circledast^{\dagger} \underbrace{\begin{pmatrix} h(1,1) & h(1,2) & h(1,3) \\ h(2,1) & h(2,2) & h(2,3) \\ h(3,1) & h(3,2) & h(3,3) \end{pmatrix}}_{\mathbf{h}} = \underbrace{\begin{pmatrix} u(1,1) & \cdots & u(1,4) \\ u(2,1) & \cdots & u(2,4) \\ u(3,1) & \cdots & u(3,4) \\ u(4,1) & \cdots & u(4,4) \end{pmatrix}}_{\text{output } \mathbf{u}}. \tag{1.72}$$

The output $\mathbf{u}$ is expected to be a solution of the following de-convolution problem

$$W' \circledast \mathbf{u} = \mathbf{h}, \tag{1.73}$$

which can be written as

$$\underbrace{\begin{pmatrix} w(1,1) & w(1,2) & 0 & 0 & w(2,1) & w(2,2) & 0 & 0 & 0 & \cdots & 0 \\ 0 & w(1,1) & w(1,2) & 0 & 0 & w(2,1) & w(2,2) & 0 & 0 & \cdots & 0 \\ 0 & 0 & w(1,1) & w(1,2) & 0 & 0 & w(2,1) & w(2,2) & 0 & \cdots & 0 \\ \vdots & & & & & & & & & & \vdots \\ 0 & \cdots & 0 & 0 & 0 & w(1,1) & w(1,2) & 0 & 0 & w(2,1) & w(2,2) \end{pmatrix}}_{\mathbb{W}:\text{Toeplitz matrix}} \underbrace{\begin{pmatrix} u(1,1) \\ u(1,2) \\ u(1,3) \\ u(1,4) \\ u(2,1) \\ u(2,2) \\ u(2,3) \\ u(2,4) \\ u(3,1) \\ \vdots \\ u(4,4) \end{pmatrix}}_{\mathbf{u}_{\text{vectorize}}} = \underbrace{\begin{pmatrix} h(1,1) \\ h(1,2) \\ \vdots \\ \vdots \\ h(3,3) \end{pmatrix}}_{\mathbf{h}_{\text{vectorize}}} \tag{1.74}$$

Then, the output $\mathbf{u}$ is computed by

$$\mathbb{W}^T \mathbf{h}_{\text{vectorize}} = \mathbf{u}_{\text{vectorize}}. \tag{1.75}$$

The up-convolution feels like an inverse of the convolution, but this is not really a mathematical inverse or de-convolution.

In general, the encoder Φ is of the form

$$\Phi(\mathbf{x}) := W^{\ell-1} \circledast \left(\sigma(W^{\ell-2} \circledast \sigma\left(\cdots pool\left(\sigma\left(W^1 \circledast \mathbf{x} + \mathbf{b}^1\right)\right)\cdots\right)\right) + \mathbf{b}^{\ell-1}, \tag{1.76}$$

where σ is $ReLU$. Similarly, the decoder Ψ is of the form:

$$\Psi(\mathbf{h}) = \tanh\left(W^{2\ell} \circledast^{\dagger} \left(\sigma(W^{2\ell-1} \circledast^{\dagger} \sigma\left(\cdots\sigma\left(W^{\ell+1} \circledast^{\dagger} \mathbf{h}\right)\cdots\right)\right)\right), \tag{1.77}$$

where tanh is the hyperbolic tangent function.

AEs aim to develop a low-dimensional latent representation of higher-dimensional images while keeping the image structural information. The dimension of the latent space (N_{latent}) and the depth of the AE architecture must be carefully selected in order not to lose important information in the input image $\mathbf{x}$. By increasing the depth of the AE and increasing the degrees of freedom, the reconstruction loss on the training data can approach zero. However, having too many degrees of freedom can lead to overfitting, making it difficult for the decoder to achieve its generative purpose. It is easy to develop AEs with little reproducing loss on the training data, if we don't care how the latent space is structured. But this is pointless, because the decoder is hoping to generate new images by randomly sampling the latent vectors.

1.4.3 Variational Autoencoders (VAEs)

Given training data $\mathcal{T}_{images} = \{\mathbf{x}^{(k)}\}_{k=1}^{N_{\text{data}}}$, AEs do their best to learn (Φ, Ψ) with as little reconstruction loss (1.63) as possible, rather than trying to organize the latent space well for generative purposes. Therefore, the AE's architecture has a limitation in constructing the latent space in an interpretable and usable form based on the characteristics of images. In other words, the architecture of AEs may not provide

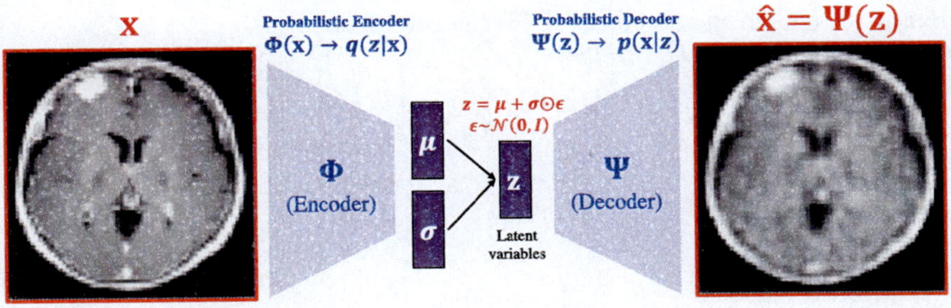

$$Loss(\Phi, \Psi) = \sum_{n=1}^{N_{data}} \left\{ \underbrace{D_{KL}[q(\mathbf{z}|\mathbf{x}^{(n)}) \parallel p(\mathbf{z})]}_{\text{Regularization}} - \underbrace{E_{q_\Phi(\mathbf{z}|\mathbf{x}^{(n)})} \log p(\mathbf{x}^{(n)}|\mathbf{z})}_{\text{Reconstruction loss}} \right\}$$

Fig. 1.17 The structure of VAE. VAE is designed to prevent overfitting and to regularize latent space to enable generative processes. Note that VAE encodes the input as $\mathbf{z} = \mu + \sigma \odot \epsilon, \epsilon \sim \mathcal{N}(0, I)$ (a distribution over the latent space)

regularity in the latent space. If the two images are close from the doctor's point of view, it is desirable that the corresponding points in the latent space are close with respect to Euclidian distance. It would be desirable to organize the latent space in such a way that the average of the two points in the latent space should be about halfway between the corresponding two images with respect to the doctor's point of view (Fig. 1.17).

Variational AE (VAE) is a variant of AE designed to prevent overfitting and to allow latent space to be regularized to enable generative processes [36]. The difference between AE and VAE is as follows: AE encodes the input as a single point, whereas VAE encodes the input as a distribution over the latent space. In the training process of VAEs, an input $\mathbf{x}$ is encoded into a latent distribution $p(\mathbf{z}|\mathbf{x})$ and then $\mathbf{z}$ is sampled from $p(\mathbf{z}|\mathbf{x})$ (i.e., $\mathbf{z} \sim p(\mathbf{z}|\mathbf{x})$).

The VAE was developed by Kingma and Welling in 2013 [35]. The goal is to find a generative model $p_{\text{model}}(\mathbf{x})$ such that $p_{\text{model}}(\mathbf{x}) \approx p_{\text{data}}(\mathbf{x})$. Here, we use the dataset $\{\mathbf{x}^{(1)}, \cdots, \mathbf{x}^{(N)}\} \sim p_{\text{data}}(\mathbf{x})$ to find a model $p_{\text{data}}(\mathbf{x})$. Let us denote $p_{\text{model}}(\mathbf{x})$ by $p_\theta(\mathbf{x})$ where θ stands for the set of paramters of $p_{\text{model}}(\mathbf{x})$. To fit $p_\theta(\mathbf{x})$ to $p_{\text{data}}(\mathbf{x})$, we optimize θ by mimimizing the KL-divergence between two distributions $p_\theta(\mathbf{x})$ and $p_\theta(\mathbf{x})$:

$$\theta^* = \underset{\theta}{\text{argmin}} \ D_{KL}(p_{\text{data}} \parallel p_\theta), \tag{1.78}$$

where $D_{KL}(q \parallel p) = \int q(x) \log \frac{q(x)}{p(x)} dx$. Note that

$$D_{KL}(p_{\text{data}} \parallel p_\theta) = \underbrace{E_{p_{\text{data}}}[\log p_{\text{data}}]}_{\text{constant w.r.t. } \theta} - E_{p_{\text{data}}}[\log p_\theta]. \tag{1.79}$$

Hence, the optimization problem (1.78) can be expressed as

$$\theta^* = \underset{\theta}{\text{argmax}}\ E_{p_{\text{data}}}[\log p_\theta], \tag{1.80}$$

Using the dataset $\{\mathbf{x}^{(1)}, \cdots, \mathbf{x}^{(N)}\} \sim p_{\text{data}}(\mathbf{x})$, the optimization problem (1.80) can be approximated as

$$\theta^* = \underset{\theta}{\text{argmax}}\ \frac{1}{N}\sum_{n=1}^{N} \log p_\theta(\mathbf{x}^{(n)}). \tag{1.81}$$

VAEs are latent variable models $p_\theta(\mathbf{x}|\mathbf{z})$, where learning the distribution $p_\theta(\mathbf{x})$ from $\{\mathbf{x}^{(1)}, \cdot, \mathbf{x}^{(n)}\} \sim p_{\text{data}}(\mathbf{x})$ means learning the dependencies among pixels. Here, $p_\theta(\mathbf{x})$ is expressed as

$$\log p_\theta(\mathbf{x}) = E_{q_\phi(\mathbf{z}|\mathbf{x})}[\log p_\theta(\mathbf{x})] = E_{p_\phi(\mathbf{z}|\mathbf{x})}[\log \frac{p_\theta(\mathbf{x}, \mathbf{z})}{q_\phi(\mathbf{z}|\mathbf{x})}] \tag{1.82}$$

$$= E_{p_\phi(\mathbf{z}|\mathbf{x})}[\log \frac{p_\theta(\mathbf{x}, \mathbf{z})}{q_\phi(\mathbf{z}|\mathbf{x})} \frac{q_\phi(\mathbf{z}|\mathbf{x})}{p_\theta(\mathbf{z}|\mathbf{x})}] \tag{1.83}$$

$$= E_{p_\phi(\mathbf{z}|\mathbf{x})}[\log \frac{p_\theta(\mathbf{x}, \mathbf{z})}{q_\phi(\mathbf{z}|\mathbf{x})}] + E_{p_\phi(\mathbf{z}|\mathbf{x})}[\log \frac{q_\phi(\mathbf{z}|\mathbf{x})}{p_\theta(\mathbf{z}|\mathbf{x})}] \tag{1.84}$$

$$= \mathscr{L}(\mathbf{x}, \phi, \theta) + D_{KL}[q_\theta(\mathbf{z}|\mathbf{x})||p_\theta(\mathbf{z}|\mathbf{x})]. \tag{1.85}$$

Since the distribution $p_\theta(\mathbf{x})$ is intractable, the goal is reduced to optimize the ELBO $\mathscr{L}(\mathbf{x}, \phi, \theta) = E_{p_\phi(\mathbf{z}|\mathbf{x})}[\log \frac{p_\theta(\mathbf{x}, \mathbf{z})}{q_\phi(\mathbf{z}|\mathbf{x})}]$, where $\mathscr{L}(\mathbf{x}, \phi, \theta)$ is tight when $q_\phi(\mathbf{z}|\mathbf{x}) \approx p_\theta(\mathbf{z}|\mathbf{x})$. Maximization of ELBO is equivalent to minimizing $KL[q_\theta(\mathbf{z}|\mathbf{x})||p_\theta(\mathbf{z}|\mathbf{x})]$ [37]. The details of the structure of the VAE are described in the next section.

1.4.3.1 Framework of VAEs

In VAE, the probabilistic encoder produces

$$q(\mathbf{z}|\mathbf{x}) = \mathcal{N}(\mathbf{z}|\underbrace{\Phi_{\text{mean}}(\mathbf{x})}_{\text{mean}}, \underbrace{\Phi_{\text{variance}}(\mathbf{x})\, I}_{\text{variance}}), \tag{1.86}$$

where Φ_{mean} outputs a vector of means

$$\Phi_{\text{mean}}(\mathbf{x}) = \mu = (\mu(1), \cdots, \mu(N_{\text{latent}})) \in \mathbb{R}^{N_{\text{latent}}} \tag{1.87}$$

and Φ_{variance} outputs a diagonal covariance matrix

$$\Phi_{\text{variance}}(\mathbf{x}) = \Sigma = \text{diag}(\sigma(1)^2, \cdots, \sigma(N_{\text{latent}})^2). \tag{1.88}$$

In practice, the encoder Φ is of the following nondeterministic form:

$$\Phi(\mathbf{x}) = \Phi_{\text{mean}}(\mathbf{x}) + \sqrt{\Phi_{\text{variance}}(\mathbf{x})} \odot \mathbf{h}_{\text{noise}}, \tag{1.89}$$

where $\mathbf{h}_{\text{noise}}$ is an auxiliary noise variable sampled from standard normal distribution $\mathcal{N}(0, I)$ and $\odot$ is the element-wise product (Hadamard product).

Denoting by $p(\mathbf{x}|\mathbf{z})$ the probabilistic decoder, the decoder Ψ is chosen to maximize

$$E_{\mathbf{z}\sim q_{\mathbf{x}}} \log p(\mathbf{x}|\mathbf{z}) = E_{\mathbf{z}\sim q_{\mathbf{x}}} \frac{|\Psi(\mathbf{z}) - \mathbf{x}|^2}{\sigma^2}. \tag{1.90}$$

Given training data $\mathcal{T}_{images} = \{\mathbf{x}^{(n)}\}_{n=1}^{N_{\text{data}}}$, the goal is to maximize the reconstruction log-likelihood:

$$\sum_{n=1}^{N_{\text{data}}} \log p(\mathbf{x}^{(n)}) = \sum_{n=1}^{N_{\text{data}}} \log \int p(\mathbf{x}^{(n)}|\mathbf{z})p(\mathbf{z})d\mathbf{z}. \tag{1.91}$$

To achieve the goal, we try to minimize the following loss function:

$$Loss(\Psi, \Phi) = \sum_{n=1}^{N_{\text{data}}} \{D_{KL}[q(\mathbf{z}|\mathbf{x}^{(n)}) \parallel p(\mathbf{z}|\mathbf{x}^{(n)})] - \log p(\mathbf{x}^{(n)})\}. \tag{1.92}$$

For compact encoding and smooth interpolation in the latent space, we additionally enforce that the distribution of the encoder output is close to a normal distribution $\mathcal{N}(0, I)$, by penalizing the KL divergence loss between $\mathcal{N}(\mu, \Sigma)$ and $\mathcal{N}(0, I)$. The KL-divergence between two distributions p and q on the same probability space is given by $D_{KL}(q \parallel p) = \int q(z) \log \frac{q(z)}{p(z)} dz$. Direct computation shows that

$$D_{KL}(\mathcal{N}(\mu, \Sigma) \parallel \mathcal{N}(0, I)) = \frac{1}{2} \sum_{j=1}^{N_{\text{latent}}} \left[(\mu(j)^2 + \sigma(j)^2 - \log \sigma(j) - 1\right] \tag{1.93}$$

Lemma 1.1 *The $Loss(\Psi, \Phi)$ in (1.92) is expressed as*

$$Loss(\Psi, \Phi) = \sum_{n=1}^{N_{\text{data}}} \{D_{KL}[q(\mathbf{z}|\mathbf{x}^{(n)}) \parallel p(\mathbf{z})] - E_{q_\Phi(z|\mathbf{x}^{(n)})} \log p(\mathbf{x}^{(n)}|\mathbf{z})\}. \tag{1.94}$$

Proof Using the Bayes' theorem ($p(\mathbf{z}|\mathbf{x}) = \frac{p(\mathbf{x}|\mathbf{z})p(\mathbf{z})}{p(\mathbf{x})}$),

$$\begin{aligned} D_{KL}[q(\mathbf{z}|\mathbf{x}^{(n)}) \parallel p(\mathbf{z}|\mathbf{x}^{(n)})] &= \int q(\mathbf{z}|\mathbf{x}^{(n)}) \log \left(\frac{q(\mathbf{z}|\mathbf{x}^{(n)})}{p(\mathbf{z})} \frac{p(\mathbf{x}^{(n)})}{p(\mathbf{x}^{(n)}|\mathbf{z})} \right) d\mathbf{z} \\ &= D_{KL}[q(\mathbf{z}|\mathbf{x}^{(n)}) \parallel p(\mathbf{z})] + \log p(\mathbf{x}^{(n)}) - E_{q_\Phi(z|\mathbf{x}^{(n)})} \log p(\mathbf{x}^{(n)}|\mathbf{z}). \end{aligned}$$

Then (1.94) follows from substituting the above identity into (1.92). □

Theorem 1.2 *The* $Loss(\Psi, \Phi)$ *in (1.94) is expressed as*

$$Loss(\Psi, \Phi) = \frac{1}{2}\sum_{n=1}^{N_{\text{data}}}\left\{D_{KL}[tr(\Sigma^{(n)}) + |\mu^{(n)}|^2 - \log|\Sigma^{(n)}| - E_{q_\Phi(\mathbf{z}|\mathbf{x}^{(n)})}\log p_\Psi(\mathbf{x}^{(n)}|\mathbf{z})\right\} \tag{1.95}$$

where $\mu^{(n)} = \Phi_{mean}(\mathbf{x}^{(n)}))$ *and* $\Sigma^{(n)} = \Phi_{variance}(\mathbf{x}^{(n)}))$.

Proof From (1.87), we have

$$\begin{aligned} 2D_{KL}[q(\mathbf{z}|\mathbf{x})) \parallel p(\mathbf{z})] &= 2\int q(\mathbf{z}|\mathbf{x}))\log\left(\frac{q(\mathbf{z}|\mathbf{x})}{p(\mathbf{z})}\right)d\mathbf{z} \\ &= -E_{q_\Phi(z|\mathbf{x})}\left[(\mathbf{z}-\mu)\Sigma^{-1}(\mathbf{z}-\mu)^T - |\mathbf{z}|^2\right] - \log|\Sigma| \\ &= -N_{\text{latent}} + \text{tr}(\Sigma) + |\mu_z|^2 - \log|\Sigma|. \end{aligned}$$

Then (1.95) follows from substituting the above identity into (1.94). □

The integral $E_{q_\Phi(\mathbf{z}|\mathbf{x}^{(n)})}\log p_\Psi(\mathbf{x}^{(n)}|\mathbf{z})$ admits no closed-form, making direct optimization over Ψ and Φ intractable. Likewise, any detailed analysis of the underlying objective function becomes problematic as well. Hence, we use Monte-Carlo stochastic approximation

$$E_{q_\Phi(\mathbf{z}|\mathbf{x}^{(n)})}\log p_\Psi(\mathbf{x}^{(n)}|\mathbf{z}) \approx \frac{1}{S}\sum_{s=1}^{S}\log p_\Psi(\mathbf{x}^{(n)}|\mathbf{z}^{n,s}) \tag{1.96}$$

Dai et al. [14] viewed VAE as the natural evolution of robust PCA models, capable of learning nonlinear manifolds of unknown dimension obscured by gross corruptions. Given data $\mathscr{T}_{images} = \{\mathbf{x}^{(k)}\}_{k=1}^{N_{\text{data}}}$, the encoder $\Phi(\mathbf{x})$ can be viewed as a conditional distribution $q(\mathbf{h}|\mathbf{x})$ that satisfies $q(\mathbf{h}|\mathbf{x}) = \mathscr{N}(\mu, \Sigma)$. The decoder Ψ can be represented by a conditional distribution $p(\mathbf{x}|\mathbf{h})$ with $p(\mathbf{h}) = \mathscr{N}(0, I)$. VAE tries to match $p(\mathbf{h}|\mathbf{x})$ and $q(\mathbf{h}|\mathbf{x})$. VAE encoder covariance can help to smooth out undesirable minima in the energy landscape of what would otherwise resemble a more traditional deterministic autoencoder [14].

1.4.3.2 AEs Versus VAEs: Comparison of Generative Models

Deep Convolution AEs have a series of convolutional layers, each with trainable convolutional filters that produce a feature map (that is an input to the next layer). Here, ReLU (Rectified Linear Unit) is commonly used to limit the input of each convolutional layer to a specific range, and the pooling layer is usually added after ReLU(the activation), which downsamples the feature map. The goal of its encoder is to extract increasingly complex features (hierarchical features) through multiple

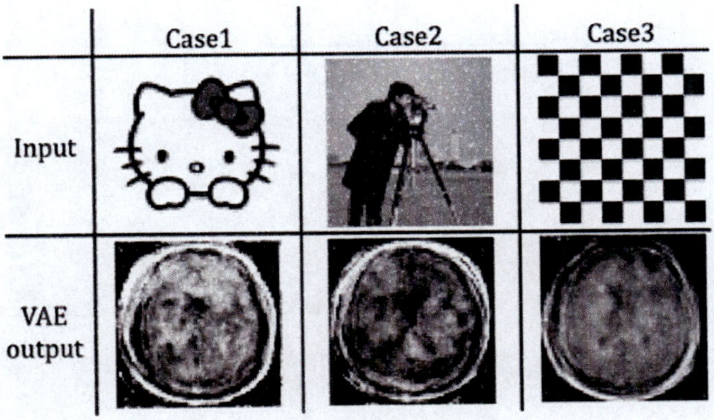

Fig. 1.18 VAE as generative model. Unlike AE, which only aims to reproduce its input, VAE encodes latent variables to be compressed and normalized to a normal distribution, which enables the generative process

hidden convolutional layers, eventually extracting high-level abstractions $\mathbf{z}$ at the bottleneck in the network. In AEs, a mirrored decoder aims to reconstruct the input $\mathbf{x}$ using the latent variable $\mathbf{z} = \Phi(\mathbf{x})$ (i.e., high-level abstractions at the bottleneck). Shortcut connections in ResNet (Residual Neural Network) can be used to propagate the finer details of the input to deeper layers.

What we hope is that the encoder Φ extracts a high-level abstraction $\mathbf{z}$ so that the decoder Ψ generates new contents. For the decoder Ψ to play a generative role, the output $\hat{\mathbf{x}} = \Psi \circ \Phi(\mathbf{x})$ is expected to have low energy in the sense that

$$Energy(\hat{\mathbf{x}}) = \begin{cases} \text{low,} & \text{near a manifold containing } \mathcal{T}_{images} \\ \text{high,} & \text{elsewhere} \end{cases}, \tag{1.97}$$

where $\mathcal{T}_{images}$ is a set of training data. AEs in Fig. 1.14 do their best to learn (Φ, Ψ) with as little reconstruction loss as possible, rather than trying to organize the latent space well for generative purposes. Hence, AE only aims to reproduce its input and is not considered a generative model. On the other hand, VAE acts as a generative model to some extent, as shown in Fig. 1.18. In VAE, the encoded latent variables are compressed and normalized to a normal distribution, to enable the generative process. VAEs can achieve somewhat successful manifold learning only in a relatively low-dimensional image space. However, we have not yet succeeded in fitting low-dimensional manifolds to real high-dimensional CT image data using VAEs (Figs. 1.19 and 1.20).

The expressive power of VAE is very limited, because the latent variable $\mathbf{z}$ has global information of $\mathbf{x}$ but may not have detailed pixel-wise information. VAEs can achieve somewhat successful manifold learning only on relatively low-dimensional data and not on high-dimensional data. For high-dimensional data, VAEs suffer from image blurring and loss of small details.

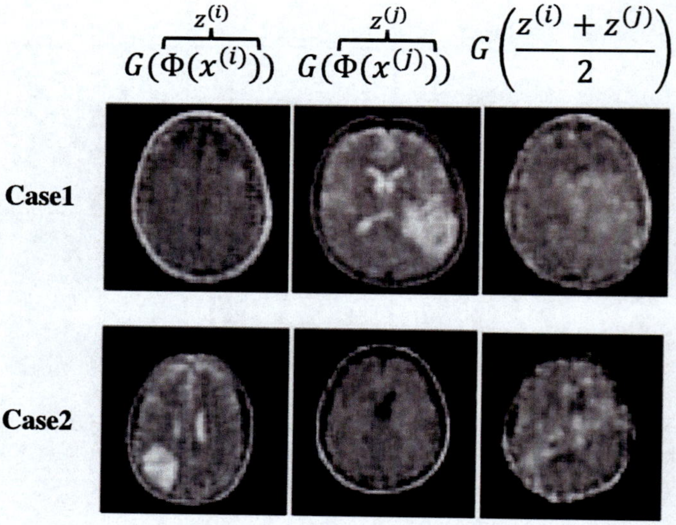

Fig. 1.19 Latent walk using VAE. The average of the two points in the latent space generates about halfway between the corresponding two images with respect to the image point of view

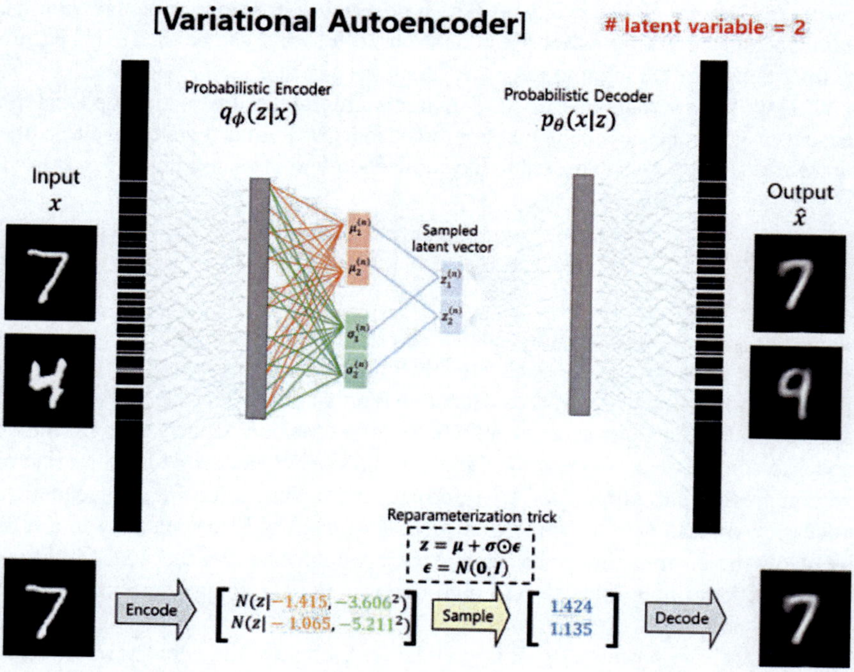

Fig. 1.20 This shows an example of VAE using MNIST data with the latent dimension given as 2. The difference with the AE is that the bottleneck structure contains a sampling process, which enables smooth interpolation in the latent space

To reproduce the fine details, the U-net use skip connections where the high-resolution features of the encoding layer are directly connected to the corresponding decoding layer [59]. In the U-net, information transmitted over long skip connections in the encoding path often contains unnecessary elements of the input image, such as noise. Due to these skip connections, the U-net cannot serve as a generative model. In our opinion, AEs without skip connections, as in the U-net, seem to continue to suffer from blurring and loss of detail when generating new synthetic medical images.

1.4.3.3 Generative Adversarial Networks

GAN is a clever way of generating realistic images by introducing the discriminative network that is trained to classify an input image as real or synthetic [20, 21]. The training method of GAN is that in order to generate synthetic images that are as realistic as possible, the generative network is subjected to the strict punishment of the discriminatory network, finally reaching a level where the discriminative network is tricked into discriminating synthetic images as real images.

The ultimate goal is to learn a generator

$$G : \mathscr{X} \to \mathscr{Y}, \tag{1.98}$$

where $\mathscr{X}$ is an input distribution of a generator G and $\mathscr{Y}$ is a target distribution of medical images to be generated. In practice, G is learned by finite samples from the input distribution $\mathscr{X}$ and $\mathscr{Y}$. Assume that $\mathscr{S}_\mathbf{x} := \{\mathbf{x}_j\}_{j=1}^N$ and $\mathscr{S}_\mathbf{y} := \{\mathbf{y}_j\}_{j=1}^M$ are training samples drawn from $\mathscr{X}$ and $\mathscr{Y}$, respectively. We denote the data distribution as $\mathbf{x} \sim p_{\text{data}}(\mathbf{x})$ and $\mathbf{y} \sim p_{\text{data}}(\mathbf{y})$. Let θ_G and θ_D be parameters of G and D, respectively.

The goal of the GANs is to simultaneously train the generator G and the discriminator D in an adversarial relationship to improve their mutual abilities; D tries to maximize its ability to distinguish between $G(\mathbf{x}), \mathbf{x} \sim \mathscr{X}$ and $\mathbf{y} \sim \mathscr{Y}$, while the generator G tries cheat the discriminator D by bringing the generated samples $G(\mathbf{x}), \mathbf{x} \sim \mathscr{X}$ close to the target distribution $\mathscr{Y}$.

We begin with explaining the original GANs made by Goodfellow et al. [20]. The input of G is the standard distribution $\mathscr{X} = \mathscr{N}(0, I)$ of a certain dimension. Learning G and D is done through training on the parameters θ_G and θ_D of the following discriminator/generator minimax game:

$$\min_{\theta_G} \max_{\theta_D} \Gamma_{GAN}(\theta_G, \theta_D), \tag{1.99}$$

where $\Gamma_{GAN}(\theta_G, \theta_D)$ is the adversarial loss function given by

$$\Gamma_{GAN}(\theta_G, \theta_D) := E_{\mathbf{y} \sim p_{\text{data}}(\mathbf{y})} \left[\log D(\mathbf{y}; \theta_D)\right] + E_{\mathbf{x} \sim p_{\text{data}}(\mathbf{x})} \left[\log(1 - D(G(\mathbf{x}; \theta_G); \theta_D)\right]. \tag{1.100}$$

Here, $E[\mathbf{y}]$ denotes the expectation of $\mathbf{y}$. In (1.100), the discriminator D is trained to minimize $\Gamma_{GAN}(\theta_G, \theta_D)$; it tries to have $D(\mathbf{y}) = 1$ for $\mathbf{y} \sim \mathscr{Y}$ and $D(G(\mathbf{x})) = 0$ for $\mathbf{x} \sim \mathscr{X}$. The generator G, on the other hand, tries to maximize $D(G(\mathbf{x}))$; it tries to generate $G(\mathbf{x})$ as if $G(\mathbf{x}) \sim \mathscr{Y}$ as much as possible, causing the discriminator D to misclassify $D(G(z)) = 1$. The learning of G and D can be achieved using alternating stochastic gradient descent:

1. Given G, train D to minimize $\Gamma_{GAN}(\theta_G, \theta_D)$.
2. Given D, train G to maximize $\Gamma_{GAN}(\theta_G, \theta_D)$.

This original GAN loss has suffered from vanishing gradient problems that prevent the generator from updating well. The term $\log(1 - D(G(\mathbf{x}))$ will saturate when $D(G(\mathbf{x}) \approx 0$ [47]. The adversarial loss can be replaced by various kinds of GAN loss function (e.g., least squares loss [45, 47] and Wasserstein loss [23]) to get high performance or training stability. Noting that the objective still makes intuitive sense if we replace log by any monotone function $\varphi : [0, 1] \to \mathbb{R}$ [4], Wasserstein GAN(WGAN) [3] used $\varphi(t) = t$.

The least squares GANs (LSGANs) [45, 47] use the following objective functions that yield minimizing the Pearson χ^2 divergence [47]:

$$\begin{cases} \min\limits_{\theta_D} \frac{1}{2} E_{\mathbf{y}\sim p_{\text{data}}(\mathbf{y})} \left[(D(\mathbf{y}; \theta_D) - 1)^2\right] + \frac{1}{2} E_{\mathbf{x}\sim p_{\text{data}}(\mathbf{x})} \left[(D(G(\mathbf{z}; \theta_G); \theta_D))^2\right], \\ \min\limits_{\theta_G} \frac{1}{2} E_{\mathbf{x}\sim p_{\text{data}}(\mathbf{x})} \left[(D(G(\mathbf{z}; \theta_G); \theta_D) - 1)^2\right], \end{cases} \tag{1.101}$$

where 1 and 0 are the labels for the fake data and the real data, respectively, and the term $(D(G(\mathbf{x}; \theta_G); \theta_D) - 1)^2$ is used when G wants D to believe the fake data. A good GAN loss function will captures the difference between the two distributions in order to update G well.

The original GANs cannot provide image-to-image translation (i.e., mapping from $\mathbf{x}$ to a target output $\mathbf{y}$). CycleGAN can be viewed as an extension of GAN and is designed to solve the problem of GAN of unpaired image-to-image conversion. It uses a cycle-consistency constraint to reliably translate an image from one domain to the corresponding image from another domain without having paired training data: Given an input $\mathbf{x}$ in one domain $\mathscr{X}$, a generator G produces $\hat{\mathbf{y}} = G(\mathbf{x})$ in another domain $\mathscr{Y}$. The corresponding fake generator F (that can be viewed as an inverse of G) maps from $\mathbf{y}$ to $\hat{\mathbf{x}} = F(\mathbf{y})$. The cycle-consistency loss is the differences $\|\mathbf{x} - F \circ G(\mathbf{x})\|$ and $\|\mathbf{y} - G \circ F(\mathbf{y})\|$. The Cycle-GAN can be used as semi-automated labeling systems to generate synthetic datasets for medical machine learning applications, which could help address the data-hungry nature of diverse contents. CycleGANs have had remarkable success in changing the style of an image while preserving contents (e.g., ultrasound image-to-MRI image synthesis). However, cycleGAN should only be used in a very limited range of applications in medicine, as it tends to reject the presence of small anomalies that are rarely seen, due to the strong punishment of the discriminator (i.e., inflexible training method that only scores on approval or rejection). The lack of flexibility of the discriminator to transform images to their own taste without leaving the anomalous parts untouched can cause serious misdiagnosis problems in medicine.

Now, let us explain the mathematical framework of cycleGAN. Throughout this section, the input $\mathbf{x}$ represents a synthesized image by the method in the previous section and the output $\mathbf{y}$ represents a practical CBCT image with metal artifacts. We assume that unpaired training samples $\{\mathbf{x}^{(n)}\}_{n=1}^N$ and $\{\mathbf{y}^{(m)}\}_{m=1}^M$ are drawn from unknown synthetic and real CBCT image distributions ($p_{\mathscr{X}}(\mathbf{x})$ and $p_{\mathscr{Y}}(\mathbf{y})$), respectively. The goal is to find an optimal generator $G_{\mathscr{Y}} : \mathscr{X} \to \mathscr{Y}$ such that the output distribution over $\widehat{\mathbf{y}} = G_{\mathscr{Y}}(\mathbf{x})$ is a good approximation of $p_{\mathscr{Y}}(\mathbf{y})$ (i.e., $\widehat{\mathscr{Y}} = G_{\mathscr{Y}}(\mathscr{X})$ is distributed almost similarly to $\mathscr{Y}$).

The model of CycleGAN comprises two mappings $G_{\mathscr{Y}} : \mathscr{X} \to \mathscr{Y}$ (generating images for $\mathscr{Y}$) and $G_{\mathscr{X}} : \mathscr{Y} \to \mathscr{X}$ (generating images for $\mathscr{X}$). Similar to the standard GAN as in (1.99), this model uses the corresponding adversarial discriminators $D_{\mathscr{Y}}$ (for distinguishing between real $\mathbf{y}$ and fake $G_{\mathscr{Y}}(\mathbf{x})$) and $D_{\mathscr{X}}$ (for distinguishing between real $\mathbf{x}$ and fake $G_{\mathscr{X}}(\mathbf{y})$).

Roughly speaking, the main objective of CycleGAN is to find $(G^*_{\mathscr{Y}}, G^*_{\mathscr{X}})$ such that

$$(G^*_{\mathscr{Y}}, G^*_{\mathscr{X}}) := \underset{G_{\mathscr{Y}}, G_{\mathscr{X}}}{\operatorname{argmin}} \underset{D_{\mathscr{X}}, D_{\mathscr{Y}}}{\operatorname{argmax}} \mathscr{L}(G_{\mathscr{Y}}, G_{\mathscr{X}}, D_{\mathscr{Y}}, D_{\mathscr{X}}), \tag{1.102}$$

where the generator $G^*_{\mathscr{Y}}$ corresponds to our desired synthetic-to-realistic image refinement map. Here, the loss function $\mathscr{L}$ is given by

$$\mathscr{L}(G_{\mathscr{Y}}, G_{\mathscr{X}}, D_{\mathscr{X}}, D_{\mathscr{Y}}) := \mathscr{L}_{\text{GAN}}(G_{\mathscr{Y}}, D_{\mathscr{Y}}) + \mathscr{L}_{\text{GAN}}(G_{\mathscr{X}}, D_{\mathscr{X}}) + \lambda \mathscr{L}_{\text{cyc}}(G_{\mathscr{Y}}, G_{\mathscr{X}}), \tag{1.103}$$

where λ is a parameter controlling the contribution of the consistency. The first term $\mathscr{L}_{\text{GAN}}(G_{\mathscr{Y}}, D_Y)$ is for estimating optimal $G_{\mathscr{Y}}$ and $D_{\mathscr{Y}}$, given by

$$\mathscr{L}_{\text{GAN}}(G_{\mathscr{Y}}, D_{\mathscr{Y}}) = \mathbb{E}_{\mathbf{y}\sim p_{\text{data}}(\mathbf{y})}[\ln D_{\mathscr{Y}}(\mathbf{y})] + \mathbb{E}_{\mathbf{x}\sim p_{\text{data}}(\mathbf{x})}[\ln(1 - D_{\mathscr{Y}}(G_{\mathscr{Y}}(\mathbf{x}))], \tag{1.104}$$

where p_{data} denotes data distribution over a given dataset. In the same sense, the second term is for estimating $G_{\mathscr{X}}$ and $D_{\mathscr{X}}$, given by

$$\mathscr{L}_{\text{GAN}}(G_{\mathscr{X}}, D_{\mathscr{X}}) = \mathbb{E}_{\mathbf{x}\sim p_{\text{data}}(\mathbf{x})}[\ln D_{\mathscr{X}}(\mathbf{x})] + \mathbb{E}_{\mathbf{y}\sim p_{\text{data}}(\mathbf{y})}\left[\ln(1 - D_{\mathscr{X}}(G_{\mathscr{X}}(\mathbf{y}))\right], \tag{1.105}$$

The last key term is the cycle consistency loss that attempts to impose invertibility on image-to-image translation (i.e., $\mathbf{x} \approx G_{\mathscr{X}}(G_{\mathscr{Y}}(\mathbf{x}))$ and $\mathbf{y} \approx G_{\mathscr{Y}}(G_{\mathscr{X}}(\mathbf{y}))$). It is given by

$$\mathscr{L}_{\text{cyc}}(G_{\mathscr{Y}}, G_{\mathscr{X}}) = \mathbb{E}_{\mathbf{x}\sim p_{\text{data}}(\mathbf{x})} \left\|\mathbf{x} - G_{\mathscr{X}} \circ G_{\mathscr{Y}}(\mathbf{x})\right\| + \mathbb{E}_{\mathbf{y}\sim p_{\text{data}}(\mathbf{y})} \left\|\mathbf{y} - G_{\mathscr{Y}} \circ G_{\mathscr{X}}(\mathbf{y})\right\|, \tag{1.106}$$

where $\|\cdot\|$ is ℓ^1 or ℓ^2 norm. Here, it is desired that the generator's fidelity $\|\mathbf{x} - G_{\mathscr{X}} \circ G_{\mathscr{Y}}(\mathbf{x})\|$ is reasonably small for every $\mathbf{x} \sim p_{\mathscr{X}}(\mathbf{x})$.

The additional training of $G^*_{\mathscr{X}} (\approx (G^*_{\mathscr{Y}})^{-1}|_{\text{on } \mathscr{Y}})$ can be viewed as a means of preventing loss of information on original data $\mathbf{x}$ during the image-to-image conversion via $G^*_{\mathscr{Y}}$. In our case, especially, it may mean that the primary metal artifacts in $\mathbf{x}$ (generated by the CT forward model) are not considerably affected while making up for the lack of reality via $G^*_{\mathscr{Y}}$.

1.5 Application of Auto-Encoder: Automatic 3D Cephalometric Annotation System

Cephalometric analysis is a clinical application of cephalometry and requires the estimation of skeletal relationships between the cranial base and the maxilla or mandible, the relationships between the teeth, jaw and the skull, and others. The annotation of cephalometric landmarks on 2D radiographs or 3D CT or CBCT is essentially performed for cephalometric analysis, which facilitates the development of morphometric guidelines for the diagnosis, planning, and treatment of craniofacial disease [41, 66]. Although the value of cephalometric analysis and the definition of landmarks remains controversial, its usefulness is recognized.

Until recently, cephalometric landmark annotation on 2D radiographs or 3D CT or CBCT was mostly performed through manual tracing. This annotation requires a high level of professionalism, time and labor. Automatic landmark identification can reduce work and eliminate operator subjectivity. Therefore, there has been a great demand for an automatic landmark annotation system that can reduce the labor intensiveness of this task and improve the workflow and operator subjectivity.

In 1986, Lévy-Mandel et al. [43] proposed the first automatic cephalometric annotation system on 2D radiographs. Since then, various algorithms have been introduced for automatic 2D cephalometry; the knowledge-based, model-based, soft computing-based, and hybrid approaches [8, 9, 18, 24, 30, 43, 51, 58, 65]. Early works focused on the knowledge-based approaches using edge detection and image processing techniques [18, 43, 51, 65], while the later works implemented model-based approaches [8, 24, 58]. However, despite these efforts, the automated annotation methods using the conventional knowledge-based or atlas-based techniques failed to reach the level of clinical application. Recently, thanks to the remarkable advance of AI, the automation on 2D cephalometry has reached the level of clinical application by using deep learning (DL) technologies [2, 44, 53]. The main advantage of these DL approaches compared to conventional image processing is that the experience of the experts can be reflected in the algorithm through learning training datasets [41, 66]. DL-based automatic annotation of 2D cephalometric landmarks has now reached the level of clinical application.

Recently, as the use of CT or CBCT has become common, there is a tendency to switch from 2D cephalometry to 3D. Since the anatomical structure of the head has a 3D geometry, 3D cephalometry has many advantages over 2D for accurate craniofacial analysis. With the recent development of deep learning technology, a lot of studies have been done on automated cephalometric landmark annotation that enables immediate 3D cephalometric analysis. Existing automatic annotation methods for 3D cephalometry still needs improvement as the error level for some landmarks does not meet the clinical application requirements (less than 2 mm).

This section explains a DL-based fully automatic landmarking system for 3D cephalometry in 3D CT, based on the articles [41, 66, 67]. For ease of explanation of this DL system, we use the following concrete example:

- **Input:** 3D CT image $\mathbf{x} \in \mathbb{R}^{512\times512\times400}$ with the number of voxels being $512 \times 512 \times 400$.
- **Output:** 93 landmarks $\mathfrak{R} = (\mathfrak{r}_1, \cdots, \mathfrak{r}_{93}) \in \mathbb{R}^{93\times3}$. Here, $\mathfrak{r}_j$ denotes the position of j-th landmark.
- **Goal:** Find a landmarking map $f : \mathbf{x} \to \mathfrak{R}$.

The main strategy is to detect only 15 easy-to-find landmarks from $\mathbf{x}$, and the remainder of the landmarks $(78 = 93 - 15)$ are estimated from the knowledge of 15 landmarks, using the variational autoencoder (VAE)-based representation of morphological similarity/dissimilarity of the normalized skull. This mimics the detection procedure of experts in that it first estimates easily detectable landmarks, and then detects the remainder of the landmarks.

This approach has the potential to alleviate experts' time-consuming workflow by dramatically reducing the time taken for landmarking, although the error level of some landmarks does not meet the requirement of clinical applications (less than 2 mm). This method can immediately help guide the operator to the approximate position and image setting for 3D cephalometry. It can also reduce the burden of moving the 3D skull object and scrolling over the multi-planar image settings during landmark pointing tasks.

1.5.1 The Overall Process of the Automatic 3D Cephalometric Annotation System

Assume that $\mathbf{x}$ (3D CT image) is defined on a voxel grid $\Omega := \{\mathbf{v} = (v_1, v_2, v_3) : v_1, v_2 \in \{1, \cdots, 512\},\ \ v_3 \in \{1, \cdots, 400\}\}$, where the number $512 \times 512 \times 400$ comes from the CT image size. The value $\mathbf{x}(\mathbf{v})$ at voxel position $\mathbf{v}$ can be regarded as the X-ray attenuation coefficient. Given an input $\mathbf{x}$, we try to detect the total 93 landmarks ($\mathfrak{R}$). Identification of $\mathfrak{R}$ is done using the following procedure (Fig. 1.21):

1. **Generate a skull image $\mathbf{x}_{seg}$ by a soft threshold-based weighting method** [34]. Roughly speaking, $\mathbf{x}_{seg}$ is

$$\mathbf{x}_{seg}(\mathbf{v}) \approx \begin{cases} 1 \text{ if } \mathbf{x}(\mathbf{v}) > \theta, \\ 0 \ \text{ otherwise}, \end{cases} \tag{1.107}$$

where θ is a threshold for segmenting bones excluding soft tissue. For ease of notational simplicity, we will use the same notation $\mathbf{x}_{seg}$ to represent the segmented region, which will be clear from the context. See Fig. 1.22.

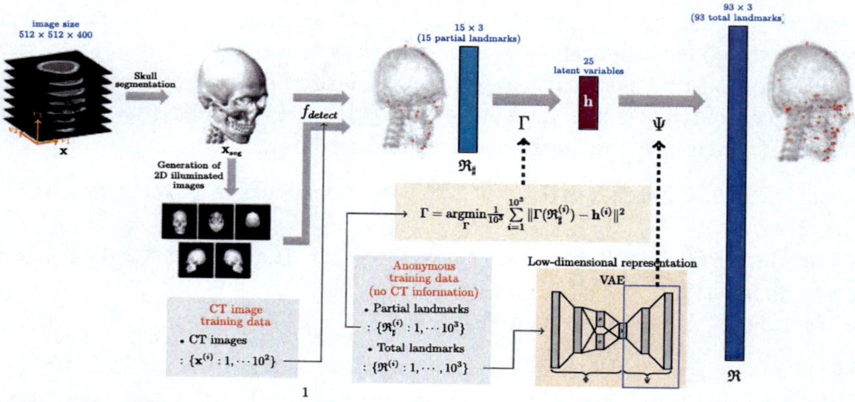

Fig. 1.21 Overall process of automatic 3D cephalometric annotation system

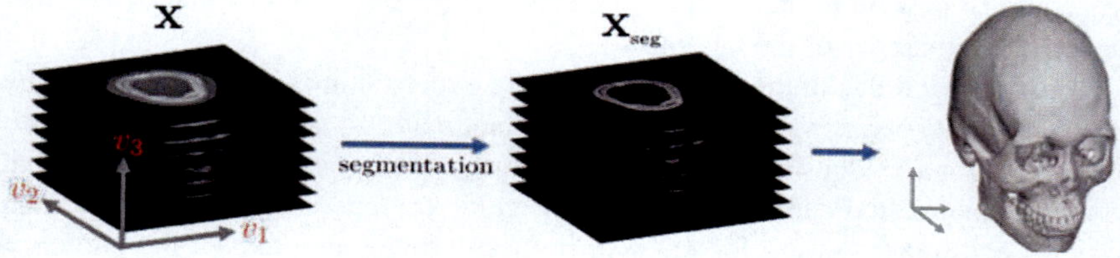

Fig. 1.22 Illustration of 3D CT image and skull image. Our landmark detection method is based on the use of 3D skull image which can be segmented from 3D CT data

2. **Generating the 2D image containing 3D geometric cues.** Given the skull geometry $\mathbf{x}_{\text{seg}}$, we generate five different 2D illuminated images $(\tilde{\mathbf{x}}_1, \cdots, \tilde{\mathbf{x}}_5)$ that contain 3D geometric features caused by using lighting and shadowing [42], as shown in Figs. 1.23 and 1.24.
3. **Find only some of the landmarks that are relatively easy to find.** Detect 15 easy-to-find landmarks $\mathfrak{R}_\sharp$ from 2D illuminated images using a deep learning method. We denote the map from $\mathbf{x}$ to $\mathfrak{R}_\sharp$ by f_{detect}, where the detailed process is

$$\underbrace{\mathbf{x} \mapsto \mathbf{x}_{\text{seg}} \mapsto (\tilde{\mathbf{x}}_1, \cdots, \tilde{\mathbf{x}}_5; \mathbf{x}_{\text{seg}}) \mapsto \mathfrak{R}_\sharp}_{f_{\text{detect}}:\mathbf{x} \mapsto \mathfrak{R}_\sharp}. \tag{1.108}$$

   Note that $\dim(\mathfrak{R}_\sharp) = 15 \times 3 < 93 \times 3 = \dim(\mathfrak{R})$.
4. **Learning a low-dimensional latent representation of $\mathfrak{R}$.** Find a low-dimensional disentanglement representation $\Psi : \mathbf{h} \to \mathfrak{R}$ with $\dim(\mathbf{h}) = 25 \ll 93 \times 3 = \dim(\mathfrak{R})$ by applying the variational autoencoder(VAE) [35]. We hope that each latent variable is sensitive to changes in individual morphological factors, while being relatively insensitive to other changes.

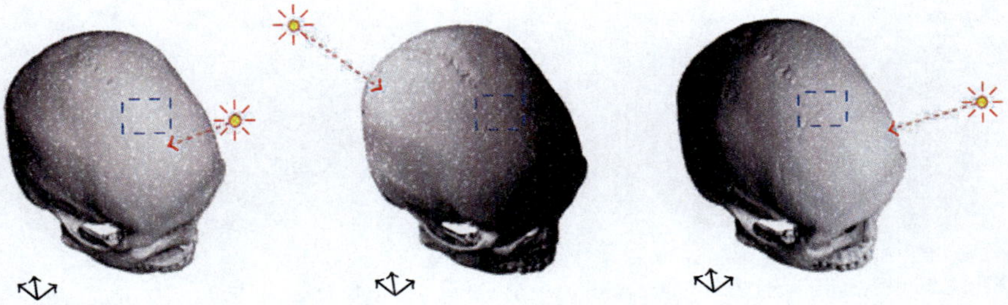

Fig. 1.23 2D images with three different light sources. This figure was extracted from [41]

5. **Connection between $\mathfrak{R}_\sharp$ and $\mathbf{h}$.** Learn a map $\Gamma : \mathfrak{R}_\sharp \to \mathbf{h}$, which connects the latent variables $\mathbf{h}$ and the fractional data $\mathfrak{R}_\sharp$.
6. **Detect the total landmark vector ($\mathfrak{R}$) from the fractional information ($\mathfrak{R}_\sharp$).** The landmark detection map f is

$$f = \Psi \circ \Gamma \circ f_{\text{detect}} : \mathbf{x} \underbrace{\mapsto}_{f_{\text{detect}}} \mathfrak{R}_\sharp \underbrace{\mapsto}_{\Gamma} \mathbf{h} \underbrace{\mapsto}_{\Psi} \mathfrak{R}. \tag{1.109}$$

1.5.2 Generating the 2D Image Containing 3D Geometric Cues

Lee et al. [41] use a 3D skull image $\mathbf{x}_{\text{seg}}$ to generate multiple 2D images containing 3D geometric features caused by using a light source to create shadows [42], as shown in Fig. 1.24. Now, let us explain how to generate the illuminated images $(\tilde{\mathbf{x}}_1, \cdots, \tilde{\mathbf{x}}_5)$. To generate the 2D image $\tilde{\mathbf{x}}_j$, we select a plane Π_j, lying outside of the skull, with its normal vector d_j. This plane Π_j can be expressed as $\Pi_j = \{\mathbf{r}_j(\mathbf{s}) : \mathbf{s} \in \{1, \cdots, 512\}^2\}$, where $\mathbf{r}_j(\mathbf{s})$ is a point on the plane Π_j at the pixel position $\mathbf{s}$. Let ℓ_j be the line passing through $\mathbf{p}_j(\mathbf{s})$ in the direction d_j. Then, each point $\mathbf{r}_j(\mathbf{s}) \in \Pi_j$ corresponds to the point $\mathbf{p}_j(\mathbf{s})$ on the surface of the skull ($\mathbf{x}_{\text{seg}}$) given by

$$\mathbf{p}_j(\mathbf{s}) = \underset{\mathbf{p} \in \ell_j \cap \mathbf{x}_{\text{seg}}}{\arg\min} \|\mathbf{p} - \mathbf{r}_j(\mathbf{s})\|. \tag{1.110}$$

By placing a light source at a point $\mathbf{q}_j$ outside of the skull, the illuminated image $\tilde{\mathbf{x}}_j$ can be defined as an intensity of light on the skull with the viewing direction d_j and the light source at $\mathbf{q}_j$. To be precise, $\tilde{\mathbf{x}}_j$ can be defined as

$$\tilde{\mathbf{x}}_j(\mathbf{s}) = \max\left(\frac{\mathbf{n}(\mathbf{p}_j(\mathbf{s})) \cdot (\mathbf{q}_j - \mathbf{p}_j(\mathbf{s}))}{||\mathbf{q}_j - \mathbf{p}_j(\mathbf{s})||}, 0\right), \tag{1.111}$$

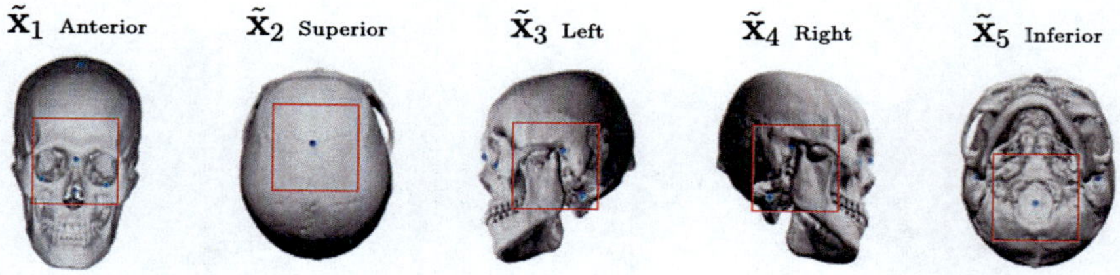

Fig. 1.24 Five different 2D images with various views and light sources. These 2D images provide 3D geometric cues

where $\mathbf{n}(\mathbf{p})$ is the unit normal vector at $\mathbf{p}$ to the surface of the skull. Figure 1.24 shows 2D images $\tilde{\mathbf{x}}_j$, $j = 1, \cdots, 5$ from different views and multiple light sources. These 2D images provide 3D geometric cues.

1.5.3 *Finding only Some of the Landmarks that are Relatively Easy to Find*

The network $f_{\text{detect}} : \mathbf{x} \to \mathfrak{R}_\sharp$ in (1.109) takes advantage of 2D illuminated images that are generated by manipulating various lighting and viewing directions. Lee et al. [41] applied VGGNet [60] to these illuminated images, and $\mathfrak{R}_\sharp$ are accurately and automatically identified (Fig. 1.25).

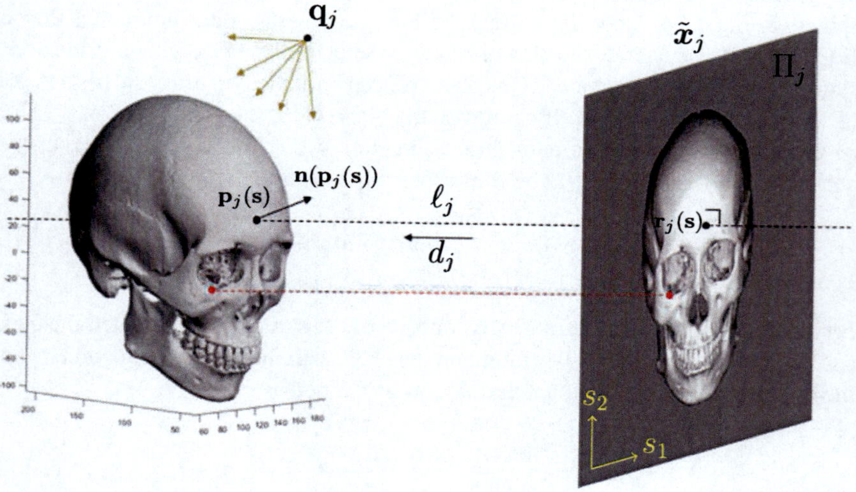

Fig. 1.25 The process of generating 2D images containing 3D geometric information by lighting and shadowing. This figure was extracted from [41]

The architecture of VGGNet is as follows. The input image $\tilde{\mathbf{x}}_j$ is convolved with the set of 64 predetermined filters $\mathbf{w}_1 = \{w_{1,k} \in \mathbb{R}^{3\times 3} : k = 1, \cdots, 64\}$. This convolution provides a set of 64 feature maps ($\mathbf{h}_1 = \{h_{1,1}, \cdots, h_{1,64}\}$), where the k-th feature map at the pixel position (m, n) is given by

$$h_{1,k}(m,n) = \text{ReLU}\left((w_{1,k} * \tilde{\mathbf{x}})(m,n) + b_{1,k}\right)$$
$$= \text{ReLU}\left(\sum_{i=1}^{3}\sum_{j=1}^{3} w_{1,k}(i,j)\tilde{\mathbf{x}}(m-i+2, n-j+2) + b_{1,k}\right).$$

Here, ReLU(x)=max$\{0, x\}$ is the rectified linear unit, acting as an activation function to solve the vanishing gradient problem [19] and $\mathbf{b}_1 = \{b_{1,k} \in \mathbb{R} : k = 1, \cdots, 64\}$ is the set of 64 biases. Similarly, at the second convolution layer, we compute a set of 64 feature maps ($\mathbf{h}_2 = \{h_{2,1}, \cdots, h_{2,64}\}$) using a set of weights ($\mathbf{w}_2 = \{w_{2,k,k'} \in \mathbb{R}^{3\times 3} : k, k' = 1, \cdots, 64\}$, $\mathbf{b}_2 = \{b_{2,k} \in \mathbb{R} : k = 1, \cdots, 64\}$), where $h_{2,k}$ is given by

$$h_{2,k}(m,n) = \text{ReLU}\left(\sum_{k'=1}^{64}(w_{2,k,k'} * h_{1,k'})(m,n) + b_{2,k}\right)$$
$$= \text{ReLU}\left(\sum_{k'=1}^{64}\sum_{i=1}^{3}\sum_{j=1}^{3} w_{2,k,k'}(i,j)h_{1,k'}(m-i+2, n-j+2) + b_{2,k}\right)$$

Next, we apply a 2×2 max pooling operation with stride 2 for down-sampling to get $\mathbf{h}_3$. By applying either convolution or max pooling to each layer, we obtain $\mathbf{h}_{17}$, a set of 512 feature maps. This feature map is vectorized to $\tilde{\mathbf{h}}_{17} \in \mathbb{R}^{(16\times 16\times 512)\times 1}$ and are passed through fully-connected layers. At the first fully-connected layer, we compute a vector $\mathbf{h}_{18} = (h_{18,1}, \cdots, h_{18,4096})$ by multiplying the feature vector $\tilde{\mathbf{h}}_{17}$ with the predetermined matrix $\mathbf{W}_{18} \in \mathbb{R}^{4096\times(16\times 16\times 512)}$, then adding the bias $\mathbf{b}_{17} = (b_{17,1}, \cdots, b_{17,4096})$, which is given by

$$\mathbf{h}_{18} = \text{ReLU}\left(\mathbf{W}_{18}\tilde{\mathbf{h}}_{17} + \mathbf{b}_{17}\right). \tag{1.112}$$

Through the remaining fully-connected layers in a similar way, we finally obtain the 2D landmark position $\mathbf{s}_{d_k} = (s_1, s_2)$.

1.5.4 *Learning a Low-Dimensional Latent Representation of $\mathfrak{R}$*

For effective feature learning of $\mathfrak{R}$, the first step normalizes the data [41]. The hexahedron made by five landmarks (center of foramen magnum (CFM), nasion, left/right porion (L/R Po)) is normalized by a suitable anisotropic scaling. The data normalization is based on facial width (the distance between the x-coordinate of L Po and R Po), facial depth (the distance between y-coordinate of L Po and nasion), and facial height (the distance between z-coordinate of CFM and bregma). We normalize the data by setting the width, depth, and height as a fixed value so that each reference hexahedron has (to some extent) fixed shape and size. This normalization reveals facial deformities, and enables efficient feature learning of similarity/dissimilarity when applying VAE.

We use the VAE to find a nonlinear expression of the concatenated landmark vector $\mathfrak{R}$ by variables $\mathbf{h}$ in the low-dimensional latent space. Here, we use $\dim(\mathbf{h}) = 25$. To be precise, the VAE uses the training datasets $\{\mathfrak{R}^{(i)}\}_{i=1}^N$ to learn two functions, the encoder $\Phi : \mathfrak{R} \to \mathbf{h}$ and the decoder $\Psi : \mathbf{h} \to \mathfrak{R}$ by the following loss minimization over the training data:

$$(\Psi, \Phi) = \underset{(\Psi,\Phi)\in\mathbb{VAE}}{\operatorname{argmin}} \frac{1}{N} \sum_{i=1}^{N} \left[\|\Psi \circ \Phi(\mathfrak{R}^{(i)}) - \mathfrak{R}^{(i)}\|^2 + \text{VAE-regularization} \right], \tag{1.113}$$

where $\mathbb{VAE}$ describes a class of functions in the form of the deep learning network described in Fig. 1.26. Note that VAE encodes the input as a distribution over the latent space. For VAE-regularization, please refer to [35]. The decoder $\Psi : \mathbf{h} \to \mathfrak{R}$ in (1.113) provides a low-dimensional disentanglement representation, so that each latent variable is sensitive to changes in individual morphological factors, while being relatively insensitive to other changes.

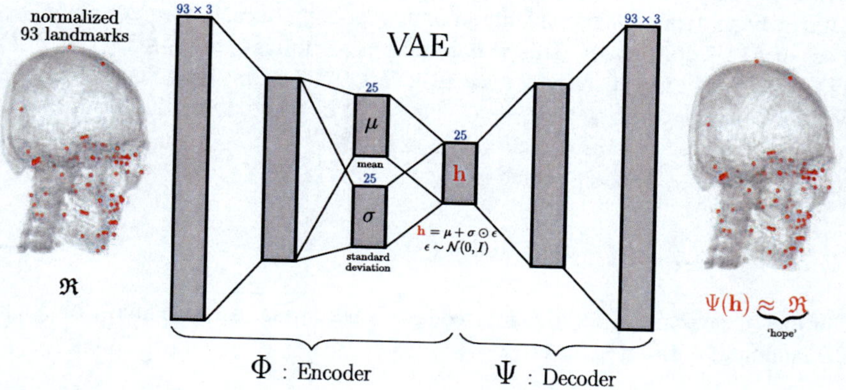

Fig. 1.26 Architecture of VAE-based low-dimensional latent representation of $\mathfrak{R}$. This low-dimensional representation can address data shortages of CT images due to legal and ethical limitations. Here, we took advantage of many anonymous landmark datasets $\{\mathfrak{R}^{(i)} : i = 1, \cdots, N\}$, where CT images were not used for the training of VAE

1.5.5 Detecting the Total Landmark Vector ($\mathfrak{R}$) from the Fractional Information ($\mathfrak{R}_\sharp$)

We connect the latent variables $\mathbf{h}$ and the fractional data $\mathfrak{R}_\sharp$ by learning a nonlinear map $\Gamma : \mathfrak{R}_\sharp \to \mathbf{h}$, where the training data $\{(\mathbf{h}^{(i)}, \mathfrak{R}_\sharp^{(i)}) : i = 1, 2, ..., N\}$ is provided from the encoder map $\mathbf{h}^{(i)} = \Phi(\mathfrak{R}^{(i)})$. See Fig. 1.27. To be precise, the nonlinear regression map Γ is obtained by minimizing the loss $\frac{1}{N}\sum_{i=1}^{N} ||\Gamma(\mathfrak{R}_\sharp^{(i)}) - \mathbf{h}^{(i)}||^2$. Then, the landmark detection map f is $f = \Psi \circ \Gamma \circ f_{\text{detect}}$.

1.5.6 Remarks

The proposed method exhibited relatively high performance, but the error level did not meet the requirement for immediate clinical application (such as in less than 2 mm of error levels). However, this approach has a lot of room for improvement and the errors can be significantly reduced by improving deep learning performance with an increased number of training data. Although our protocol cannot intuitively determine the exact location set to achieve the expert human standards, it can immediately help

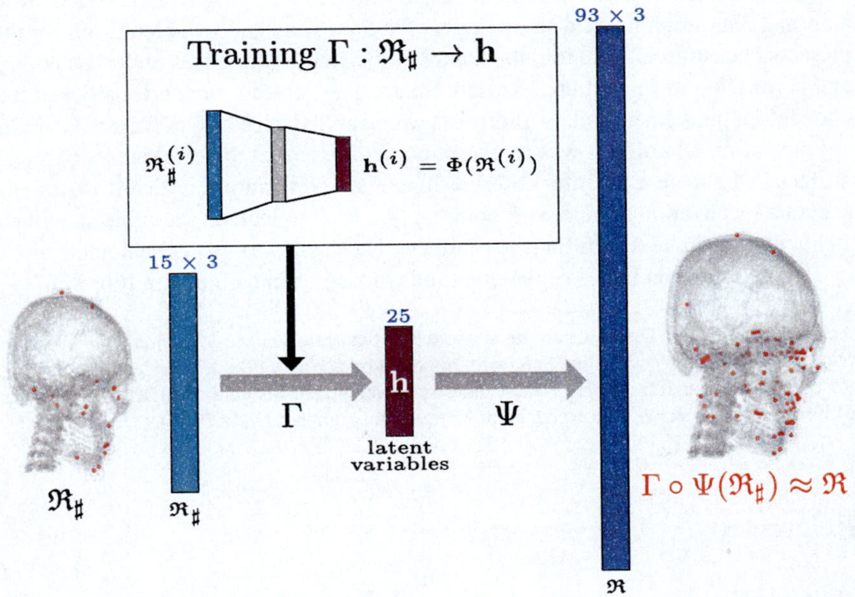

Fig. 1.27 Connection between $\mathfrak{R}_\sharp$ (fractional information) and $\mathbf{h}$ (latent variables in VAE). This connection $\Gamma : \mathfrak{R}_\sharp \to \mathbf{h}$ allows us to detect $\mathfrak{R}$ (the total landmarks) from its fractional information. The map Γ is trained using the dataset $\{(\mathbf{h}^{(i)} = \Phi(\mathfrak{R}^{(i)}), \mathfrak{R}_\sharp^{(i)}) : i = 1, \cdots, N\}$, where Φ is obtained from the VAE

guide the operator to the approximate position and image setting for 3D cephalometry. It can also reduce the burden of moving the 3D skull object and scrolling over the multi-planar image settings during landmark pointing tasks. Finally, it can be applied prior to data processing for segmentation, thus assisting in the orientation of the head to the calibrated posture.

It is pertinent to consider the radiation dose for 3D cephalometry. The recommendations from the WHO Global Initiative on Radiation Safety in Health Care Settings, American guidelines such as the Appropriateness Criteria of the American College of Radiology, and European guidelines such as the Royal College of Radiologists Referral Guidelines are widely known sources of evidence-based guidelines.

All diagnostic imaging basically needs to adhere to three basic principles of radiation dose, including justification, optimization and application of dose limits. Though the imaging guidelines for the craniomaxillofacial-dental region in 3D cephalometry have not yet been set up completely, cone-beam CT (CBCT) is a frequently employed dental radiograph for 3D cephalometry. However, one distinct limit in CBCT for 3D cephalometry lies in its limited field of view (FOV). We needed the whole skull-sized FOV of CBCT for Delaire's 3D cephalometric analysis. Unfortunately, we did not know of any CBCT machine that covers the whole skull and face in a single scan.

Although the organ and effective doses can vary based on field size, acquisition angle, and positioning of the beam in relation to the radiosensitive organs , the radiation effective dose of CBCT is known to be 30$\sim$1073 μSv, while multi-slice CT comes to 280$\sim$1410 μSv . In addition, the CT images were originally produced from a normal 3D morphological study project for class I skeletal volunteers, not for this project. The subjects were told the nature of the study, which was under the control and permission of Institutional Review Board. The subjects were critically selected under strict guidelines, and the protocols were applied to reduce radiation dose. The absence of metal artifacts was an essential requirement of the original study. Metal artifacts, which are frequently observed in modern populations, critically disturb the accurate localization of landmark pointing. We tried to include the subjects without any dental restoration or fillings to optimize the accuracy of 3D cephalometry.

Detailed information of cephalometric landmarks can be found in [66, 67].

Acknowledgements This research was supported by Samsung Science & Technology Foundation (No. SRFC-IT1902-09). Yun and Seo were supported by a grant of the Korea Health Technology R&D Project through the Korea Health Industry Development Institute (KHIDI), funded by the Ministry of Health & Welfare, Republic of Korea (grant number : HI20C0127).

References

1. Adams G.L., et al.: Comparison between traditional 2-dimensional cephalometry and a 3-dimensional approach on human dry skulls. Am. J. Orthod. Dentofac. Orthop. 126(4), 397–409 (2004)
2. Arik, S.Ö., Ibragimov, B., Xing, L.: Fully automated quantitative cephalometry using convolutional neural networks. J. Med. Imag. 4(1), 014501 (2017)

3. Arjovsky, M., Chintala, S., Bottou, L.: Wasserstein generative adversarial networks. In: International Conference on Machine Learning. PMLR (2017)
4. Arora, S., et al.: Generalization and equilibrium in generative adversarial nets (gans). In: International Conference on Machine Learning. PMLR (2017)
5. Baur, C., et al.: Deep autoencoding models for unsupervised anomaly segmentation in brain MR images. In: International MICCAI Brainlesion Workshop. Springer, Cham (2018)
6. Candès, E.J., Romberg, J., Tao, T.: Robust uncertainty principles: exact signal reconstruction from highly incomplete frequency information. IEEE Trans. Inf. Theory **52**(2), 489–509 (2006)
7. Candès, E.J., Tao, T.: Reflections on compressed sensing. IEEE Inf. Theory Soc. Newsl. **58**(4), 20–23 (2008)
8. Cardillo, J., Sid-Ahmed, M.A.: An image processing system for locating craniofacial landmarks. IEEE Trans. Med. Imag. **13**(2), 275–289 (1994)
9. Chakrabartty, S., et al.: Robust cephalometric landmark identification using support vector machines. In: 2003 International Conference on Multimedia and Expo. ICME'03. Proceedings (Cat. No. 03TH8698), vol. 3. IEEE (2003)
10. Codari, M., et al.: Computer-aided cephalometric landmark annotation for CBCT data. Int. J. Comput. Assist. Radiol. Surgery 12(1), 113–121 (2017)
11. Chang Rick, J.H., et al.: One network to solve them all–solving linear inverse problems using deep projection models. In: Proceedings of the IEEE International Conference on Computer Vision (2017)
12. Donoho, D.L., Elad, M.: Optimally sparse representation in general (nonorthogonal) dictionaries via $\ell 1$ minimization. Proc. Natl. Acad. Sci. **100**(5), 2197–2202 (2003)
13. Donoho, D.L.: Compressed sensing. IEEE Trans. Inf. Theory 52(4), 1288–1306 (2006)
14. Dai, B., et al.: Hidden talents of the variational autoencoder (2017). arXiv:1706.05148
15. Diker, B., Tak, Ø.: Comparing the accuracy of six intraoral scanners on prepared teeth and effect of scanning sequence. J. Adv. Prosthodont. **12**(5), 299 (2020)
16. Elnagar, M.H., Aronovich, S., Kusnoto, B.: Digital workflow for combined orthodontics and orthognathic surgery. Oral Maxillofac. Surg. Clin. North Am. **32**(1), 1–14 (2020)
17. Esteva, A, et al.: Dermatologist-level classification of skin cancer with deep neural networks. Nature 542(7639), 115–118 (2017)
18. Giordano, D., et al.: Automatic landmarking of cephalograms by cellular neural networks. In: Conference on Artificial Intelligence in Medicine in Europe. Springer, Berlin, Heidelberg (2005)
19. Glorot, X., Bordes, A., Bengio, Y.: Deep sparse rectifier neural networks. In: Proceedings of the Fourteenth International Conference on Artificial Intelligence and Statistics. JMLR Workshop and Conference Proceedings (2011)
20. Goodfellow, I., et al.: Generative adversarial nets. In: Advances in Neural Information Processing Systems, vol. 27 (2014)
21. Goodfellow, I., Bengio, Y., Courville, A.: Deep learning. MIT Press (2016)
22. Gondara, L.: Medical image denoising using convolutional denoising autoencoders. In: 2016 IEEE 16th International Conference on Data Mining Workshops (ICDMW), IEEE, pp. 241–246 (2016)
23. Gulrajani, I., et al.: Improved training of wasserstein gans (2017). arXiv:1704.00028
24. Hutton, T.J., Cunningham, S., Hammond, P.: An evaluation of active shape models for the automatic identification of cephalometric landmarks. Eur. J. Orthodont. **22**(5), 499–508 (2000)
25. Hinton, G.E., Salakhutdinov, R.R.: Reducing the dimensionality of data with neural networks. Science **313**(5786), 504–507 (2006)
26. Hinton, G.E., Krizhevsky, A., Wang, S.D.: Transforming auto-encoders. In: International Conference on Artificial Neural Networks. Springer, Berlin, Heidelberg (2011)
27. Hyun, C.M., Kim, H.P., Lee, S.M., Lee, S., Seo, J.K.: Deep learning for undersampled MRI reconstruction. Phys. Med. Biol. 63(13) (2018)
28. Hyun, C.M., Kim, K.C., Cho, H.C., Choi, J.K., Seo, J.K.: Framelet pooling aided deep learning network: the method to process high dimensional medical data. Mach. Learn. Sci. Technol. **1**, 015009 (2020)

29. Hyun, C.M., Baek, S.H., Lee, M., Lee, S.M., Seo, J.K.: Deep learning-based solvability of underdetermined inverse problems in medical imaging. In: Medical Image Analysis (2021)
30. Innes, A., et al.: Landmark detection for cephalometric radiology images using pulse coupled neural networks. In: Proceedings of International Conference on Artificial Intelligence, vol. 2 (2002)
31. Jolliffe, I.T.: Principal components in regression analysis. In: Principal Component Analysis, pp. 167–198 (2002)
32. Jalali, S., Yuan, X.: Using auto-encoders for solving ill-posed linear inverse problems (2019). arXiv:1901.05045
33. Kramer, M.A.: Nonlinear principal component analysis using auto associative neural networks. AIChE J. 37(2), 233–243 (1991)
34. Kyriakou, Y., et al.: Empirical beam hardening correction (EBHC) for CT. Med. Phys. 37(10), 5179–5187 (2010)
35. Kingma, D.P., Welling, M.: Auto-encoding variational bayes (2013). arXiv:1312.6114
36. Kingma, D.P., Ba, J.: Adam: a method for stochastic optimization (2014). arXiv:1412.6980
37. Kingma, D.P., Welling, M.: An introduction to variational autoencoder. Found. Trends Mach. Learn. **12**(4), 307–392 (2019)
38. Kang, S.H., et al.: Automatic three-dimensional cephalometric annotation system using three-dimensional convolutional neural networks: a developmental trial. Comput. Methods Biomech. Biomed. Eng. Imag. Vis. 8(2), 210–218 (2020)
39. Kernen, F., et al.: A review of virtual planning software for guided implant surgery-data import and visualization, drill guide design and manufacturing. BMC Oral Health 20(1), 1–10 (2020)
40. Lee, S.-H., et al.: Three-dimensional architectural and structural analysis-a transition in concept and design from Delaire's cephalometric analysis. Int. J. Oral Maxillofac. Surg. **43**(9), 1154–1160 (2014)
41. Lee, S.M., et al.: Automatic 3D cephalometric annotation system using shadowed 2D image-based machine learning. Phys. Med. Biol. 64(5), 055002 (2019)
42. Lengyel, E.: Mathematics for 3D game programming and computer graphics. Charles River Media, Inc. (2003)
43. Levy-Mandel, A.D., Venetsanopoulos, A.N., Tsotsos, J.K.: Knowledge-based landmarking of cephalograms. Comput. Biomed. Res. **19**(3), 282–309 (1986)
44. Lindner, C., et al.: Fully automatic system for accurate localisation and analysis of cephalometric landmarks in lateral cephalograms. Sci. Rep. 6(1), 1–10 (2016)
45. Mao, X., et al.: Least squares generative adversarial networks. In: Proceedings of the IEEE International Conference on Computer Vision (2017)
46. Montúfar, J., Romero, M., Scougall-Vilchis R.J.: Automatic 3-dimensional cephalometric landmarking based on active shape models in related projections. Am. J. Orthodont. Dentofac. Orthoped. 153(3), 449–458 (2018)
47. Mao, Q., et al.: Mode seeking generative adversarial networks for diverse image synthesis. In: Proceedings of the IEEE/CVF Conference on Computer Vision and Pattern Recognition (2019)
48. Moon, Y.-G., Lee, K.-M.: Comparison of the accuracy of intraoral scans between complete-arch scan and quadrant scan. Prog. Orthod. **21**(1), 1–6 (2020)
49. Nalcaci, R., Ozturk, F., Sokucu, O.: A comparison of two-dimensional radiography and three-dimensional computed tomography in angular cephalometric measurements. Dentomaxillofacial Radiol. 39(2), 100–106 (2010)
50. Nagy, Z., et al.: Comparing the trueness of seven intraoral scanners and a physical impression on dentate human maxilla by a novel method. BMC Oral Health 20(1), 1–10 (2020)
51. Parthasarathy, S., et al.: Automatic landmarking of cephalograms. Comput. Biomed. Res. **22**(3), 248–269 (1989)
52. Paszke, A., et al.: Pytorch: an imperative style, high-performance deep learning library. Adv. Neural Inf. Process. Syst. 32, 8026–8037 (2019)
53. Pei, Y., et al.: Anatomical structure sketcher for cephalograms by bimodal deep learning. BMVC (2013)

54. Pittayapat, P., et al.: Three-dimensional cephalometric analysis in orthodontics: a systematic review. Orthodont. Craniofac. Res. 17(2), 69–91 (2014)
55. Pauwels, R., et al.: Technical aspects of dental CBCT: state of the art. Dentomaxillofacial Radiol. 44(1), 20140224 (2015)
56. Pittayapat, P., et al.: A new mandible-specific landmark reference system for three-dimensional cephalometry using cone-beam computed tomography. Eur. J. Orthodont. 38(6), 563–568 (2016)
57. Proffit, W.R., et al.: Contemporary orthodontics-e-book. Elsevier Health Sciences (2018)
58. Rudolph, D.J., Sinclair, P.M., Coggins, J.M.: Automatic computerized radiographic identification of cephalometric landmarks. Am. J. Orthod. Dentofac. Orthop. **113**(2), 173–179 (1998)
59. Ronneberger, O., Fischer, P., Brox, T.: U-net: convolutional networks for biomedical image segmentation. In: International Conference on Medical Image Computing and Computer-Assisted Intervention. Springer, Cham (2015)
60. Simonyan, K., Zisserman, A.: Very deep convolutional networks for large-scale image recognition (2014). arXiv:1409.1556
61. Seo, J.K., Kim, K.C., Jargal, A., Lee, K., Harrach, B.: A learning-based method for solving ill-posed nonlinear inverse problems: a simulation : a simulation study of Lung EIT. SIAM J. Imag. Sci. (2019)
62. Tenti, F.V.: Cephalometric analysis as a tool for treatment planning and evaluation. Eur. J. Orthodont. **3**(4), 241–245 (1981)
63. Tezcan, K.C., et al.: MR image reconstruction using deep density priors. IEEE Trans. Med. Imag. 38(7), 1633–1642 (2018)
64. Vallabh, R., et al.: The morphology of the human mandible: a computational modelling study. Biomech. Model. Mechanobiol. 19(4) (2020)
65. Vucinic, P., Trpovski, Z., Scepan, I.: Automatic landmarking of cephalograms using active appearance models. Eur. J. Orthodont. **32**(3), 233–241 (2010)
66. Yun, H.S., et al.: Learning-based local-to-global landmark annotation for automatic 3D cephalometry. Phys. Med. Biol. 65(8), 085018 (2020)
67. Yun, H.S., et al.: Automated 3D cephalometric landmark identification using computerized tomography (2020). arXiv:2101.05205
68. Zeiler, M.D., Fergus, R.: Visualizing and understanding convolutional networks. In: European Conference on Computer Vision. Springer, Cham (2014)

Chapter 2
Deep Learning Techniques for Medical Image Segmentation and Object Recognition

Kang Cheol Kim, Tae Jun Jang, and Jin Keun Seo

Abstract Segmentation of a target object in the form of closed curves has many potential applications in medical imaging because it provides quantitative information related to its size and shape. Over the last few decades, many innovative methods of performing segmentation have been proposed, and these segmentation techniques are based on the basic recipes using thresholding and edge-based detection. Segmentation and classification in medical imaging are in fact experiencing a paradigm shift due to a marked and rapid advance in deep learning (DL) techniques. DL methods have nonlinear representability to extract and utilize global spatial features and local spatial features simultaneously, showing amazing overall performance in medical image segmentation. DL methods mostly lack transparency due to the black-box output, so clinicians cannot trace the output back to present the causal relationship of the output diagnosis. Therefore, in order to safely utilize DL algorithms in the medical field, it is desirable to design the models to transparently explain the reason for making the output diagnosis rather than a black-box. For explainable DL, a systematic study is needed to rigorously analyze which input characteristics affect the output of the network. Despite the lack of rigorous analysis in DL, recent rapid advances indicate that DL algorithms will improve their performance as training data and experience accumulate over time.

2.1 Introduction

Medical image segmentation plays an important role for a quantitative diagnosis using medical images such as X-ray, CT, MRI, and ultrasound. It is the process of separating a target object from a medical image and has several applications

K. C. Kim · T. J. Jang · J. K. Seo (✉)
School of Mathematics and Computing (Computational Science and Engineering), Yonsei University, Seoul, Korea
e-mail: seoj@yonsei.ac.kr

T. J. Jang
e-mail: taejunjang@yonsei.ac.kr

J. K. Seo (ed.), *Deep Learning and Medical Applications*, Mathematics in Industry 40,
https://doi.org/10.1007/978-981-99-1839-3_2

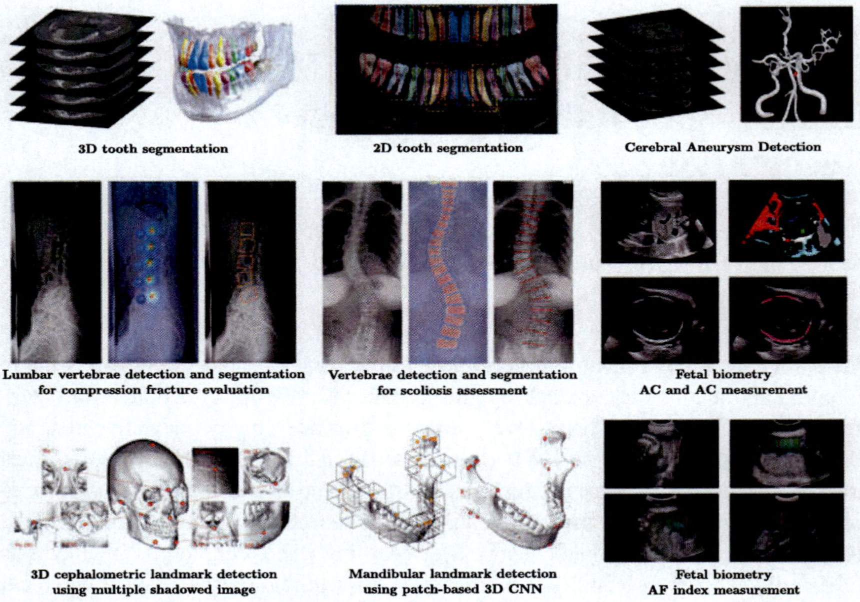

Fig. 2.1 Examples of medical image detection and segmentation

including tumor detection, 3D tooth segmentation from 3D CT images, lumbar vertebrae segmentation for compression fracture evaluation in lateral radiographs, and fetal organ segmentation for fetal biometry from ultrasound images as shown in Fig. 2.1. Since manual segmentations are cumbersome and time consuming, numerous computer-aided methods for automated segmentation have been developed by several researchers during the last four decades. Unfortunately, automated medical image segmentation suffers from various difficulties such as variability of organ shape, neighboring multiple objects, image artifacts, noise, low contrast of images, and weak boundaries.

Over the last four decades, various image segmentation methods have been proposed for extracting information related to the size or shape of searching objects. However, in medical image segmentation, conventional segmentation techniques have difficulty in reaching the desired segmentation in the form of a simple closed curve due to uncertainty of segmentation boundaries in medical images caused by noise, low SNR, and low contrast. Conventional segmentation methods such as level set [30], active contour [19], region-growing [1], histogram-based methods, and graph partition method have fundamental limitation to achieve fully automated segmentation. For example, the level set method essentially relies on an initial guess and stopping criterion determined by the local structure of the image. Moreover, it is difficult to reflect the doctor's thoughts.

Recently, medical image segmentation is experiencing a paradigm shift due to a marked and rapid advance in deep learning (DL) techniques. This advancement is somehow motivated from the remarkable performance of convolution neural networks (CNNs) equipped with rectified linear unit (ReLU) and dropout in 2012. These

CNN-based DL algorithms include Region with CNN (R-CNN), fast R-CNN, faster R-CNN, U-net, and YOLO. DL techniques can see both local and global structures at the same time, and can properly reflect the doctor's thoughts. DL algorithms appear to overcome limitations of conventional mathematical methods and have reached the level of clinical applications in segmentation and classification. Figure 2.1 shows the various applications of medical image detection and segmentation such as spine segmentation [23], scoliosis assessment [22], 3D cephalometry [24, 39], fetal abdomen segmentation [17, 20], fetal head circumference [21], amniotic fluid measurement [7], and 3D tooth segmentation [18].

In this chapter, we discuss the limitations of the conventional methods and the comparative analysis of deep learning methods. For clear and practical explanation, we will use special models in digital dentistry such as tooth segmentation where the goal is to develop a fully automated method of segmenting individual teeth from panoramic X-ray images or CT images. Currently, most of the deep learning methods have limitations in gaining the trust of doctors due to the black box characteristics. In order to gain the confidence of doctors and patients, it is important for deep learning methods to be able to transparently describe which image features of the input were used to determine the output segmentation. For explainable AI, it is necessary to understand the basic structure and working mechanisms of deep learning methodologies. We hope that this chapter provides a basis for deep learning methods for developing explainable AIs.

2.2 Segmentation Problem

In image segmentation, the goal is to find a segmentation map f such that

$$f(\text{input data}) = \text{useful output segmentation.} \tag{2.1}$$

Here, examples of the input are X-ray, CT, MRI, ultrasound images and outputs are tooth, lung CT lesion, spine, and skull segmentations. For a successful segmentation, the map f must contain the ability to extract key filters from the input associated with the target output. See Fig. 2.2.

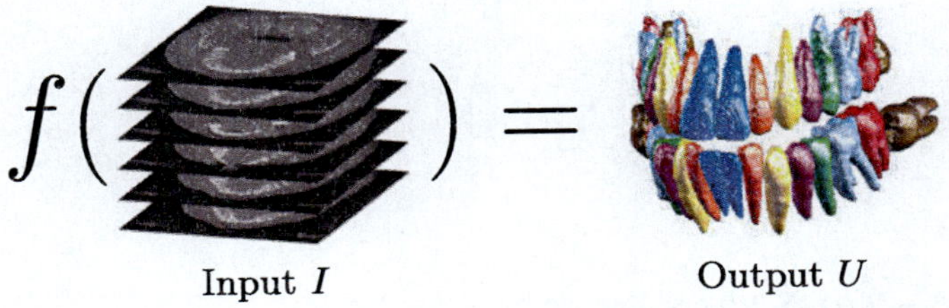

Fig. 2.2 A segmentation function that maps from CBCT image to tooth segmentation

Throughout this chapter, we use the following notations for image segmentation:

- **(Input image I)** Let I denote a 2D (or 3D) medical image with pixel grid Ω. For simplicity, we use a image pixel size of $n_{x_1} \times n_{x_2}$ (or $n_{x_1} \times n_{x_2} \times n_{x_3}$, that is, $\Omega := \{\mathbf{x} = (x_1, x_2) : x_i = 1, \ldots, n_{x_i}, i = 1, 2\}$ (or $\Omega := \{\mathbf{x} = (x_1, x_2, x_3) : x_i = 1, \ldots, n_{x_i}, i = 1, 2, 3\}$). Here, I can be viewed as a map $I : \Omega \mapsto \mathbb{R}$, where $\mathbb{R}$ is the set of real numbers. The $I(\mathbf{x})$ can be a gray-level intensity value at pixel position $\mathbf{x}$.
- **(Output U)** In the case of segmentation, the output U is a binary map $U : \Omega \to \{0, 1\}$. In the case of classification, the output U is a vector representing classifications where each component corresponds to a probability of a target class.

⋆⋆⋆ *Readers interested in only deep learning can skip Sect.* 2.3 *and jump to Sect.* 2.4.

2.3 Conventional Segmentation Methods

2.3.1 Thresholding Methods

Thresholding is the simplest method that separates a gray-scale image by grouping all pixels with intensity greater than a fixed constant (called a threshold value) into one class and all other pixels into another class. Since the thresholding approach does not reflect spatial features of tomographic images, it is sensitive to various image quality factors including noise, artifacts, and intensity non-uniformities. Hence, thresholding is often used as an initial guess for segmentation. The thresholding of I with a threshold value ϑ, denoted by I_ϑ, is defined by

$$I_\vartheta(\mathbf{x}) = \begin{cases} 0 & \text{if } \; I(\mathbf{x}) < \vartheta \\ 1 & \text{otherwise.} \end{cases} \tag{2.2}$$

To determine the threshold value ϑ automatically, Otsu's thresholding method [31], named after Nobuyuki Otsu, is widely used. It automatically determines the threshold value ϑ using histogram $h(t) = \sharp\{\mathbf{x} : I(\mathbf{x}) = t\}$, the distribution of I expressing the number of pixels of each gray-level value t. Otsu's threshold value is given by

$$\vartheta = \operatorname*{argmax}_t \left[\sum_{i<t} p(i) \left(\frac{\sum\limits_{i<t} ip(i)}{\sum\limits_{i<t} p(i)} \sum_i ip(i) \right)^2 + \sum_{t \le i} p(i) \left(\frac{\sum\limits_{t\le i} ip(i)}{\sum\limits_{t \le i} p(i)} - \sum_i ip(i) \right)^2 \right], \tag{2.3}$$

where $p(t) = h(t)/\sum\limits_i h(i)$. See Fig. 2.3.

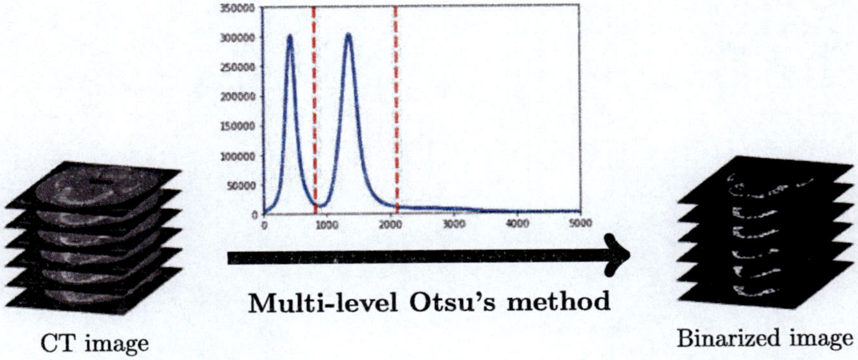

Fig. 2.3 Otsu's thresholding method that determines threshold value ϑ using histogram

2.3.2 *Region-Growing Method*

The region-growing method is a pixel-based image segmentation technique that generates connected regions created by growing from initial seed points. This method starts extracting neighboring pixels of initial seed points by examining the similarity of the intensity values of each neighboring pixels with the initial seed point. This process is repeated in the same way to determine the final region. This method may not be automatic due to the requirement of manual selection of initial seed points. Like thresholding, this approach is also sensitive to various image quality factors including noise, artifacts, and weak boundaries. This is basically a local method which does not reflect the global structure of images. Implementation details of region-growing method are given in Algorithm 1.

Algorithm 1 Region-growing method

Step 1. Select a set of seed points
Set an initial point $\mathbf{x}_0, \mathbf{x}_1, \ldots, \mathbf{x}_n$ and a set $\mathscr{R} = \{\mathbf{x}_0, \mathbf{x}_1, \ldots, \mathbf{x}_n\}$

Step 2. Given a region $\mathscr{R}$, find candidate points by checking the neighboring pixels
The set of candidate points can be $C := \{\mathbf{x} \notin \mathscr{R} | \aleph(\mathbf{x}) \cap \mathscr{R} \neq \emptyset\}$, where $\aleph(\mathbf{x})$ is a neighboring points of $\mathbf{x}$.

Step 3. Measure how different $\mathbf{x} \in C$ is from a neighboring area.

if $\left| I(\mathbf{x}) - \underset{\mathbf{x}' \in \mathscr{R}}{\text{mean}} I(\mathbf{x}') \right| < \epsilon$ for $\mathbf{x} \in C$, **then**
  append $\mathbf{x}$ to $\mathscr{R}$ and go to Step 2
else
  return output
end if

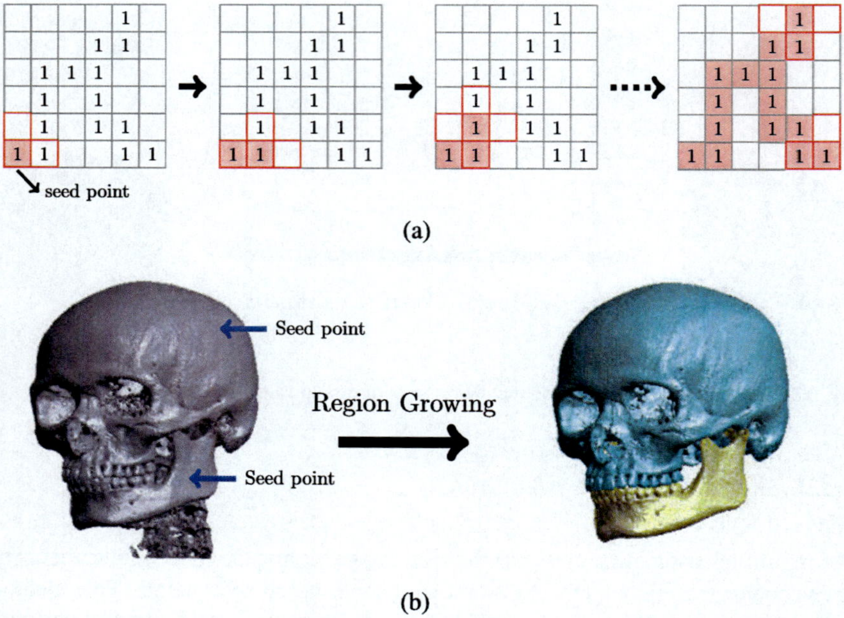

Fig. 2.4 Skull segmentation result using the region-growing method. **a** Region-growing process from a seed point to its neighbors. **b** The left image is a binary skull segmentation obtained from 3D CBCT image I with Otsu's thresholding. Here, locations pointing to blue arrows indicate selected seed points. The right image is the result of segmentation of the upper and lower jaw of skull generated from the upper and lower seed point, respectively

Figure 2.4 shows the skull segmentation result using the region-growing method. We applied the region-growing method to binary skull images obtained from Otsu's thresholding method.

2.3.3 Energy-Based Deformable Models

Deformable models are an energy-based segmentation technique that forces a closed curve defined within an image area G toward the target segmentation by minimizing an energy consisting of internal energy defined on the curve and external energy from the image I. Numerous energy-based deformable models such as active contour and level set have been developed in the last four decades. Various application-dependent energy functionals have been developed to evolve an active contour toward the boundary of the target region. The energy functional is designed to obtain a local minimum through iterative contour evolution through energy minimization process. Hence, it requires a good choice of initial contour and a suitable stopping criterion to terminate the iterative process at the target boundary. Representing a contour by $\mathscr{C}$, the common form of energy-based segmentation methods is as follows:

$$\mathscr{C} = \underset{\mathscr{C}}{\text{argmin}}\ \mathbf{Fit}(\mathscr{C}) + \lambda \mathbf{Reg}(\mathscr{C}), \tag{2.4}$$

where $\mathbf{Fit}(\mathscr{C})$ is a fitting term for attracting the contour $\mathscr{C}$ toward the target object in the image; $\mathbf{Reg}(\mathscr{C})$ is a regularization term to reflect prior knowledge of contour structure; and λ is a tunable fitting parameter which controls the balance between those two terms. There exist a variety of fitting models such as edge-based methods [3, 4, 13, 28], region-based methods [5], etc., and these fitting models are mostly combined with the standard regularization term penalizing arc length of the contour C_φ.

2.3.3.1 Active Contour Method

The active contour method (or snakes) uses an energy functional $\Phi(\mathscr{C})$ to evolve an active contour $\mathscr{C}^t$ (representing a contour at time t) toward the boundary of the target region. To be precise, let $I(x, y)$ be a given medical image. A desired segmentation can be achieved by finding a closed curve $\mathscr{C}$ that minimizes the energy functional:

$$\Phi(\mathscr{C}) = \mathbf{Fit}(\mathscr{C}) + \lambda \mathbf{Reg}(\mathscr{C}). \tag{2.5}$$

We begin with the following simplest form of the energy functional:

$$\Phi(\mathscr{C}) := \int_{\mathscr{C}} \frac{1}{1 + |\nabla I|^2}\, ds. \tag{2.6}$$

For computation of a local minimum $\mathscr{C}$ of the functional, we may start from an initial contour $\mathscr{C}^0$ and get a minimizing sequence $\mathscr{C}^1, \mathscr{C}^2, \ldots$ which converges to the local minima $\mathscr{C}$:

$$\Phi(\mathscr{C}^0) \geq \Phi(\mathscr{C}^0) \geq \cdots \Phi(\mathscr{C}^n) \geq \Phi(\mathscr{C}^{n+1}) \cdots \to \Phi(\mathscr{C}).$$

See Fig. 2.5 for the sequence $\{\mathscr{C}^n : n = 1, 2, \ldots\}$. See also Fig. 2.6 for its 3D version.

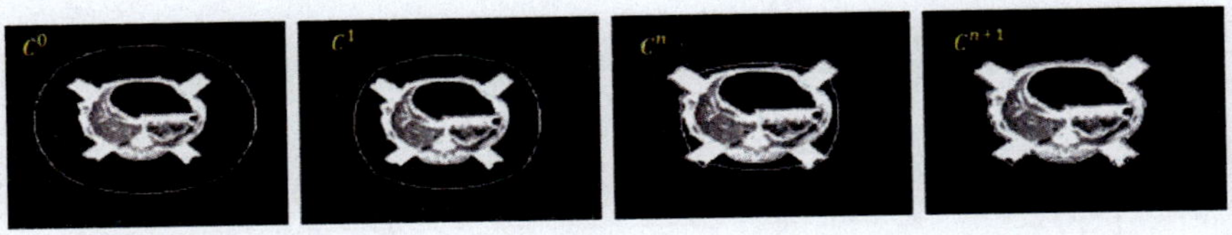

Fig. 2.5 Image segmentation using 2D active contour

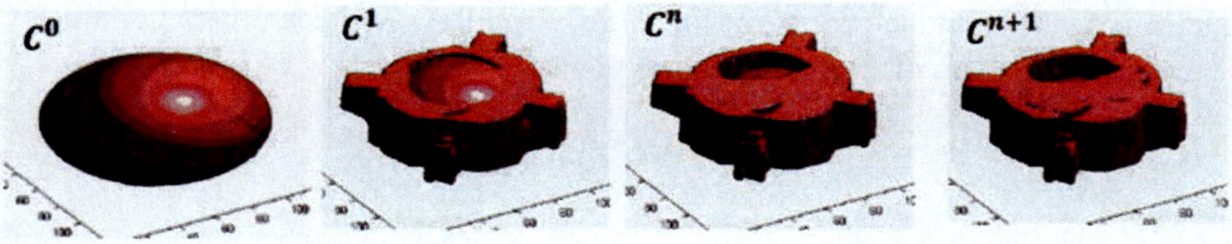

Fig. 2.6 Image segmentation by 3D level surface

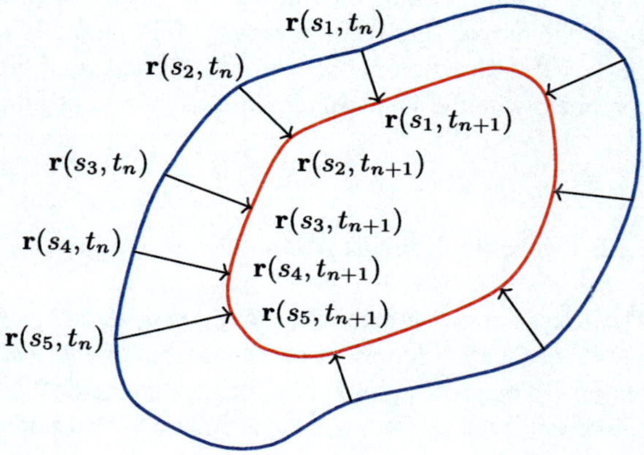

Fig. 2.7 Active contour model at time t

To calculate the next contour $\mathscr{C}^{n+1}$ from $\mathscr{C}^n$, the gradient descent method based on the Fréchet gradient $\nabla\Phi(\mathscr{C}^n)$ is widely used:

$$\mathscr{C}^{n+1} = \mathscr{C}^n \quad \underbrace{-\alpha\nabla\Phi(\mathscr{C}^n)}_{\text{steepest descent direction}} \tag{2.7}$$

What does $\nabla\Phi(\mathscr{C}^n)$ (the gradient with respect to the contour parameters) mean? To explain $\nabla\Phi(\mathscr{C}^n)$, it is convenient to use the time-varying contour $\mathscr{C}^t$ instead of the sequence $\{\mathscr{C}^n\}$. Then, (2.7) is expressed as

$$\mathscr{C}^{t+\Delta t} = \mathscr{C}^t - \alpha\Delta t\nabla\Phi(\mathscr{C}^t) \tag{2.8}$$

To calculate $\mathscr{C}^{t+\Delta t}$, let us parameterize $\{\mathscr{C}^t\}$ by $\mathbf{r}(s,t) = x(s,t)(1,0) + y(s,t)(0,1)$, $0 < s < 1$:

$$\mathscr{C}^t = \{\mathbf{r}(s,t) = x(s,t)(1,0) + y(s,t)(0,1) \mid 0 < s < 1\}. \tag{2.9}$$

See Fig. 2.7 for $\mathscr{C}^t$ and $\mathscr{C}^{t+\Delta t}$.

Setting $\Phi(t) = \Phi(\mathscr{C}^t)$, we have

$$\Phi(t) := \int_{\mathscr{C}^t} \frac{1}{1+|\nabla I|^2}\, ds = \int_0^1 g_I(\mathbf{r}(s,t))\, |\mathbf{r}_s(s,t)| ds, \tag{2.10}$$

where $g_I(\mathbf{r}) = \frac{1}{1+|\nabla I(\mathbf{r})|^2}$. The variation of the energy functional Φ is

$$\begin{aligned}
\Phi'(t) &= \int_0^1 |\mathbf{r}_s| \, [\nabla g_I \cdot \mathbf{r}_t] \, ds + \int_0^1 g_I \left[\frac{\mathbf{r}_s}{|\mathbf{r}_s|} \cdot \mathbf{r}_{ts} \right] ds \\
&= \int_0^1 |\mathbf{r}_s| \, [\nabla g_I \cdot \mathbf{r}_t] \, ds - \int_0^1 \langle g_I \left[\frac{\mathbf{r}_s}{|\mathbf{r}_s|} \right]_s + [\nabla g_I \cdot \mathbf{r}_s] \frac{\mathbf{r}_s}{|\mathbf{r}_s|} \, , \mathbf{r}_t \rangle ds \\
&= \int_0^1 |\mathbf{r}_s| \, \mathbf{r}_t \cdot [\nabla g_I - \kappa g_I \mathbf{n} - \langle T, \nabla g_I \rangle T] \, ds,
\end{aligned}$$

where $\mathbf{r}_t = \frac{\partial \mathbf{r}}{\partial t}$, $\mathbf{r}_s = \frac{\partial \mathbf{r}}{\partial s}$, $\mathbf{n} = \mathbf{n}(s, t)$ the unit normal to the curve $\mathscr{C}^t$, and $T = T(s, t)$ the unit tangent vector. Hence, the direction for which $\Phi(t)$ decreases most rapidly is given by

$$\mathbf{r}_t = -\left[\nabla g - \kappa g \mathbf{n} - \langle T, \nabla g \rangle T\right]. \tag{2.11}$$

Decomposing $\nabla g_I = \langle \nabla g_I, \mathbf{n} \rangle \, \mathbf{n} + \langle \nabla g_I, T \rangle \, T$, the Eq. (2.11) becomes

$$\mathbf{r}_t \; = \; \underbrace{(\kappa \tilde{g} - \langle \nabla g_I, \mathbf{n} \rangle) \; \mathbf{n}}_{\pm \, speed \;\; |\mathbf{r}_t|} \qquad (or \quad \frac{\partial}{\partial t} \mathscr{C}^t = (\kappa g_I - \nabla g_I \cdot \mathbf{n}) \, \mathbf{n} \;) \tag{2.12}$$

which means that the curve $\mathbf{r}(s, t)$ moves along its normal with the speed

$$F = \kappa g - \langle \nabla g, \mathbf{n} \rangle. \tag{2.13}$$

Since

$$\frac{\mathbf{r}(s, t + \Delta t) - \mathbf{r}(s, t)}{\Delta t} \quad \approx \quad F(\mathbf{r}(s, t)) \, \mathbf{n}(s, t), \tag{2.14}$$

we can determine the update $\mathscr{C}^{n+1}$ by

$$\mathbf{r}(s, n + 1) = \mathbf{r}(s, n) + \Delta t \; F(\mathbf{r}(s, n)) \, \mathbf{n}(s, n). \tag{2.15}$$

2.3.3.2 Level Set Method

The explicit expression $\mathscr{C}^t = \{\mathbf{r}(s, t) \; : \; 0 \le s \le 1\}$ is not appropriate when $\mathscr{C}^t$ changes its topology (splitting multiple closed curves). When segmenting multiple targets whose locations are unknown, it is desirable to use the implicit expression of $\mathscr{C}^t$ in order to allow topological change to track multiple targets. The level set method [30] is based on the implicit expression of the zero level set

$$\mathscr{C}^t = \{\mathbf{r} \; : \; \phi(\mathbf{r}, t) = 0\}. \tag{2.16}$$

For the ease of explanation, let $\{\mathbf{r}(s,t) \ : \ s \in [0,1]\} \subset \mathscr{C}^t$ (a part of the contour). Then, ϕ satisfies

$$\phi(\mathbf{r}(s,t),t) = 0, \quad \forall t > 0, \ s \in [0,1]. \tag{2.17}$$

Taking the derivative $\frac{d}{dt}$ leads to the equation of ϕ containing the embedded motion of $\mathscr{C}^t$:

$$0 = \frac{d}{dt}\phi(\mathbf{r}(s,t),t) = \frac{\partial}{\partial t}\phi(\mathbf{r},t) + \frac{\partial}{\partial t}\mathbf{r}(s,t) \cdot \nabla\phi(\mathbf{r}(s,t),t). \tag{2.18}$$

The above equation explains that $\mathscr{C}^t$ evolves according to the velocity $\mathbf{V} = \frac{\partial}{\partial t}\mathbf{r}(s,t)$. Note that $\mathbf{n}(s,t) = \frac{\nabla\phi(\mathbf{r}(s,t),t)}{|\nabla\phi(\mathbf{r}(s,t),t)|}$ is the unit normal vector of the contour $\mathscr{C}^t$. Writing $F(\mathbf{r}(s,t),t) = V \cdot n$ as the motion in the normal direction, the above identity can be rewritten as

$$\frac{\partial}{\partial t}\phi(\mathbf{r},t) + \underbrace{\frac{\partial}{\partial t}\mathbf{r}(s,t)\frac{\nabla\phi(\mathbf{r}(s,t),t)}{|\nabla\phi(\mathbf{r}(s,t),t)|}}_{F(\mathbf{r},t)}|\nabla\phi(\mathbf{r},t)| = 0. \tag{2.19}$$

In (2.13), the normal speed F can be expressed as

$$F = \kappa g \nabla \cdot \left[\frac{\nabla\phi}{|\nabla\phi|}\right] - \langle \nabla g, \frac{\nabla\phi}{|\nabla\phi|}\rangle, \tag{2.20}$$

where the convection term $\langle \nabla g, \nabla\phi \rangle$ increases the attraction of the deforming contour toward the boundary of objects.

Even with a smooth initial contour, $\mathscr{C}^t$ may develop irregularities (or singularities) during its evolution. For the stability during its evolution, the level set function ϕ is desirable to satisfy $|\nabla\phi| \approx 1$. To maintain $|\nabla\phi| \approx 1$ during evolution, the following distant regularized level set method was proposed [25]:

$$\Phi(\phi) = \int \frac{1}{1+|\nabla I|}\delta(\phi)|\nabla\phi|d\mathbf{r} + \int (|\nabla\phi| - 1)^2 d\mathbf{r}, \tag{2.21}$$

where δ is a smeared Dirac delta function.

2.4 Deep Learning-Based Segmentation Methods

Medical image segmentation requires an appropriate balance between the semantics of viewing the entire image structure and the local information to determine the boundaries of the expected segmentation region. It would be desirable to mimic a clinician's procedure for the segmentation. Compared to the conventional methods described earlier, a deep learning approach with training data can be an effective way to view global and local information simultaneously.

Conventional segmentation techniques including energy-based segmentation methods (using active contour and level-set) have fundamental drawbacks in achieving fully automated segmentations. These methods start from an initial contour and use the iterative contour evolution through an energy minimization process. Therefore, these approaches require a good choice of initial contour so that user intervention through manual initialization is necessary and the results often are affected by initialization. The critical drawback of the conventional approaches is the inability to reflect contextual (global) information, and the stopping criterion for curve evolution tends to depend mostly on local image intensity. To overcome these problems, it would be desirable to segment on the basis of local patterns of the target boundary as well as the global image structure.

DL methods are advantageous over the conventional methods in medical image semantic segmentation where there are various uncertain factors about the borders of the target domain. Using training data, the DL methods effectively take into account prior knowledge of anatomical structures. This section explains various DL methods for segmentation and identification.

For ease of explanation, DL methods will be described using a digital dental model that aims to automatically recognize and segment individual teeth from cone-beam computerized tomography (CBCT) images. In 3D individual tooth segmentation, conventional methods have fundamental limitations because of the difficulty in separating an individual tooth from adjacent teeth and its surrounding alveolar bone. On the other hand, DL methods appear to have a strong capability to explore the prior information of the borders of target segmentation via training data. In particular, U-net has shown remarkable overall performance in medical image segmentations by simultaneously making use of global features and local spatial information [21, 26, 27, 35].

This section is based on the recent paper written by Jang et al. [18]. The final goal is to find a map $f : I \to U$, where I is a 3D CBCT image and U is the corresponding 3D segmentation of individual tooth and identification of them into four types (e.g., incisor, canine, premolar, molar) according to tooth morphology. DL methods use training data $\{(I^{(n)}, U^{(n)})\}_{n=1}^{N}$ to learn f by minimizing the distance between $f(I^{(n)})$ and $U^{(n)}$ for all n. To be precise, the learning objective is as follows:

$$f = \underset{f \in \mathbb{NN}}{\text{argmin}} \frac{1}{N} \sum_{n=1}^{N} \text{dist}(f(I^{(n)}), U^{(n)}), \tag{2.22}$$

where $\mathbb{NN}$ denotes a set of functions described in a special form of neural network and $\text{dist}(f(I^{(n)}), U^{(n)})$ is the distance between $f(I^{(n)})$ and $U^{(n)}$.

A DL network's performance depends not only on its architecture, but also on the quality and quantity of training data. In medical areas, the high dimensionality of the input data and limited number of training data are the main factors that hinder the training of deep learning networks for learning the tooth segmentations from 3D CBCT data. Moreover, due to the current legal and ethical restrictions on medical data, it is very difficult to utilize CBCT data from patients. Hence, we need to overcome the above-mentioned learning problems caused by the high input dimensions and training data deficiencies.

To circumvent the high-dimensionality problem associated with CT images, we use 2D panoramic images that are generated from 3D CBCT images. From a 2D panoramic image, we can develop a tooth detection method that localizes bounding boxes that enclose each tooth and classifies them into four types according to tooth morphology. Next, we can perform 2D segmentation for individual teeth. Finally, we can achieve 3D segmentation by utilizing the above-mentioned 2D segmentation and identification.

Before explaining the whole process of 3D tooth segmentation, which will be discussed in the final subsection of this chapter, let's start with understanding the basic mechanisms of a convolutional neural networks (CNNs).

2.4.1 Convolutional Neural Networks (CNNs)

To explain CNNs intuitively, we use the following simplified model of the tooth identification network $f : I \to U$.

- The input I is a image patch from a bounding box (that enclose each tooth) in a panoramic image. This panoramic image is generated from 3D CBCT image. In Sect. 2.4.5, we will explain how to detect this bounding box.
- The output U is a vector representing tooth classification, where the first, second, third, and fourth component of $U = f(I)$ correspond to the probability of I being incisor, canine, premolar, and molar, respectively. To be precise,

$$f(\text{incisor}) = \begin{bmatrix} 1 \\ 0 \\ 0 \\ 0 \end{bmatrix}, \quad f(\text{canine}) = \begin{bmatrix} 0 \\ 1 \\ 0 \\ 0 \end{bmatrix}, \quad f(\text{premolar}) = \begin{bmatrix} 0 \\ 0 \\ 1 \\ 0 \end{bmatrix}, \quad f(\text{molar}) = \begin{bmatrix} 0 \\ 0 \\ 0 \\ 1 \end{bmatrix}.$$

See Fig. 2.8 for visual understanding.

The CNNs consist of four components: (i) convolutional layer, (ii) pooling layer, (iii) fully connected layer, and (iv) element-wise nonlinear activation function such as the sigmoid function or rectifier linear unit (ReLU). The tooth identification function f can be expressed as $f(I) = f^{(6)}(f^{(5)} \cdots (f^{(1)}(I)))$, where $f^{(j)}$ denotes the j-th layer. Here, $f^{(j)}$ can be a convolutional layer, pooling layer, or fully connected layer. Figure 2.9 shows the network architecture of the CNN.

In the following sections, we will go through the basic components one by one to understand the CNN architecture.

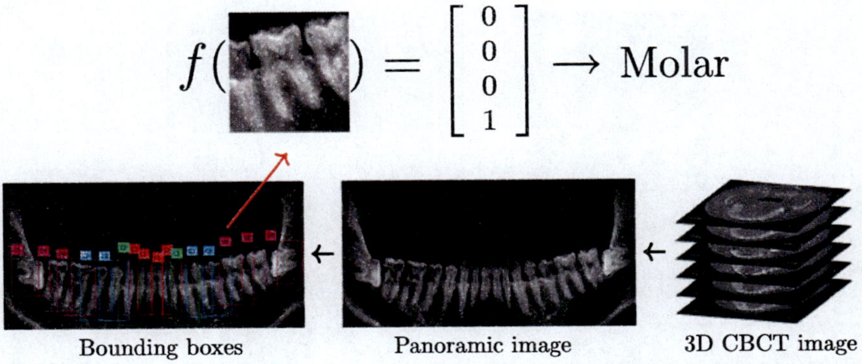

Fig. 2.8 Tooth identification model. The teeth are classified as incisor (class 1), canine (class 2), premolar (class 3), and molar (class 4)

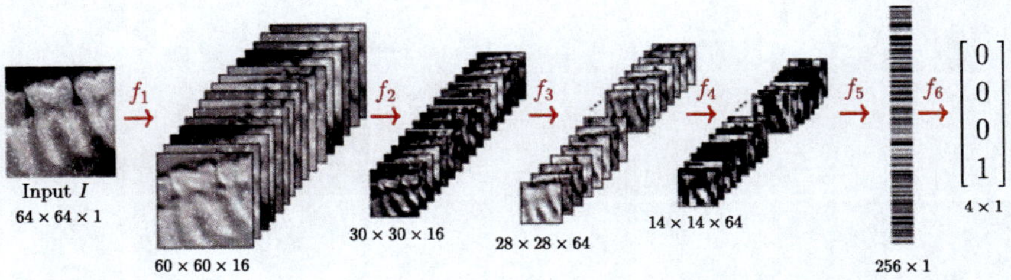

Fig. 2.9 A simple architecture of CNN. This CNN consists of two convolution layers (f_1, f_3), two pooling layers (f_2, f_4), one fully connected layer (f_5), and one classification layer (f_6) with 4 units

2.4.1.1 Forward Mapping of CNN

[Input layer] In Fig. 2.10, the input I has one image of dimensions 64×64 which contains a tooth.

[Layer 1: $f_1(I)$] We use 16 convolution filters of size 5×5, denoted by $\mathbf{W}^1$, to generate $f_1(I)$. Here, the $\mathbf{W}^1 = [\mathbf{w}_1^1, \ldots, \mathbf{w}_{16}^1]$ is given by

$$\mathbf{w}_i^1 = \begin{bmatrix} w_{i,(1,1)}^1 & \cdots & w_{i,(1,5)}^1 \\ \vdots & \ddots & \vdots \\ w_{i,(5,1)}^1 & \cdots & w_{i,(5,5)}^1 \end{bmatrix}, \; i = 1, \ldots, 16. \tag{2.23}$$

Given a filter $\mathbf{w}$ of size 5×5, the convolution $\mathbf{w} \circledast_1 I$ with stride 1 produces an image of dimension 60×60 that is given by

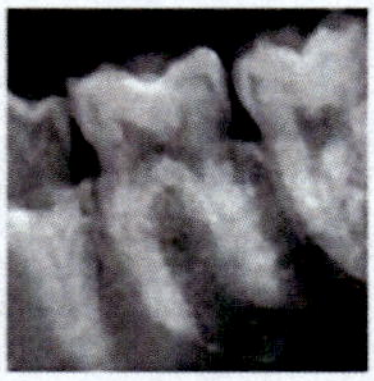

$$\text{Input } I = \begin{bmatrix} I_{1,1} & \cdots & I_{1,64} \\ \vdots & \ddots & \vdots \\ I_{64,1} & \cdots & I_{64,64} \end{bmatrix}$$

Fig. 2.10 The input I of the CNN is an image of 64 × 64 pixels

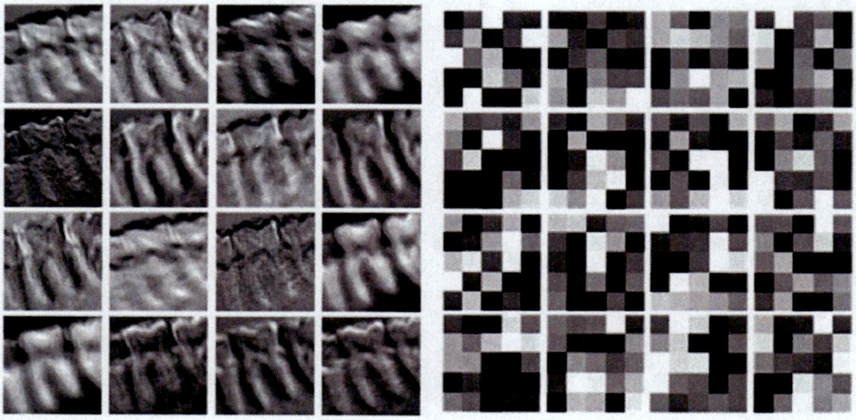

(a) (b)

Fig. 2.11 **a** Feature extraction by the $\mathbf{W}^1 \circledast_1 I$. **b** 16 convolution filters $\mathbf{W}^1 = [\mathbf{w}_1^1, \ldots, \mathbf{w}_{16}^1]$

$$\mathbf{w} \circledast_1 I = \underbrace{\begin{bmatrix} \sum_{i,j=0}^{5} w(i,j)I(i,j) & \sum_{i,j=0}^{5} w(i,j)I(i,j+1) & \cdots & \sum_{i,j=0}^{5} w(i,j)I(i,j+59) \\ \sum_{i,j=0}^{5} w(i,j)I(i+1,j) & \sum_{i,j=0}^{5} w(i,j)I(i+1,j+1) & \cdots & \sum_{i,j=0}^{5} w(i,j)I(i+1,j+59) \\ \vdots & \vdots & \ddots & \vdots \\ \sum_{i,j=0}^{5} w(i,j)I(i+59,j) & \sum_{i,j=0}^{4} w(i,j)I(i+59,j+1) & \cdots & \sum_{i,j=0}^{5} w(i,j)I(i+59,j+59) \end{bmatrix}}_{\text{This is a } 60\times 60 \text{ matrix, where 60 comes from 60=64-5+1.}} . \quad (2.24)$$

Figure 2.11 shows the convolved output $\mathbf{W}^1 \circledast_1 I$ with the convolution filter $\mathbf{W}^1$. The output of the first hidden layer is given by

$$f_1(I) = \sigma(\mathbf{W}^1 \circledast I + \mathbf{b}^1) = \left[\sigma(\mathbf{w}_1^1 \circledast I + \mathbf{b}_1^1), \ldots, \sigma(\mathbf{w}_{16}^1 \circledast I + \mathbf{b}_{16}^1)\right], \quad (2.25)$$

where $\mathbf{b}^1$ is a bias term and σ is a nonlinear activation function. Throughout this chapter, the nonlinear activation function σ is ReLU given by:

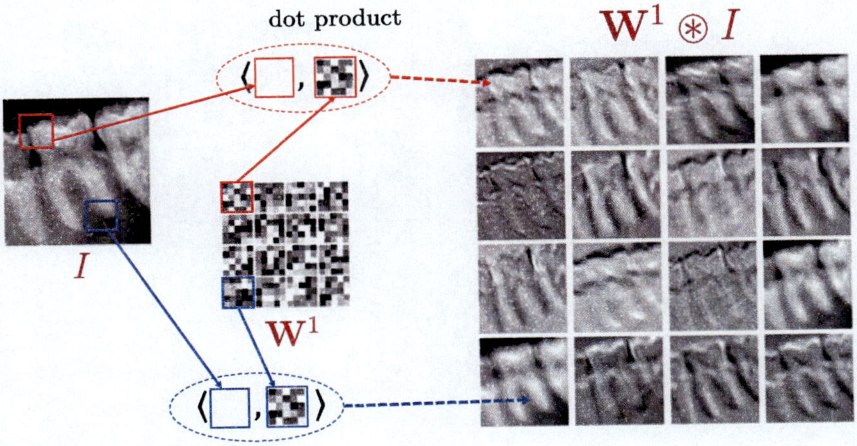

Fig. 2.12 Visualization of features by the convolution $\mathbf{W}^1 \circledast_1 I$

$$\sigma(\mathbf{h}) = ReLU(\mathbf{h}) = \max(0, \mathbf{h}),$$
$$\sigma(h_1, \ldots, h_m) = (\sigma(h_1), \ldots, \sigma(h_m)). \tag{2.26}$$

This map f_1 is used to extract useful feature information from images as shown in Fig. 2.12.

[Layer 2: $f_2(\mathbf{h}^1)$ **]** This layer applies the max-pooling operator to the $\mathbf{h}^1 = f_1(I)$. The output of $\mathbf{h}^2 = f_2(\mathbf{h}^1)$ 16 feature images of dimension 30×30. The i-th feature image in $\mathbf{h}^2$ is given by

$$\mathbf{h}_i^2 = \begin{bmatrix} \max\begin{pmatrix} h^1_{i,(1,1)} & h^1_{i,(1,2)} \\ h^1_{i,(2,1)} & h^1_{i,(2,2)} \end{pmatrix} & \cdots & \max\begin{pmatrix} h^1_{i,(1,59)} & h^1_{i,(1,60)} \\ h^1_{i,(2,59)} & h^1_{i,(2,60)} \end{pmatrix} \\ \vdots & \ddots & \vdots \\ \max\begin{pmatrix} h^1_{i,(59,1)} & h^1_{i,(59,2)} \\ h^1_{i,(60,1)} & h^1_{i,(60,2)} \end{pmatrix} & \cdots & \max\begin{pmatrix} h^1_{i,(59,59)} & h^1_{i,(59,60)} \\ h^1_{i,(60,59)} & h^1_{i,(60,60)} \end{pmatrix} \end{bmatrix}. \tag{2.27}$$

This pooling helps to make a translation invariant identification system by producing representations that are invariant to small shifts in the input. The pooling function summarizes the outputs of neighboring group of chosen size. There are other types of pooling, such as min pooling and average pooling.

[Layer 3: $f_3(\mathbf{h}^2)$ **]** In this layer, we used 64 filters of size $3 \times 3 \times 16$, denoted by $\mathbf{W}^3$ to generate $\mathbf{h}^3 = f_3(\mathbf{h}_2)$. As shown in Fig. 2.13, $\mathbf{W}^3 = [\mathbf{w}_1^3, \mathbf{w}_2^3, \ldots, \mathbf{w}_{64}^3]$, where each $\mathbf{w}_i^3$ is given by

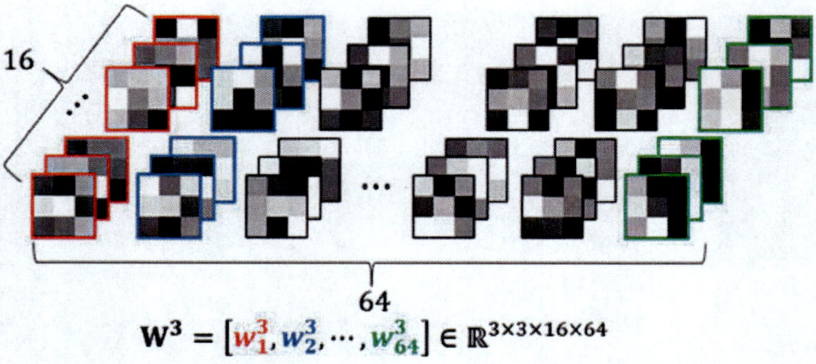

Fig. 2.13 64 convolution filters $\mathbf{W}^3 = [\mathbf{w}_1^3, \mathbf{w}_2^3, \ldots, \mathbf{w}_{64}^3]$. Each filter $\mathbf{w}_j^3$ has the size of $3 \times 3 \times 16$

$$\mathbf{w}_i^3 = \left[\underbrace{\begin{bmatrix} w^3_{i,(1,1,1)} & \cdots & w^3_{i,(1,3,1)} \\ \vdots & \ddots & \vdots \\ w^3_{i,(3,1,1)} & \cdots & w^3_{i,(3,3,1)} \end{bmatrix}}_{\mathbf{w}^3_{i,1}}, \ldots, \underbrace{\begin{bmatrix} w^3_{i,(1,1,16)} & \cdots & w^3_{i,(1,3,16)} \\ \vdots & \ddots & \vdots \\ w^3_{i,(3,1,16)} & \cdots & w^3_{i,(3,3,16)} \end{bmatrix}}_{\mathbf{w}^3_{i,16}} \right]. \tag{2.28}$$

The output $\mathbf{h}^3 = f_3(\mathbf{h}_2)$ (ignoring the bias term) is given by

$$\mathbf{h}_i^3 = \sigma(\mathbf{w}_i^3 \circledast \mathbf{h}^2) = \sigma\left(\sum_{j=1}^{16} \mathbf{w}_{i,j} \circledast \mathbf{h}_j^2\right) \quad i = 1, \ldots, 64. \tag{2.29}$$

Figure 2.14 shows the output $\mathbf{h}_1^3$.

[Layer 4: $f_4(\mathbf{h}^3)$**]** In this layer, we apply another pooling layer to reduce the size of $\mathbf{h}^3$ by factor 2. Maximum pooling summarizes the information, producing a smaller output, and helping it to be invariant to the displacement of target in the input image. The i-th feature image in $\mathbf{h}^4$ is given by the following:

$$\mathbf{h}_i^4 = \begin{bmatrix} \max\begin{pmatrix} h^3_{i,(1,1)} & h^3_{i,(1,2)} \\ h^3_{i,(2,1)} & h^3_{i,(2,2)} \end{pmatrix} & \cdots & \max\begin{pmatrix} h^3_{i,(1,27)} & h^3_{i,(1,28)} \\ h^3_{i,(2,27)} & h^3_{i,(2,28)} \end{pmatrix} \\ \vdots & \ddots & \vdots \\ \max\begin{pmatrix} h^3_{i,(27,1)} & h^3_{i,(27,2)} \\ h^3_{i,(28,1)} & h^3_{i,(28,2)} \end{pmatrix} & \cdots & \max\begin{pmatrix} h^3_{i,(27,27)} & h^3_{i,(27,28)} \\ h^3_{i,(28,27)} & h^3_{i,(28,28)} \end{pmatrix} \end{bmatrix}. \tag{2.30}$$

[Layer 5: $f_5(\mathbf{h}^4)$ **]** This is a fully connected layer immediately before the classifier. This generates a feature vector as a barcode for the classifier. The fully connected layer connects all neurons in one layer to all neurons in another layer. If the output

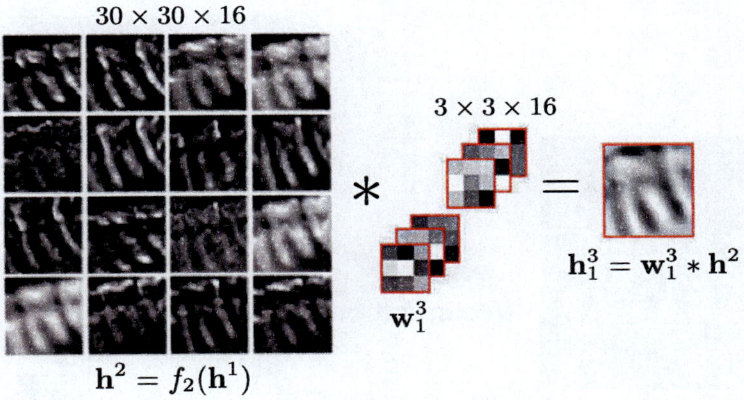

Fig. 2.14 Application of one $3 \times 3 \times 16$ filter to the input $\mathbf{h}^2$ (of size $30 \times 30 \times 16$) with stride 1. The output has the size of 28×28. Since we are using 64 filters, the output $\mathbf{h}^3 = f_3(\mathbf{h}_2)$ has the size of $28 \times 28 \times 64$

$f_5(\mathbf{h}^4)$ has the size of 256×1, the weight $\mathbf{W}^5$ is a 256×12544 matrix. To be precise, the output $f_5(\mathbf{h}^4)$ is given by

$$f_5(\mathbf{h}^4) = \sigma(\mathbf{W}^5\mathbf{h}^4), \tag{2.31}$$

where $\mathbf{W}^5\mathbf{h}^4$ is the standard matrix multiplication and the input of the fully connected layer (output of layer 4) is vectorized for the matrix multiplication, as shown in Fig. 2.15. Here, the input image I is encoded as barcode $\mathbf{h}^5$, so that we are ready to classify I using $\mathbf{h}^5$.

[Output layer: $f_6(\mathbf{h}^5)$] This $f_6(\mathbf{h}^5)$ is a linear classifier that aims to classify the input I into 4 classes (e.g., incisor, canine, premolar, molar teeth). For the classification, we apply a 4×256 matrix $\mathbf{W}^6$ to $\mathbf{h}^5$, where the number of the column of $\mathbf{W}^6$ is equal to the dimension of $\mathbf{h}^5$ and the number of the row of $\mathbf{W}^6$ is equal to the number of classes being classified. The output is given by

$$f_6(\mathbf{h}^5) = \sigma_{\text{softmax}}(\mathbf{W}^6\mathbf{h}^5), \tag{2.32}$$

where $\mathbf{W}^6 = [\mathbf{w}_1^6, \mathbf{w}_2^6, \mathbf{w}_3^6, \mathbf{w}_4^6] \in \mathbb{R}^{4\times 256}$ is a weight and σ_{softmax} is softmax activation function which has a value

$$\sigma_{\text{softmax},i} = \frac{\exp(h_i^6)}{\sum_{k=1}^{4}\exp(h_k^6)}. \tag{2.33}$$

The j-th component of $f_6(\mathbf{h}^5)$ can be viewed as the posterior probability of the j-th class. See Fig. 2.16.

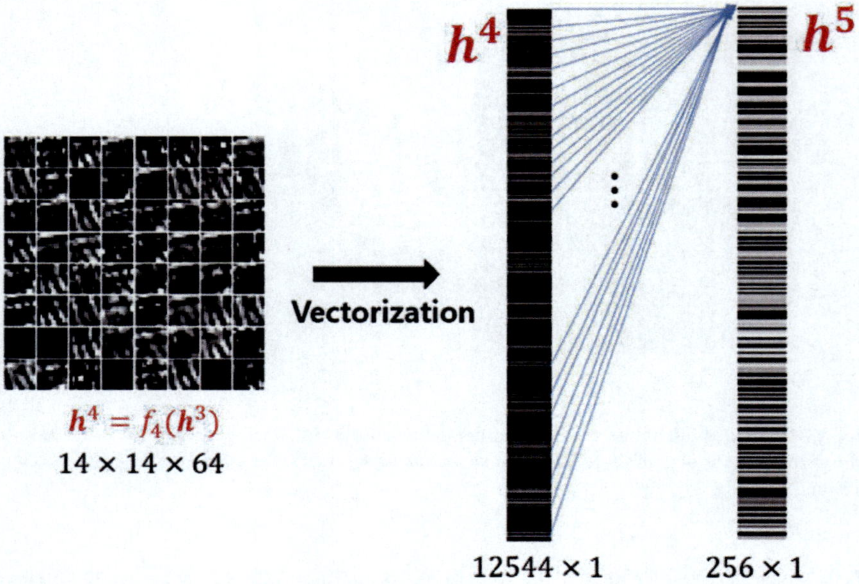

Fig. 2.15 Fully connected layer generates a low-dimensional feature vector

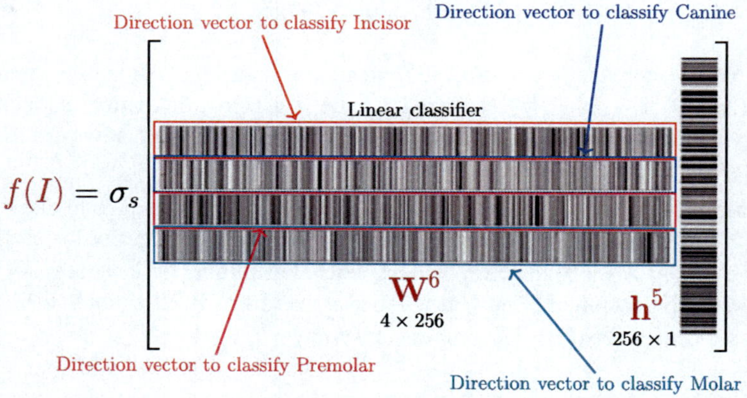

Fig. 2.16 The last layer is the 4-way softmax classifier. There are four barcodes to classify tooth class

2.4.1.2 Loss Function and Backpropagation

The goal is learn the map $f : I \to U$ which is defined by

$$f(I) = \sigma_s \left(\mathbf{W}^6 \sigma \left(\mathbf{W}^5 P_4 \left(\sigma \left(\mathbf{W}^3 P_2 \left(\sigma \left(\mathbf{W}^1 I \right) \right) \right) \right) \right) \right). \tag{2.34}$$

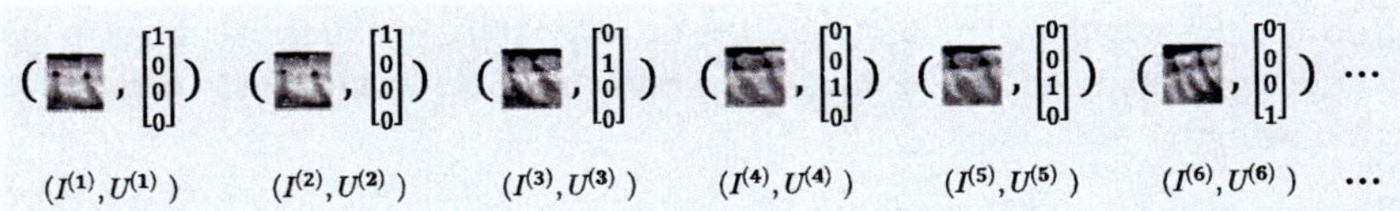

Fig. 2.17 Labeled training data for supervised learning

Hence, f is determined by the weights $\mathbf{W} = [\mathbf{W}^1, \mathbf{W}^3, \mathbf{W}^5, \mathbf{W}^6]$. We will learn $f : I \to U$ using labeled training data $\{(I^{(1)}, U^{(1)}), \ldots, (I^{(N)}, U^{(N)})\}$ as shown in Fig. 2.17. In this supervised learning, f is obtained in a way that $f(I^{(n)}) \approx U^{(n)}$ for all n. In other ward, the network f is estimated by minimizing the discrepancy between the predicted output $f(I^{(n)})$ and the reference target $U^{(n)}$ for all n.

To achieve this goal, we need to define a loss function $Loss(\mathbf{W})$. Generally, there are two types of loss functions:

$$Loss(\mathbf{W}) = \frac{1}{N}\sum_{n=1}^{N} \|f(I^{(n)}) - U^{(n)}\|^2 \quad L^2 \text{ minimization,} \tag{2.35}$$

and

$$Loss(\mathbf{W}) = \frac{1}{N}\sum_{n=1}^{N}\sum_{i=1}^{4} y_i^{(n)} \left(\log(f(I^{(n)})_i)\right) \quad \text{cross entropy.} \tag{2.36}$$

In this section, we only consider the L^2-minimization (2.35). A learning neural network $f(I) = \sigma_s\left(\mathbf{W}^6\sigma\left(\mathbf{W}^5 P_4\left(\sigma\left(\mathbf{W}^3 P_2\left(\sigma\left(\mathbf{W}^1 I\right)\right)\right)\right)\right)\right)$ is achieved by minimizing the loss function $Loss(\mathbf{W})$. The value of $Loss(\mathbf{W})$ measures how far the network performance deviates from perfection for the given training data set $\{(I^{(1)}, U^{(1)}), \ldots, (I^{(N)}, U^{(N)})\}$. For optimizing the objective function $Loss(\mathbf{W})$, we use gradient descent method that is an iterative method for finding a local minimum of $Loss(\mathbf{W})$:

$$\mathbf{W} \leftarrow \mathbf{W} - \eta \nabla_{\mathbf{W}} Loss(\mathbf{W}), \tag{2.37}$$

where η is learning rate. The gradient $-\nabla_{\mathbf{W}} Loss(\mathbf{W})$ has the direction in which the loss function $Loss(\mathbf{W})$ has the steepest descent.

About the dimension of $\nabla_{\mathbf{W}} Loss(\mathbf{W})$, we should note that the number of parameters in the $\mathbf{W}$ is huge:

- $\mathbf{w}_i^1$ for $i = 1, \ldots, 16 \;\to\; (5 \times 5) \times 16$ unknowns.
- $\mathbf{w}_i^3$ for $i = 1, \ldots, 64 \;\to\; (3 \times 3 \times 16) \times 64$ unknowns.
- $\mathbf{w}_i^5$ for $i = 1, \ldots, 256 \;\to\; (14 \times 14 \times 64) \times 256$ unknowns.
- $\mathbf{w}_i^6$ for $i = 1, \ldots, 4 \;\to\; (256) \times 4$ unknowns.
- $25 \times 16 + 576 \times 64 + 12544 \times 256 + 256 \times 4$ parameters $> 3M$.

Moreover, the $Loss(\mathbf{W}) = \frac{1}{N}\sum_{n=1}^{N} \|f(I^{(n)}) - U^{(n)}\|^2$ is hardly convex and there may be infinitely many minima. Fortunately, a good local minimum ($Loss(\mathbf{W}) \approx 0$)

can do as much as a global minimum. Since there are some misunderstandings in the scientific community about the issue of the optimal f, we would like to emphasize the following:

- There may be a lot of **W**s that provide a good network f.
- It is fruitless (or impossible) to find a good f in the entire high-dimensional ambient space, due to the curse of dimensionality. The probability of any point in the ambient space to be a medical image is approximately zero. We only need to focus on a suitable probability distribution associated with the training data.
- A well-trained function f seems to work only in the immediate vicinity of a nonlinear data manifold associated with a training set. As a result, even if the input deviates even slightly from the training data manifold, deep learning models may produce incorrect or unwanted results.
- In practice, the gradient often reaches a plateau ($Loss(\mathbf{W}) \approx 0$), making it difficult to improve our training loss.

There are variants of gradient descent such as Batch, Stochastic, and Mini-batch. The difference between these methods depends on how much data we use to compute the gradient of $Loss(\mathbf{W})$. Depending on the amount of training data, we need to find a balance between time for updating and the accuracy of the update. Various methods have been developed to get out of the plateus, including adjusting the learning rate η.

For optimization, we need to understand the derivative of $Loss(\mathbf{W})$ with respect to any weight in the network $f(I)$. For simplicity, we express $Loss(\mathbf{W})$ in term of $\{I^1, \ldots, I^N\}$:

$$Loss(\mathbf{W}) = \frac{1}{N}\sum_{n=1}^{N} \mathscr{L}^{(n)}(\mathbf{W}), \tag{2.38}$$

where

$$\mathscr{L}^{(n)}(\mathbf{W}) = ||f(I^{(n)}) - U^{(n)}||^2. \tag{2.39}$$

We express $\mathscr{L}^{(n)}(\mathbf{W}) = \left\| f(I^{(n)}) - U^{(n)} \right\|^2$ with respect to $\mathbf{W}$ as follows:

$$\begin{aligned}
\mathscr{L}(\mathbf{W}) &= \|f(I) - U\|^2 && (2.40)\\
&= \left\|\sigma_s\left(\mathbf{W}^6\mathbf{h}^5\right) - U\right\|^2 && (2.41)\\
&= \left\|\sigma_s\left(\mathbf{W}^6\sigma\left(\mathbf{W}^5\mathbf{h}^4\right)\right) - U\right\|^2 && (2.42)\\
&= \left\|\sigma_s\left(\mathbf{W}^6\sigma\left(\mathbf{W}^5 P_4\left(\mathbf{h}^3\right)\right)\right) - U\right\|^2 && (2.43)\\
&= \left\|\sigma_s\left(\mathbf{W}^6\sigma\left(\mathbf{W}^5 P_4\left(\sigma\left(\mathbf{W}^3\mathbf{h}^2\right)\right)\right)\right) - U\right\|^2 && (2.44)\\
&= \left\|\sigma_s\left(\mathbf{W}^6\sigma\left(\mathbf{W}^5 P_4\left(\sigma\left(\mathbf{W}^3 P_2\left(\mathbf{h}^1\right)\right)\right)\right)\right) - U\right\|^2 && (2.45)\\
&= \left\|\sigma_s\left(\mathbf{W}^6\sigma\left(\mathbf{W}^5 P_4\left(\sigma\left(\mathbf{W}^3 P_2\left(\sigma\left(\mathbf{W}^1 I\right)\right)\right)\right)\right)\right) - U\right\|^2. && (2.46)
\end{aligned}$$

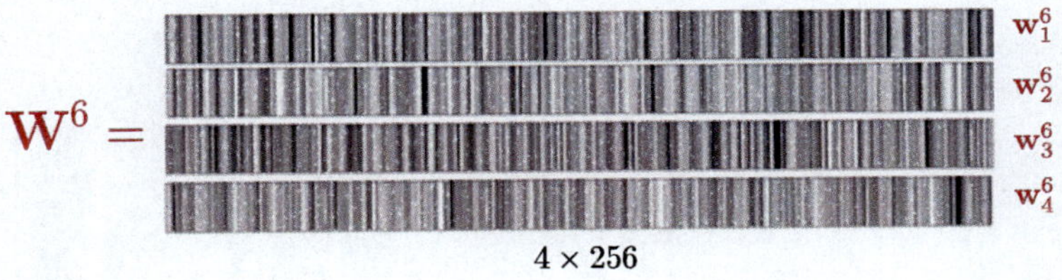

Fig. 2.18 W^6 is a 4×256 matrix. Hence, $\frac{\partial \mathscr{L}(\mathbf{W})}{\partial \mathbf{w}_i^6}$ is a 256-dimensional vector

[Gradient of $\mathscr{L}(\mathbf{W})$: output layer] The partial gradient of $\mathscr{L}(\mathbf{W})$ with respect to $\mathbf{w}_i^6$ is given by

$$\frac{\partial \mathscr{L}(\mathbf{W})}{\partial \mathbf{w}_i^6} = \frac{\partial \mathscr{L}(\mathbf{W})}{\partial \mathbf{h}^6} \cdot \frac{\partial \mathbf{h}^6}{\partial \mathbf{w}_i^6} \tag{2.47}$$

$$= \left(\sigma_s\left(\mathbf{w}_i^6 \mathbf{h}^5\right) - y_i\right) \cdot \sigma_s'\left(\mathbf{w}_i^6 \mathbf{h}^5\right) \mathbf{h}^5. \tag{2.48}$$

Note that $\mathscr{L}(\mathbf{W}) = \left\|\sigma_s\left(\mathbf{W}^6 \mathbf{h}^5\right) - U\right\|^2$ and $\frac{\partial \mathscr{L}(\mathbf{W})}{\partial \mathbf{w}_i}$ should be understood as a "partial gradient" which is a 256=dimensional vector $\frac{\partial \mathscr{L}(\mathbf{W})}{\partial \mathbf{w}_i} = \left(\frac{\partial \mathscr{L}(\mathbf{W})}{\partial w_{i,1}}, \ldots, \frac{\partial \mathscr{L}(\mathbf{W})}{\partial w_{i,256}}\right)$. See Fig. 2.18.

The term $\frac{\partial \mathscr{L}(\mathbf{W})}{\partial \mathbf{w}_3^6}$ can be viewed as a 256-dimensional-partial gradient:

$$\frac{\partial \mathscr{L}(\mathbf{W})}{\partial \mathbf{w}_3^6} = \frac{\text{the maximal change of } \mathscr{L}(\mathbf{W})}{\text{w.r.t. perturbation of } \mathbf{w}_3^6} = \nabla_{\mathbf{w}_3^6} \mathscr{L}(\mathbf{W}). \tag{2.49}$$

[Gradient of $\mathscr{L}(\mathbf{W})$: layer 5] The partial gradient of $\mathscr{L}(\mathbf{W})$ with respect to $\mathbf{w}_i^5$ is given by

$$\frac{\partial \mathscr{L}(\mathbf{W})}{\partial \mathbf{w}_i^5} = \frac{\partial}{\partial \mathbf{w}_i^5} \left\|\sigma_s\left(\mathbf{W}^6 \sigma\left(\mathbf{W}^5 \mathbf{h}^4\right)\right) - U\right\| \tag{2.50}$$

$$= \left(\frac{\mathscr{L}}{\partial \mathbf{h}^6} \cdot \frac{\partial \mathbf{h}^6}{\partial \mathbf{h}_i^5}\right) \frac{\partial \mathbf{h}_i^5}{\partial \mathbf{w}_i^5} \tag{2.51}$$

$$= \left[\left(\sigma_s\left(\mathbf{w}^6 \mathbf{h}^5\right) - y\right) \cdot \left(\sigma_s'\left(\mathbf{w}^6 \mathbf{h}^5\right) \mathbf{w}_{i,c}^6\right)\right] \sigma'\left(\mathbf{w}_i^5 \mathbf{h}^4\right) \mathbf{h}^4. \tag{2.52}$$

Here, $\mathbf{w}_{i,c}^6$ is i-th column of W^6. See Fig. 2.19.

[Backpropagation: Layer 4. max-pooling layer] The partial gradient of $\mathscr{L}(\mathbf{W})$ with respect to $\mathbf{h}_i^4$ is given by

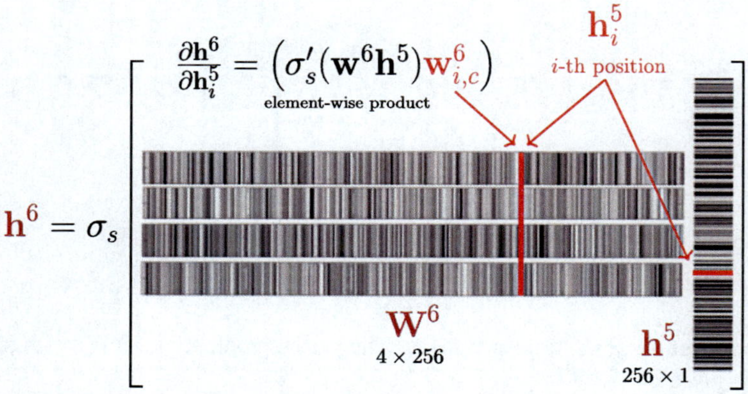

Fig. 2.19 The derivative $\frac{\partial \mathbf{h}^6}{\partial \mathbf{h}_i^5}$ (i.e., the change of $\mathbf{h}^6$ with respect to perturbations of $\partial \mathbf{h}_i^5$) is affected by $\mathbf{w}_{i,c}^6$ (i-th column of W^6)

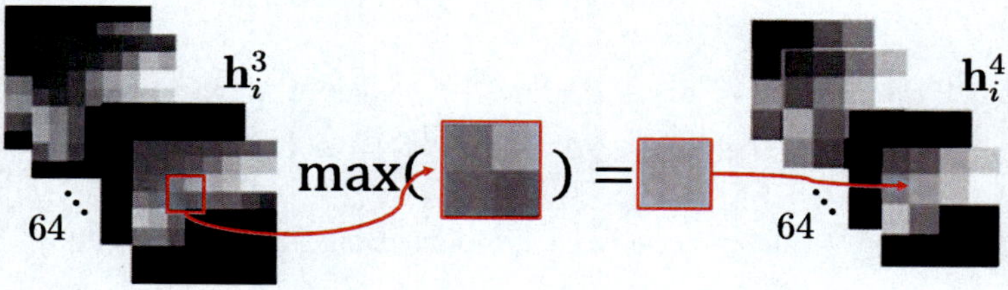

Fig. 2.20 Since max-pooling chooses the largest from a pool of four elements, the derivative is 1 at the biggest element and 0 otherwise

$$\frac{\partial \mathcal{L}(\mathbf{W})}{\partial \mathbf{h}_i^4} = \left(\frac{\partial \mathcal{L}}{\partial \mathbf{h}^6} \cdot \frac{\partial \mathbf{h}^6}{\partial \mathbf{h}_i^5}\right) \frac{\partial \mathbf{h}_i^5}{\partial \mathbf{h}_i^4} \tag{2.53}$$

$$= \left[\left(\sigma_s\left(\mathbf{w}^6\mathbf{h}^5\right) - y\right) \cdot \left(\sigma_s'\left(\mathbf{w}^6\mathbf{h}^5\right)\mathbf{w}_{i,c}^6\right)\right] \sigma'\left(\mathbf{w}_i^5\mathbf{h}^4\right)\mathbf{w}_i^5 \tag{2.54}$$

Next, we need the derivative of the pooling P_4 for $\frac{\partial \mathcal{L}}{\partial \mathbf{h}_i^3} = \frac{\partial \mathcal{L}}{\partial \mathbf{h}_i^4}\frac{\partial \mathbf{h}_i^4}{\partial \mathbf{h}_i^3}$. See Fig. 2.20.

The derivative of the max-pooling function in terms of backpropagation is

$$\frac{\partial}{\partial \mathbf{h}_{i,(a,b)}^3} \max \begin{pmatrix} \mathbf{h}_{i,(1,1)}^3 & \mathbf{h}_{i,(1,2)}^3 \\ \mathbf{h}_{i,(2,1)}^3 & \mathbf{h}_{i,(2,2)}^3 \end{pmatrix} = \begin{cases} 1, & \text{if } \mathbf{h}_{i,(a,b)}^3 = \max \begin{pmatrix} \mathbf{h}_{i,(1,1)}^3 & \mathbf{h}_{i,(1,2)}^3 \\ \mathbf{h}_{i,(2,1)}^3 & \mathbf{h}_{i,(2,2)}^3 \end{pmatrix} \\ 0, & \text{otherwise.} \end{cases} \tag{2.55}$$

This computation requires us to keep track of the index of the max-pooling. See Figs. 2.21 and 2.22.

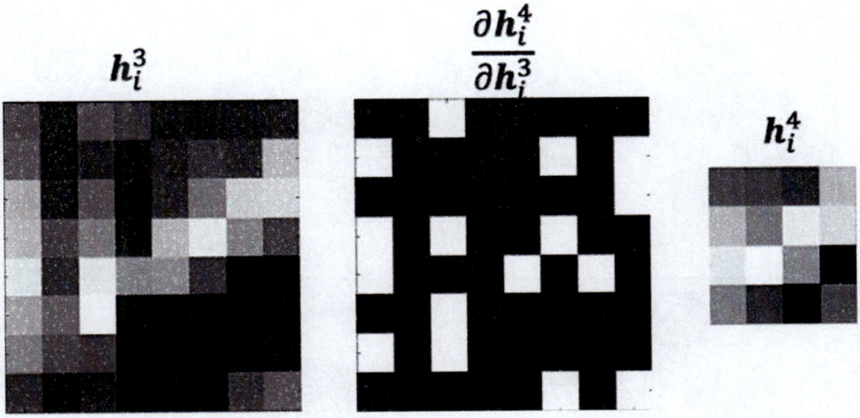

Fig. 2.21 Max-pooling passes a gradient flow through the neuron with the largest response at the input

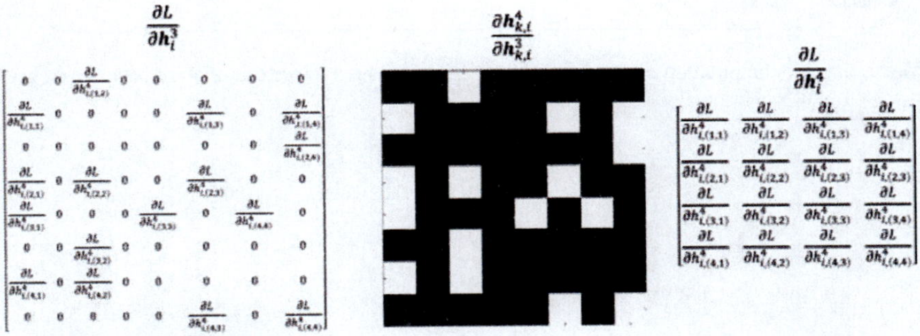

Fig. 2.22 Simplified version of the derivative $\frac{\partial \mathscr{L}(\mathbf{W})}{\partial \mathbf{h}_i^3}$

[Gradient of $\mathscr{L}(\mathbf{W})$ (Layer 3: convolutional layer)] We need to compute

$$\frac{\partial \mathscr{L}(\mathbf{W})}{\partial \mathbf{w}_{i,j,(1,1)}^3}, \ldots, \frac{\partial \mathscr{L}(\mathbf{W})}{\partial \mathbf{w}_{i,j,(3,3)}^3}, \quad i = 1, \ldots, 64, \ j = 1, \ldots, 16. \tag{2.56}$$

The partial gradient of $\mathscr{L}(\mathbf{W})$ with respect to $\mathbf{w}_{i,j,(1,1)}^3$ is given by

$$\frac{\partial \mathscr{L}(\mathbf{W})}{\partial \mathbf{w}_{i,j,(1,1)}^3} = \frac{\partial \mathscr{L}(\mathbf{W})}{\partial \mathbf{h}_i^3} \cdot \frac{\partial \mathbf{h}_i^3}{\partial \mathbf{w}_{i,j,(1,1)}^3} = \frac{\partial \mathscr{L}(\mathbf{W})}{\partial \mathbf{h}_i^3} \cdot \mathbf{h}_{j,(1,1)}^2. \tag{2.57}$$

Here, $\mathbf{w}_{i,j}^3$ relates between $\mathbf{h}_j^2$ and $\mathbf{h}_i^3$. See Figs. 2.23 and 2.24.

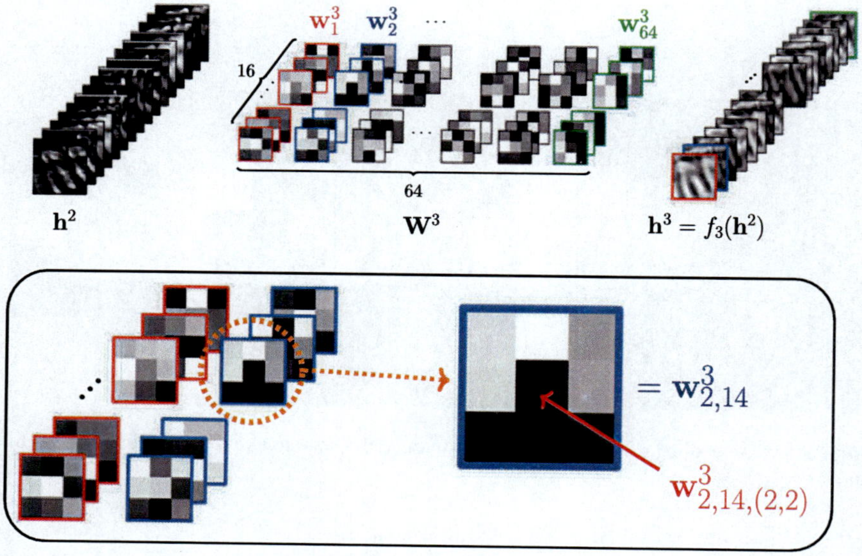

Fig. 2.23 For computation of the derivative $\frac{\partial \mathscr{L}(\mathbf{W})}{\partial \mathbf{w}_{2,14}^3}$, we need to see the relation between $\mathbf{h}_j^2$ and $\mathbf{h}_i^3$ carefully

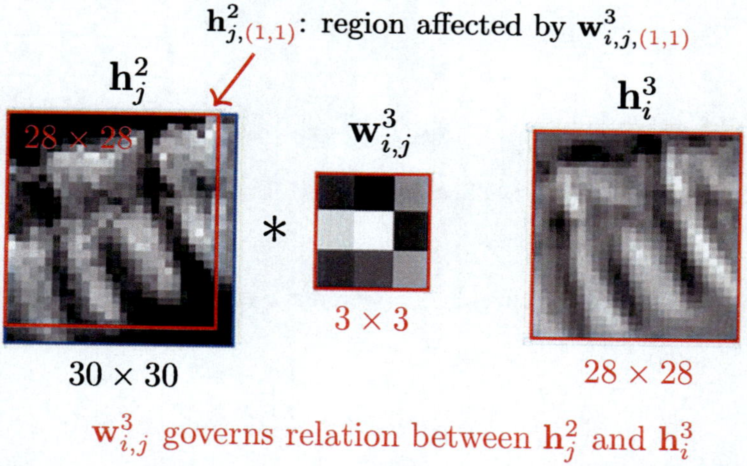

Fig. 2.24 The derivative $\frac{\partial \mathbf{h}_i^3}{\partial \mathbf{w}_{i,j,(1,1)}^3}$ is $\mathbf{h}_{j,(1,1)}^2$, which is the 28×28 matrix image in the left most red box. The derivative $\frac{\partial \mathbf{h}_i^3}{\partial \mathbf{w}_{i,j,(1,2)}^3}$ is $\mathbf{h}_{j,(1,2)}^2$, which is the box shifted to the right from the box for $\mathbf{h}_{j,(1,1)}^2$

2.4.2 Fully Convolutional Networks

In the previous section, we studied convolutional neural networks (CNNs) that allow to provide image classification by building a hierarchical structure of image features through training over labeled images. In this section, we will study fully convolutional networks (FCNs) for tooth segmentation. For the example of tooth segmentation, we use the following simplified model of a segmentation map $f : I \to U$:

- The input I is the image in the bounding box described in the previous subsection.
- The output U is a binary image segmentation, as shown in Fig. 2.25.

Figure 2.26 shows a simple FCN architecture, consisting of a feature encoder with a CNN architecture and a decoder performing upsampling. Hence, compared to CNNs, FCNs add an expanding path consisting of unpooling layers to CNN structure. For segmentation purposes, we need to keep the spatial information in the last layer of the encoder. In CNN architecture for classification purposes, CNNs consist of a contracting path and the last fully connected layer discards all spatial information and retains semantic information. See Fig. 2.27. When segementation is the aim, this is replaced by a 1×1 convolution layer to maintain spatial information. The decoder uses upsampling to produce a segmented output from the semantic information extracted by the encoder.

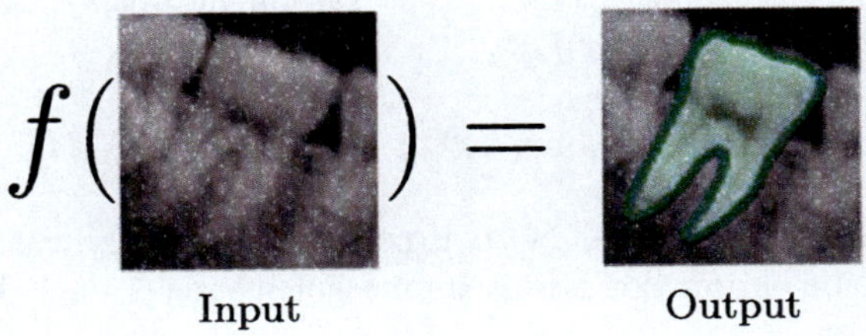

Fig. 2.25 Segmentation map

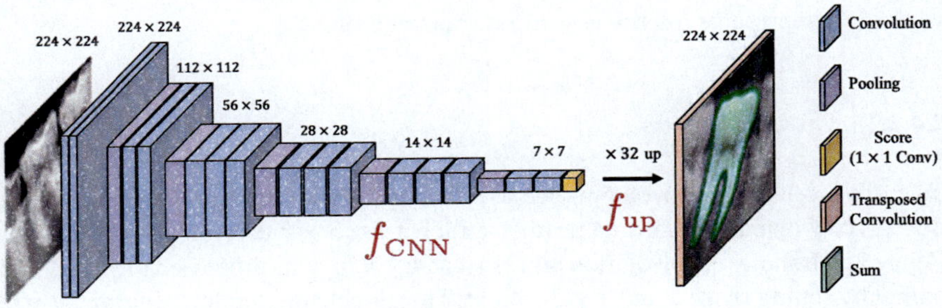

Fig. 2.26 Fully convolutional network for segmentation

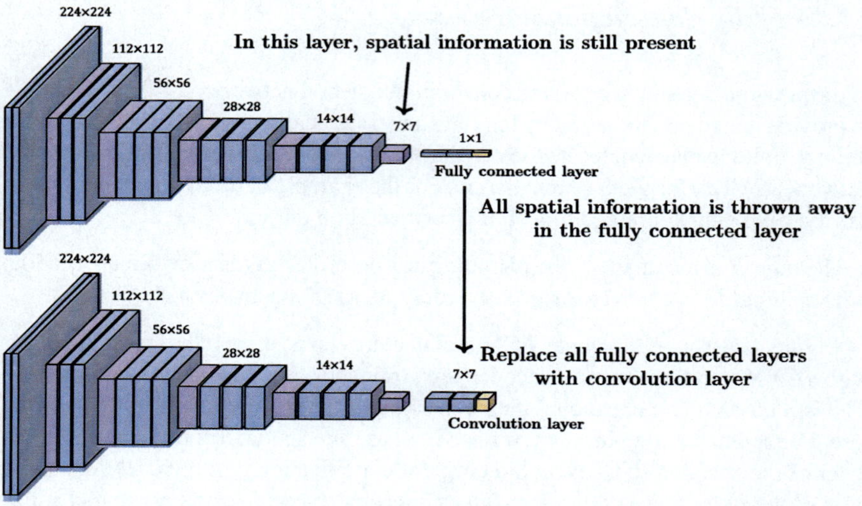

Fig. 2.27 Image classification to semantic segmentation

2.4.2.1 Forward Mapping of FCNs

In the FCN architecture shown in Fig. 2.26, the feed-forward map $f : I \to U$ consists of a encoder f_{CNN} and a decoder f_{up}:

$$f(I) = f_{\text{up}} \circ f_{\text{CNN}} = f_{\text{up}}(\mathbf{h}) \quad \text{where} \quad \mathbf{h} = f_{\text{CNN}}(I). \tag{2.58}$$

Given an input data I, $\mathbf{h} = f_{\text{CNN}}(I)$ is expected to contain semantic information of I. The semantic information $\mathbf{h}$ is used for segmentation through the upsampling decoder $f_{\text{up}}(\mathbf{h})$:

$$f_{\text{up}}(\mathbf{h}) = \sigma_s\left(\mathbf{h} \circledast \uparrow \mathbf{W}\right) \tag{2.59}$$

where $\circledast \uparrow$ represents fractionally-strided convolution.

2.4.2.2 Up-Convolution

To easily explain up-convolution, we use a simple model in which an input of size 2×2 is convolved with a learnable filter kernel of size 3×3 to form the 4×4 output. Figure 2.29b shows a convolution process that 4×4 input is processed to get a 2×2 output by setting stride 1 and 3×3 weight. The up-convolution feels like an inverse of the convolution. But this is not really a mathematical inverse or deconvolution.

Now, we explain how to generate the transposed matrix in Fig. 2.28 from the filter in Fig. 2.29a. We first use the notation for vectorization of 2D image. The

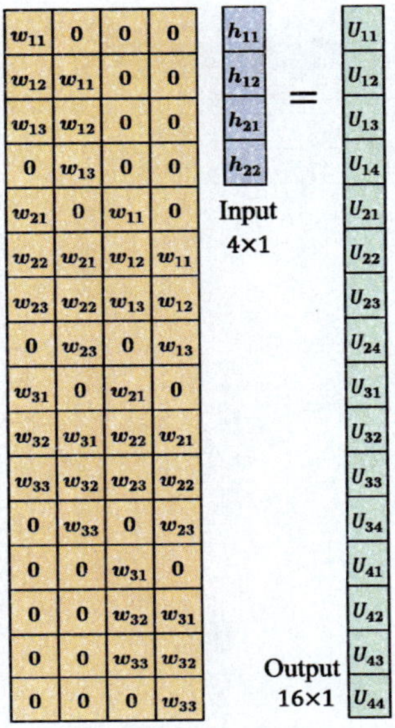

Fig. 2.28 An example of fractionally-strided convolution

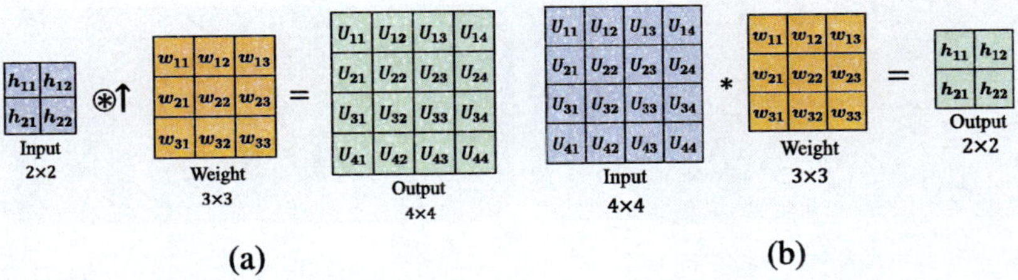

Fig. 2.29 The relationship between **a** up-convolution and **b** convolution

vectorization of the image matrix $I = \begin{pmatrix} I_{1,1} & \cdots & I_{1,n} \\ \vdots & \ddots & \vdots \\ I_{m,1} & \cdots & I_{m,n} \end{pmatrix}$ is obtained by stacking the rows of the matrix I:

$$I = (I_{1,1}, \ldots, I_{1,n}, I_{2,1}, \ldots, I_{2,n}, \ldots, \ldots, \ldots, I_{m,1}, \ldots, I_{m,n}). \tag{2.60}$$

For simplicity, we sometimes abuse the notation of I as follows:

$$I = (I_1, \ldots, x_n, I_{n+1}, \ldots, I_{2n}, \ldots, \ldots, \ldots, I_{(m-1)n+1}, \ldots, I_{mn}). \tag{2.61}$$

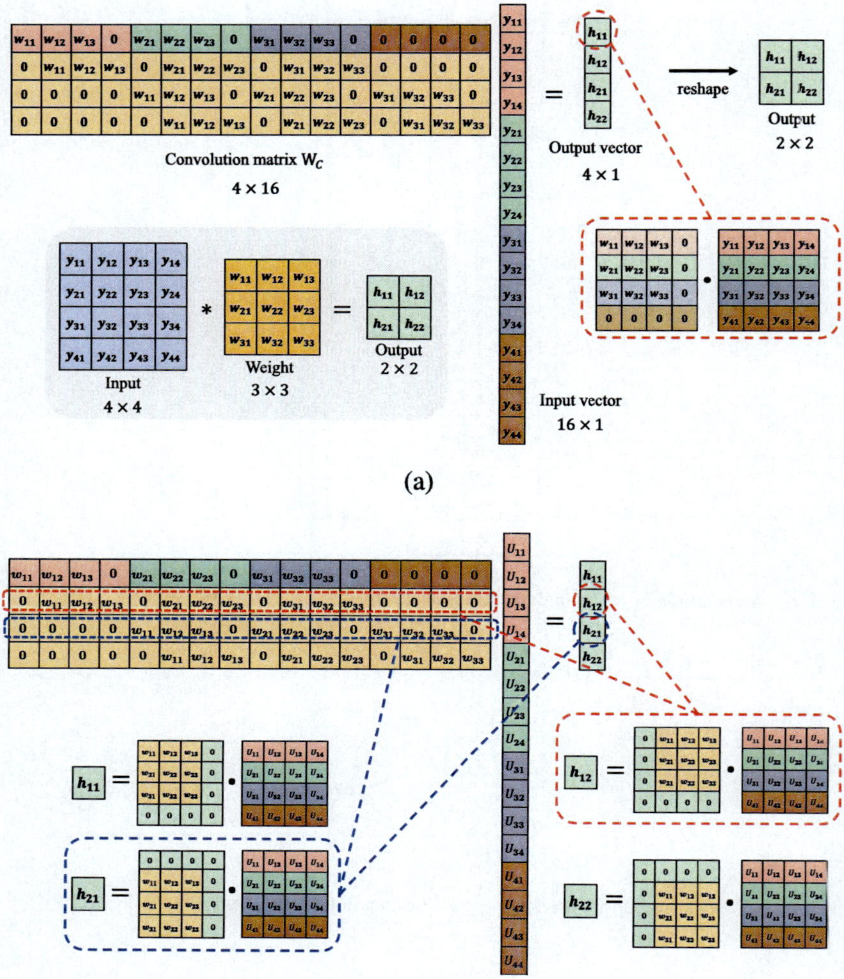

(b)

Fig. 2.30 Convolution can be viewed as a multiplication of a Toeplitz matrix

The regular 2D convolution operation can be represented as a matrix multiplication with the Toeplitz matrix, as shown in Fig. 2.30. The transpose convolution operation conducts up-sampling that is expressed by the transpose of the Toeplitz matrix associated with the regular convolution, as shown in Fig. 2.31.

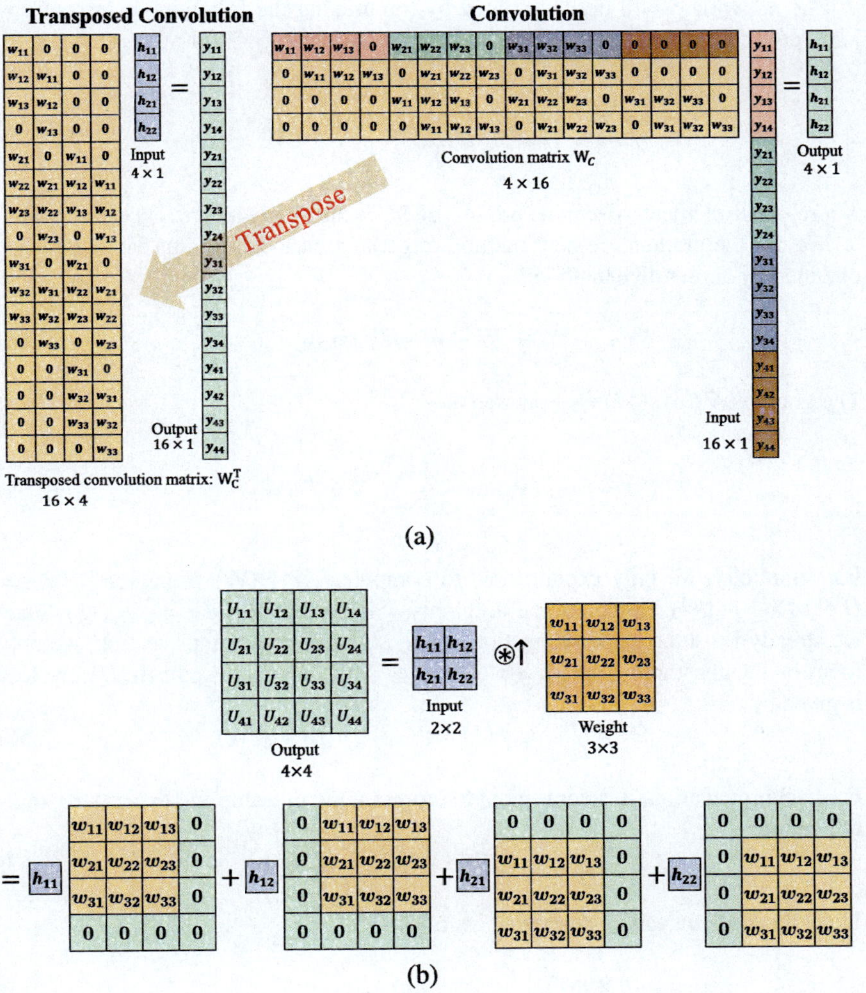

Fig. 2.31 Transposed convolution is like the transpose of a Toeplitz convolution matrix

2.4.2.3 Backpropagation and Loss Function

Now, we are ready to discuss backpropagation. We use supervised learning to learn $f = f_{\text{up}} \circ f_{\text{CNN}}$ that is given by

$$f(I) = \sigma_s \left(\mathbf{W}^{k+1} \circledast \uparrow f_{\text{CNN}}(I) \right). \tag{2.62}$$

The network f will be determined by minimizing the following cross entropy loss between the recovered $f(I^{(n)})$ and the ground-truth $U^{(n)}$ for all n:

$$Loss(\mathbf{W}) = -\frac{1}{N}\sum_{n=1}^{N} U^{(n)} \cdot \log f(I^{(n)}), \tag{2.63}$$

where $\cdot$ is the element-wise inner product and U is the binary image for segmentation.

We use the gradient descent method to get a good local minima $\mathbf{W}$ that can be obtained by an iterative method:

$$\mathbf{W} \leftarrow \mathbf{W} - \eta \nabla_{\mathbf{W}} Loss(\mathbf{W}). \tag{2.64}$$

The gradient $\nabla Loss(\mathbf{W})$ is expressed as

$$\nabla Loss(\mathbf{W}) = -\frac{1}{N}\sum_{n=1}^{N} \nabla \mathscr{L}^{(n)}(\mathbf{W}). \tag{2.65}$$

For simplicity, we only explain how to compute $\nabla \mathscr{L}^{(n)}(\mathbf{W})$ from $\nabla \mathscr{L}^{(n)}(\mathbf{W}) = U^{(n)} \odot \log f(I^{(n)})$. For notational simplicity, we fix $I = I^{(n)}$ and $\mathbf{h} = f_{\text{CNN}}(I)$. Since we already learn the backpropagation for $f_{\text{CNN}}(I)$ in the previous section, we only focus on the upsampling part $f_{\text{up}}(\mathbf{h}) = \sigma_s(\mathbf{h} \circledast \uparrow \mathbf{W})$. Given the pair $(\mathbf{h}, U)$, the loss is given by

$$\mathscr{L} = U \cdot \log f(I) = U \cdot \log \sigma_s(\mathbf{h} \circledast \uparrow \mathbf{W}). \tag{2.66}$$

For backpropagation, it is convenient to represent the up-sampling by a matrix multiplication:

$$f\text{up}(\mathbf{h}) = \sigma_s(\mathbf{h} \circledast \uparrow \mathbf{W}) = \sigma_s(\mathbf{W}_C^T \mathbf{h}). \tag{2.67}$$

Then, the gradient can be computed as follows:

$$\frac{\partial \mathscr{L}(\mathbf{W})}{\partial \mathbf{w}_{i,j}} = \frac{\partial \mathscr{L}}{\partial \mathbf{h}^\sharp} \cdot \frac{\partial \mathbf{h}^\sharp}{\partial \mathbf{w}_{i,j}} \tag{2.68}$$

$$= \left(y \frac{\sigma_s'(\mathbf{W}_c^T \mathbf{h})}{\sigma_s(\mathbf{W}_c^T \mathbf{h})} \right) \cdot (\mathbf{W}_c^T)_{i,j} \mathbf{h}. \tag{2.69}$$

Here, $\mathbf{h}^\sharp = \mathbf{h} \circledast \uparrow \mathbf{W}$ and $y \frac{\sigma_s'(\mathbf{W}_c^T \mathbf{h})}{\sigma_s(\mathbf{W}_c^T \mathbf{h})}$ can be computed using the Hadamard product. See Fig. 2.32 for the meaning of $(\mathbf{W}_c^T)_{i,j}$.

$$
\mathbf{W}_c^T = \begin{bmatrix}
w_{11} & 0 & 0 & 0 \\
w_{12} & w_{11} & 0 & 0 \\
w_{13} & w_{12} & 0 & 0 \\
0 & w_{13} & 0 & 0 \\
w_{21} & 0 & w_{11} & 0 \\
w_{22} & w_{21} & w_{12} & w_{11} \\
w_{23} & w_{22} & w_{13} & w_{12} \\
0 & w_{23} & 0 & w_{13} \\
w_{31} & 0 & w_{21} & 0 \\
w_{32} & w_{31} & w_{22} & w_{21} \\
w_{33} & w_{32} & w_{23} & w_{22} \\
0 & w_{33} & 0 & w_{23} \\
0 & 0 & w_{31} & 0 \\
0 & 0 & w_{32} & w_{31} \\
0 & 0 & w_{33} & w_{32} \\
0 & 0 & 0 & w_{33}
\end{bmatrix}
\qquad
(\mathbf{W}_c^T)_{1,3} = \begin{bmatrix}
0 & 0 & 0 & 0 \\
0 & 0 & 0 & 0 \\
1 & 0 & 0 & 0 \\
0 & 1 & 0 & 0 \\
0 & 0 & 0 & 0 \\
0 & 0 & 0 & 0 \\
0 & 0 & 1 & 0 \\
0 & 0 & 0 & 1 \\
0 & 0 & 0 & 0 \\
0 & 0 & 0 & 0 \\
0 & 0 & 0 & 0 \\
0 & 0 & 0 & 0 \\
0 & 0 & 0 & 0 \\
0 & 0 & 0 & 0 \\
0 & 0 & 0 & 0 \\
0 & 0 & 0 & 0
\end{bmatrix}
$$

Fig. 2.32 The matrix of $(\mathbf{W}_c^T)_{1,3}$ in the computation of the derivative $\frac{\partial \mathbf{h} \circledast \uparrow \mathbf{W}}{\partial \mathbf{w}_{i,j}} = (\mathbf{W}_c^T)_{i,j}\mathbf{h}$

Variants of FCNs

Segmentation requires recognizing the global features of the target, while the boundaries of the object must be finely divided into pixels. Deeper layers tend to have deeper features while losing spatial location information. This means that the shallower the layer, the more sophisticated the location information.

For fine segmentation, there have been developed variants of the FCN architecture, such as FCN-32, FCN-16 and FCN-8. See Fig. 2.33 for variants of FCNs, where the skip connections are used to enhance the output segmentation performance.

FCN-8s can improve segmentation performance by properly combining deep and shallow layers, in order to make use of global features and location information simultaneously. In FCN-8s, upsampled feature maps are combined with feature maps at less deep levels from the encoding path. These skip connections have been shown to help recover the full spatial resolution at the network's output, making fully convolutional methods suitable for semantic segmentation.

The usefulness of the skip connections is related to the receptive field. The receptive field of a node in j-th layer is defined by the area of the region in the input I that affects the node through the forward propagation. If the receptive field of the network is large, accurate segmentation is difficult due to loss of location information. On the other hand, if the receptive field is small, the network may not properly recognize large objects.

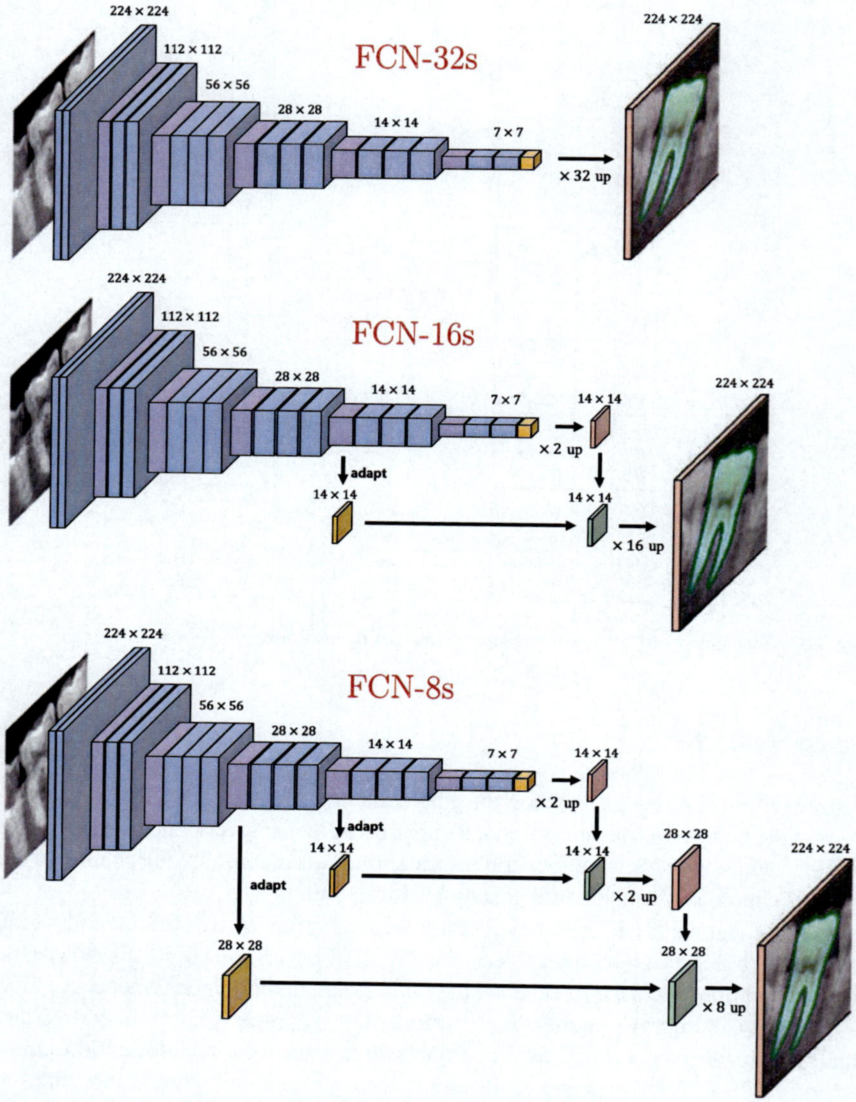

Fig. 2.33 Variants of the FCN architecture; FCN-32, FCN-16 and FCN-8

The standard convolution-based networks use a stack of transformations consisting of convolution-ReLU-convolution-ReLU. To be precise, the feature map $\mathbf{h}^{k+2}$ is obtained by the following transform $\mathscr{H} : \mathbf{h}^k \to \mathbf{h}^{k+2}$ given by

$$\mathbf{h}^{k+2} = \mathscr{H}(\mathbf{h}^k) = \sigma(W^{k+1} \circledast \underbrace{\sigma(W^k * h^k + b^k)}_{\mathbf{h}^{k+1}} + b^k). \tag{2.70}$$

Taking account of vanishing gradient problem associated with ReLU, these networks become more difficult to optimize as depth increases. To deal with vanishing gradient problem in deep networks, ResNets use the following simple modification:

$$\mathbf{h}^{k+2} = \mathbf{h}^k + \mathscr{H}(\mathbf{h}^k), \tag{2.71}$$

which allows it to train deep networks.

2.4.3 U-net and M-net

U-net is one of the most popular networks for medical image segmentation. U-net consists of two main parts: the encoding and decoding paths. The basic idea of the U-net architecture is somewhat similar to FCN-8s. Hence, U-net can be regarded as a U-shaped FCN. U-net is symmetric (for taking advantage of information transfer) and has the hierarchical skip connections for concatenation with the corresponding feature in the decoding path. See Fig. 2.34.

The encoding path of U-net is based on a series of convolutions followed by pooling to reliably recognize image features so that the resulting output is fairly resilient to variations in position and scale on a target structure. Also, it utilizes concatenative skip connections between the same-scale feature maps obtained from the encoder and decoder subnetworks, which play the important role in restoring the spatial information lost during downsampling. U-net has shown remarkable overall performance in medical image segmentations, by making use of global features and local spatial information simultaneously. However, we still do not know what the optimal depth is, and there are criticisms that the same-scale skip-connections are unnecessarily restrictive. Deeper networks may learn more complex features of images, but various experiments also show that deeper is not always better. The optimal network depth may vary depending on a number of factors, including the size of the input image, the amount of training data, the variance in the size of the target features in the image, and the difficulty of the task. Instead of increasing the network depth and the filter size, spatially-adaptive filters such as atrous convolution can be used to increase receptive field size, and this technique has been used to handle the large differences in target size properly. Various modifications or auxiliary means of U-net have been developed to supplement the limitations of the classical U-net architecture.

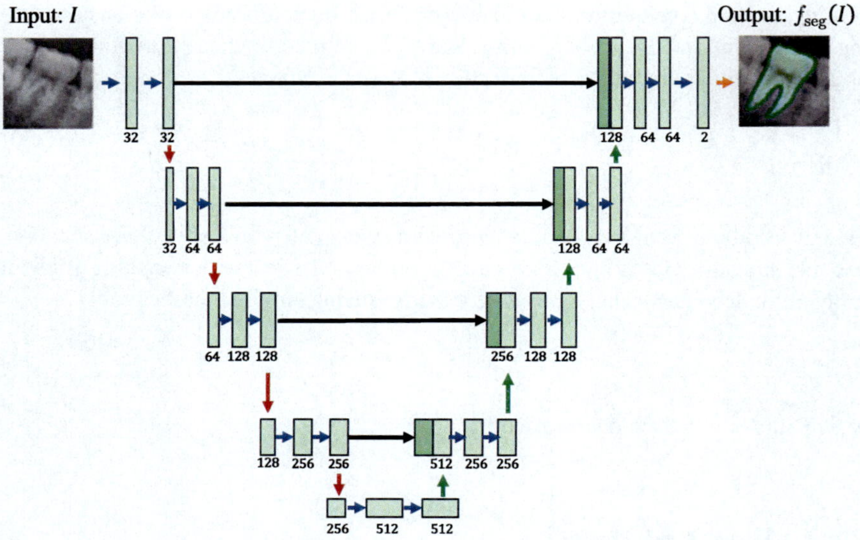

Fig. 2.34 U-net consists of two main parts: encoding path and decoding path

At the first convolution layer in the encoding path, a set of feature maps (denoted by $\mathbf{h}^1$) is computed using a set of filters (denoted by $\mathbf{W}^1$): $\mathbf{h}^1 = \sigma(I \circledast \mathbf{W}^1 + \mathbf{b}^1)$ where $\sigma = ReLU$. Similarly, at the second layer, the second feature map of $\mathbf{h}^2$ is obtained by $\mathbf{h}^2 = \sigma(\mathbf{h}^1 \circledast \mathbf{W}^2 + \mathbf{b}^2)$. After two convolution layers, max-pooling operation is applied to reduce the dimensions of the feature maps. The above three consecutive processes are repeated up to the end of the encoding path to extract feature maps. The decoding path is a reverse process of the encoding path, replacing the pooling operator by the average un-pooling to restore the size of the output. The un-pooled output is concatenated with the corresponding feature from the encoding path. See Fig. 2.35. At the last layer, we take the pixel-wise softmax activation function after applying a 1×1 convolution.

As in FCNs, f_{seg} can be determined by minimizing the cross-entropy loss

$$Loss(\mathbf{W}) = -\frac{1}{N}\sum_{n=1}^{N} U^{(n)} \odot \left(\log f_{seg}(I^{(n)}),\right), \qquad (2.72)$$

where $\odot$ is element-wise inner product. As before, we can find a good local minima $\mathbf{W}$ by updating

$$\mathbf{W} \leftarrow \mathbf{W} - \eta \nabla_{\mathbf{W}} Loss. \qquad (2.73)$$

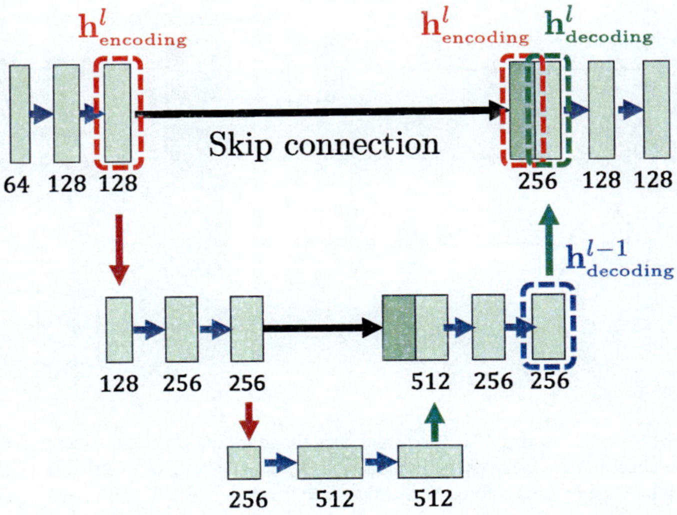

Fig. 2.35 Each un-pooled output is concatenated with the corresponding feature from the encoding path

2.4.3.1 M-net

The architecture of the M-net is based on U-net and two major parts are added in the input and output layers, as shown in Fig. 2.36. In the input layer, an image pyramid constructed by multi-scale images is used to integrate a multi-level receptive field. Here, the image is down sampled by the average pooling and convolution with ReLU applied to the down sampled image. In the output layer, a side-output layer is used to learn local and global information at the same time. A multi-label loss function with a side output is used to deal with the vanishing gradient problem by replenishing the back-propagated gradients. At the output layer, a 1×1 convolution and an element-wise softmax activation function are applied.

The final segmentation f_{seg} is obtained by averaging 4 side output maps (f^k_{seg}, $k = 1, 2, 3, 4$).

As in FCNs, f_{seg} can be determined by minimizing the cross-entropy loss

$$Loss(\mathbf{W}) = -\frac{1}{N}\sum_{k=1}^{4}\sum_{n=1}^{N} U^{(n)} \odot \left(\log f^k_{\text{seg}}(I^{(n)}),\right), \tag{2.74}$$

where $\odot$ is the element-wise inner product. As before, we can find good local minima $\mathbf{W}$ by updating

$$\mathbf{W} \leftarrow \mathbf{W} - \eta \nabla_{\mathbf{W}} Loss. \tag{2.75}$$

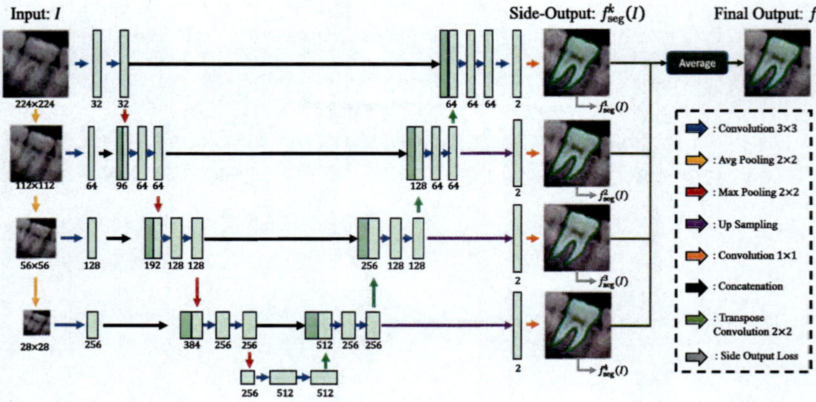

Fig. 2.36 The M-net is based on U-net and has advantages by adding two major parts: (i) multi-scale layer used to construct an image pyramid input and (ii) multi-label loss function with side-output layer to learn local and global information at the same time

2.4.3.2 R-CNN

As seen in Sect. 2.4.1, the CNN's architecture has been applied to object classification and detection, and has a remarkable performance. This architecture has been generalized and evolved in various ways, which include R-CNN (Regional CNN) [12], Faster R-CNN [34], and Mask R-CNN [15]. In these R-CNNs, the input is an image having various objects to be detected and the output is bounding boxes with classifications. Given an input image, R-CNNs roughly propose multiple boxes and next check which of them actually contain the object properly. R-CNNs use a class-specific Support Vector Machine (SVM) to know what the object in the box is. Then, R-CNNs also perform a linear regression on boxes, making the bounding box coordinates tighter to fit the object's dimensions.

2.4.4 Confidence Map

Given the 2D medical image, the confidence map is used to provide a spatial representation of the likelihood of a particular object being located. In this section, we will use a confidence map to localize individual teeth. Given a panoramic X-ray image, the confidence map for all teeth is given by

$$\Psi(\mathbf{p}) = \{\psi_1(\mathbf{p}), \psi_2(\mathbf{p}), \ldots, \psi_{N_{\text{tooth}}}(\mathbf{p})\}, \tag{2.76}$$

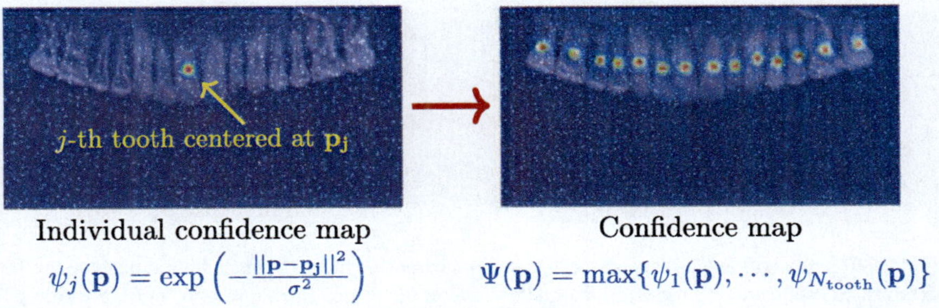

Fig. 2.37 Confidence map to achieve tooth localization. The individual confidence map ψ_j represents the belief of a tooth being located. Then, the confidence map for all teeth is computed from individual confidence maps

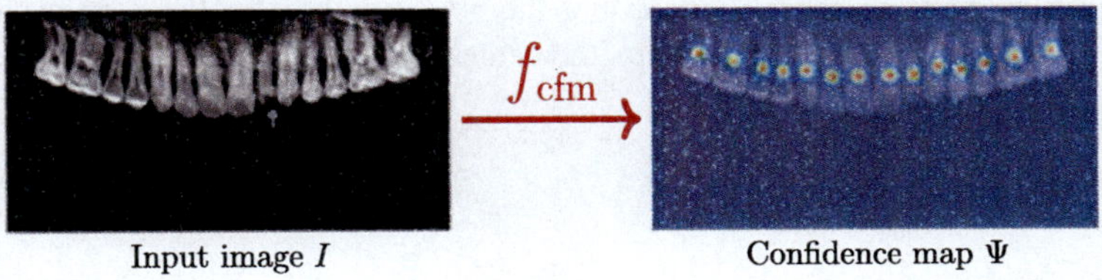

Fig. 2.38 Deep learning for localization of teeth

where $\mathbf{p}$ denotes the pixel position and N_{tooth} is the number of teeth. The individual confidence map $\psi_j(\mathbf{p}) = \exp\left(\frac{\|\mathbf{p}-\mathbf{p}_j\|^2}{\sigma^2}\right)$ for localization of $\mathbf{p}_j$ represents the belief of the centroids at each pixel position $\mathbf{p}$. See Fig. 2.37 for the confidence map.

We try to learn the network $f_{\text{cfm}} : I \to \Psi$ for estimating the confidence map as shown in Fig. 2.38. A well-trained network f_{cfm} allows us to estimate the centroids of the all teeth given by

$$\mathbf{P} = (\mathbf{p}_1, \mathbf{p}_2, \ldots, \mathbf{p}_{N_{\text{tooth}}}) \in \mathbb{R}^{2\times N_{\text{tooth}}}. \tag{2.77}$$

In practice, a simple network f_{cfm} may not provide an accurate confidence map. Then, f_{cfm} could consist of a double pipeline, where the first estimation of the confidence map can be used as a good initial guess. Then, the final confidence map provides the refinement of the initial prediction where the initial confidence map is concatenated with the intermediate feature map as an input of the refinement network.

Using the estimate of the center positions, we can detect bounding box coordinates for all individual teeth which are expressed as

$$\mathbf{B} = (\mathbf{b}_1, \mathbf{b}_2, \ldots, \mathbf{b}_{N_{\text{tooth}}}) \in \mathbb{R}^{4\times N_{\text{tooth}}}, \tag{2.78}$$

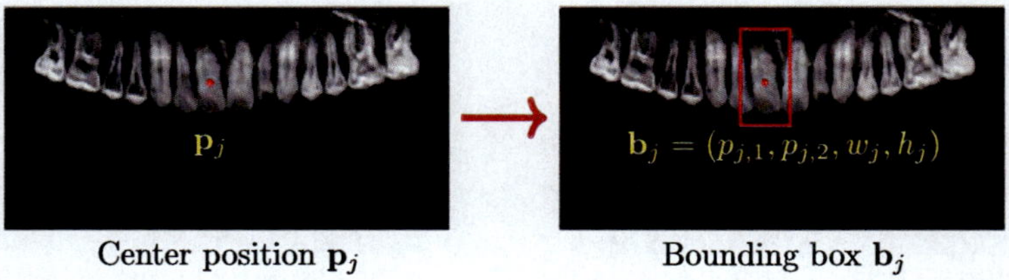

Fig. 2.39 The confidence map only estimates the center position. Then, a bounding box can be determined by estimating the width and height. Here, the left image shows the center position $\mathbf{p}_j$ of j-th tooth estimated by the confidence map. The right image shows the bounding box for j-th tooth described by the 4D vector $\mathbf{b}_j = (p_{j,1}, p_{j,2}, w_j, h_j)$

where $\mathbf{b}_j = (p_{j,1}, p_{j,2}, w_j, h_j)$ describes the bounding box for the j-th tooth. See Fig. 2.39. Here, $(p_{j,1}, p_{j,2})$ represents the center of the bounding box and (w_j, h_j) indicates its width and height.

2.4.5 YOLO

In this section, we explain a one-stage object detection method based on YOLO (You Only Look Once) [33]. YOLO was developed for real-time object detection and classification. The advantage of YOLO is that it applies a single neural network to the entire image and provides directly the location of the bounding box with the related category in the output layer as shown in Fig. 2.40.

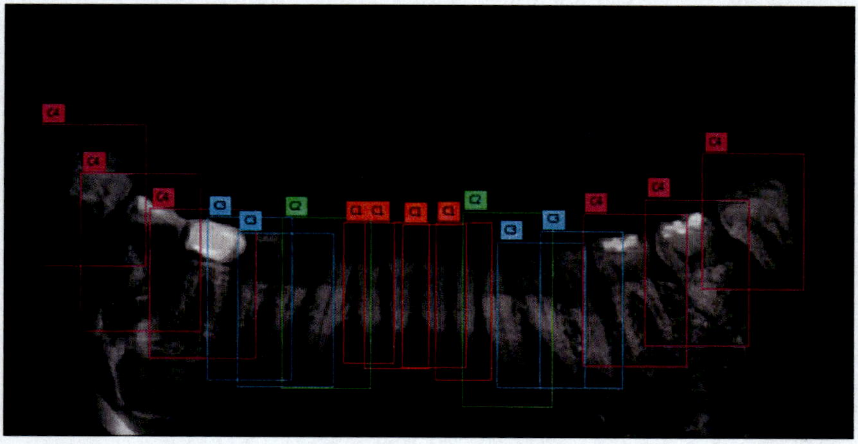

Fig. 2.40 YOLO can detect and classify multiple teeth with as single neural network

YOLO's pipeline is roughly as follows:

- The input image is divided into grid cells. Each grid has tasked with detecting whether the center of an object belongs to the grid.
- For each grid cell, it predicts a bounding box, confidence score, and its class probability.
- A non-max suppression (NMS) technique is used to remove boxes with low confidence scores.

Let the input I be a panoramic image with size of $N_s \times N_z$ (e.g., $N_s \times N_z = 640 \times 320$). We divide the image I into uniform cell of size $k \times k$ (e.g., $k = 16$) as shown in Fig. 2.41. Let Q_{ij} be a (i, j) grid cell where $(i, j) \in \{(i, j) : i = 1, 2, \ldots, \frac{N_s}{k}, j = 1, 2, \ldots, \frac{N_z}{k}\}$. We estimate a confidence score c_{ij} which represents belief of existence of the center of the tooth in grid cell Q_{ij}, where it has a value 1 if the central point lies in Q_{ij}, otherwise 0. And we estimate bounding box coordinates, denoted by

$$\mathbf{b}_{ij} = (s_{ij}, z_{ij}, w_{ij}, h_{ij}). \tag{2.79}$$

Here, (s_{ij}, z_{ij}) is the center position of the bounding box in Q_{ij} and (w_{ij}, h_{ij}) indicates its width and height. This bounding box is designed to completely enclose one tooth. For a tooth in the bounding box corresponding to $\mathbf{b}_{ij}$, we estimate a class probability

$$\mathbf{p}_{ij} = (p_{ij,1}, p_{ij,2}, p_{ij,3}, p_{ij,4}), \tag{2.80}$$

where $p_{ij,k}$ represent the probability of being tooth class k.

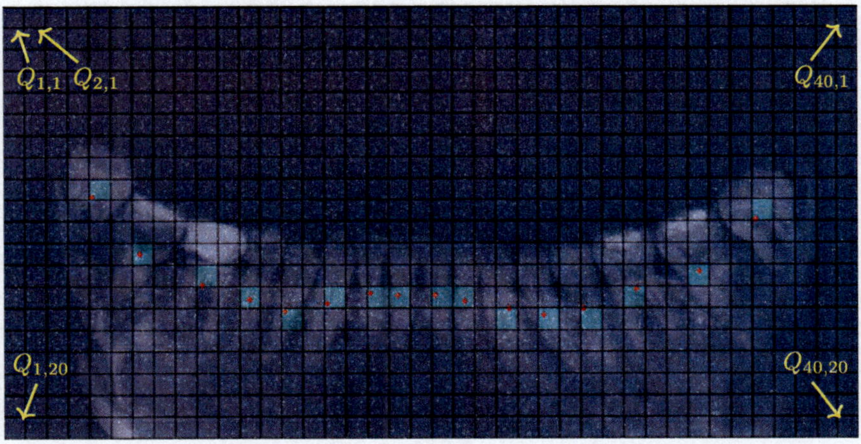

Fig. 2.41 YOLO divides the panoramic input image into 40×20 grid cells. Each grid cell $Q_{i,j}$ has a size of 16×16. Here, a red cross represents a tooth center and cyan-colored square represents a grid cell containing a tooth center

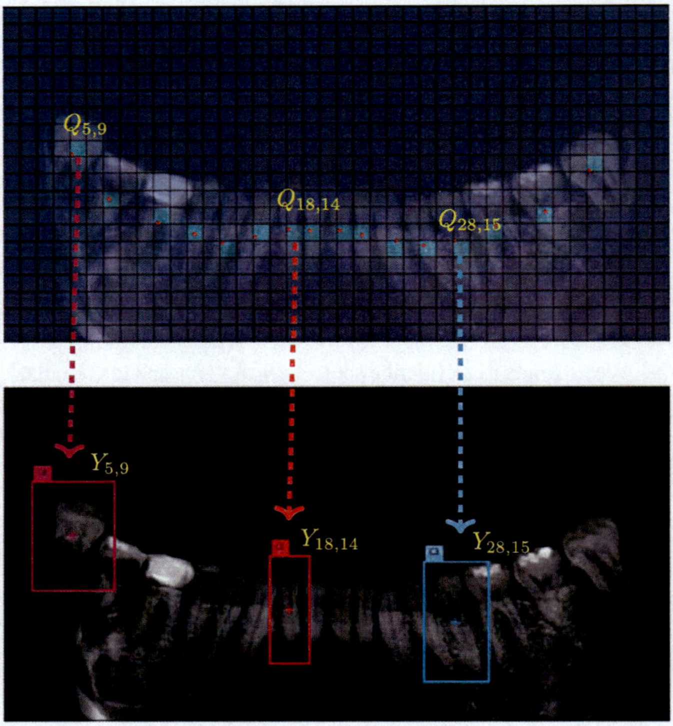

Fig. 2.42 For each grid cell, YOLO predicts a bounding box, confidence score, and its class probability

To be more precise, the above pipeline is to find a tooth detection map $f_{det} : I \mapsto Y$ that is given by

$$f_{det}(I) = \begin{pmatrix} Y_{1,1} & Y_{2,1} & \cdots & Y_{40,1} \\ Y_{1,2} & Y_{2,2} & \cdots & Y_{40,2} \\ \vdots & & \ddots & \vdots \\ Y_{1,20} & \cdots & \cdots & Y_{40,20} \end{pmatrix}, \tag{2.81}$$

where Y_{ij} is a vector representing confidence score c_{ij}, bounding box components $\mathbf{b}_{ij} = (s_{ij}, z_{ij}, w_{ij}, h_{ij})$ in (2.79), and class probability $\mathbf{p}_{ij} = (p_{ij,1}, p_{ij,2}, p_{ij,3}, p_{ij,4})$ in (2.80) corresponding to Q_{ij}:

$$Y_{ij} = (c_{ij}, \mathbf{b}_{ij}, \mathbf{p}_{ij}) \in \mathbb{R}^9. \tag{2.82}$$

See Fig. 2.42 for Q_{ij} and Y_{ij} and Fig. 2.43 for the network architecture of f_{det}.

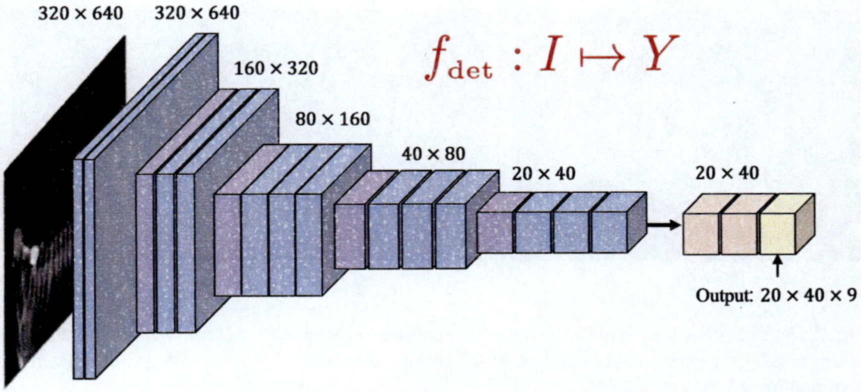

Fig. 2.43 Simplified version of network architecture for one-stage detection based on YOLO

Using a labeled training dataset $\{(I^{(n)}, Y^{*(n)})\}_{n=1}^{N}$ where Y^* is a ground truth, f_{det} is learned by minimizing the loss between the output $Y = f_{det}(I)$ and the ground truth Y^* as follows:

$$\sum_{n=1}^{N} \left[\mathcal{L}_{obj}(Y, Y^*) + \lambda_1 \mathcal{L}_{noobj}(Y, Y^*) + \lambda_2 \mathcal{L}_{box}(Y, Y^*) + \mathcal{L}_{cls}(Y, Y^*)\right], \tag{2.83}$$

where

$$\mathcal{L}_{obj}(Y, Y^*) := \sum_{(i,j)\in\{(i,j)|c_{ij}^*=1\}} (1 - c_{ij})^2, \tag{2.84}$$

$$\mathcal{L}_{noobj}(Y, Y^*) := \sum_{(i,j)\in\{(i,j)|c_{ij}^*=0\}} (0 - c_{ij})^2, \tag{2.85}$$

$$\mathcal{L}_{box}(Y, Y^*) := \sum_{(i,j)\in\{(i,j)|c_{ij}^*=1\}} |\mathbf{b}_{ij}^* - \mathbf{b}_{ij}|^2, \tag{2.86}$$

$$\mathcal{L}_{cls}(Y, Y^*) := -\sum_{(i,j)\in\{(i,j)|c_{ij}^*=1\}} \sum_{k=1}^{4} \mathbf{p}_{ij,k}^* \log \mathbf{p}_{ij,k}. \tag{2.87}$$

Now, we need to filter overlapping boxes. To find exact bounding boxes among the predicted boxes for all Q_{ij}, we use the score $e_{ij} = c_{ij} * (\max_k p_{ij,k})$ (score = confidence value × class probability). We also adopt the non-maximal suppression (NMS) technique [2] to eliminate highly overlapped bounding boxes of the same tooth. The process of filtering overlapping boxes is as follows:

Step 1 Setting threshold score = 0.1, we filter overlapping boxes. (More than 90% of the boxes can be removed at this stage.)

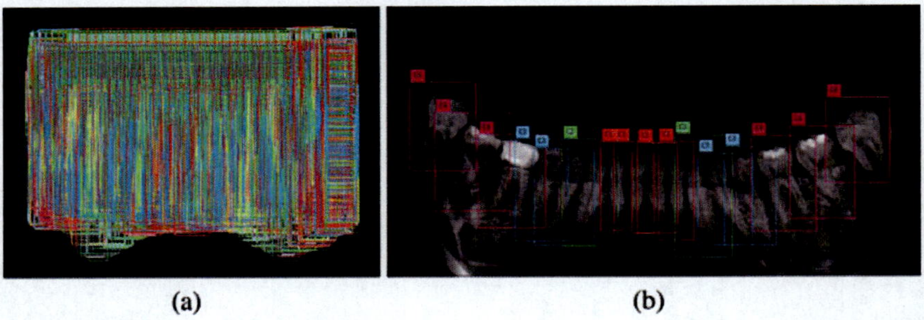

Fig. 2.44 The non-max suppression technique is used to remove boxes with low box confidence scores and highly overlapped bounding boxes of the same tooth. **a** 800 bounding boxes from Y. **b** 16 bounding boxes after NMS

Step 2 For each object class, we first select the box with the highest score. See the red box of the premolar class.

Step 3 Compute IoU (intersection over union) between a box selected in Step 2 and a box with the second highest score. IOU is a metric that measures the similarity between two boxes [8] given by

$$IoU = \frac{|box_1 \cap box_2|}{|box_1 \cup box_2|}, \tag{2.88}$$

Step 4 Discard the boxes with IOU larger than 0.6 (high overlap).

Step 5 For the remaining boxes, select a box with the second highest score as in Step 2. Repeat the same process (Steps 3–4) for the box with second highest score among the remaining boxes.

Through this elimination process, only one bounding box is left per tooth as shown in Fig. 2.44.

For stable learning of bounding box regression [12], $\mathbf{b}_{ij}$ is replaced by $\hat{\mathbf{b}}_{ij} = (\hat{s}_{ij}, \hat{z}_{ij}, \hat{w}_{ij}, \hat{h}_{ij})$ which satisfies the following:

$$\begin{aligned} s_{ij} &= 16(\hat{s}_{ij} + i - 1), \quad z_{ij} = 16(\hat{z}_{ij} + j - 1), \\ w_{ij} &= a_w \exp(\hat{w}_{ij}), \quad h_{ij} = a_h \exp(\hat{h}_{ij}), \end{aligned} \tag{2.89}$$

where a_w and a_h are the width and height of an anchor box. We set the size of the anchor box as the mean size of the ground truth bounding boxes.

2.4.5.1 Remarks on CNNs

Although it is somewhat unclear how deep learning works, there is a way to indirectly visualize which parts of I play an important role in the classification $f(I)$. Let $f_{class1}(I)$ denote the first component of f (the classifier for incisor). Note that the

f_{class1} can be regarded as a logistic regression. Given a well-trained f, the gradient $\nabla f_{class1}(I)$ is used to find the features. To visualize features that f_{class1} responds to, one may use the following maximaization method:

$$I = \text{argmax}_I f(I) - \lambda \text{Reg}(I), \tag{2.90}$$

where $\text{Reg}(I)$ is a regularization mentioned in the previous section.

CNNs generate feature vectors for an input image I in the layer immediately before the classifier.

2.4.6 Dentistry Application: 3D Tooth Segmentation from 3D CBCT Image

This section explains a fully automated method of identifying and segmenting 3D individual teeth from dental CBCT images, which was developed by Jang et al. [18]. Automatic and accurate 3D individual tooth segmentation from CBCT images is a difficult task for the following reasons: (i) similar intensities between teeth roots and their surrounding alveolar bone; (ii) the attached boundary between adjacent teeth in the crown parts.

To deal with the above-mentioned difficulties, Jang et al. [18] use the following hierarchical multi-step model.

Step 1 *Panoramic image reconstruction of the upper and lower jaws from a 3D CBCT image.* This step is to circumvent the high-dimensionality problem associated with CT images. This step automatically generates panoramic images of the upper and lower jaws from CT images where its size is smaller than the original CT image. The low-dimensional 2D panoramic image is designed to find accurate 3D tooth regions of interest (ROIs) and identify individual teeth. The panoramic images of the upper and lower jaws are separated to reduce overlaps between adjacent teeth. See Fig. 2.45.

Step 2 *Bounding box detection, identification, and 2D segmentation of individual teeth in the reconstructed panoramic images.* This step is to automatically identify individual teeth by two-digit numbers relative to their quadrant and location, as shown in Fig. 2.46a. We developed a tooth detection method that localizes bounding boxes that enclose each tooth and classifies them into four types according to tooth morphology. This method solves misclassification problems caused by similar adjacent teeth. The individual teeth are then identified using the results of tooth detection. Additionally, we performed 2D segmentation for individual teeth. See Fig. 2.46b.

Step 3 *Extraction of loose and tight 3D tooth ROIs using the detected bounding boxes and segmented tooth regions.* The third step extracts loose and tight 3D tooth ROIs from the detected boxes and segmented tooth regions for

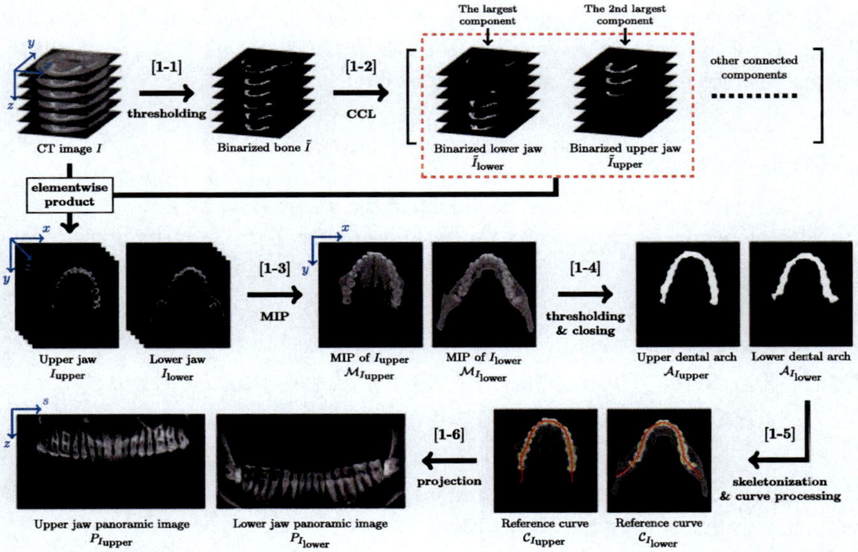

Fig. 2.45 Workflow of step 1. This shows the reconstruction process of upper jaw panoramic image $P_{I_{\text{upper}}}$ and low jaw panoramic image $P_{I_{\text{lower}}}$ from a 3D CT image I

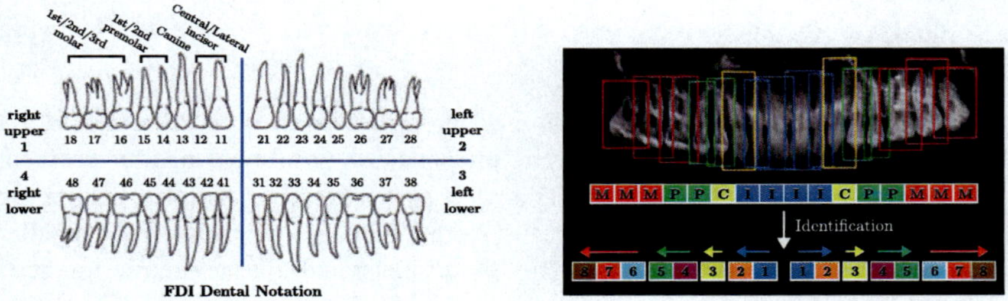

(a) (b)

Fig. 2.46 **a** Fédération Dentaire Internationale (FDI) dental notation using a two-digit numbering system. The first digit (quadrant code) represents a quadrant of teeth, and the second digit (tooth code) represents the order of the tooth from the central incisor in a quadrant. **b** Tooth identification process using the classification results in step 2. The capital letters represent the first letters of the tooth type and the numbers are tooth codes

accurate 3D individual tooth segmentation in the final step. Tight 3D tooth ROIs improve the segmentation accuracy. See Fig. 2.47.

Step 4 *3D segmentation for individual teeth from the 3D tooth ROIs.* In this final step, 3D individual tooth segmentation is performed using the loose ROI and tight ROI, as shown in Fig. 2.48. The tight ROI is crucial for improving the segmentation accuracy at the attached boundaries between a target tooth and its neighboring teeth.

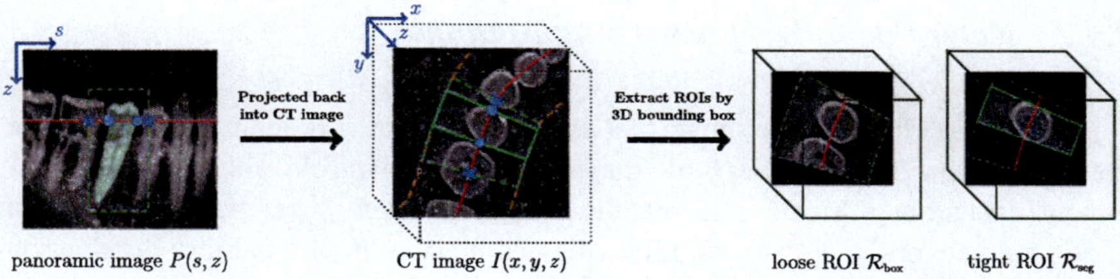

Fig. 2.47 Extraction of loose and tight 3D tooth ROIs. A loose ROI domain (green dotted line in I) is determined by the domain of projection between points (blue stars in I) on the reference curve corresponding to points (blue stars in P) on the bounding box. A 3D bounding box is then obtained by closely fitting the loose ROI domain. A loose 3D tooth ROI $\mathscr{R}_{box}$ is extracted by cropping the CT image I by the 3D bounding box and by changing values of voxels outside the loose ROI domain to 0. Similarly, a tight 3D tooth ROI $\mathscr{R}_{seg}$ is obtained from the 2D segmented tooth region

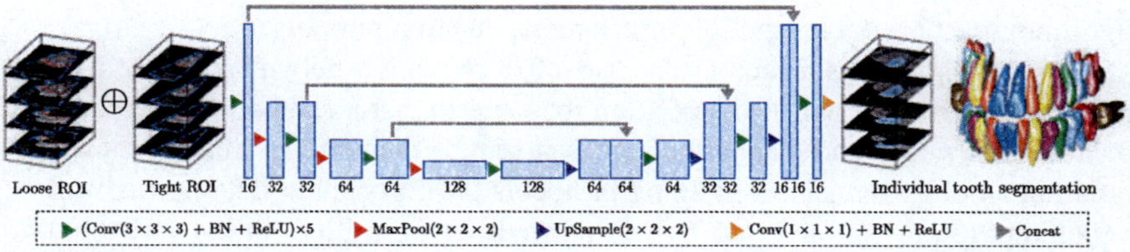

Fig. 2.48 Architecture diagram of the proposed network in step 4, which is a 3D version of U-net [35]. Individual tooth segmentation is performed using loose and tight 3D ROIs of a target tooth obtained in step 3. The tight ROI, which provides the boundary of the target tooth, is used for accurate segmentation at the boundary between the target tooth and its neighboring teeth

Individual tooth segmentation can be formulated as an instance segmentation problem. Mask R-CNN is the state-of-the-art deep learning framework for instance segmentation. However, Mask R-CNN cannot be applied to large size 3D CBCT images directly because of the computational limit.

The method by Jang et al. [18] has the advantage of using loose and tight ROIs that provide a considerable background region in advance. In particular, the tight ROI excludes structures on sides (adjacent teeth, jaw, etc.) of the target tooth. To evaluate the effectiveness of the use of both loose and tight ROIs, we implemented experiments using either loose or tight ROI, or both on the same 3D segmentation network. When using only tight ROIs, the recall is the lowest because loss of tooth information may occur where the tight ROI boundary intersects the tooth boundary. The use of only loose ROI containing a tooth boundary shows a higher recall. However, HD tends to be high because there is no information on tooth boundaries. A combination of the two ROIs enhances the segmentation performance, as the tight ROI provides detailed information on the target tooth and the loose ROI compensates for the disadvantage of the tight ROI.

2.4.7 Remarks on Deep Learning Methods

Numerous experiments have shown that a well-trained network seems to work only in the immediate vicinity of a very low-dimensional data manifold embedded in a high-dimensional ambient space $\mathbb{R}^n$. Here, the dimension of $\mathscr{M}_{\text{image}}$ is much smaller than n, and the network f appears to utilize the geometry of the probability distribution associated with the training data. As a result, even if the input deviates even slightly from the training data manifold, DL models may produce incorrect or unwanted results. Several experiments regarding adversarial classifications (e.g., false positive output of cancer) have shown that deep neural networks are vulnerable to various noise-like perturbations, resulting in incorrect output (which might be critical in medical environments) [6, 9, 14, 38]. In practice, the measured data are exposed to various noise sources such as machine dependent noise; therefore, the developed algorithm must be stable against perturbations due to input noise.

Normalizing data is an important part of improving a network's generalization ability (by enhancing out-of-distribution robustness), but it can be very challenging. Data normalization and standardization can reduce diversity in images caused by variation among scanners or imaging protocols [11]. For input data normalization, we attempt to project the input I to its normalized form $\mathfrak{N}(I)$ in the way that two images I and $\mathfrak{N}(I)$ are almost the same from the viewpoint of radiologists. However, definition of a similarity distance between images that reflects the radiologist's view is somewhat complicated. Regarding adversarial attacks due to noise-like perturbations mentioned above, a possible solution is the use of denoising networks that project a noisy image to the corresponding noise-suppressed image while preserving its salient features. Several DL methods have been developed for image denoising, such as CycleGANs [32, 41], denoising autoencoders [36], and denoising blocks [37].

A DL network's performance depends not only on its architecture, but also on the quality and quantity of training data. The implementation of U-net-based denoising [32] showed that its performance depends significantly on whether the training data set contains images with very small features. Let us briefly discuss U-net as one of the most popular networks. Its encoding path is based on a series of convolutions followed by pooling to reliably recognize image features so that the resulting output is fairly resilient to variations in position and scale of a target structure. It also utilizes concatenative skip connections between the same-scale feature maps obtained from the encoder and decoder subnetworks, which play an important role in restoring spatial information lost during downsampling. U-net has shown remarkable overall performance in medical image segmentations by simultaneously making use of global features and local spatial information [26, 27, 35]. However, the optimal depth remains unknown, and same-scale skip-connections have been criticized as unnecessarily restrictive. Deeper networks may learn more complex features of images, but various experiments have shown that deeper is not always better [40]. The optimal network depth may depend on a number of factors, including the size of the input image, the amount of training data, the variance in the size of the target features in the image, and the difficulty of the task. Instead of increasing the network depth and the filter size, spatially-adaptive filters such as atrous convolution can be used to increase the receptive field size. This technique has been used to properly handle large differences in a target's size. Various modifications or auxiliary means of U-net have been developed to supplement the limitations of the classical U-net architecture. These include Attention U-net [29], M-net [10], U-net++ [40], and MultiResUnet [16].

Acknowledgements This research was supported by Samsung Science & Technology Foundation (No. SRFC-IT1902-09). Jang and Seo were supported by a grant of the Korea Health Technology R&D Project through the Korea Health Industry Development Institute (KHIDI), funded by the Ministry of Health & Welfare, Republic of Korea (grant number : HI20C0127). We are deeply grateful to HDXWILL for their help and collaboration.

References

1. Adams, R., Bischof, L.: Seeded region growing. IEEE Trans. Pattern Anal. Mach. Intell. **16**(6), 641–647 (1994)
2. Alexe, B., Thomas, D., Ferrari, V.: Measuring the objectness of image windows. IEEE Trans. Pattern Anal. Mach. Intell. **34**(11), 2189–2202 (2012)
3. Caselles, V., Catte, F., Francine, C., Tomeu, D., Dibos, F.: A geometric model for active contours in image processing. Numerische Mathematik **66**(1), 1–31 (1993)
4. Caselles, V., Kimmel, R., Sapiro, G.: Geodesic active contours. Int. J. Comput. Vis. **22**(1), 61–79 (1997)
5. Chan, T.F., Vese, L.A.: Active contours without edges. IEEE Trans. Image Proc. **10**(2), 266–277 (2001)
6. Ching, T., Himmelstein, D.S., Beaulieu-Jones, B.K., Kalinin, A.A., Do, B.T., Way, G.P., Ferrero, Paul-Michael Agapow, E., Zietz, M., Hoffman, M.M. et al.: Opportunities and obstacles for deep learning in biology and medicine. J. R. Soc. Interf. **15**(141), 20170387 (2018)
7. Cho, H.C., Sun, S., Hyun, C.M., Kwon, J.-Y., Kim, B., Park, Y., Seo, J.K.: Automated ultrasound assessment of amniotic fluid index using deep learning. Med. Image Anal. 101951
8. Everingham, M., Van Gool, L., Williams, C.K., Winn, J., Zisserman, A.: The pascal visual object classes (VOC) challenge. Int. J. Comput. Visi. **88**(2), 303–338 (2010)
9. Finlayson, S.G., Bowers, J.D., Ito, J., Zittrain, J.L., Beam, A.L., Kohane, I.S.: Adversarial attacks on medical machine learning. Science **363**(6433), 12871289 (2019)
10. Huazhu, F., Cheng, J., Yanwu, X., Wong, D.W.K., Liu, J., Cao, X.: Joint optic disc and cup segmentation based on multi-label deep network and polar transformation. IEEE Trans Med Imaging **37**(7), 1597–1605 (2018)
11. Gao, Y., Liu, Y., Wang, Y., Shi, Z., Jinhua, Y.: A universal intensity standardization method based on a many-to-one weak-paired cycle generative adversarial network for magnetic resonance images. IEEE Trans. Med. Imaging **38**(9), 20592069 (2019)
12. Girshick, R., Donahue, J., Darrell, T., Malik, J.: Rich feature hierarchies for accurate object detection and semantic segmentation. In: Proceedings of the IEEE conference on computer vision and pattern recognition, pp 580–587 (2014)
13. Goldenberg, R., Kimmel, R., Rivlin, E., Rudzsky, M.: Fast geodesic active contours. IEEE Trans. Image Proc. **10**(10), 1467–1475 (2001)
14. Goodfellow, I.J., Shlens, J., Szegedy, C.: Explaining and Harnessing Adversarial Examples (2014). arXiv:1412.6572
15. He, K., Gkioxari, G., Dollar, P., Girshick, R.: Mask r-CNN. In: Proceedings of the IEEE International Conference on Computer Vision, pp. 2961–2969 (2017)

16. Ibtehaz, N., Rahman, M.S.: Multiresunet: rethinking the u-net architecture for multimodal biomedical image segmentation. Neural Netw. **121**, 74–87 (2020)
17. Jang, J., Park, Y., Kim, B., Lee, S.M., Kwon, J.-Y., Seo, J.K.: Automatic estimation of fetal abdominal circumference from ultrasound images. IEEE J. Biomed. Health Inf. **22**(5), 1512–1520 (2017)
18. Jang, T.J., Kim, K.C., Cho, H.C., Seo, J.K.: A fully automated method for 3d individual tooth identification and segmentation in dental CBCT. Submitted to IEEE Transactions on Pattern Analysis and Machine Intelligence (2021)
19. Kass, M., Witkin, A., Terzopoulos, D.: Snakes: active contour models. Int. J. Comput. Vis. **1**(4), 321–331 (1988)
20. Kim, B., Kim, K.C., Park, Y., Kwon, J.-Y., Jang, J., Seo, J.K.: Machine-learning-based automatic identification of fetal abdominal circumference from ultrasound images. Physiol. Measurem **39**(10), 105007 (2018)
21. Kim, H.P. Lee, S.M., Kwon, J.-Y., Park, Y., Kim, K.C., Seo, J.K.: Automatic evaluation of fetal head biometry from ultrasound images using machine learning. Physiol. Measur. **40**(6), 065009 (2019)
22. Kim, K.C., Cho, H.C., Jang, T.J., Choi, J.M., Seo, J.K.: Automatic detection and segmentation of lumbar vertebrae from x-ray images for compression fracture evaluation. Computer Methods and Programs in Biomedicine, p. 105833 (2020)
23. Kim, K.C., Yun, H.S., Kim, S., Seo, J.K.: Automation of spine curve assessment in frontal radiographs using deep learning of vertebral-tilt vector. IEEE Access **8**, 84618–84630 (2020)
24. Lee, S.M., Kim, H.P., Jeon, K., Lee, S.-H., Seo, J.K.: Automatic 3d cephalometric annotation system using shadowed 2d image-based machine learning. Phys. Med. Biol. **64**(5):055002 (2019)
25. Li, C., Xu, C., Gui, C., Fox, M.D.: Distance regularized level set evolution and its application to image segmentation. IEEE Trans. Image Proc. **19**(12), 3243–3254 (2010)
26. Litjens, G., Kooi, T., Bejnordi, B.E., Setio, A.A.A., Ciompi, F., Ghafoorian, M., Van Der Laak, J.A., Van Ginneken, B., Sanchez, C.I.: A survey on deep learning in medical image analysis. Med. Image Anal. **42**, 60–88 (2017)
27. Livne, M., Rieger, J., Aydin, O.U., Taha, A.A., Akay, E.M., Kossen, T., Sobesky, J., Kelleher, J.D., Hildebrand, K., Frey, D. et al.: A u-net deep learning framework for high performance vessel segmentation in patients with cerebrovascular disease. Front. Neurosci. **13**, 97 (2019)
28. Malladi, R., Sethian, J.A., Vemuri, B.C.: Shape modeling with front propagation: a level set approach. IEEE Trans. Pattern Anal. Mach. Intell. **17**(2), 158–175 (1995)
29. Oktay, O., Schlemper, J., Folgoc, L.L., Lee, M., Heinrich, M., Misawa, K., Mori, K., McDonagh, S., Hammerla, N.Y., Kainz, B. et al.: Attention u-net: learning where to look for the pancreas (2018). arXiv:1804.03999
30. Osher, S., Fedkiw, R.P.: Level set methods: an overview and some recent results. J. Comput. Phys. **169**(2), 463–502 (2001)
31. Otsu, N.: A threshold selection method from gray-level histograms. IEEE Trans. Syst. Man Cybern. **9**(1), 62–66 (1979)
32. Park, H.S., Baek, J., You, S.K., Choi, J.K., Seo, J.K.: Unpaired image denoising using a generative adversarial network in x-ray ct. IEEE Access **7**, 110414110425 (2019)
33. Redmon, J., Divvala, S., Girshick, R., Farhadi, A.: You only look once: unified, real-time object detection. In: Proceedings of the IEEE Conference on Computer Vision and Pattern Recognition, pp. 779–788 (2016)
34. Ren, S., He, K., Girshick, R., Sun, J.: Faster r-CNN: towards real-time object detection with region proposal networks. In: Advances in Neural Information Processing Systems, pp. 91–99 (2015)
35. Ronneberger, O., Fischer, P., Brox, T.: U-net: convolutional networks for biomedical image segmentation. In: International Conference on Medical Image Computing and Computer-Assisted Intervention, pp. 234–241. Springer (2015)
36. Vincent, P., Larochelle, H., Lajoie, I., Bengio, Y., Manzagol, P.-A., Bottou, L.: Stacked denoising autoencoders: learning useful representations in a deep network with a local denoising criterion. J. Mach. Learn. Res. **11**(12) (2010)

37. Xie, C., Wu, Y., van der Maaten, L., Yuille, A.L., He, K.: Feature denoising for improving adversarial robustness. In: Proceedings of the IEEE Conference on Computer Vision and Pattern Recognition, pp. 501–509 (2019)
38. Yuan, X., He, P., Zhu, Q., Li, X.: Adversarial examples: attacks and defenses for deep learning. IEEE Trans. Neural Netw. Learn. Syst. **30**(9), 2805–2824 (2019)
39. Yun, H.S., Jang, T.J., Lee, S.M., Lee, S.-H., Seo, J.K.: Learning-based local-to-global landmark annotation for automatic 3d cephalometry. Phys. Med. Biol. **65**(8), 085018 (2020)
40. Zhou, Z., Siddiquee, M., Rahman, M., Nima, T., Jianming, L.: Unet++: redesigning skip connections to exploit multiscale features in image segmentation. IEEE Trans. Med. Imaging **39**(6), 1856–1867 (2019)
41. Zhu, J.-Y., Park, T., Isola, P., Efros, A.A.: Unpaired image-to-image translation using cycle-consistent adversarial networks. In Proceedings of the IEEE International Conference on Computer Vision, pp. 2223–2232 (2017)

Chapter 3
Deep Learning for Dental Cone-Beam Computed Tomography

Chang Min Hyun, Taigyntuya Bayaraa, Sung Min Lee, Hong Jung, and Jin Keun Seo

Abstract This chapter reviews metal artifact reduction (MAR) methods for low-dose cone-beam computed tomography (CBCT). MAR is of vital significance because the number of aged people with artificial prostheses and metallic implants is swiftly increasing with the rapidly aging population. Metallic objects present in the CBCT field of view produce streaking artifacts that highly degrade the reconstructed CT images, resulting in a loss of information on the teeth and other anatomical structures. Metallic object-related artifacts are associated with beam hardening, scattering, partial volume effects, and a high degree of inhomogeneous attenuation to name a few. As metal-induced artifacts are complex and nonlinearly intertwined, MAR has remained a challenging problem over the last four decades. Metal artifacts are caused mainly due to a mismatch in the forward model of the filtered back-projection (FBP) algorithm. The presence of metallic objects in an imaging subject violates the model's assumption that the CT sinogram data is equal to the Radon transform of an image. FBP ignores the polychromatic nature of the X-ray data **P**, which has non-linear dependence on the distribution of the metallic object. Various MAR methods have been suggested, but the existing MAR methods do not reduce the metal artifacts effectively in low-dose CBCT environments and may introduce new streaking artifacts that did not previously exist. We hope that this chapter will help develop

---

C. M. Hyun · S. M. Lee · J. K. Seo (✉)
School of Mathematics and Computing (Computational Science and Engineering), Yonsei University, Seoul, Korea
e-mail: seoj@yonsei.ac.kr

C. M. Hyun
e-mail: chammyhyun@yonsei.ac.kr

T. Bayaraa
Department of Mathematics, School of Applied Sciences, Mongolian University of Science and Technology, Ulaanbaatar, Mongolia
e-mail: taigiintuya@must.edu.mn

H. Jung
HDXWILL, Seoul, Korea
e-mail: jh21star@iwillmed.com

J. K. Seo (ed.), *Deep Learning and Medical Applications*, Mathematics in Industry 40,
https://doi.org/10.1007/978-981-99-1839-3_3

new MAR algorithms that overcome the limitations of existing MAR methods and effectively reduce metal artifacts to facilitate diagnosis, preoperative and presurgical assessments, surgical navigation, and workflows for rapid prototyping.

3.1 Introduction

This chapter aims to review existing results of metal artifact reduction (MAR) methods for low-dose cone-beam computed tomography (CBCT). Metallic objects present in the CBCT field of view produce streaking artifacts that highly degrade the reconstructed CT image, resulting in a loss of information on the teeth and other anatomical structures. Metal artifacts are common in dental CBCT. As the number of aged patients with metallic implants is increasing, metal-induced artifacts are becoming a major factor degrading the CBCT diagnostic performance. Figure 3.1 shows that a dental CBCT image after applying state-of-the-art MAR still contains serious metal artifacts caused by gold crowns.

The presence of highly attenuating materials such as metallic objects complicates reconstruction techniques by violating the forward model assumption of the sinogram data being equal to the Radon transform of an image. The metallic objects-related artifacts are associated with beam hardening-induced sinogram inconsistency, scattered radiation, nonlinear partial volume effect (NLPV), high degree of attenuation inhomogeneity (i.e., bone, tissue, air), and others. Since metal-induced artifacts are complex and nonlinearly intertwined, MAR has remained a challenging problem over the last three decades.

MAR in the dental CBCT is more difficult than that in the standard multi-detector CT (MDCT). In the era of low-dose CBCT protocols requiring the reduction of either the tube voltage or tube current or both, artifacts related to high-attenuation materials could be accentuated. This occurs because the reduced X-ray tube voltage or current causes more severe beam hardening, scattering, photon starvation, and photon noise. Moreover, the field of view (FOV) size in dental CBCT is usually small compared

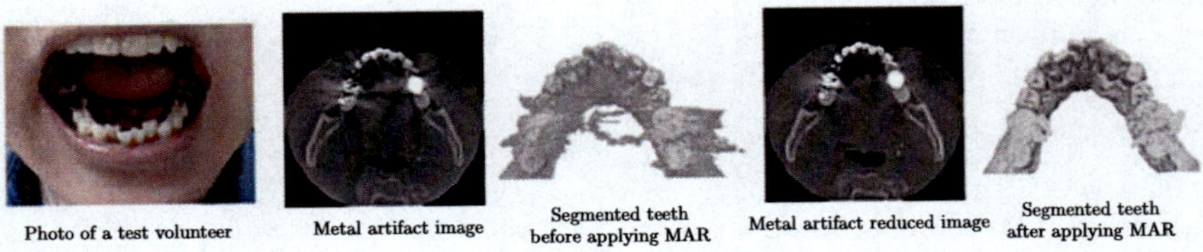

Fig. 3.1 Metal artifacts caused by gold crowns in a dental CBCT image. (left) Photo of a test volunteer; (middle) CBCT and segmented image before applying MAR; (right) CBCT and segmented image obtained by applying a state-of-the-art MAR algorithm

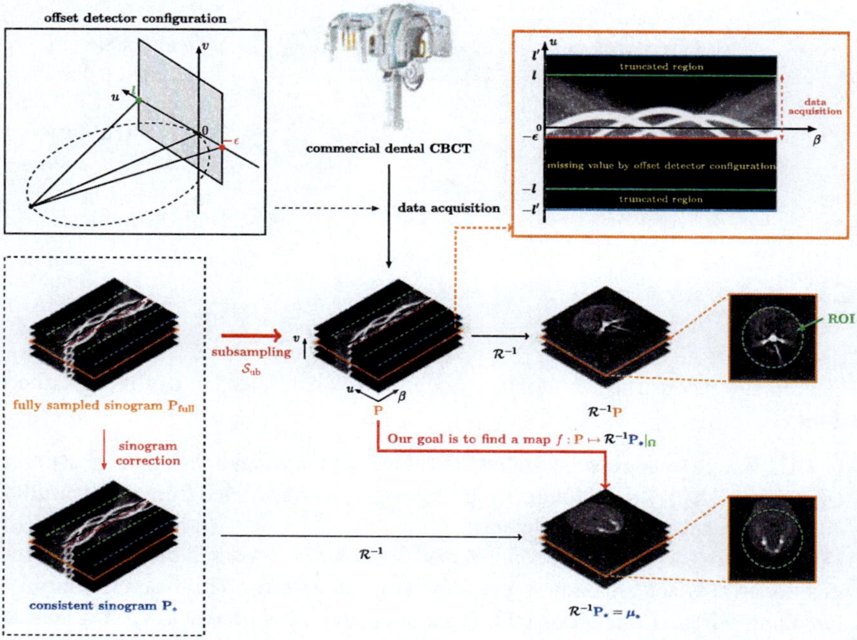

Fig. 3.2 [10] Dental CBCT system. Owing to the offset detector configuration and interior-ROI-oriented scan, the measured sinogram data can be viewed as a subsampled sinogram from a fully sampled sinogram acquirable in the standard MDCT. This subsampling causes sinogram truncation and asymmetry. This system will be explained in detail in Sect. 3.2.4. This figure was extracted from [10]

to the size of a patient's head because a small detector is employed to reduce system costs. This FOV truncation produces additional artifacts combined with teeth and high-attenuation materials. See Fig. 3.2.

Dental CBCT is advantageous over MDCT in the following aspects: (i) The price of CBCT is much cheaper than the price of MDCT; (ii) The X-ray dose of CBCT is much lower than the MDCT dose. On the other hand, CBCT is more disadvantageous than MDCT in the following aspects: (a) The step-and-shoot CBCT machine cannot obtain sufficient projection data to obtain accurate reconstruction due to violation of Tuy's condition, whereas hellical MDCT is designed to satisfy Tuy's condition which requires that every plane intersecting the object under study must intersect the focal trajectory; (b) the scan time of CBCT is much longer than that of MDCT. Despite these shortcomings, the CBCT system has seen a rapid increase in demand based on qualifications as a promising diagnostic and therapeutic procedure. This is because the dental CBCT system significantly reduces the patient's radiation dose, enhancing the confidence of the equipment's healthcare worker. Figure 3.3 summarizes the machine specifications of MDCT and dental CBCT.

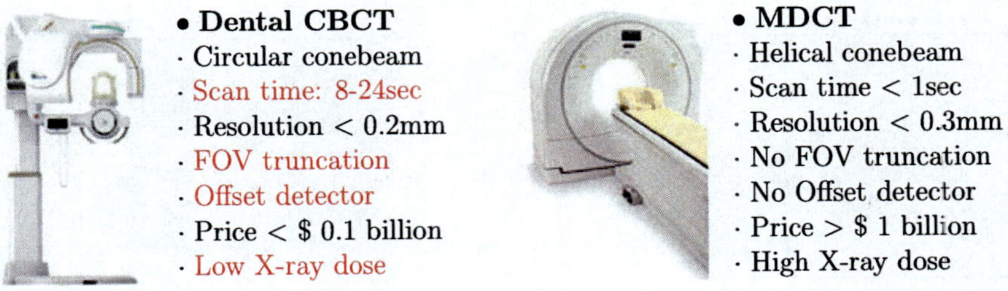

Fig. 3.3 Machine specifications of MDCT and dental CBCT

The image reconstruction problem in CBCT (planar detector) can be described as follows:

- **(CBCT image to be reconstructed)** The three-dimensional image $\mu(\mathbf{x}, z)$ at point $(\mathbf{x}, z) = (x_1, x_2, z) \in \mathbb{R}^3$ (or the tissue density) is reconstructed from the measured X-ray beam attenuation coefficients.
- **(X-ray beam for CBCT)** A cone-shaped X-ray beam is passed through a patient positioned between an X-ray source and flat-panel detector. This beam is transmitted along different directions by rotating the gantry that houses the X-ray source and detector. Thus, data are acquired for CBCT reconstruction.
- **(CBCT projection data)** CBCT projection data $\mathbf{P}(\beta, u, v)$ are acquired using a planar detector after passing the X-ray beams in several directions, where $\beta \in [0, 2\pi)$ is the projection angle and (u, v) is the planar detector position. See Fig. 3.4 for details of the CBCT geometry.
  - The relationship between the data location (β, u, v) and the image location $(\mathbf{x}, z)$ is the following:

$$u = u_{\beta,\mathbf{x}} = R\frac{\mathbf{x} \cdot \Theta_\beta}{U_{\beta,\mathbf{x}}} \quad \text{and} \quad v = v_{\beta,\mathbf{x},z} = \frac{zR}{U_{\beta,\mathbf{x}}}, \tag{3.1}$$

  where $\Theta_\beta = (\cos\beta, \sin\beta)$, R is the distance between the source point and the origin, and

$$U_{\beta,\mathbf{x}} = R + \mathbf{x} \cdot \Theta_\beta^\perp. \tag{3.2}$$

- **(Inverse problem)** Recover $\mu(\mathbf{x}, z)$ from the series of the X-ray data $\mathbf{P}(\beta, u, v)$.

The standard CBCT reconstruction algorithm is FDK method [31], which can be expressed as

$$\mu(\mathbf{x}, z) = \frac{1}{4\pi}\int_0^{2\pi} \frac{R^2}{U_{\beta,\mathbf{x}}^2} \int_{\mathbb{R}} \mathbf{P}(\beta, u, v_{\mathbf{x},z,\beta}) \frac{R}{\sqrt{R^2 + u^2 + v_{\beta,\mathbf{x},z}^2}} \hbar(u_{\beta,\mathbf{x}} - u)\,du\,d\beta, \tag{3.3}$$

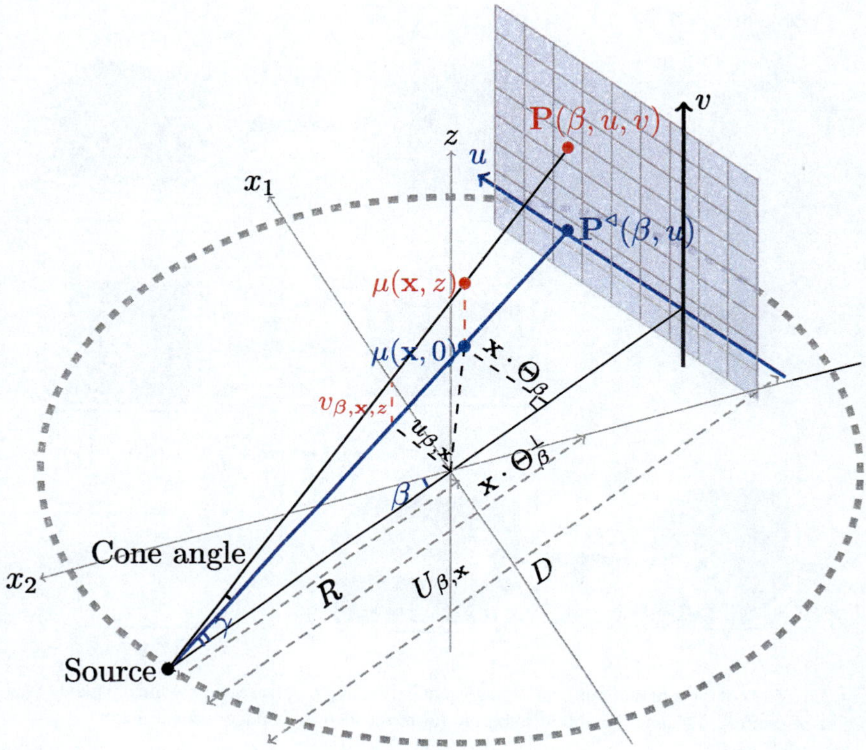

Fig. 3.4 Dental CBCT geometry

where $\hbar(s)$ is a 1-D ramp filter given by $\hbar(s) = \frac{1}{2\pi}\int_{\mathbb{R}} |\omega| e^{i\omega s} d\omega$. Given an object location $(\mathbf{x}, z)$, the identification of the attenuation value $\mu(\mathbf{x}, z)$ requires knowing $\mathbf{P}(\beta, u, v)$ with the data location (β, u, v) being related with $(\mathbf{x}, z)$ according to (3.1), which takes account of all beam lines passing though the image location $(\mathbf{x}, z)$. The data position (β, u, v) contains the information on the incident angle of detected photons.

An unavoidable mismatch between the real projection data and mathematical model exists, due to undesired noise and the polychromatic nature of the incident X-ray beam [84]. In the presence of metals or high-attenuation materials in the field of view of clinical CT, this mismatch produces serious streaking artifacts associated with metal geometries, which are caused mainly by the beam-hardening factor [25, 30, 119]. See Fig. 3.5.

Numerous metal artifact reduction (MAR) methods have been suggested. Classical MAR methods have very limited performance [7, 78, 79]. Dual-energy CT [5] provides somehow satisfactory reconstruction via beam-hardening correction through dual-energy blending, but it requires a longer postprocessing time and a

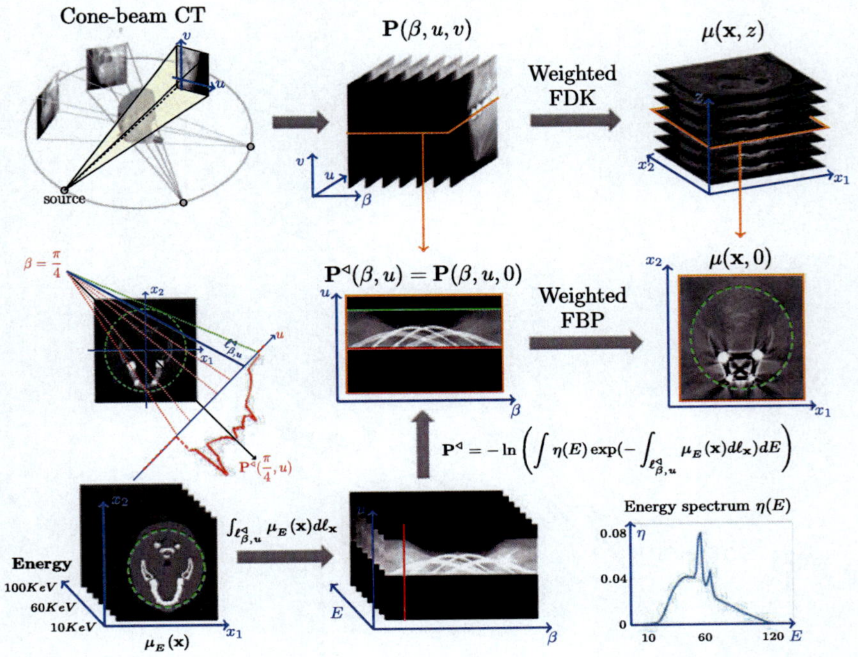

Fig. 3.5 The cone-beam and fan-beam image acquisition and reconstruction, which explains how region of interest is scanned to obtain sinogram for cross-sectional image reconstruction

higher dose of radiation compared with single-energy CT [13]. Currently, commercially available MAR algorithms include SEMAR (Toshiba Medical Systems) [135], O-MAR (Philips Healthcare) [86], iMAR (Siemens Healthcare) [64], Smart MAR (GE Healthcare) [44]. However, the existing MAR methods do not reduce the metal artifacts effectively in low-dose CBCT environments, and may introduce new streaking artifacts that did not previously exist.

Various experiences/studies have shown that the performance of MAR depends on the material, geometry, size, and location of the metal. The development of a unified MAR method, which includes all cases such as small/medium/large metal objects and complete photon starvation, is difficult. Thus, we need to develop a case-based adaptive MAR algorithm instead of a unified MAR method. The purpose of this chapter is to review existing results of MAR, to analyze the problems mathematically, and to suggest future research directions.

We begin with reviewing the basics of CT, existing MAR methods, and their limitations. Next, we explain mathematical theories on describing characteristics of metal artifacts and a beam-hardening correction formula, which accounts for the measurement nonlinearities arising from the energy dependence of the attenuation coefficients of metallic materials. Then, we discuss challenging issues and suggest future research directions. We review various existing MAR methods and their limitations.

3.2 Basics of CT

3.2.1 *History of CT*

X-ray computed tomography (CT) is one of the most established diagnostic tools for medical and dental imaging. CT provides cross-sectional images (2D slice images) of the human body (3D object) by assigning an X-ray attenuation coefficient to each voxel that characterizes how easily a medium can be penetrated by X-ray beam. X-rays are produced by accelerating charged particles. X-rays are electromagnetic waves with wavelengths ranging from 10^{-8} to 10^{-12} m. X-ray radiation was discovered by Röntgen, who was awarded the first Nobel Prize in Physics in 1901. In CT, X-ray beams are passed from an anode X-ray generator in a tube to a multiple set of detectors. The idea underlying CT is to use X-ray beams passing through the body to acquire information about the CT image's projection values (2D X-ray images) in various directions. CT images are reconstructed from X-ray images at different angles, which are acquired by rotating the X-ray tube.

In 1917, Johann Radon [102] found that the CT images could be reconstructed from the X-ray information at all directions. The first clinical CT scanner was invented by Hounsfield [52] in England at the end of 1960s where data acquisition is based on parallel-beam geometry, and Hounsfield and Cormack [19] shared the Nobel prize in Medicine in 1979. The idea of CT reconstruction is the general concept of filtered back-projection, back-projecting filtered X-ray data to a reconstruction volume according to the beam's path. Using Hounsfield's scanner, the first clinical CT scan of a female patient with suspected frontal lobe was performed at Atkinson Morley's Hospital in London, in 1971. Each scan took about 5 minutes to collect data to produce an 80×80 pixel image.

In 1973, the first whole-body CT scanner was developed by Robert S. Ledley. The paper [113] explained the interesting story behind the development of the first whole-body CT scanner. Ledley was supposed to be paid a big NIH grant around 1970, when Nixon was president of the USA. However, after a while, Nixon cut back the medical research funds, and as a result, his NIH research funding was cut completely. So Ledley had to quickly find sa way to pay his workers. When he heard that Dr. Lussenhop, the chief neurosurgeon, was interested in buying the CT scanner developed by Hounsfield, Ledley went to see him. Ledley said to Dr. Luessenhop: "I can make it, and it'll be half the price". At that time, he didn't even know what the price was. Ironically, this funding crisis was the driving force behind the focus on the development of the first whole-body CT scanner [113].

The introduction of helical CT scanners and multi-detector computed tomography (MDCT) has led to technological advances in CT scanning. Currently, CT scanners take less than 0.3 s to collect data to produce a 1024×1024 pixel image and can simultaneously acquire 64 slices per gantry rotation.

In 1999, the first commercial CBCT was introduced. Since then, CBCT has continuously evolved to provide high-resolution image while reducing its cost, size, and radiation dose exposure. Now, CBCT is widely used for dental/maxillofacial applications. In clinical dentistry, dental CBCT has been gaining great attention as a crucial supplementary radiographic technique to aid diagnosis, treatment planning, and prognosis assessment such as diagnosis of dental caries, reconstructive cranio-facial surgery planning, and evaluation of the patient's face [40, 77, 114, 115].

3.2.2 Parallel-Beam CT: Basic Principles

The underlying concept of CT is based on the mathematical model proposed by Cormack in 1963 (previously by Radon in 1917 [103]), which aimed to provide an image of a 3D subject by taking X-ray data from all angles around the subject. Each pixel in the CT image is assigned by an X-ray attenuation coefficient, which indicates how easily the media can be penetrated by X-ray beam.

For convenience, the basic principle of CT will be explained using a two-dimensional parallel-beam CT model (although impractical). The following symbols are used in this section:

- $\mu(\mathbf{x})$ describes the attenuation coefficient of a medium at position $\mathbf{x} = (x_1, x_2) \in \mathbb{R}^2$.
- $\mathbf{P}^{\ddagger}(\varphi, s)$ represents the projection data at projection angle $\varphi \in [0, 2\pi)$ and detector position $s \in \mathbb{R}$. The superscript $\ddagger$ stands for the parallel beam.
- $\ell^{\ddagger}_{\varphi,s} := \{\mathbf{x} \in \mathbb{R}^2 : \mathbf{x} \cdot \Theta_\varphi = s\}$ represent the beam line, where $\Theta_\varphi = (\cos\varphi, \sin\varphi)$ represents the projection direction.
- $\mathscr{R}$ denotes the Radon transform given by

$$\mathscr{R}\mu(\varphi, s) := \int_{\mathbb{R}^2} \mu(\mathbf{x})\delta(\mathbf{x} \cdot \Theta_\varphi - s)d\mathbf{x} = \int_{\ell^{\ddagger}_{\varphi,s}} \mu(\mathbf{x})dl_{\mathbf{x}}, \tag{3.4}$$

  where $dl_{\mathbf{x}}$ is the line element and $\delta(\cdot)$ is the Dirac delta function.

In parallel-beam CT, X-ray projection data $\mathbf{P}^{\ddagger}(\varphi, s)$ are collected for a slice after passing X-ray beams in different directions through an object, as shown in Fig. 3.6.

The FBP algorithm is based on the assumption that the X-ray projection data $\mathbf{P}^{\ddagger}(\varphi, s)$ is in the range of the Radon transform, that is, there exist μ such that

$$\mathbf{P}^{\ddagger}(\varphi, s) = \mathscr{R}\mu(\varphi, s) \quad \text{for all projection angle } \varphi \in [0, 2\pi). \tag{3.5}$$

In practice, the above assumption is not true in general, that is, $\mathbf{P}^{\ddagger}(\varphi, s)$ is *NOT* in the range of the Radon transform. This will be discussed in Sect. 3.3.2.

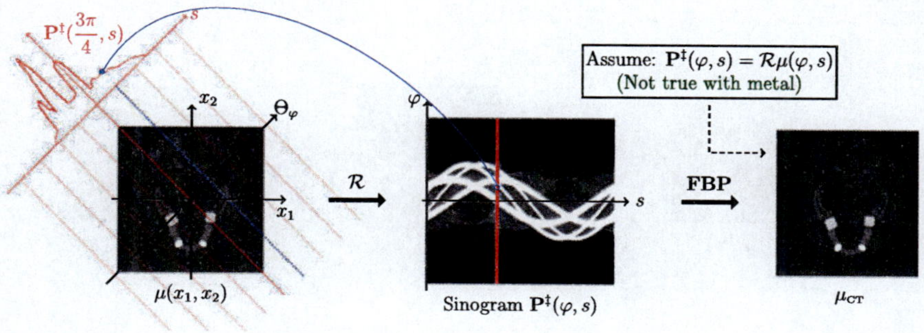

Fig. 3.6 Data acquisition and image reconstruction in the parallel-beam CT system

3.2.2.1 Filtered Back-Projection (FBP)

The concept of CT reconstruction was based on a variant of the following mathematical formula:

$$\mu(\mathbf{x}) = \frac{1}{8\pi^2} \int_{-\pi}^{\pi} \int_{\mathbb{R}} |\omega| \mathscr{F}_1[\mathscr{R}\mu(\varphi, \cdot)](\omega) e^{i\omega \mathbf{x}\cdot\Theta_\varphi} d\omega d\varphi, \tag{3.6}$$

where $\mathscr{F}_1$ is 1D-Fourier transform(FT) with respect to the s-variable. Hence, under the linear assumption (3.5), we have the following FBP algorithm:

$$\mu(\mathbf{x}) = \mathscr{R}^{-1}\mathbf{P}^{\ddagger}(\mathbf{x}) = \underbrace{\frac{1}{8\pi^2} \int_{-\pi}^{\pi} \underbrace{\int_{-\infty}^{\infty} \underbrace{|\omega| \mathscr{F}_1[\mathbf{P}^{\ddagger}(\varphi, \cdot)](\omega)}_{\text{FT of } \mathbf{P}^{\ddagger} \text{ at each } \varphi \text{ multiplied by } |\omega|} e^{i\omega \mathbf{x}\cdot\Theta_\varphi} d\omega}_{\text{Inverse FT}}\, d\varphi}_{\text{Back-project for each } \varphi} . \tag{3.7}$$

See Fig. 3.7 for imaging principles in parallel-beam CT.

The FBP formula (3.7) is based on the following two well-known facts:

1. Fourier slice theorem: $\mathscr{F}_2\mu(\omega\Theta_\varphi) = \mathscr{F}_1[\mathscr{R}\mu(\varphi, \cdot)](\omega)$, where $\mathscr{F}_2$ is the 2D-Fourier transform. To be precise,

$$\underbrace{\mathscr{F}_1[\mathscr{R}\mu(\varphi, \cdot)](\omega)}_{=\int_{\mathbb{R}} \mathscr{R}_{\Theta_\varphi}\mu(\varphi, s) e^{-i2\pi\omega s} ds} = \underbrace{\iint \mu(s\Theta_\varphi + t\Theta^{\perp})\, e^{-i2\pi\omega s}\, ds dt}_{=\iint \mu(\mathbf{x}) \exp[-i2\pi\omega \underbrace{\mathbf{x}\cdot\Theta_\varphi}_{s}]\, d\mathbf{x}} = \mathscr{F}_2(\mu)(\omega\Theta_\varphi). \tag{3.8}$$

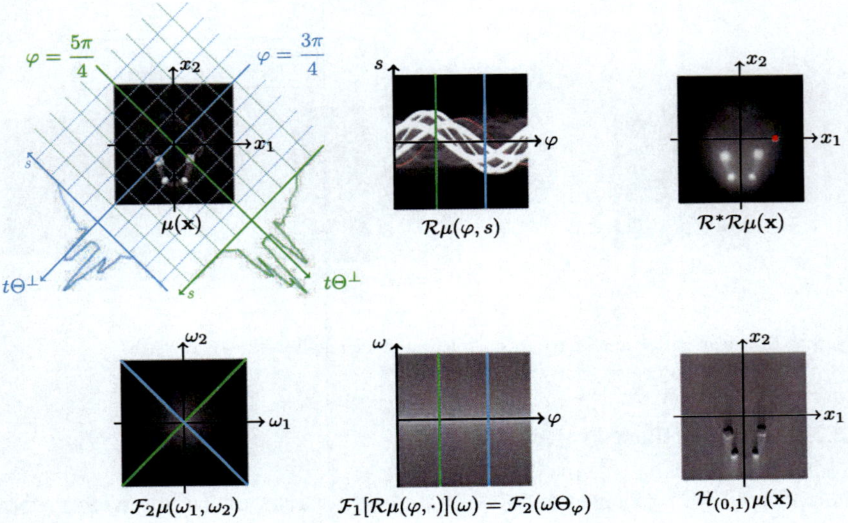

Fig. 3.7 Imaging principle in parallel-beam CT

Here, we use

$$\mathscr{R}\mu(\varphi, s) = \int_{\mathbb{R}} \mu(s\Theta_\varphi + t\Theta^\perp)dt = \int_{\mathbb{R}^2} \mu(\mathbf{x})\, \delta(\mathbf{x} \cdot \Theta_\varphi - s)d\mathbf{x}. \tag{3.9}$$

See Fig. 3.8.

2. Filtered back-projection: Back-project the filtered data, which is the inverse Fourier transform of $|\omega|\mathscr{F}_1[\mathscr{R}\mu(\varphi, \cdot)](\omega)$.

This FBP method works well for CT imaging because for most human tissues, the X-ray data $\mathbf{P}^\ddagger(\varphi, s)$ approximately satisfy the linear assumption that $\mathbf{P}^\ddagger = \mathscr{R}\mu$. However, the X-ray data $\mathbf{P}^\ddagger(\varphi, s)$ fail to satisfy the assumption when metallic objects are present in the imaging slice. This is because the incident X-ray beams comprise a number of photons of different energies, and the X-ray attenuation coefficients vary with E. In particular, the attenuation coefficient μ of a metallic object varies greatly with the energy E. Hence, the presence of metal objects in the scan slice causes the data $\mathbf{P}^\ddagger(\varphi, s)$ to violate the assumption of linearity.

3.2.2.2 Other Reconstruction Methods

There are several alternative expressions of (3.7) for analytic CT reconstruction. Various expressions of $\mathscr{R}^{-1}$ are the following:

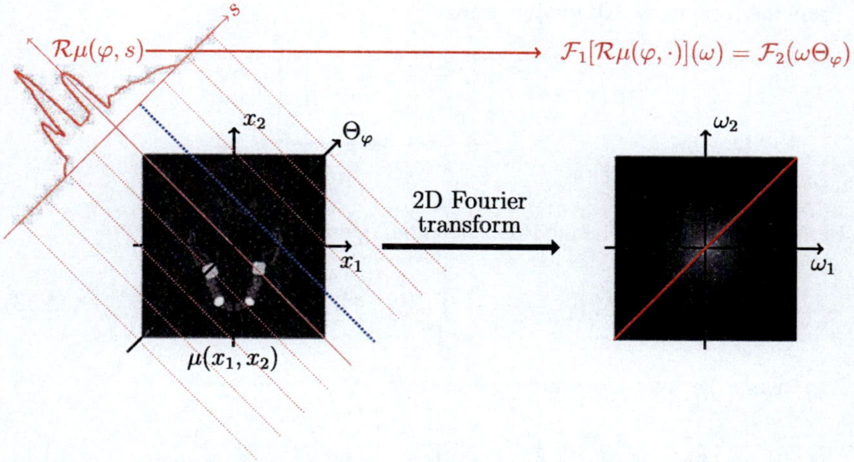

Fig. 3.8 Graphical illustration of the Fourier slice theorem

- According to the formula (3.7), $\mathscr{R}^{-1}$ can be expressed as

$$\mathscr{R}^{-1}\mathbf{P}^{\ddagger} = \frac{1}{2}\mathscr{R}^{*}\left(\mathscr{H}[\frac{\partial}{\partial s}\mathbf{P}^{\ddagger}]\right), \tag{3.10}$$

where $\mathscr{H}$ is the Hilbert transform. Let us briefly explain how to get the formula (3.10). We have

$$\mu(\mathbf{x}) \doteq \underbrace{\iint_{\mathbb{R}^2} \mathscr{F}_2\mu(\boldsymbol{\xi})\, e^{i\mathbf{x}\cdot\boldsymbol{\xi}} d\boldsymbol{\xi}}_{\int_0^{2\pi}\int_0^{\infty} \underbrace{\mathscr{F}_2\mu(\omega\Theta_\varphi)}_{\mathscr{F}_1(\mathscr{R}\mu)(\varphi,\omega)} e^{i\omega\mathbf{x}\cdot\Theta_\varphi}\omega d\omega d\varphi}$$

$$\doteq \int_0^{\pi}\int_{\mathbb{R}} \underbrace{\mathscr{F}_1\left(\mathscr{H}[\frac{d}{ds}\mathscr{R}\mu]\right)(\varphi,\omega)}_{\omega\frac{\omega}{|\omega|}\mathscr{F}_1(\mathscr{R}\mu)(\varphi,\omega)} e^{i\omega\mathbf{x}\cdot\Theta_\varphi} d\omega d\varphi. \tag{3.11}$$

Here, we use the following identities.

- $\frac{1}{2\pi i}\mathscr{F}_1\left(\frac{\partial}{\partial s}\mathscr{R}\mu(\varphi,\cdot)\right)(\omega) = \omega\,\mathscr{F}_1[\mathscr{R}\mu(\varphi,\cdot)](\omega)$.
- $\mathscr{F}_1(\mathscr{H}g)(\omega) = \frac{\omega}{|\omega|}\mathscr{F}_1 g(\omega)$.
- $\mathscr{R}^*\mathscr{R}\mu(\mathbf{x}) \doteq \int_0^{\pi}\mathscr{R}\mu(\mathbf{x}\cdot\Theta_\varphi,\varphi)d\varphi$.

Then, the formula (3.10) follows from

$$\mu(\mathbf{x}) \doteq \int_0^{\pi} \underbrace{\left[\int_{\mathbb{R}} \frac{\frac{\partial}{\partial s}\mathscr{R}\mu(\varphi, s)}{\mathbf{x}\cdot\Theta_{\varphi} - s} ds \right]}_{\mathscr{H}\frac{\partial}{\partial s}\mathscr{R}\mu(\mathbf{x}\cdot\Theta_{\varphi},\varphi)} d\varphi. \tag{3.12}$$

- Let $\mathbf{P}^{\ddagger} = \mathscr{R}\mu$. Let $\mathscr{I}_a$ be the Riesz potential operator in 1-D given by

$$\mathscr{I}_a^n g(s) = \frac{1}{\sqrt{2\pi}} \int_R e^{is\omega} |\omega|^{-a} \mathscr{F} g(\omega) d\omega. \tag{3.13}$$

The inverse Radon transform is expressed by

$$\mathscr{R}^{-1}\mathbf{P}^{\ddagger}(\mathbf{x}) \doteq \int_{-\pi}^{\pi} \underbrace{\left[\int_{\mathbb{R}} \int_{\mathbb{R}} |\omega| \mathbf{P}^{\ddagger}(\varphi, s) e^{i\omega(\mathbf{x}\cdot\Theta_{\varphi} - s)} ds d\omega \right]}_{\mathscr{I}_{-1}\mathscr{R}\mu(\mathbf{x}\cdot\Theta_{\varphi},\varphi)} d\varphi. \tag{3.14}$$

This follows from

$$\begin{aligned} \mu(\mathbf{x}) \doteq & \underbrace{\int_{\mathbb{R}^2} \mathscr{F}_2(\mu)(\boldsymbol{\xi}) e^{i\boldsymbol{\xi}\cdot\mathbf{x}} d\boldsymbol{\xi}}_{\int_{-\pi}^{\pi}\int_0^{\infty} \underbrace{\mathscr{F}_2\mu(\omega\Theta_{\varphi})}_{\mathscr{F}_1[\mathscr{R}\mu(\varphi,\cdot)](\omega)} e^{i\omega\mathbf{x}\cdot\Theta_{\varphi}} \omega d\omega d\varphi} \\ \doteq & \int_{-\pi}^{\pi} \underbrace{\left[\int_{-\infty}^{\infty} |\omega| \mathscr{F}_1(\mathscr{R}(\varphi, \cdot)\mu)(\omega) e^{i\omega\mathbf{x}\cdot\Theta_{\varphi}} d\omega \right]}_{\mathscr{I}_{-1}\mathscr{R}\mu(\mathbf{x}\cdot\Theta_{\varphi},\varphi)} d\varphi. \end{aligned} \tag{3.15}$$

- $\mathscr{R}^{-1}$ can be expressed by using the directional Hilbert transforms: The sketch of the proof is

$$\mu(\mathbf{x}) \doteq \underbrace{\mathscr{H}_{\Theta_{\varphi_0}} \underbrace{\mathscr{R}^*_{\Theta_{\varphi_0}} \frac{\partial}{\partial s} \mathscr{R}\, \mu\, (\mathbf{x})}_{\frac{1}{\pi}\int_{|\varphi-\varphi_0|<\frac{\pi}{2}} \frac{\partial}{\partial s}\mathscr{R}\mu(\varphi,s)|_{s=\mathbf{x}\cdot\Theta_{\varphi}} d\varphi}}_{\frac{1}{\pi}\int \frac{1}{s}\mathscr{R}^*_{\Theta_{\varphi_0}} \frac{\partial}{\partial s}\mathscr{R}\, \mu(\mathbf{x}-s\Theta_{\varphi_0})\, ds} \tag{3.16}$$

$$\Rightarrow \mathscr{H}_{\Theta_{\varphi_0}} \mu(\mathbf{x}) \doteq \mathscr{R}^*_{\Theta_{\varphi_0}} \frac{\partial}{\partial s} \mathscr{R}\, \mu(\mathbf{x}). \tag{3.17}$$

The proof is based on the following identities:

- $\mathscr{F}\{\mathscr{H}_{\Theta_{\varphi_0}}\mu\}(\boldsymbol{\xi}) = -i\ \mathrm{sgn}(\Theta_{\varphi_0}\cdot\mathbf{k})\mathscr{F}\mu(\boldsymbol{\xi})$.
- $\mathscr{F}\{\mathscr{R}^*_{\Theta_{\varphi_0}}\mathscr{R}\mu\}(\boldsymbol{\xi}) = \frac{1}{|\boldsymbol{\xi}|}\mathscr{F}_1\mathscr{R}\mu(\mathrm{sgn}(\Theta_{\varphi_0}\cdot\boldsymbol{\xi})\varphi, \mathrm{sgn}(\Theta_{\varphi_0}\cdot\boldsymbol{\xi})|\boldsymbol{\xi}|)$.
- $\mathscr{F}\{\mathscr{R}^*_{\Theta_{\varphi_0}}\mathscr{R}f + \mathscr{R}^*_{-\Theta_{\varphi_0}}\mathscr{R}f\}(\boldsymbol{\xi}) = \frac{1}{|\boldsymbol{\xi}|}\left(\mathscr{F}_1\mathscr{R}\mu(\varphi, |\boldsymbol{\xi}|) + \mathscr{F}_1\mathscr{R}\mu(-\varphi, -|\boldsymbol{\xi}|)\right)$.

3.2.3 Fan-Beam CT: Reconstruction Algorithm

Figure 3.9 shows the fan-beam imaging geometry, which explains how patients are scanned to obtain projection data for cross-sectional image reconstruction. CBCT can be viewed as a wide fan-beam CT (FBCT) due to its wide collimator openings.

Fan-beam multiple detector CT (MDCT) is widely used in clinical fields. Although CT theory was initiated for parallel beams, parallel CT is not used in clinical field. In MDCT, the X-ray source moves in a tomographic plane along a circle of radius R. The fan-beam reconstruction algorithm is similar to parallel-beam reconstruction algorithm except "data filtering" and weighting in the back-projection.

In order to clarify the data acquisition difference between parallel-beam and fan-beam, we use the following notations of projection data:

- $\mathbf{P}^{\ddagger}(\varphi, s)$ denotes the projection data in parallel-beam CT, where the data acquisition is based on a beam line $\ell^{\ddagger}_{\varphi,s} := \{\mathbf{x} \in \mathbb{R}^2 \,:\, \mathbf{x} \cdot \Theta_\varphi = s\}$ passing though the body. Here, $\Theta_\varphi = (\cos\varphi, \sin\varphi)$ represents the projection angle and s represents the position of detector.
- As shown in Fig. 3.9, the projection data of fan-beam CT is denoted by $\mathbf{P}^{\triangleleft}(\beta, u)$, where the data acquisition is based on a fan-beam line

$$\ell^{\triangleleft}_{\beta,u} := \{\mathbf{x} \in \mathbb{R}^2 \,:\, \mathbf{x} \cdot \Theta_{\beta+\gamma_u} = R \sin \gamma_u\}. \tag{3.18}$$

Here, γ_u is the fan angle given by $u = R \tan \gamma$; R is the distance between the source point and the origin; $\beta + \frac{\pi}{2}$ is the beam-source angle; the source point of the beam line $\ell^{\triangleleft}_{\beta,u}$ is located at $R\Theta_{\beta+\pi/2}$; and the superscript $\triangleleft$ stands for the fan-beam. See Fig. 3.11 for a visual explanation of the fan-beam line.

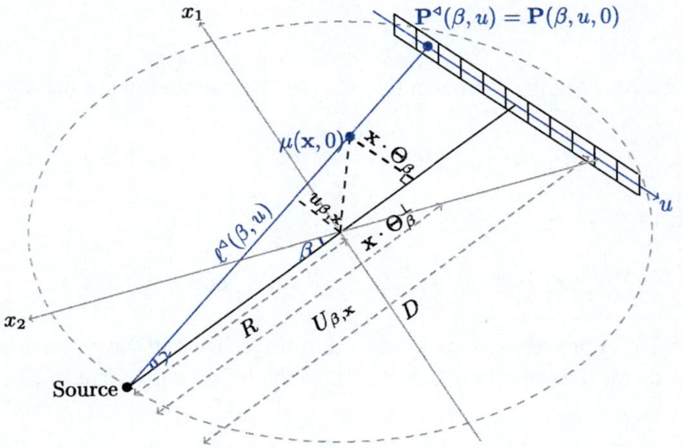

Fig. 3.9 Fan-beam CT geometry

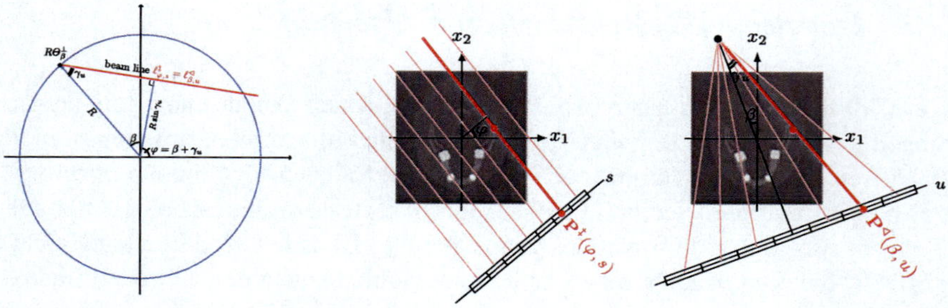

Fig. 3.10 The relation between parallel-beam and fan-beam CT geometry

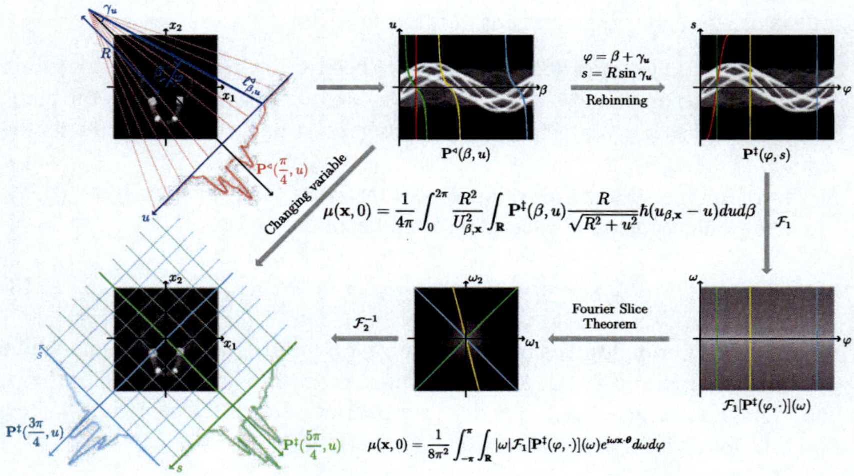

Fig. 3.11 Fan-beam CT image reconstruction process

The relation between the fan-beam line $\ell^{\triangleleft}_{\beta,u}$ and the parallel-beam line $\ell^{\ddagger}_{\varphi,s}$ is

$$\ell^{\ddagger}_{\varphi,s} = \ell^{\triangleleft}_{\beta,u} \iff s = R\sin\gamma_u, \quad \varphi = \gamma_u + \beta. \tag{3.19}$$

Hence, we have

$$\mathbf{P}^{\ddagger}(\varphi, s) = \mathbf{P}^{\triangleleft}(\beta, u) \;\text{ if }\; s = R\sin\gamma_u \;\text{ and }\; \varphi = \gamma_u + \beta. \tag{3.20}$$

See Fig. 3.10 for the relation between the fan-beam line and the parallel-beam line.
Under the ideal assumption (3.5), we have the following variant of FBP (3.7):

$$\mu(\mathbf{x}) = \underbrace{\frac{1}{4\pi}\int_0^{2\pi}\int_{\mathbb{R}} \mathbf{P}^{\ddagger}(\varphi, s)\hbar(s_{\mathbf{x},\varphi} - s)\,ds\,d\varphi}_{:=\mathscr{R}^{-1}\mathbf{P}^{\ddagger}(\mathbf{x})}. \tag{3.21}$$

A fan-beam CT version of the FBP (3.21) can be expressed as

$$\mu(\mathbf{x}) = \frac{1}{4\pi} \int_0^{2\pi} \frac{R^2}{U_{\beta,\mathbf{x}}^2} \int_{\mathbb{R}} \mathbf{P}^{\triangleleft}(\beta, u) \frac{R}{\sqrt{R^2 + u^2}} \hbar(u_{\beta,\mathbf{x}} - u) du d\beta, \tag{3.22}$$

where $U_{\beta,\mathbf{x}} = (\mathbf{x} - R\Theta_\beta^\perp) \cdot \Theta_\beta^\perp$ and $u_{\beta,\mathbf{x}} = R\tan(\gamma_{\beta,\mathbf{x}})$.

In practice, the projection data $\mathbf{P}^{\triangleleft}(\beta, u)$ contains various artifact sources including scattering, nonlinear effects due to the polychromatic beam, and patient movement. Analytic reconstruction methods such as (3.22) can not deal with these artifact sources effectively. To improve image quality, it is recommended to develop a hybrid method combining the analytic method, image processing techniques, and iterative method to solve an optimization problem. Due to serious difficulties in handling nonlinear effects, it would be desirable to develop application-specific reconstruction techniques.

3.2.4 Cone-Beam CT

Cone-beam CT (CBCT) machines have been widely used for dental CT, where a cone-shaped X-ray beam rotates around the patient's head to obtain many 2D projections using a 2D area detector. Figure 3.12 shows the difference between CBCT and fan-beam CT. As an MDCT tends to increase z-axis coverage of the beam with increasing numbers of rows in detector arrays, it is becoming increasingly difficult to set a standard for distinguishing between MDCT and CBCT.

This section focuses only on the step-and-shoot cone-beam acquisition as shown in Fig. 3.12, where the source trajectory is a circle and the patient is stationary during CT scanning [54]. In a circular CBCT, the projection data is insufficient for exact analytic reconstruction. The pros and cons of CBCT over MDCT are the followings:

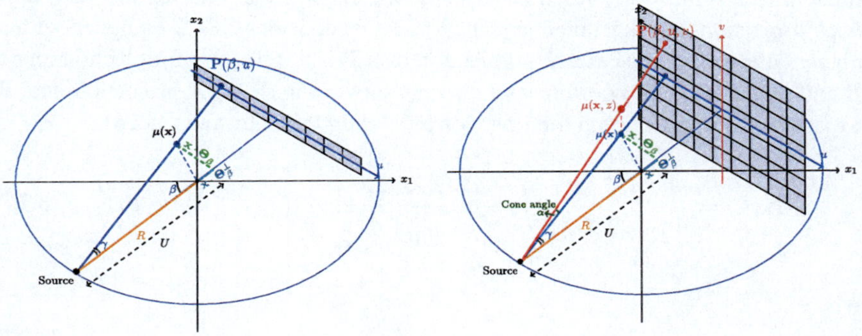

Fig. 3.12 The relation between fan-beam and cone-beam CT geometry

- *Price*: The price of CBCT is much less than that of MDCT.
- *X-ray dose*: The X-ray dose for CBCT is much lower than that of MDCT.
- *CT reconstruction*: The step-and-shoot CBCT machine cannot obtain sufficient projection data to get an exact reconstruction due to violation of Tuy's condition. On the other hand, hellical MDCT is designed to satisfy Tuy's condition which requires that every plane intersecting the object under study must intersect the focal trajectory.
- *Artifacts*: Artifacts are common in CBCT due to additional data defects including FOV truncation.
- *Scan time*: The scan time of CBCT is much longer than that of MDCT.

Let us begin with considering the ideal case where the measured projection data is the full 3D Radon transform $\mathscr{R}\mu$. Here, the 3D Radon transform is given by

$$\mathscr{R}\mu(\rho, \mathbf{n}) = \iint_{\Pi_{\rho,\mathbf{n}}} \mu dS, \forall \mathbf{n} \in S^2, \rho, \tag{3.23}$$

where S^2 denotes the unit sphere and $\Pi_{\rho,\mathbf{n}}$ is the plane given by

$$\Pi_{\rho,\mathbf{n}} = \{(\mathbf{x}, z) \in \mathbb{R}^3 : \ \rho = (\mathbf{x}, z) \cdot \mathbf{n}\}. \tag{3.24}$$

Using the 3D Fourier slice theorem as in 2D FBP, the inverse of Radon transform can be expressed as

$$\mu(\mathbf{x}, z) = -\frac{1}{8\pi^2} \iint_{|\mathbf{n}|=1} \frac{\partial^2}{\partial \rho^2} \mathscr{R}\mu(\rho, \mathbf{n}) dS_{\mathbf{n}}. \tag{3.25}$$

However, we cannot use this inversion method directly because the full projection data covering all direction $\mathbf{n}$ is not available.

One of popular CBCT reconstruction algorithms is the FDK method, developed by Feldkamp, Davis, and Kress [31]. The nature of the FDK algorithm is similar to filtered back-projection, which consists of two major parts: *filtering* and *weighted back-projection*. For a detailed explanation, let us consider CBCT equipped with a planar detector. In this case, the projection data $\mathbf{P}$ is a function of projection-angle β and planar detector coordinates (u, v), as shown Fig. 3.13. A projection data $\mathbf{P}$ experiences weighted ramp filtering, corresponding to the formula (3.26):

$$\mu(\mathbf{x}, z) = \frac{1}{4\pi} \int_0^{2\pi} \frac{R^2}{U_{\beta,\mathbf{x}}^2} \int_{\mathbb{R}} \frac{R\mathbf{P}(\beta, u, v_{\mathbf{x},z,\beta})}{\sqrt{R^2 + u^2 + v_{\mathbf{x},z,\beta}^2}} \hbar(u_{\beta,\mathbf{x}} - u) du d\beta. \tag{3.26}$$

where

$$v_{\beta,\mathbf{x},z} = \frac{zR}{R + \mathbf{x} \cdot \Theta_\beta}. \tag{3.27}$$

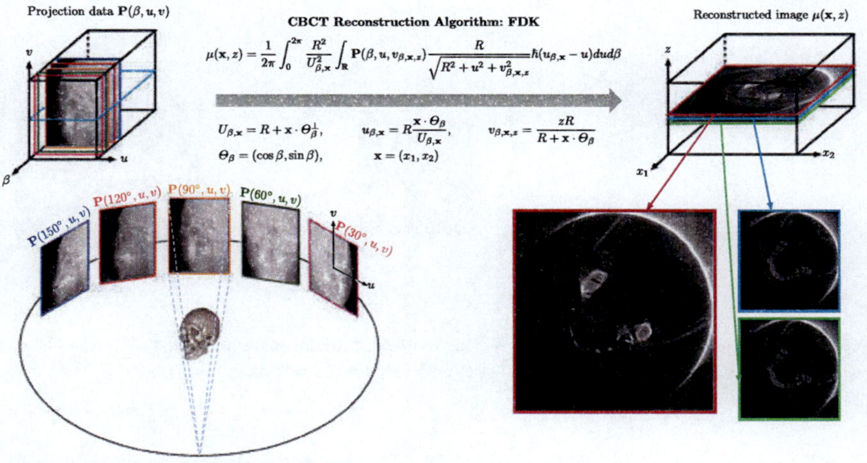

Fig. 3.13 CBCT reconstruction algorithm

Remark 3.1 Within only two weeks after the announcement of the discovery of "Roentgen rays", Friedrich Otto Walkhoff made the first dental radiograph in 1896, for which a long exposure time of 25 min was used [92].

3.2.5 *Dental CBCT*

In clinical dentistry, the most widely used dental CBCT systems use an offset detector configuration and an interior-ROI-oriented scan, as shown in Fig. 3.14. Dental CBCT is designed to provide high-quality 3D maxillofacial images with the lowest possible X-ray dose exposure to radiation and minimal cost.

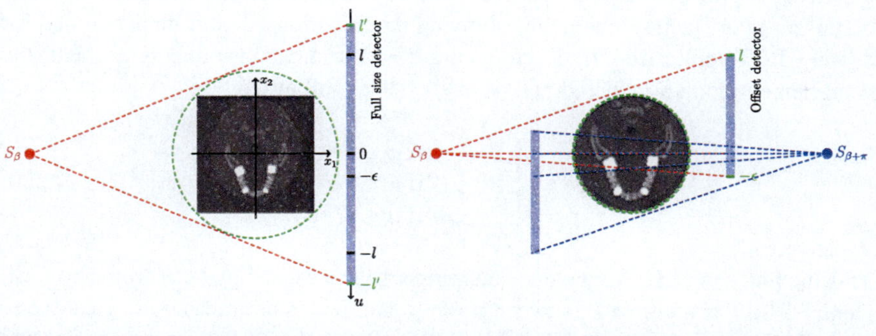

Fig. 3.14 Detector configuration for MDCT and dental CBCT. MDCT uses a symmetric detector that covers the entire FOV. On the other hand, dental CBCT uses a small detector with offset array to minimize production cost

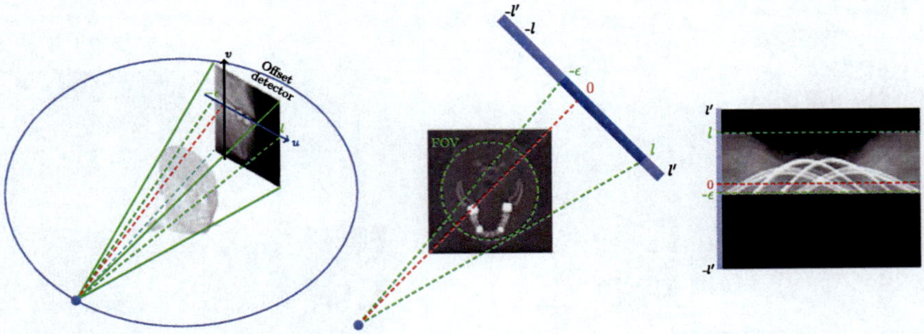

Fig. 3.15 Data acquisition in dental CBCT. Due to offset detector configuration, projection data is only acquired in the asymmetric region $[-\epsilon, l]$ with respect to the u axis

Because the effective FOV of dental CBCT does not cover the entire region of an object to be scanned and offset detector geometry is used, the sinogram $\mathbf{P}$ can be expressed by

$$\mathbf{P} = \mathscr{S}_{\text{ub}}(\mathbf{P}_{\text{full}}), \tag{3.28}$$

where $\mathbf{P}_{\text{full}}$ is the corresponding sinogram acquirable with non-offset and wide-detector CBCT providing complete information of a sinogram and $\mathscr{S}_{\text{ub}}$ is a subsampling operator determined by the size and offset configuration of a detector. More precisely, let a 2D flat-panel detector be aligned in $[-\epsilon, l]$ with respect to the u-axis. As shown in Fig. 3.15, sinogram $\mathbf{P}$ is truncated by

$$\mathbf{P} = \mathscr{S}_{\text{ub}}(\mathbf{P}_{\text{full}}) = \begin{cases} \mathbf{P}_{\text{full}} & \text{if } u \in [-\epsilon, l], \\ 0 & \text{if } u \in [-l', -\epsilon] \cup [l, l'], \end{cases} \tag{3.29}$$

where $[-l', l']$ is the support of $\mathbf{P}_{\text{full}}$ with respect to the u-axis. This missing information in $\mathbf{P}$ along the u-axis makes the application of existing methods difficult.

Owing to the subsampling $\mathscr{S}_{\text{ub}}$ (i.e., FOV truncation and offset detector array), the direct application of the standard FDK algorithm in (3.26) can produce additional image artifacts. Hence, the following modifications should be performed for dental CBCT application. (i) To compensate FOV truncation-inducing artifacts, the sinogram extrapolation method [124] is applied as follows:

$$\mathscr{E}_{\text{xpol}}(\mathbf{P}) = \begin{cases} \mathbf{P} & \text{if } u \in [-\epsilon, \ell], \\ \mathbf{P}|_{u=-\epsilon} & \text{if } u \in [-\ell', -\epsilon), \\ \mathbf{P}|_{u=\ell} & \text{if } u \in (\ell, \ell']. \end{cases} \tag{3.30}$$

This method can effectively reduce cupping artifacts caused by FOV truncation. (ii) Dental CBCT sinograms $\mathbf{P}$ between $u = -\epsilon$ and $u = \epsilon$ are measured twice due to the offset detector geometry. To address this partial data redundancy, a weighting

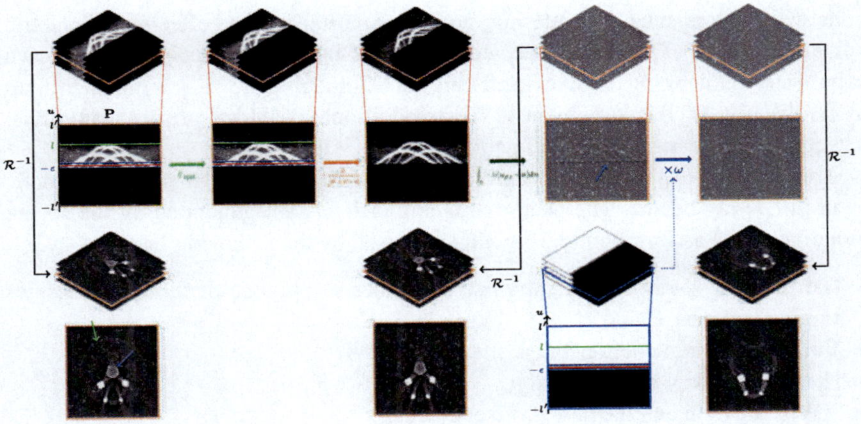

Fig. 3.16 Dental CBCT reconstruction algorithm. The standard FDK algorithm is modified to deal with the problems caused by offset detector and FOV truncation. Sinogram extrapolation is applied to handle the cupping artifact caused by FOV truncation. The weighting function is applied to address the offset detector-inducing partial redundancy in $[-\epsilon, \epsilon]$ with respect to the u-axis

function ω is used before taking back-projection in a similar sense to [121]. See Fig. 3.16. The modified FDK algorithm for dental CBCT is expressed as follows:

$$\mu(\mathbf{x}, z) = \frac{1}{4\pi}\int_0^{2\pi} \frac{R^2\omega(u_{\beta,\mathbf{x}})}{U^2_{\beta,\mathbf{x}}} \int_{\mathbb{R}} \frac{R\mathscr{E}_{\text{xpol}}(\mathbf{P})(\beta, u, v_{\mathbf{x},z,\beta})}{\sqrt{R^2 + u^2 + v^2_{\mathbf{x},z,\beta}}} \hbar(u_{\beta,\mathbf{x}} - u)du d\beta, \quad (3.31)$$

where ω is given by

$$\omega(u) = \frac{1 - \cos(\pi(-u + \epsilon)/(2\epsilon))}{2}. \quad (3.32)$$

3.3 Dental CBCT Artifacts: Beam-Hardening

Maxillofacial CBCT imaging still suffers from various artifacts that significantly degrade the image quality regarding bone and teeth. Compared to standard multi-detector CT (MDCT), the additional difficulty of artifact reduction in most dental CBCT is caused by the use of low X-ray irradiation and a small-size flat-panel detector in which the center axis of rotation is offset relative to the source–detector axis to maximize the transaxial FOV [14, 15].

As the number of patients with metallic implants and dental filling is increasing, metal-induced artifacts are common in dental CBCT [27, 107, 112, 116]. These metal-related artifacts are generated by the effects of beam-hardening-induced sinogram inconsistency and different types of complicated metal–bone–tissue interactions with factors such as scattering, nonlinear partial volume effects, and electric noise [8, 53, 70, 117]. Furthermore, reducing metal-induced artifacts, which is known to be a very challenging problem in all kinds of CT imaging [23, 37], is much difficult in the dental CBCT environment owing to additional problems arising from offset detectors, FOV truncations, and low X-ray doses.

This section explains the beam-hardening artifacts caused by polychromatic nature of X-ray beams. The beam-hardening artifacts are generated by the severe violation of the assumption (3.5), which is caused by the following factors:

- The incident X-ray beams comprise a number of photons of different energies ranging between $\underline{E}$ and $\overline{E}$.
- X-ray attenuation coefficient (μ_E) varies with E.
- The use of low X-ray irradiation.
- FOV truncation and the use of offset detector.

3.3.1 Lambert–Beer Law and Beam-Hardening Artifacts

For simplicity, the parallel-beam CT model is used to explain the nonlinear beam-hardening effects. In parallel-beam CT, the projection data $\mathbf{P}^\ddagger(\varphi, s)$ for a polychromatic X-ray is given by the Lambert–Beer law [11, 67]:

$$\mathbf{P}^\ddagger(\varphi, s) = -\ln\left(\int_{\mathbb{R}} \eta(E)\exp\left\{-\mathscr{R}\mu_E(\varphi, s)\right\}dE\right), \tag{3.33}$$

where $\eta(E)$ represents the fractional energy at photon energy E in the spectrum of the X-ray source [49, 95] with its support being the interval $[E_{min}, E_{max}]$ and $\int_{\mathbb{R}} \eta(E)dE = 1$. See Fig. 3.17 for attenuation coefficient μ_E and fractional energy $\eta(E)$.

More precisely, the projection data $\mathbf{P}^\ddagger(\varphi, s)$ is related to the following two terms:

- $I_{\text{in}}(\varphi, s)\eta(E)$ represents the intensity distribution of the incoming beam of the energy E along the beam's line $\ell_{\varphi,s}$.
- $I_{\text{out}}(\varphi, s)\eta(E)$, measured by the detector, represents the intensity of X-ray photons that are transmitted through the body along $\ell_{\varphi,s}$.

The Beer–Lambert law is

$$I_{\text{out}}(\varphi, s)\eta(E) = I_{\text{in}}(\varphi, s)\eta(E)\exp\{-\int_{\ell_{\varphi,s}} \mu(\mathbf{x}, E)dl_{\mathbf{x}}\}, \tag{3.34}$$

where $\mu(\mathbf{x}, E)$ is the distribution of linear attenuation coefficients at position $\mathbf{x} = (x_1, x_2)$ and at energy level E. The value of μ of a metal object varies greatly with E, whereas μ of a soft tissue varies little with E. For example, μ(material, E) are:

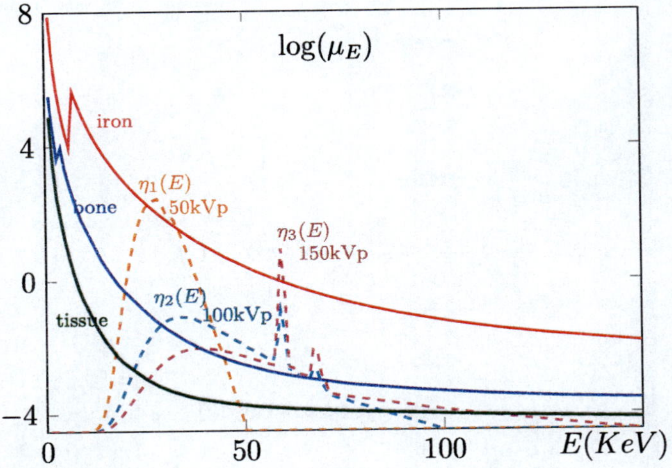

Fig. 3.17 Illustration of attenuation coefficient μ_E and fractional energy $\eta(E)$. The value of μ of a metal object varies greatly with E, whereas μ of a soft tissue varies little with E

- μ(soft tissue, 30 KeV) $\approx 0.38(\text{cm}^{-1})$, μ(soft tissue, 60 KeV) $\approx 0.21(\text{cm}^{-1})$
- μ(water, 10 KeV) $\approx 5(\text{cm}^{-1})$, μ(water, 100 KeV) $\approx 0.17(\text{cm}^{-1})$
- μ(bone, 10 KeV) $\approx 144(\text{cm}^{-1})$, μ(bone, 100 KeV) $\approx 0.40(\text{cm}^{-1})$

Integrating the Beer–Lambert law (3.34) over E, we have

$$I_{\text{out}}(\varphi, s) = I_{\text{in}}(\varphi, s) \int \eta(E) \exp\{-\int_{\ell_{\varphi,s}} \mu(\mathbf{x}, E) dl_{\mathbf{x}}\} dE \tag{3.35}$$

Hence, the logarithm of the transmission loss $I_{\text{in}}(\varphi, s)/I_{\text{out}}(\varphi, s)$ provides information of the distribution of μ from the following formula:

$$\underbrace{\ln \frac{I_{\text{in}}(\varphi, s)}{I_{\text{out}}(\varphi, s)}}_{\text{measured data}} = -\ln \int \eta(E) \exp\{-\int_{\ell_{\varphi,s}} \mu(\mathbf{x}, E) dl_{\mathbf{x}}\} dE. \tag{3.36}$$

The projection data $\mathbf{P}(\varphi, s)$ in (3.33) is derived from

$$\mathbf{P}(\varphi, s) = \ln \left(\frac{I_{\text{in}}(\varphi, s)}{I_{\text{out}}(\varphi, s)} \right). \tag{3.37}$$

Figure 3.18 shows the beam-hardening effect in the simplest case of a bi-chromatic X-ray beam, where X-ray beam consists of two different energy levels $E_1 = 40$ keV and $E_2 = 80$ keV. Let $\Omega = \{\mathbf{x} : x_1 + x_2 < a \ , \ 0 < x_1, x_2 < a \ \}$ represent the triangular-shaped homogeneous imaging object, as shown in Fig. 3.18. Assume that

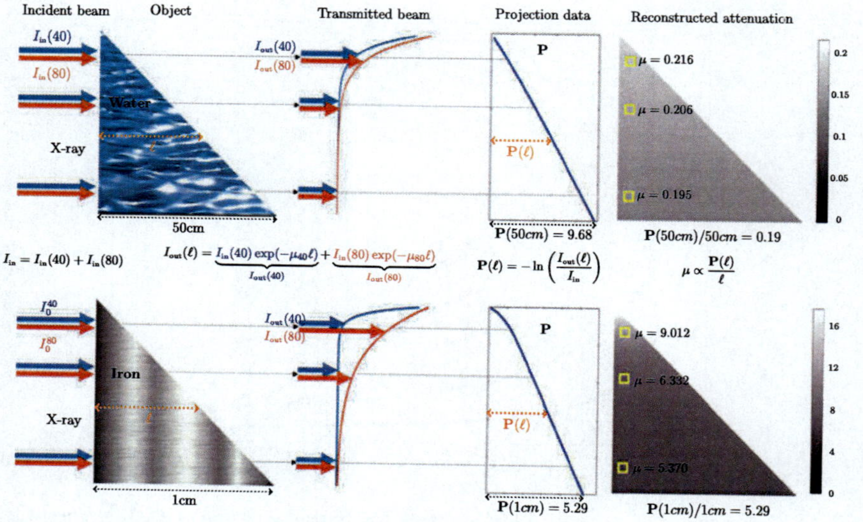

Fig. 3.18 Beam-hardening effect for bichromatic energy in two different materials. Beam hardening in high-attenuation materials is relatively strong, relative to that in low attenuation ones

$\eta(E) = \frac{1}{2}(\delta(E - E_1) + \delta(E - E_2))$ (bi-chromatic) and the attenuation coefficient at the energy level E_j is μ_j. Then, according to (3.33), the projection data $\mathbf{P}^{\ddagger}(s) = \mathbf{P}^{\ddagger}(\pi/2, s)$ is

$$\mathbf{P}^{\ddagger}(s) = -\ln\Big(\underbrace{\int_{\mathbb{R}} \eta(E)\exp\{-\mathscr{R}\mu_E(s)\}dE}_{\frac{1}{2}\sum_{j=1}^{2}\exp\{-\mu_j\mathscr{R}\chi_\Omega(s)\}}\Big). \tag{3.38}$$

Since $\mathscr{R}\chi_\Omega(s) = a - s$,

$$\mathbf{P}^{\ddagger}(s) = -\ln\Big(\frac{1}{2}\sum_{j=1}^{2} e^{-\mu_j(a-s)}\Big) \tag{3.39}$$

Hence, if we back-project the projection $\mathbf{P}^{\ddagger}(s)$ into the object domain Ω along the horizontal direction, then we get the reconstructed image μ^* depending on vertical direction $x_2 = s$ that is given by

$$\mu_*(s) = \frac{1}{a-s}\mathbf{P}^{\ddagger}(s) = -\frac{1}{a-s}\ln\Big(\frac{1}{2}\sum_{j=1}^{2} e^{-\mu_j(a-s)}\Big) \tag{3.40}$$

for $0 < s < a$. It is easy to observe the following:

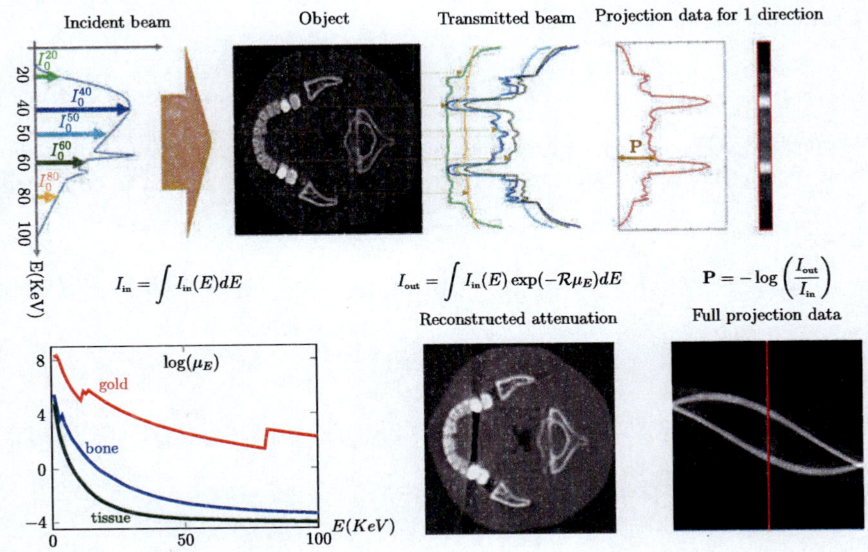

Fig. 3.19 A material has a different attenuation coefficient value depending on energy level. The linear assumption is violated in the presence of metallic objects in the CT scan field

- If $\mu_1 = \mu_2$, then $\mu_*(s) = \mu_1$. Hence, if $\mu_1 \approx \mu_2$, then $\mu_*(s) \approx \mu_1$
- In the case when $\mu_1 > \mu_2$, we have
  - $\mu_*(a-) = \frac{\mu_1+\mu_2}{2}$ by applying L'Hospital's Rule.
  - $\mu_*(s) = \mu_2 - \frac{1}{a-s}\ln\left(\frac{1}{2}(e^{(\mu_2-\mu_1)(a-s)} + 1)\right)$ for $0 < s < a$.
  - $\lim_{a\to\infty} \mu_*(0+) = \mu_2$ by applying L'Hospital's Rule.

This shows a beam-hardening effect in that the lower-energy photons tend to be absorbed more rapidly than higher-energy photons.

Figure 3.19 shows the violation of the linear assumption (3.5) due to beam-hardening. This can be explained by the mismatch between the data $\mathbf{P}^\ddagger$ and the range space of $\mathscr{R}$. In Fig. 3.19, we decompose the body into three regions; tissue region Ω_{tissue}, bone region Ω_{bone}, and metal region Ω_{metal}. Then, μ_E can be decomposed into

$$\mu_E(\mathbf{x}) = \mu_E^t(\mathbf{x}) + \mu_E^b(\mathbf{x}) + \mu_E^m(\mathbf{x}), \tag{3.41}$$

where $\mu_E^t = \mu_E \chi_{\Omega_{\text{tissue}}}$, $\mu_E^b = \mu_E^b \chi_{\Omega_{\text{bone}}}$, $\mu_E^m = \mu_E^m(\mathbf{x})\chi_{\Omega_{\text{metal}}}$. Given $\mathbf{P}^\ddagger$, let

$$(E_t, E_b, E_m) := \underset{(E_1,E_2,E_3)}{\text{argmin}} \|\mathscr{R}(\mu_{E_1}^t + \mu_{E_2}^b + \mu_{E_3}^m) - \mathbf{P}^\ddagger\|_{L^2((0,2\pi]\times\mathbb{R})}. \tag{3.42}$$

Let

$$\mu^* = \mu_{E_t}^t + \mu_{E_b}^b + \mu_{E_m}^m,$$

The mismatch can be written as

$$[\mathbf{P}^{\ddagger} - \mathscr{R}\mu^*](\varphi, s) = \Upsilon_{\Omega_{\text{tissue}}} + \Upsilon_{\Omega_{\text{bone}}} + \Upsilon_{\Omega_{\text{metal}}}, \tag{3.43}$$

where $\Upsilon_{\Omega_{\text{tissue}}}, \Upsilon_{\Omega_{\text{bone}}}, \Upsilon_{\Omega_{\text{metal}}}$ are mismatches induced mainly by the tissue region, the bone region, and metal region, respectively. Metal-induced mismatch can be written as

$$\Upsilon_{\Omega_{\text{metal}}} = -\ln\left(\int \eta(E) \exp\left\{-\mathscr{R}\left[\mu_E \chi_{\Omega_{\text{metal}}} - \mu^m_{E_m}\right](\varphi, s)\right\} dE\right). \tag{3.44}$$

The mismatch $\mathbf{P}^{\ddagger} - \mathscr{R}\mu^*$ is approximated by

$$[\mathbf{P}^{\ddagger} - \mathscr{R}\mu^*] \approx \beta_t \left(\mathscr{R}\chi_{\Omega_{\text{tissue}}}\right)^2 + \beta_b \left(\mathscr{R}\chi_{\Omega_{\text{bone}}}\right)^2 + \beta_m \left(\mathscr{R}\chi_{\Omega_{\text{metal}}}\right)^2, \tag{3.45}$$

where $\beta_t, \beta_b, \beta_t$ are optimal constants.

3.3.2 Effect of Sinogram Discrepancies

To illustrate the effect of sinogram discrepancies, let us consider the CT scan of a head containing a simulated metallic object. The projection data $\mathbf{P}^{\ddagger}$ are approximately consistent in all the regions except the diamond-shaped area $A_\diamond := \{(\varphi, s) : \mathscr{R}\chi_{D_1}(\varphi, s)\mathscr{R}\chi_{D_2}(\varphi, s) \neq 0\}$, where D_1, D_2 are two disks occupying the cross-sectional areas of the dental filling. A serious inconsistency exists in $\mathbf{P}^{\ddagger}$ in this small diamond-shaped region $A_\diamond$ due to beam-hardening as shown in Fig. 3.20. Due to the inherent nature of the pseudo inverse of the Radon transform $\mathscr{R}$, the local inconsistency of $\mathbf{P}^{\ddagger}$ on $A_\diamond$ generates severe global artifacts in $\mathscr{R}^{-1}\mathbf{P}^{\ddagger}$, which appear as streaking and shading artifacts. Denoting the back-projection by $\mathscr{R}^*$ (the dual of $\mathscr{R}$), its mathematical structure can be explained in Fig. 3.21.

This inconsistency is mapped to the streaking artifacts between the boundaries of two metallic objects in $\mathscr{R}^{-1}\mathbf{P}^{\ddagger}$. Moreover, the gradually varying inconsistency in $A_\diamond$ causes shading artifacts near or between the two metallic objects. It is important to note that the mismatch differs from the mathematical inconsistency $\mathbf{P}^{\ddagger}_\diamond := \mathscr{R}(\mathscr{R}^{-1}\mathbf{P}^{\ddagger}) - P$.

$\mathbf{P}^{\ddagger}_\diamond$ is not a practical mismatch to be corrected in MAR because $\mathscr{R}^{-1}(\mathbf{P}^{\ddagger} + \mathbf{P}^{\ddagger}_\diamond)$ equals $\mathscr{R}^{-1}\mathbf{P}^{\ddagger}$, which represents artifacts introduced by the discrepancy $P^\diamond$. We can briefly explain the reason as follows. Based on the inherent nature of the pseudo inverse $\mathscr{R}^{-1}$, $\mathscr{R}(\mathscr{R}^{-1}\mathbf{P}^{\ddagger})$ is the closest sinogram in the range space of $\mathscr{R}$ from P. According to Hilbert's projection theorem, the discrepancy $\mathbf{P}^{\ddagger}_\diamond$ is orthogonal to any sinogram in the range space, and therefore

$$\mathscr{R}^*\mathbf{P}^{\ddagger} = \mathscr{R}^*\left(\mathbf{P}^{\ddagger} + t\mathbf{P}^{\ddagger}_\diamond\right) \quad \text{and} \quad \mathscr{R}^{-1}\mathbf{P}^{\ddagger} = \mathscr{R}^{-1}(\mathbf{P}^{\ddagger} + t\mathbf{P}^{\ddagger}_\diamond) \quad \text{for all } t \in \mathbb{R}. \tag{3.46}$$

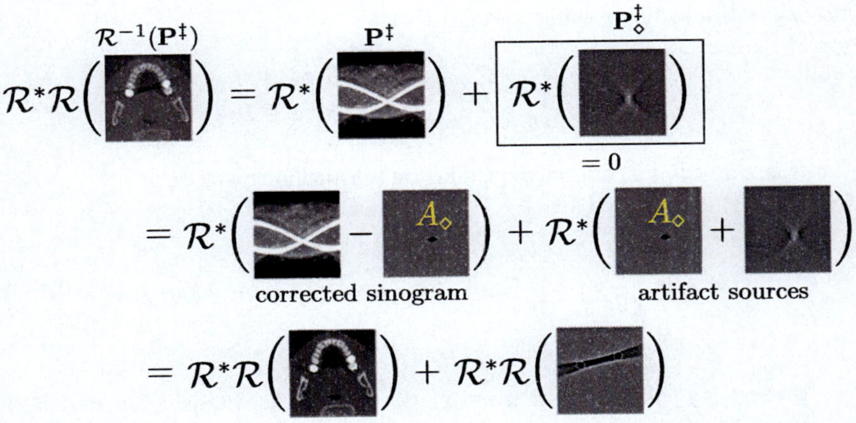

Fig. 3.20 Cause of metal artifact, which is inconsistency on the intersecting area ($A_\diamond$) of two metal traces

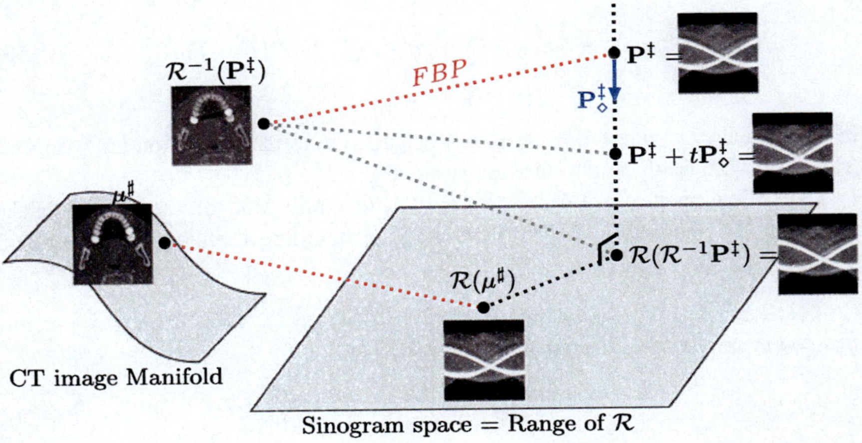

Fig. 3.21 The inherent nature of $\mathscr{R}^{-1}$ when the projection data $\mathbf{P}^\ddagger$ is not in the range space of $\mathscr{R}$

The discrepancy is mapped to the bright and dark streaking artifacts between the boundaries of any two metallic objects.

3.3.3 *Mathematical Analysis of Streaking Artifacts*

For a rigorous analysis and simplicity's sake, we assume the following [87] for a moment:

A1. $D = D_1 \cup D_2 \cup \cdots \cup D_J$ is a metal region, where $D_1, \cdots, D_J$ are disjoint simply connected smooth domains in $\mathbb{R}^2$.

A2. μ_E is differentiable with respect to E and

$$\left.\frac{\partial \mu_E}{\partial E}\right|_{E=E_0}(\mathbf{x}) = \begin{cases} 0 & \text{if } \mathbf{x} \notin D_j \\ \alpha \neq 0 & \text{if } \mathbf{x} \in D_j, \end{cases} \tag{3.47}$$

where $\alpha < 0$ is a constant depending on the metallic material.

A3. $\mathbf{P}^\ddagger(\varphi, s)$ is expressed as:

$$\mathbf{P}^\ddagger(\varphi, s) = -\ln\left(\frac{1}{2\epsilon}\int_{E_0-\epsilon}^{E_0+\epsilon} \exp\{-\mathscr{R}\mu_{E_0}(\varphi, s) - \alpha(E - E_0)\mathscr{R}\chi_D(\varphi, s)\}dE\right), \tag{3.48}$$

where ϵ is a positive constant and χ_D denotes the characteristic function of D; $\chi_D = 1$ in D and 0 otherwise.

From the second assumption [A2], we have

$$\mu_E(\mathbf{x}) = \mu_{E_0}(\mathbf{x}) + (E - E_0)\sum_{j=1}^{J} \alpha\chi_{D_j}(\mathbf{x}) + O(|E - E_0|^2), \tag{3.49}$$

In the special case when $\eta(E) = \frac{1}{2\epsilon}\chi_{[E_0-\epsilon, E_0+\epsilon]}$, the third assumption [A3] follows from [A2] with (3.49) and a sufficiently small ϵ.

The following proposition expresses the decomposition of the filtered back-projected CT image $\mathscr{R}^{-1}[\mathbf{P}^\ddagger]$ into the metal artifact-free term μ_{E_0} and the metal artifact term $\Upsilon_{\mathbf{P}^\ddagger}$.

Proposition 3.1 *[87] Under the assumptions [A1]–[A3], the* $\mathscr{R}^{-1}[\mathbf{P}^\ddagger]$ *(the reconstructed image using FBP) can be decomposed into*

$$\mathscr{R}^{-1}[\mathbf{P}^\ddagger](\mathbf{x}) = \mu_{E_0}(\mathbf{x}) + \Upsilon_{\mathbf{P}^\ddagger}(\mathbf{x}), \tag{3.50}$$

where $\Upsilon_{\mathbf{P}^\ddagger}$ *represents the metal artifact term given by*

$$\Upsilon_{\mathbf{P}^\ddagger}(\mathbf{x}) = -\frac{1}{8\pi^2}\int_{-\pi}^{\pi}\int_{-\infty}^{\infty} |\omega|\mathscr{F}\left[\ln\left(\frac{\sinh(\alpha\epsilon\mathscr{R}\chi_D(\varphi, \cdot))}{\alpha\epsilon\mathscr{R}\chi_D(\varphi, \cdot)}\right)\right](\omega)e^{i\omega\mathbf{x}\cdot\Theta_\varphi}d\omega d\varphi, \tag{3.51}$$

where $\ln\left(\frac{\sinh\beta}{\beta}\right)$ *is understood to be zero when* $\beta = 0$.

The artifact $\Upsilon_{\mathbf{P}^\ddagger}$ is generated by the mismatch:

$$\mathbf{P}^\ddagger_{\text{MA}}(\varphi, s) := \begin{cases} -\ln\left(\frac{\sinh(\alpha\epsilon\mathscr{R}\chi_D(\varphi,s))}{\alpha\epsilon\mathscr{R}\chi_D(\varphi,s)}\right) & \text{if } (\varphi, s) \in G, \\ 0 & \text{if } (\varphi, s) \notin G, \end{cases} \tag{3.52}$$

where $G := \{(\varphi, s) \in (-\pi, \pi] \times \mathbb{R} \ : \ \mathscr{R}\chi_D(\varphi, s) \neq 0\}$. It follows from Taylor's expansion that

$$\mathbf{P}^{\ddagger}_{\mathrm{MA}} = -\ln M + \sum_{k=1}^{\infty} \frac{(-1)^k}{k} \frac{1}{M^k} \left[\frac{\sinh(\alpha\epsilon\mathscr{R}\chi_D)}{\alpha\epsilon\mathscr{R}\chi_D} - M \right]^k \tag{3.53}$$

where M is the positive number given by

$$M := \left\| \frac{\sinh(\alpha\epsilon\mathscr{R}\chi_D)}{\alpha\epsilon\mathscr{R}\chi_D} \right\|_{L^\infty(G)}. \tag{3.54}$$

Note that $M \geq 1$. We define $\mathbf{P}^{\ddagger}_{\mathrm{MA},N}$ as

$$\mathbf{P}^{\ddagger}_{\mathrm{MA},N}(\varphi, s) := \begin{cases} -\ln M + \sum_{k=1}^{N} \frac{(-1)^k}{k} \left[\frac{h_N(\varphi,s)}{M} \right]^k & \text{if } (\varphi, s) \in G, \\ 0 & \text{if } (\varphi, s) \notin G. \end{cases} \tag{3.55}$$

where h_N is given by

$$h_N(\varphi, s) = \sum_{n=0}^{N} \frac{(\alpha\epsilon)^{2n}}{(2n+1)!} (\mathscr{R}\chi_D(\varphi, s))^{2n} - M. \tag{3.56}$$

$$\begin{aligned} \mathbf{P}^{\ddagger}(\varphi, s) &= -\ln \left(\frac{1}{2\epsilon} \int_{E_0-\epsilon}^{E_0+\epsilon} \exp\{-\alpha(E - E_0)\mathscr{R}\chi_D(\varphi, s)\} dE \right) \\ &= -\ln \left(\frac{1}{2\epsilon} \int_{-\epsilon}^{\epsilon} \exp\{-\alpha E \mathscr{R}\chi_D(\varphi, s)\} dE \right) \\ &= -\ln \left(\frac{1}{2} \int_{-1}^{1} \exp\{-\alpha\epsilon t \mathscr{R}\chi_D(\varphi, s)\} dt \right) \\ &\approx \beta(\mathscr{R}\chi_D)^2(\varphi, s). \end{aligned} \tag{3.57}$$

3.3.3.1 Analysis of Streaking Artifacts Using Wavefront Set

This subsection explains rigorous mathematical analysis of streaking artifacts using a wavefront set. This material can be skipped by anyone who does not focus on rigorous mathematical analysis.

The first mathematical analysis to characterize the structure of metal streaking artifacts was given in the paper [87]. Based on this mathematical analysis, the authors first found the mathematical formula (3.51) for the beam-hardening metal artifacts, which has been experimentally validated using industrial CT [88].

In the paper [87], the metal artifacts are viewed as the singularities in an image, which are closely related to the interrelation between the structure of the data $\mathbf{P}^{\ddagger}$ and the FBP $\mathscr{R}^{-1}[\mathbf{P}^{\ddagger}]$. This can be interpreted effectively using the Fourier integral operator and the wavefront set [22, 50, 51, 94, 122].

The wavefront set is a useful tool to describe simultaneously the locations and orientations of singularities.

Definition 3.1 Let $f \in \mathscr{D}'(\mathbb{R}^2)$, $g \in \mathscr{E}'(\mathbb{R}^2)$, and $\mathbf{x} \in \mathbb{R}^2$.

1. The singular support of f, denoted as sing-supp(f), is the smallest closed subset in $\mathbb{R}^2$ outside of which f is C^∞.
2. $\Sigma(g)$ is the smallest closed conic subset of $\mathbb{R}^2 \setminus \{\mathbf{0}\}$ such that

$$\emptyset = \Sigma(g) \cap \{\boldsymbol{\xi} : \exists \text{conic nbd } V \text{ of } \boldsymbol{\xi} \text{ s.t. } \sup_{\boldsymbol{\xi}' \in V} (1 + |\boldsymbol{\xi}'|)^N |\mathscr{F}[g](\boldsymbol{\xi}')| < \infty \,, N \in \mathbb{Z}^+\}. \tag{3.58}$$

   Note that $V \subseteq \mathbb{R}^2 \setminus \{\mathbf{0}\}$ is called a conic set if $r\boldsymbol{\xi} \in V$ whenever $r > 0$ and $\boldsymbol{\xi} \in V$. If V is an open conic set which contains $\boldsymbol{\xi} \neq \mathbf{0}$, we say that it is a conic neighborhood of $\boldsymbol{\xi}$.
3. For $\mathbf{x} \in \mathbb{R}^2$, $\Sigma_{\mathbf{x}}(f)$ is a closed conic subset in $\mathbb{R}^2 \setminus \{\mathbf{0}\}$ defined as

$$\Sigma_{\mathbf{x}}(f) = \bigcap \{\Sigma(\eta f) : \eta \in C_c^\infty(\mathbb{R}^2),\ \eta(\mathbf{x}) \neq 0\}. \tag{3.59}$$

4. The wavefront set of f, denoted as WF(f), is a closed conic subset in $\mathbb{R}^2 \times (\mathbb{R}^2 \setminus \{\mathbf{0}\})$ defined as

$$\mathrm{WF}(f) = \{(\mathbf{x}, \boldsymbol{\xi}) \in \mathbb{R}^2 \times (\mathbb{R}^2 \setminus \{\mathbf{0}\}) : \boldsymbol{\xi} \in \Sigma_{\mathbf{x}}(f)\}. \tag{3.60}$$

Note that for $g \in \mathscr{E}'(\mathbb{R}^2)$, $g \in C_0^\infty(\mathbb{R}^2)$ if and only if $\Sigma(g) = \emptyset$. This means that

$$\text{sing-supp}(f) = \{\mathbf{x} \in \mathbb{R}^2 : \Sigma_{\mathbf{x}}(f) \neq \emptyset\} \quad \text{for} \quad f \in \mathscr{D}'(\mathbb{R}^2). \tag{3.61}$$

If $(\mathbf{x}, \boldsymbol{\xi}) \in \mathrm{WF}(f)$, then $\boldsymbol{\xi} \in \Sigma(f)$ for $f \in \mathscr{E}'(\mathbb{R}^2)$ [51].

Definition 3.2 A straight line $L_{\varphi,s} = \{\mathbf{x} = s(\cos\varphi, \sin\varphi) + t(-\sin\varphi, \cos\varphi) : t \in \mathbb{R}\}$ is called a *streaking artifact of* $\mathscr{R}^{-1}[\mathbf{P}^{\ddagger}]$ in the sense of a wavefront set if it satisfies

$$\Sigma_{\mathbf{x}}(\mathscr{R}^{-1}[\mathbf{P}^{\ddagger}]) \neq \emptyset \quad \text{for all } \mathbf{x} \in L_{\varphi,s} \setminus \text{sing-supp}(\mu_{E_0}). \tag{3.62}$$

There are various works about the wavefront set for Radon transforms [26, 32, 33, 35, 62, 63, 97–101, 104, 105] and metal artifacts [1, 7, 17, 25, 61, 68, 79, 90, 118, 128, 132].

Theorem 3.1 *Let $D_1, D_2, \ldots, D_J$ be strictly convex and disjoint bounded domains in $\mathbb{R}^2$ with connected boundaries of the class C^∞. Let $D = \cup_{j=1}^J D_j$ be the metal region. Given $\mathbf{P}^\ddagger$, assume that $\mathscr{R}^{-1}[\mathbf{P}^\ddagger]$ is represented as*

$$\mathscr{R}^{-1}[\mathbf{P}^\ddagger](\mathbf{x}) = \mu_{E_0}(\mathbf{x}) + \Upsilon_{\mathbf{P}^\ddagger}(\mathbf{x}), \tag{3.63}$$

where

$$\Upsilon_{\mathbf{P}^\ddagger} = \frac{1}{4\pi}\mathscr{R}^*\mathscr{I}^{-1}\left[\sum_{k=1}^{N}\frac{(-1)^k}{k}\left[\sum_{n=1}^{N}\frac{(\alpha\epsilon)^{2n}}{(2n+1)!}(\mathscr{R}\chi_D)^{2n}\right]^k\right]. \tag{3.64}$$

If a line $L_{\varphi,s}$ is a streaking artifact of $\mathscr{R}^{-1}[\mathbf{P}^\ddagger]$ in the sense of wavefront set (3.62), then (φ, s) satisfies

$$\dim\left(\mathrm{Span}[\Sigma_{(\varphi,s)}(\mathscr{R}\chi_D)]\right) = 2, \tag{3.65}$$

where $\dim(\mathrm{Span}[A])$ is the dimension of the span of the set A.

The wavefront set of $\Upsilon_{\mathbf{P}^\ddagger}$ satisfies

$$\mathrm{WF}(\Upsilon_{\mathbf{P}^\ddagger}) \subseteq \bigcup_{k=1}^{N^2}\mathrm{WF}\left(\mathscr{R}^*\mathscr{I}^{-1}(\mathscr{R}\chi_D)^{2k}\right). \tag{3.66}$$

Since $\mathscr{I}^{-1}$ is an elliptic pseudodifferential operator [94, 122],

$$\mathrm{WF}(\Upsilon_{\mathbf{P}^\ddagger}) \subseteq \bigcup_{k=1}^{N^2}\mathrm{WF}\left(\mathscr{R}^*(\mathscr{R}\chi_D)^{2k}\right). \tag{3.67}$$

The wavefront set $\mathrm{WF}(\mathscr{R}\chi_D)$ can be decomposed into

$$\mathrm{WF}(\mathscr{R}\chi_D) = \mathrm{WF}_1(\mathscr{R}\chi_D) \cup \mathrm{WF}_2(\mathscr{R}\chi_D), \tag{3.68}$$

where

$$\mathrm{WF}_k(\mathscr{R}\chi_D) = \left\{((\varphi, s), \eta) \in \mathrm{WF}(\mathscr{R}\chi_D) : \dim(\mathrm{Span}[\Sigma_{(\varphi,s)}(\mathscr{R}\chi_D)]) = k, k \in \{1, 2\}\right\} \tag{3.69}$$

If $\mathrm{WF}_2(\mathscr{R}\chi_D) = \emptyset$, the reconstructed image $\mathscr{R}^{-1}[\mathbf{P}^\ddagger]$ does not have streaking artifacts in the sense of a wavefront set because

$$\mathrm{WF}(\Upsilon_{\mathbf{P}^\ddagger}) \subseteq \bigcup_{k=1}^{N^2}\mathrm{WF}\left(\mathscr{R}^*(\mathscr{R}\chi_D)^{2k}\right) \subseteq \mathrm{WF}(\chi_D). \tag{3.70}$$

If $L_{\varphi,s} \subseteq \text{sing-supp}(\mathscr{R}^{-1}[\mathbf{P}^{\ddagger}])$ (a streaking artifact), then for some $k = 1, \cdots, N^2$, we have

$$((\varphi, s), (-t, 1)) \in \text{WF}((\mathscr{R}\chi_D)^{2k}) \quad \text{for all} \quad t \in \mathbb{R}, \tag{3.71}$$

which is possible only when $\dim(\text{Span}[\Sigma_{(\varphi,s)}(\mathscr{R}\chi_D)]) = 2$.

Theorem 3.2 *Let $D \subseteq \mathbb{R}^2$ denote a metal region with the connected C^∞ boundary ∂D. If D is strictly convex, then the CT image $\mathscr{R}^{-1}[\mathbf{P}^{\ddagger}]$ does not have the streaking artifacts in the sense of wavefront set*

$$\text{WF}(\mathscr{R}^{-1}[\mathbf{P}^{\ddagger}]) \subseteq \text{WF}(\mu_{E_0}). \tag{3.72}$$

3.3.3.2 Metal Induced Beam-Hardening Corrector

Recently, Park et al.[88] developed a novel geometric corrector, which can handle the inconsistency of projection data $\mathbf{P}^{\ddagger}$. The geometric corrector is a function of metal geometries and a control parameter associated with all energy-dependent factors including attenuation coefficients and the spectrum of the X-ray source. The novelty of the geometric corrector is to extract the metal-induced streaking and shadow artifacts selectively without affecting intact anatomical images when prior knowledge about the shapes of metallic object is available.

The key observation is the decomposition of the projection data $\mathbf{P}^{\ddagger}$ given by

$$\mathbf{P}(\varphi, s) = \underbrace{\mathscr{R}\mu^{\ddagger}(\varphi, s)}_{\text{target}} + \underbrace{\ln\left(\frac{\sinh(\lambda \mathscr{R}\chi_D(\varphi, s))}{\lambda \mathscr{R}\chi_D(\varphi, s)}\right)}_{\text{model mismatch}}, \tag{3.73}$$

where D is the metal region and λ is a constant depending on the energy spectrum of the X-ray beam and absorption property of the subject. The reconstructed artifact image $\phi_{D,\lambda}(\mathbf{x})$ represented by

$$\phi_{D,\lambda}(\mathbf{x}) = -\frac{1}{8\pi^2}\int_{-\pi}^{\pi}\int_{-\infty}^{\infty} |\omega| \mathscr{F}\left[\ln\left(\frac{\sinh(\lambda \mathscr{R}\chi_D(\varphi, \cdot))}{\lambda \mathscr{R}\chi_D(\varphi, \cdot)}\right)\right](\omega) e^{i\omega \mathbf{x}\cdot\Theta_\varphi} d\omega d\varphi, \tag{3.74}$$

where $\ln\left(\frac{\sinh\beta}{\beta}\right)$ is understood to be zero when $\beta = 0$.

To illustrate the effect of sinogram discrepancies and to demonstrate the validity of the corresponding mathematical theories, both real and numerical experiments are performed in the same environment and compared with each other.

$\mathbf{P}^{\ddagger}$ is generated using a human skull phantom with three cylinders filled with high-attenuating fluid (iodinated contrast media (Pamiray 370, iopamidol 370 mgI/mL; Dongkook Pharma, Seoul, Korea) with saline dilution), as shown in Fig. 3.22. The

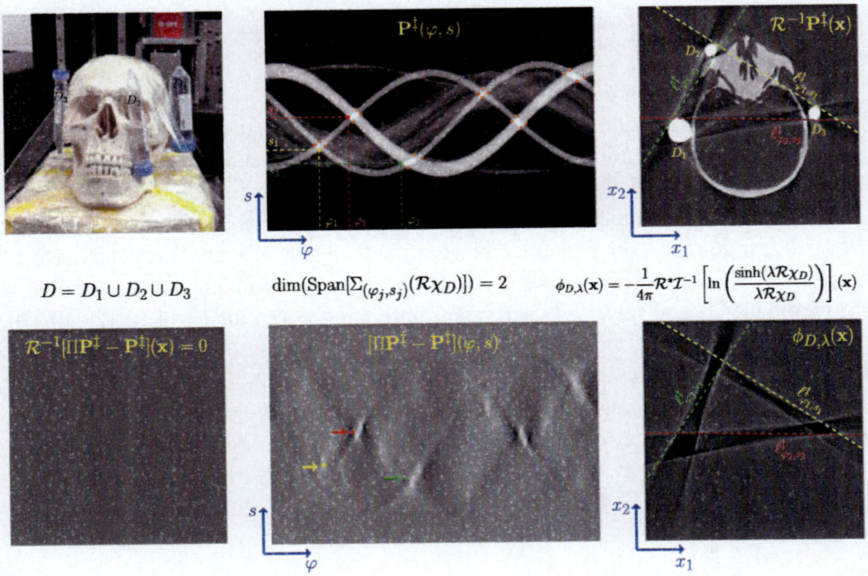

Fig. 3.22 Metal-induced beam-hardening corrector, which can handle the inconsistency of projection data. The corrector is a function of metal geometries $R\chi_D$ and a parameter associated with energy dependent factor λ

data $\mathbf{P}^{\ddagger}$ was acquired from the midplane in CBCT at a tube voltage of 120 kV and tube current of 1.3 mA with 2 mm Al filtration.

This $\mathbf{P}^{\ddagger}$ is approximately consistent in all the regions except the diamond-shaped area

$$A_{\diamond} := \{(\varphi, s) : \mathscr{R}\chi_{D_1}(\varphi, s)\mathscr{R}\chi_{D_2}(\varphi, s)\mathscr{R}\chi_{D_3}(\varphi, s) \neq 0\}, \tag{3.75}$$

where D_1, D_2, D_3 are three disks occupying the cross-sectional areas of three cylinders filled with high-attenuating fluid. This local inconsistency of $\mathbf{P}^{\ddagger}$ on $A_{\diamond}$ corresponds to a beam-hardening factor on the interval $A_{\diamond,1} := \{\varphi : (\varphi, s) \in A_{\diamond}\}$. From the inherent nature of the pseudo inverse of the Radon transform $\mathscr{R}$, the local inconsistency of $\mathbf{P}^{\ddagger}$ on $A_{\diamond}$ generates severe global artifacts in $\mathscr{R}^{-1}(\mathbf{P}^{\ddagger})$, which appear as streaking and shading artifacts. In the reconstructed image $\mathscr{R}^{-1}(\mathbf{P}^{\ddagger})$ by FBP, cupping artefacts only disturb the image in the regions of problematic objects, whereas streaking artefacts corrupt the tomographic image outside the problematic region [87, 89].

3.3.4 Data Consistency Conditions for CBCT

This section discusses data consistency conditions (DCCs) for CBCT, which are mathematical descriptions explaining data redundancy that exists between CBCT projections. DCCs are tools for characterizing projection data on the range space of the CT forward model associated with the Radon transform. In MAR applications, DCCs can be used to deal with the metal-induced inconsistency, which makes **P** not lie on the range space. For example, metal artifacts can be mitigated in the direction of imposing DCCs with a proper correction model (e.g., metal-induced beam-hardening corrector (3.73) and physics-based beam-hardening corrector (3.112)) [2, 10, 70, 131].

There have been extensive studies on DCCs for CBCT [2, 3, 21, 21, 34, 96, 130], but there is still a gap regarding DCCs for dealing with the practical dental CBCT environment that uses a single circular scan trajectory, local-ROI-oriented scanning, and offset detector array. We focus on two DCCs [21, 36], which might be a basis for understanding DCCs for practical dental CBCT.

3.3.4.1 Helgason–Ludwig Consistency-Based Data Consistency Condition for Cone-Beam Projections

We begin with brief review of the Helgason–Ludwig consistency condition (HLCC) [74], which is a well-known DCC for the parallel-beam CT. HLCC refers to the following theorem.

Theorem 3.3 ([20, 84]) *Three statements are equivalent.*

- $\mathbf{P}^{\ddagger} = \mathscr{R}\mu$ *for some* μ.
- *For a non-negative integer n and some* $\{A_k\}$,

$$\int_{\mathbb{R}} \mathbf{P}^{\ddagger}(\varphi, s)s^n ds = \sum_{k=0}^{n} A_k \cos^{n-k} \varphi \sin^k \varphi$$

.

- *For non-negative integers n and k satisfying that* $k - n$ *is odd and* $k > n$,

$$\int_0^{2\pi} e^{ik\varphi} \int_{\mathbb{R}} \mathbf{P}^{\ddagger}(\varphi, s)s^n ds d\varphi = 0$$

.

The proof can be found in [84]. Let us observe the case of $n = 0$:

$$\frac{\partial}{\partial\varphi} \int_{\mathbb{R}} \mathbf{P}^{\ddagger}(\varphi, s)ds = \frac{\partial}{\partial\varphi} \int_{\mathbb{R}} \int_{\mathbb{R}} \mu(s\Theta_\varphi + t\Theta_\varphi^{\perp})dtds = \frac{\partial}{\partial\varphi}(\int_{\mathbb{R}^2} \mu(\mathbf{x})d\mathbf{x}) = 0. \tag{3.76}$$

This corresponds to the zero-th order HLCC (i.e., $\int_{\mathbb{R}} \mathbf{P}^{\ddagger}(\varphi, s)ds$ is a constant with respect to φ if $\mathbf{P}^{\ddagger}$ is consistent).

The study [21] extended HLCC into the case of CBCT with the circular source trajectory. The consistency condition can be represented by the following theorem.

Theorem 3.4 ([21]) *If projection data* $\mathbf{P}$ *is represented by a cone-beam projection of some* μ, *i.e.,*

$$P(\beta, u, v) = \int_{\mathbb{R}^+} \mu(R\Theta_\beta + t\gamma_{\beta,u,v})dt, \tag{3.77}$$

the following equality holds. For a nonnegative integer n *and some* $\{A_k\}$,

$$\int_{\mathbb{R}}\int_{\mathbb{R}} \frac{D}{\sqrt{u^2+v^2+D^2}} P(\beta, u, v)\frac{u^n}{v^{n+2}} dudv = \sum_{k=0}^{n} A_k \cos^{n-k}\beta \sin^k \beta, \tag{3.78}$$

where $\gamma_{\beta,u,v} = (u\Theta_\beta^\perp + ve_3 - D\Theta_\beta)/\sqrt{u^2+v^2+D^2}$. *Here,* e_3 *is the unit vector in the direction of the* z *axis, and* D *is the distance between the source and detector at the* $z = 0$ *plane.*

The proof is given as follows. Let ζ be $\zeta = \sqrt{u^2+v^2+D^2}$.

$$\begin{aligned}
&\int_{\mathbb{R}}\int_{\mathbb{R}} \frac{D}{\zeta} \int_{\mathbb{R}^+} \mu(R\Theta_\beta + t\gamma_{\beta,u,v})dt \frac{u^n}{v^{n+2}} dudv && (3.79)\\
&= \int_{\mathbb{R}}\int_{\mathbb{R}}\int_{\mathbb{R}^+} \mu(R\Theta_\beta + t\gamma_{\beta,u,v})dt (\frac{\zeta}{v})^{n+2}(\frac{u}{\zeta})^n \frac{D}{\zeta^3} dudv && (3.80)\\
&= \int_{S^2}\int_{\mathbb{R}^+} \mu(R\Theta_\beta + t\gamma_{\beta,u,v}) \frac{(\gamma_{\beta,u,v}\cdot\Theta_\beta^\perp)^n}{(\gamma_{\beta,u,v}\cdot e_3)^{n+2}} dtd\gamma && (3.81)\\
&= \int_{\mathbb{R}^3} \mu(\mathbf{x}) \frac{(\frac{1}{t}(\mathbf{x}-R\Theta_\beta)\cdot\Theta_\beta^\perp)^n}{(\frac{1}{t}(\mathbf{x}-R\Theta_\beta)\cdot e_3)^{n+2}} \frac{1}{t^2} d\mathbf{x} = \int_{\mathbb{R}^3} \mu(\mathbf{x}) \frac{(\mathbf{x}\cdot\Theta_\beta^\perp)^n}{(\mathbf{x}\cdot e_3)^{n+2}} d\mathbf{x} && (3.82)\\
&= \int_{\mathbb{R}^3} \mu(\mathbf{x}) \frac{1}{(x_3)^{n+2}} \underbrace{(-x_1\sin\beta + x_2\cos\beta)^n}_{n\text{-th order polynomial}} d\mathbf{x}. && (3.83)
\end{aligned}$$

where S^2 is the unit 2-sphere. Here, we use a change of variables given by

$$d\gamma = \frac{D}{\zeta^3} dudv, \; d\mathbf{x} = t^2 dtd\gamma, \tag{3.84}$$

These relations can be derived from the change of variables from the spherical to the Cartesian coordinate system. To be precise, the first equality comes from

$$d\gamma = \cos\phi d\theta d\phi = \frac{\sqrt{D^2+v^2}}{\zeta}\frac{\partial(\phi,\theta)}{\partial(u,v)}dudv \tag{3.85}$$

$$= \frac{\sqrt{D^2+v^2}}{\zeta}\det(\begin{bmatrix} \frac{\sqrt{D^2+v^2}}{\zeta^2} & \frac{\partial\phi}{\partial v} \\ 0 & \frac{D}{D^2+v^2} \end{bmatrix})dudv \tag{3.86}$$

$$= \frac{D}{\zeta^3}dudv, \tag{3.87}$$

where

$$\phi = \tan^{-1}\frac{u}{\sqrt{D^2+v^2}}, \theta = \tan^{-1}\frac{v}{D}. \tag{3.88}$$

The second equality simply comes from

$$d\mathbf{x} = t^2\cos\phi dt d\theta d\phi = t^2 dt(\cos\phi d\theta d\phi) = t^2 dt d\gamma \tag{3.89}$$

In practical implementation, the singularity at $v = 0$ should be addressed to evaluate the DCC (3.78). In the distribution sense, the function $1/(v^{n+2})$ including the singularity can be replaced by

$$h_n(v) = \int_{\mathbb{R}} \frac{(-i)^{n+2}}{2(n+1)!}|\sigma|\sigma^n e^{i\sigma v}d\sigma. \tag{3.90}$$

Here, we note that $h_n(\lambda v) = h_n(v)/\lambda^{n+2}$ for any non-zero constant λ, which implies a homogeneous polynomial with degree $-(n+2)$. Figure 3.23 shows implementation results for testing the DCC.

3.3.4.2 Grangeat-Based Consistency Condition

This section deals with a DCC that is based on Grangeat's CBCT reconstruction formula via the first derivative of the 3D Radon transform. This DCC covers the case of CBCT with a general source trajectory.

The Grangeat-based DCC can be expressed as follows.

Theorem 3.5 ([39, 72]) *If projection data* $\mathbf{P}$ *is represented by a cone-beam projection of some* μ, *i.e.,*

$$\mathbf{P}(\beta, u, v) = \int_{\mathbb{R}+} \mu(s_\beta + t\gamma_{\beta,u,v})dt, \tag{3.91}$$

we have the following equality: For any β_1 *and* β_2 *satisfying* $(\mathbf{s}_{\beta_1} - \mathbf{s}_{\beta_2})\cdot\mathbf{n}_\nu = 0$,

$$\frac{\partial}{\partial v}\int_{\mathbb{R}} \frac{R\mathbf{P}(\beta_1, u, v)}{\sqrt{R^2+u^2+v^2}}du\Big|_{v=R\tan\nu} = \frac{\partial}{\partial v}\int_{\mathbb{R}} \frac{R\mathbf{P}(\beta_2, u, v)}{\sqrt{R^2+u^2+v^2}}du\Big|_{v=R\tan\nu}, \tag{3.92}$$

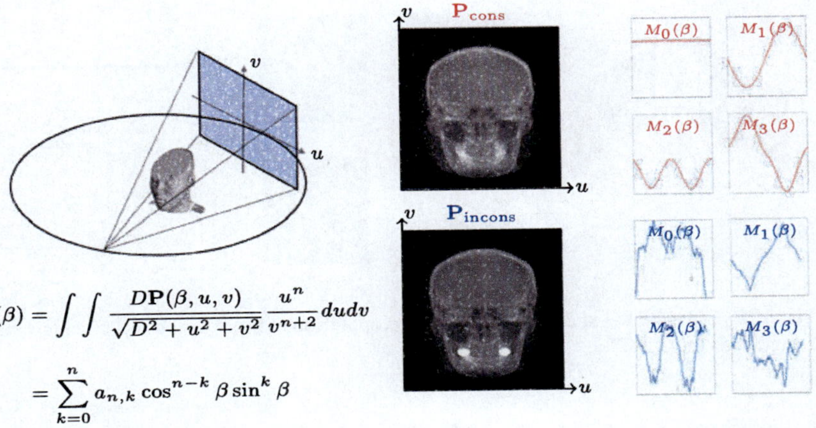

Fig. 3.23 Helgason–Ludwig consistency-based data consistency condition application for projections acquired in a circular source trajectory CBCT. The data consistency condition is checked by using consistent and inconsistent sinogram data ($\mathbf{P}_{\mathrm{cons}}$ and $\mathbf{P}_{\mathrm{incons}}$)

Fig. 3.24 Detector coordinate system of CBCT. s_β represents the source position, ν is angle between the integration plane and $\nu = 0$ plane and $\mathbf{n}_\nu$ is the normal vector of integration plane

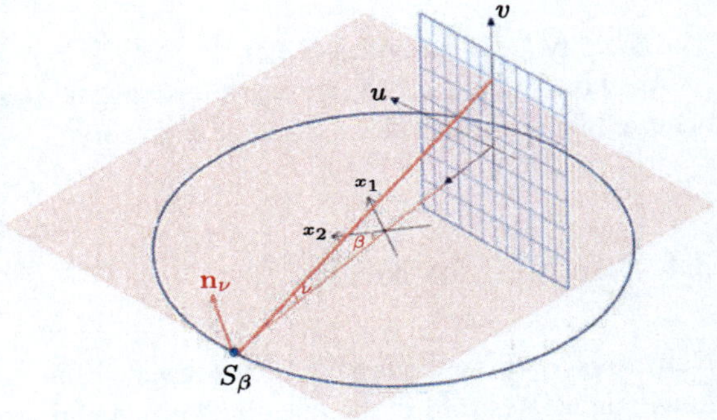

where $\mathbf{s}_\beta$ is a source position for $\beta \in \Lambda$ and $\mathbf{n}_\nu = (0, \sin\nu, \cos\nu)$ is a normal vector characterizing the plane of the incident beam with an angle of ν, as shown in Fig. 3.24. Here, Λ is a curve of the source trajectory.

The proof follows from the fundamental relation of the 3D Radon transform [39] given by

$$\frac{\partial}{\partial\rho} \int_{\Pi_{\rho,\mathbf{n}_\nu}} \mu dS \Bigg|_{\rho=\mathbf{s}_\beta\cdot\mathbf{n}_\nu} = \frac{1}{\cos^2\nu} \frac{\partial}{\partial v} \int_{\mathbb{R}} \frac{R\mathbf{P}(\beta,u,v)}{\sqrt{R^2+u^2+v^2}} du \Bigg|_{v=R\tan\nu}. \tag{3.93}$$

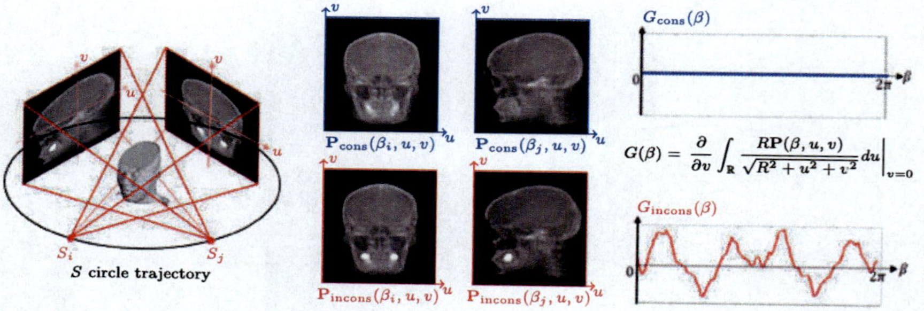

Fig. 3.25 Grangeat-based consistency condition application for projections acquired in a circular source trajectory CBCT

See [39, 72] for the proof of (3.93). Since $\Theta_{\beta_1} \cdot \mathbf{n}_\nu = \Theta_{\beta_2} \cdot \mathbf{n}_\nu$,

$$\frac{\partial}{\partial \rho} \int_{\Pi_{\rho,\mathbf{n}}} \mu dS \Big|_{\rho=\mathbf{s}_{\beta_1} \cdot \mathbf{n}_\nu} = \frac{\partial}{\partial \rho} \int_{\Pi_{\rho,\mathbf{n}}} \mu dS \Big|_{\rho=\mathbf{s}_{\beta_2} \cdot \mathbf{n}_\nu} . \tag{3.94}$$

The equality (3.94) provides (3.92).

As shown in Fig. 3.25, we test the Grangeat-based DCC with sinogram data acquired from a single circular source trajectory.

3.4 Methods for MAR

Numerous MAR methods have been suggested since the first method introduced by Lewitt and Bates [68] in the late 1970s. As mentioned in the previous section, the presence of a metal implant in the imaging domain increases the inconsistency of the projection data, mainly due to the beam-hardening. (As the polychromatic X-ray beam passes through the metal implant, more low-energy photons are absorbed, increasing the average beam energy.) This beam-hardened projection data generates streaking artifacts, and in addition, metal-induced artifacts are generated by interacting with other factors including scattering, noise, and partial volume effects.

The problem of MAR is nonlinear, because metal artifacts depend nonlinearly on the size, geometry, and material of the metal. We should note that metal artifacts are generated not only by the mismatch between the real projection data and mathematical model but also by the orthogonal projection process of the projection data to the range space of the Radon transform during CT reconstruction.

In dental CBCT, the desired MAR method should address the problems caused by "*Offset detector, FOV truncation, Low X-ray dose*". The FOV size in dental CBCT is usually small as compared to the size of a patient's head because a small detector is employed to reduce system costs. The small detector size leads to a small area of the scanner FOV, which causes the patient's head to be missing from the projection data in the transversal direction. This incomplete projection data can be combined with beam hardening of the teeth, creating streaked artifacts around the teeth even in the absence of metal. Photon starvation is very common in dental low-dose X-ray CBCT, especially when the patient has many implants.

One of the state-of-the-art MAR techniques is dual energy CT [5, 69, 125], which is designed to provide synthesized virtual monochromatic images. It uses two separate polychromatic beam spectra to generate nearly monoenergetic data to alleviate beam-hardening effects. However, dual energy CT is not appropriate for low-dose dental CT, because the use of an additional scan with higher energy increases radiation exposure to patients. Hence, we will not deal with dual-energy CT in this section. For similar reasons, we will not be dealing with photon counting detectors that can provide monochromatic data by differentiating individual photon energies. This technique is not suitable for low-dose dental CT, because low-dose exposure lowers the signal-to-noise ratio.

3.4.1 Conventional MAR Methods

Conventional MAR methods can be classified roughly into the following three categories:

- **Raw data correction approach**: Unreliable background data due to the presence of metallic objects can be recovered using various inpainting techniques.
  - Interpolation: Aims to obtain the corrected sinogram data by recovering the metal trace from the raw data, which has been affected by metallic objects. Figure 3.26 shows the overall procedure of the linear interpolation method.
  - Reprojection: This method involves identifying the metal objects in an original reconstructed image and filling those regions with the CT number of water. The image is then reprojected and subtracted from the original image projections. The corrected sinogram is reconstructed to obtain the final image.
  - Normalization: This technique uses the X-ray path length through a comparable cross-section for data normalization to ensure that the sinogram becomes comparatively flat, which enables straightforward interpolation over corrupted metal traces.

  Drawbacks: The above-mentioned methods might introduce new artifacts that did not previously exist. Moreover, these techniques tend to impair the morphological information in the areas around the metal objects in the reconstructed images.
- **Iterative reconstruction method**: An iterative reconstruction method generates an initial estimate of the target image, projects the guessed/estimated image back to the raw data space, compares it with the original raw data to generate a modified image, and repeats the process. Various iterative reconstruction methods have been developed for MAR.

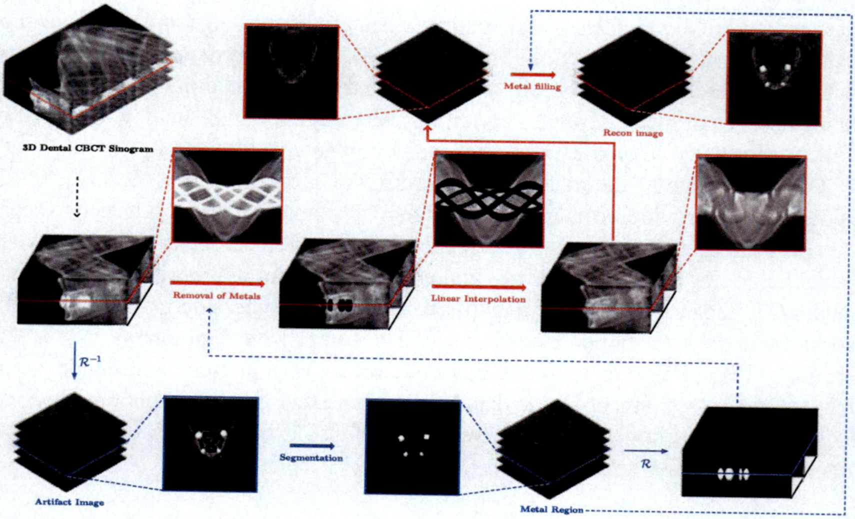

Fig. 3.26 Schematic diagram of linear interpolation method

- Corruption avoidance: This treats the MAR problem as an exterior problem, and uses the data outside the metal trace to arrive at a reconstruction result.
- Statistical compensation: Instead of explicitly omitting the metal-affected data, a statistical objective function that down-weights the data containing the metal object could be employed.
- Knowledge utilization: Iterative reconstruction methods involving incomplete data can be improved using sparse techniques such as compressed sensing (total variation) and known component models.

Drawbacks: An iterative reconstruction approach requires extensive knowledge about the CT system configuration, even though it can theoretically deal with the streaking problems very well. Additionally, the associated computation time for full iterative reconstructions can be clinically prohibitive.

- **Hybrid methods**: These combine the above two methods.

3.4.1.1 Raw Data Correction Approach

In raw data correction methods, unreliable data on the metal trace in Radon space is amended or replaced using various filling-in techniques such as interpolation [1, 9, 61, 68, 109] and sinogram inpainting [7, 78, 90, 132, 133]. Here, the metal trace is the region of the projection data that is affected by metal. In the case when data on the metal trace are completely corrupt (e.g. photon starvation), data corrections are performed either by neighboring projections or by using mathematical models [37].

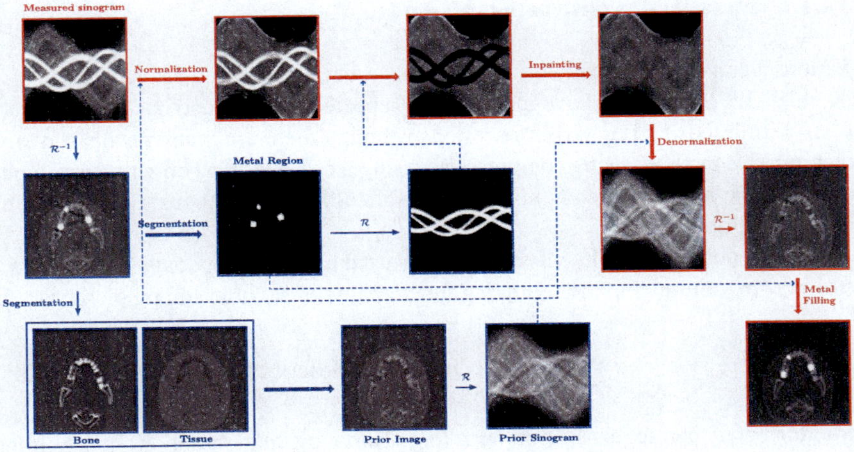

Fig. 3.27 Schematic diagram of normalized metal artifact reduction method

Helgason–Ludwig coherence conditions [48, 74] can be used to address projection data incompleteness.

We briefly explain the normalization MAR(NMAR) [79], that is designed to handle streaking artifacts caused by the non-smooth transition problem between the original and interpolated data. The main idea of NMAR is to convert the sinogram to a nearly flat sinogram and then perform interpolation on this plat sinogram for a smooth transition. The rough process of NMAR is as follows. See Fig. 3.27.

1. Given the original raw data, compute the uncorrected CT image using FBP.
2. Segment the metal region by a thresholding operation.
3. Compute a prior image by segmenting the uncorrected image (or pre-corrected image) into air, tissue, bone through multi-thresholds.
4. Apply the forward projection of the prior image and the metal region to get the corresponding sinogram and the metal trace.
5. Normalization step: Divide the original data by the projection data of the prior image.
6. Interpolation step: Apply an interpolation to the normalized projection data, in order to repair the unreliable data on the metal trace.
7. Denormalization step: Multiply the interpolated sinogram by the projection data of the prior image.
8. Compute the corrected image by applying FBP to the denormalized sinogram, and then the metal is inserted back into the image.

The performance of the NMAR method depends on the accuracy of object segmentation. In the case when metal artifacts are very strong, some image pixels are classified as the wrong tissue type, resulting in an inaccurate prior image. This is a downside because the segmentation by thresholds is not perfectly accurate.

3.4.1.2 Iterative Reconstruction Methods

Various iterative reconstruction methods have been developed for MAR [25, 30, 80, 119, 129]. These iterative methods use repeated forward and backward projections with imposing regularization prior. Although there are many results showing that iterative methods using compressed sensing techniques [16] effectively mitigate artifacts, the associated computation time for full iterative reconstructions can be clinically prohibitive.

Let us briefly explain iterative MAR. It is used to find a proper attenuation distribution μ by minimizing the following functional with suitable stopping criteria:

$$\Phi(\mu) = \frac{1}{2}\|\mathscr{R}\mu - \mathbf{P}\|^2 + \lambda \text{Reg}(\mu), \tag{3.95}$$

where Reg(μ) is a regularization term (e.g. Tikhonov, total variation) constraining prior knowledge of CBCT image and λ is the regularization parameter that controls the trade-off between the fidelity and the regularization terms. This regularization serves to suppress artifacts in the reconstructed image. Here, $\|\mathscr{R}\mu - \mathbf{P}\|$ is the L^2-norm of $\mathscr{R}\mu - \mathbf{P}$ restricted on the sinogram region unaffected by the metal. Since $-\nabla\Phi(\mu)$ is the direction of steepest decent at μ, we have the following iterative scheme:

$$\mu_{n+1} = \mu_n - \alpha \nabla \Phi(\mu_n), \tag{3.96}$$

where α is a step size. The direct computation of $\nabla\Phi(\mu_n)$ yields

$$\nabla\Phi(\mu) = \mathscr{R}^*[\mathscr{R}\mu - \mathbf{P}] + \lambda \nabla \text{Reg}(\mu). \tag{3.97}$$

In the case of Tikhonov regularization (i.e.,$\nabla\text{Reg}(\mu) = \mu$), the iterative scheme (3.96) can be expressed as

$$\mu_{n+1} = (\beta I - \alpha\mathscr{R}^*\mathscr{R})\mu_n - \alpha\mathscr{R}^*\mathscr{R}\mu_n + \alpha\mathscr{R}^*\mathbf{P}, \tag{3.98}$$

where $\beta = 1 - \alpha\lambda$.

Larry Zeng developed one-step iterative reconstruction in the case of Tikhonov regularization. He rewrote (3.98) as

$$\mu_{n+1} = \alpha \sum_{k=0}^{n} (\beta I - \alpha\mathscr{R}^*\mathscr{R})^k \mathscr{R}^*\mathbf{P} + (\beta I - \alpha\mathscr{R}^*\mathscr{R})^n \mu_0. \tag{3.99}$$

The above identity can expressed as

$$\mu_{n+1} = \alpha \left(I - (\beta I - \alpha\mathscr{R}^*\mathscr{R})\right)^{-1} \left(I - (\beta I - \alpha\mathscr{R}^*\mathscr{R})^n\right) \mathscr{R}^*\mathbf{P} + (\beta I - \alpha\mathscr{R}^*\mathscr{R})^n \mu_0. \tag{3.100}$$

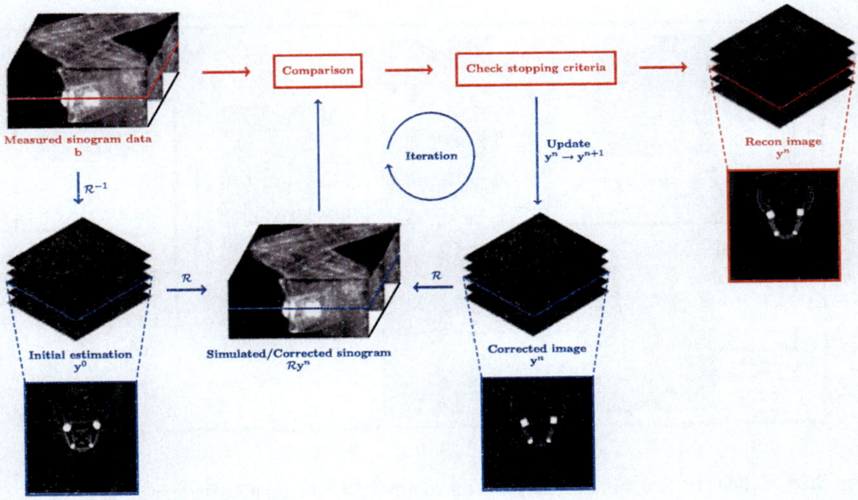

Fig. 3.28 Schematic diagram of iterative metal artifact reduction method

The key observation is that the combined operator $\mathscr{R}^*\mathscr{R}$ (i.e. back-projector· projector) has a point spread function ψ with 1/distance so that

$$\mathscr{R}^*\mathscr{R}\delta = \psi \quad \to \quad \mathscr{R}^*\mathscr{R}u\mu = \psi * \mu, \tag{3.101}$$

where $\psi * \mu$ is the convolution of ψ and μ. Thanks to this expression, we get the following simple form (changing the complicated operator $\mathscr{R}^*\mathscr{R}$ to a simple symbol$\hat{\psi}$, the Fourier transform of ψ):

$$\hat{\mu}_{n+1} = \alpha\left(1-(\beta-\alpha\hat{\psi})\right)^{-1}\left(I-(\beta-\alpha\hat{\psi})^n\right)\widehat{\mathscr{R}^*g} + (\beta I-\alpha\hat{\psi})^n\hat{\mu}_0. \tag{3.102}$$

We will briefly discuss conventional regularization techniques such as compressed sensing using ℓ^1-regularization. These approaches can effectively remove noise, but discard detailed information (e.g., small details containing clinically useful information for diagnosis could be removed) and hence have limited applications in computational medical imaging. Figure 3.28 shows the flow chart of the iterative reconstruction method.

3.4.1.3 Hybrid Methods

Various studies have shown that the MAR performance depends on the material, geometry, size, and location of the metal. The development of a universal MAR method, which includes all cases such as small/medium/large metal objects and

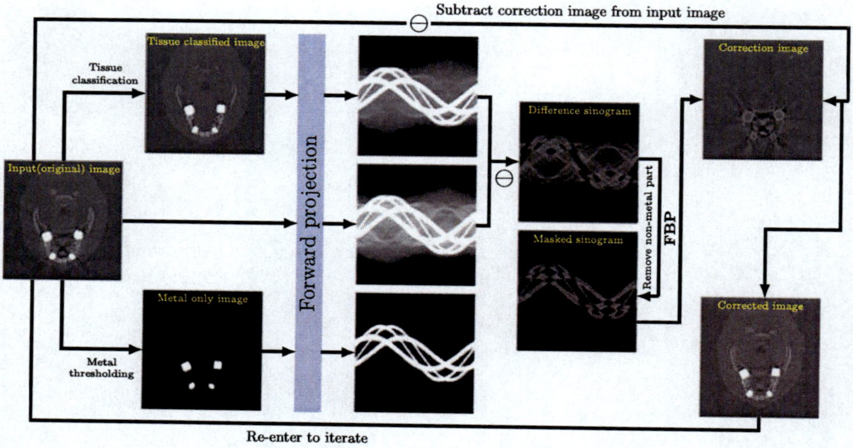

Fig. 3.29 Schematic diagram of orthopedic metal artifact reduction (O-MAR) method

complete photon starvation, is difficult. Thus, it would be desirable to develop a case-based adaptive MAR algorithm instead of a unified MAR method.

Several researchers have developed hybrid MAR methods by combining the aforementioned MAR techniques [71, 134]. This approach usually generates a pre-corrected initial image and based on it, produces a good prior image through iterative reconstruction from the unaffected projection data.

Most commercial MAR algorithms can be considered as hybrid methods. They include SEMAR (Toshiba Medical Systems) [135], O-MAR (Philips Healthcare) [86], iMAR (Siemens Healthcare) [64], Smart MAR (GE Healthcare) [44]. Figures 3.29 and 3.30 show the flowcharts of the O-MAR and SEMAR algorithms.

3.4.1.4 A Physics-Based Beam-Hardening Correction

Recently, Lee et al.[70] developed a direct sinogram correction method that does not require metal segmentation and prior knowledge to reduce metal-related artifacts in polychromatic CT. The method attempts to project the beam-hardening affected data to the range space of the Radon transform so that the mathematical model is more consistent with the corrected data, keeping a part of the data where beam-hardening effects are small.

For simplicity, we use the parallel beam CT system. The polychromatic X-ray data $\mathbf{P}^{\ddagger}$ is nonlinear with respect to the length of the object and its quantity of nonlinearity depends on the energy spectrum η. The non-linearity of projection data introduces some streaking and shadow artifacts in the reconstructed CT image. According to Alvarez and Macovski [5], μ can be decomposed approximately as:

$$\mu(\mathbf{x}; E) \approx p(E)\mu^{p}(\mathbf{x}) + q(E)\mu^{q}(\mathbf{x}), \tag{3.103}$$

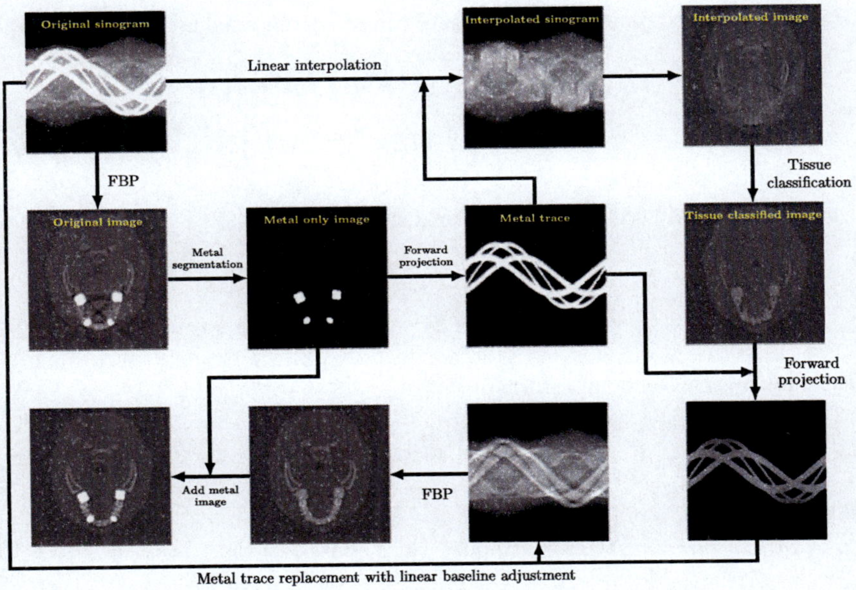

Fig. 3.30 Schematic diagram of single energy metal artifact reduction (SEMAR) method

where μ^p is the spatial-dependent photoelectric component, μ^q is the spatial-dependent Compton scattering component, $p(E) \approx E^{-3}$ (approximating the energy dependence of the photoelectric interaction), and $q(E)$ is the Klein-Nishina function [66, 120], given by

$$q(c\,\tau) = \frac{1+\tau}{\tau^2}\left[\frac{2(1+\tau)}{1+2\tau} - \frac{1}{\tau}\ln(1+2\tau)\right] + \frac{1}{2\tau}\ln(1+2\tau) - \frac{(1+3\tau)}{(1+2\tau)^2}, \tag{3.104}$$

with $c = 510.975$ keV and $\tau = E/c$. Let $\overline{E}_p$ and $\overline{E}_q$ denote effective energies for the photoelectric and Compton scattering component, respectively, given by

$$\overline{E}_a = \underset{\overline{E}_a}{\operatorname{argmin}} \left|\ln\left(\int \eta(E) e^{-a(E)+a(\overline{E}_a)} dE\right)\right|, \quad a = p, q. \tag{3.105}$$

Using this approximation (3.103), the projection data can be expressed as

$$\mathbf{P}^{\ddagger} \approx -\ln\left(\int \eta(E) e^{-p(E)\mathscr{R}\mu^p - q(E)\mathscr{R}\mu^q} dE\right). \tag{3.106}$$

This approximated projection data (3.106) can be decomposed as:

$$\begin{aligned}\mathbf{P}^{\ddagger}(\varphi, s) =&\mathscr{R}\left[p(\overline{E}_p)\mu^p + q(\overline{E}_q)\mu^q\right](\varphi, s)\\ &- \ln\left(\int \eta(E)e^{\Upsilon(E,\varphi,s)}dE\right),\end{aligned} \tag{3.107}$$

where Υ can be viewed as a source of artifacts, given by

$$\begin{aligned}\Upsilon(E, \varphi, s) =& -(p(E) - p(\overline{E}_p))\mathscr{R}\mu^p(\varphi, s)\\ &- (q(E) - q(\overline{E}_q))\mathscr{R}\mu^q(\varphi, s).\end{aligned} \tag{3.108}$$

Then, the inverse Radon transform of $\mathbf{P}$ is

$$\mathscr{R}^{-1}\mathbf{P}^{\ddagger}(\mathbf{x}) = \mu^{\sharp}(\mathbf{x}) + Err(\mathbf{x}), \tag{3.109}$$

where

$$\mu^{\sharp}(\mathbf{x}) = p(\overline{E}_p)\mu^p(\mathbf{x}) + q(\overline{E}_q)\mu^q(\mathbf{x}) \tag{3.110}$$

and $Err(\mathbf{x})$ represents beam-hardening artifact term, given by

$$Err(\mathbf{x}) = \mathscr{R}^{-1}\left[\ln\left(\int \eta(E)e^{\Upsilon(E,\varphi,s)}dE\right)\right](\mathbf{x}). \tag{3.111}$$

The goal is to reconstruct $\mu^{\sharp}$ by alleviating artifacts, given by $Err(\mathbf{x})$. Lee et al. observed the following: Given projection data $\mathbf{P}^{\ddagger}$, the corresponding target image $\mu^{\sharp}$ can be approximated by

$$\mu^{\sharp}(\mathbf{x}) \approx \mathscr{R}^{-1}[\Psi_{\boldsymbol{\lambda},t_*}(\mathbf{P}^{\ddagger})](\mathbf{x}), \tag{3.112}$$

where $\Psi_{\boldsymbol{\lambda},t_*} : \mathbb{R} \to \mathbb{R}$ is a sinogram correction function given by

$$\Psi_{\boldsymbol{\lambda},t_*}(t) = \begin{cases} t & \text{for } \ t < t_* \\ h_{t_*}(t) + \sum_{k=2}^{K} \lambda_k (t - t_*)^k & \text{for } \ t \geq t_* \end{cases} \tag{3.113}$$

where t_*, K, and $\boldsymbol{\lambda} = (\lambda_1, \lambda_2, \cdots, \lambda_K)$ are suitably chosen constants and

$$h_{t_*}(t) = \frac{\lambda_1 t_* - 1}{2\lambda_1 e^{-\lambda_1 t_*}} e^{-\lambda_1 t} + \frac{\lambda_1 t_* + 1}{2\lambda_1 e^{\lambda_1 t_*}} e^{\lambda_1 t}. \tag{3.114}$$

The parameter $\boldsymbol{\lambda}$ is determined by minimizing the following objective function:

$$\boldsymbol{\lambda} = \underset{\boldsymbol{\lambda}}{\operatorname{argmin}} \int_0^{2\pi} \left(\frac{\partial}{\partial \varphi} \int_{\mathbb{R}} \Psi_{\boldsymbol{\lambda},t_*} \mathbf{P}^{\ddagger}(\varphi, s) ds\right)^2 d\varphi, \tag{3.115}$$

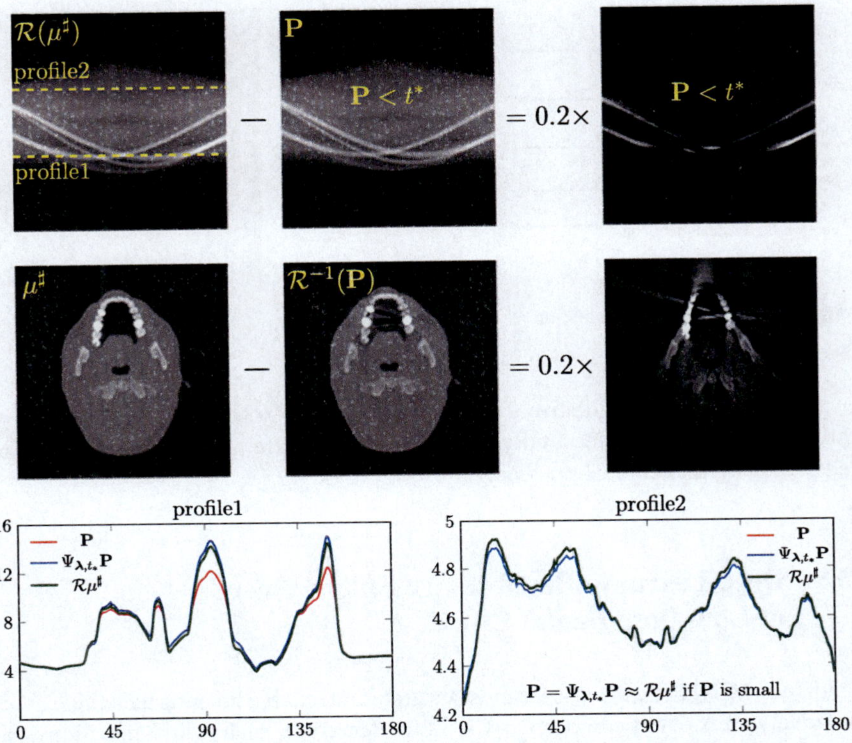

Fig. 3.31 [70] The top three images show $\mathscr{R}\mu^\sharp$, $\mathbf{P}^\ddagger$, and $\mathscr{R}\mu^\sharp - \mathbf{P}^\ddagger$, respectively. The middle three images show $\mu^\sharp$, $\mathscr{R}^{-1}\mathbf{P}^\ddagger$, and $\mu^\sharp - \mathscr{R}^{-1}\mathbf{P}^\ddagger$, respectively. $\mathscr{R}\mu^\sharp - \mathbf{P}^\ddagger \approx 0$ in the region of $\{\mathbf{P}^\ddagger < t_*\}$ for a small t_*. The bottom images show that the proposed sinogram correction works well on the profiles 1 and 2. This figure was extracted from [70]

where $\boldsymbol{\lambda} = (\lambda_1, \lambda_2, \cdots, \lambda_K)$. Figure 3.31 illustrates characteristics of beam-hardening correction function.

3.4.2 Phantom for Image Quality Evaluation

The performance of the method can be evaluated using cylindrical phantoms (manufactured by QRM GmbH, Möhrendorf, Germany) with a height and diameter of 10 cm. Each consists of five separate sections for evaluating IQ parameters in terms of CT value uniformity and linearity, image noise and contrast, 3D resolution, and artifact behavior (Fig. 3.32).

We can perform quantitative evaluation by a certificate authority to ensure that the Hounsfield unit (HU) value satisfies the error range of 140 HU in the ROI in the above figure.

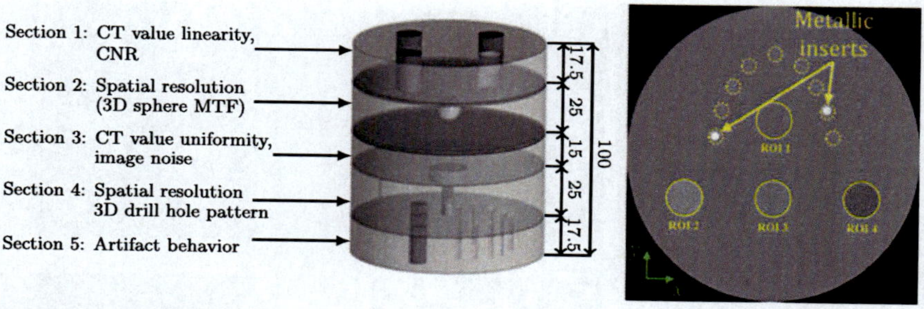

Fig. 3.32 Image quality evaluation

We can perform qualitative evaluation by several experts in the dental college hospital to confirm the feasibility of applying MAR to real data with metal artifact and applying it to clinical practice.

3.5 Deep Learning-Based Image Enhancement of Low-Dose Dental CBCT

Although MDCT is known as the most accurate and reliable imaging technique, it has relatively weak points compared to CBCT for dental use, such as high radiation dose, high equipment cost, and large space required for use. Most dental CBCT devices are designed to allow the patient to be scanned while sitting or standing, requiring less space in the dental office. In response to concerns about CT radiation, dental CBCT is being developed in the direction of minimizing radiation exposure while maintaining image quality.

The main parameters influencing radiation dose in a given CBCT device are tube current, tube voltage, and collimation. To reduce radiation exposure in CBCT, it is recommended to use as small a field of view (FOV) as possible, the lowest tube current setting, and the shortest exposure time. Most dental CBCTs use a small FOV, resulting in significant truncation in the signogram. For patients with metal implants, conventional CT image reconstruction methods using iterative reconstruction algorithms may not be suitable for application to highly truncated sinograms from low-dose CBCT scans. Collimation restricts the X-ray beam to the area of interest using lead shutters within the X-ray tube. See Fig. 3.33. A secondary beneficial effect of collimation is reduction of off-focus radiation making it to the film. Because a smaller volume of tissue is being irradiated, less scatter radiation is produced. In dental CBCT, an additional copper filter is used to reduce patient dose, as shown in Fig. 3.34.

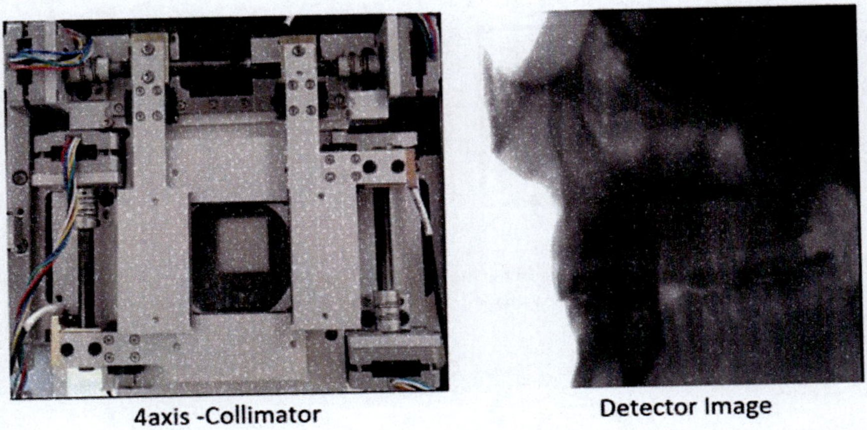

Fig. 3.33 Collimation reduces radiation exposure. Collimation limits the cross-sectional area of the X-ray beam to the size of the image receptor

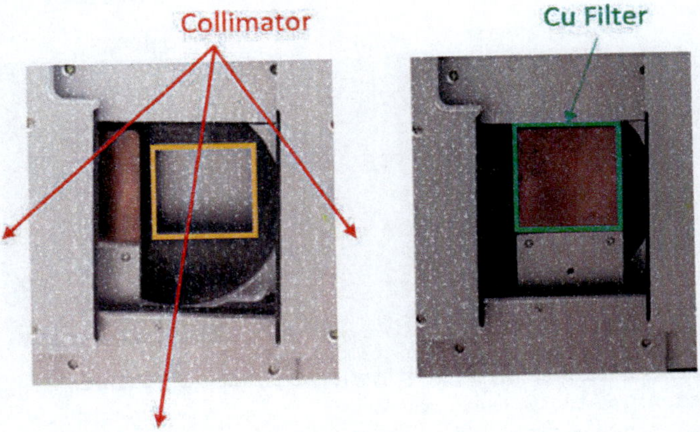

Fig. 3.34 Copper filtration for dental CBCT for reducing radiation exposure

3.5.1 Dental CBCT Data Acquisition Protocol

Specifications for commercial dental CBCT are: circular cone-beam scan, scan time 8–24 s, resolution less than 0.2 mm, FOV truncation, offset detector, low X-ray dose, and cost less than 0.1 billion dollars. Meanwhile, the specifications of MDCT are helical cone-beam scan, scan time less than 1 s, resolution less than 0.3 mm, no FOV truncation, no offset detector, high X-ray dose, and cost over 1 billion dollars. See Fig. 3.35 for FOV truncation, where the size of the FOV is determined by the detector size and shape, the beam projection geometry and the beam collimation function.

Since a significant portion of the manufacturing cost of a CBCT device is the cost of an X-ray detector, the smallest detector available is used to obtain the desired

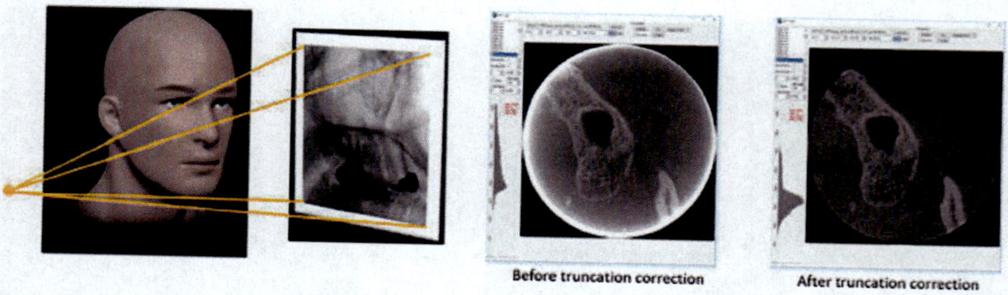

Fig. 3.35 FOV truncation. FOV represents the area of the patient's anatomy to be imaged. As the size of the detector of CBCT becomes smaller, the device becomes cheaper and takes up less space, while FOV becomes smaller

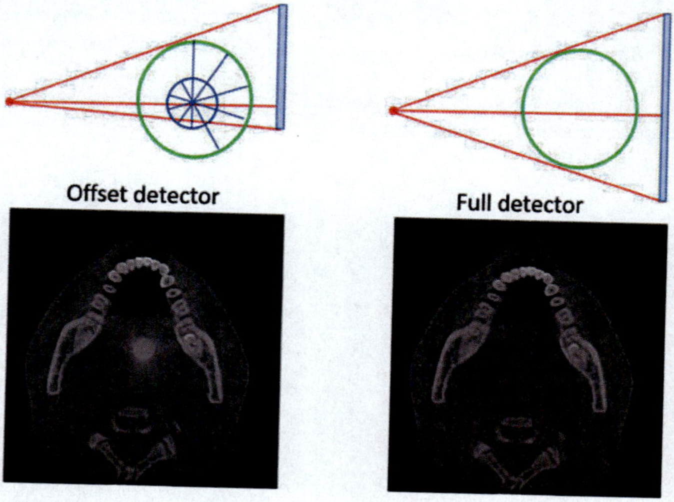

Fig. 3.36 Offset detector in dental CBCT. The center axis of rotation in a small-size flat-panel detector is offset relative to the source–detector axis to maximize the transaxial FOV

image. Due to the use of the small detector, two techniques are used: stitching and offset-scanning methods. The offset-scanning method increases the FOV by offsetting the position of the small detector, where the beam is collimated asymmetrically and the asymmetric detector covers only half of the scanning FOV. See Fig. 3.36 for the offset detector.

The stitching method is used to obtain data from two or more separate scans, and stitch adjacent image volumes to obtain a full head image acquisition, as shown in Fig. 3.37.

The most commonly used detectors in CBCT systems require a wait time after each projection in order to account for the afterglow of the scintillator and the dose delivered in each scan should be limited. In CBCT, the number of projections, together with the associated changes in the total scan time, provides a trade-off between image quality and the delivered dose that is influenced directly by user-selected parameters [59].

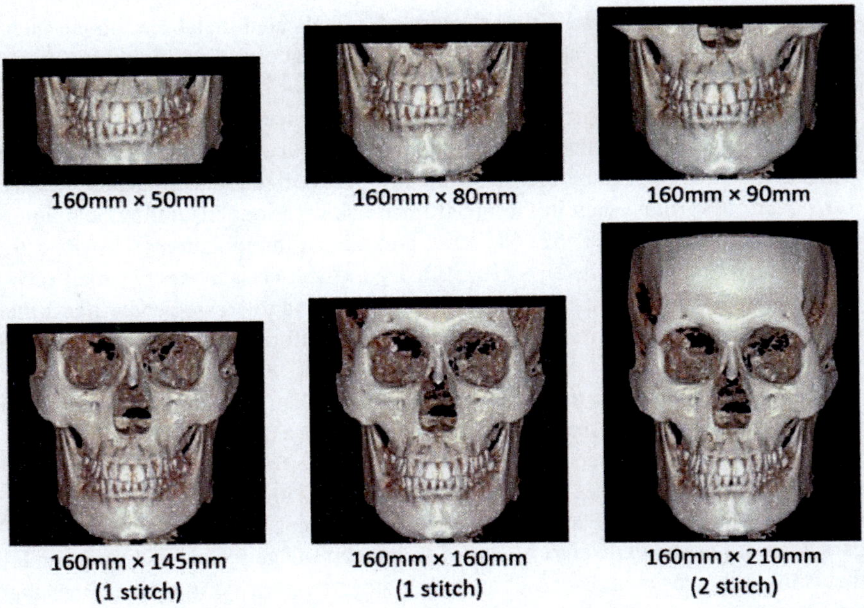

Fig. 3.37 Stitched images using a HDXWILL CBCT device

In order to reduce the patient's overall dose, tube current (mA) and tube voltage (kVp) settings should be kept as low as possible while maintaining image quality. Assuming that kVp and all other parameters are fixed, the radiation dose is directly proportional to [the applied tube current] × [duration of scan rotation], and this parameter greatly affects image quality. The dependence of radiation dose and image quality on kVp settings is very complex. Photons with higher energy have less interaction with tissue. The correct kVp and mAs settings depend greatly on several design factors, and therefore it is difficult to provide absolute guidelines [59].

3.5.2 Towards Metal Artifact Reduction in Low-Dose Dental CBCT

Reducing metal-induced artifacts in the above-mentioned dental CBCT is more difficult than in MDCT because of the additional problems caused by the offset detector, FOV truncation, and low X-ray dose. Since many patients use metal implants and dental fillings in dental CBCT, it is very important to improve image quality by reducing metal-induced artifacts. In dental CBCT, the metal-related artifacts are very difficult to handle as they are generated by the effects of beam hardening-induced sinogram inconsistency and different types of complicated metal-teeth-bone interactions with other factors such as FOV truncations, scattering, and nonlinear partial volume effects.

In the dental CBCT environment, traditional MAR methods such low data correction methods do not effectively reduce metal artifacts and tend to create new streak artifacts that did not exist before. Here, the low data correction approaches are used to recover background data in the metal trace using various inpainting techniques such as interpolation [1, 9, 61, 68, 109], normalized interpolation (NMAR) [79], Poisson inpainting [90], wavelets [78, 132, 133], tissue-class model [7], total variation [29], and Euler's elastica [46]. These methods tend to create new artifacts that did not exist before, and impair the morphological information in the area around the metallic object.

Let us briefly describe the differences between MDCT and CBCT in terms of tomographic image reconstruction. MDCT basically uses a variant of the FBP algorithm (fan-beam image reconstruction formula) that performs in slice-by-slice from sinograms. In MDCT, slices of 2D tomography images are stacked to generate a 3D CT image. Hence, the FBP algorithm is two-dimensional. On the other hand, CBCT uses a variant of the FDK algorithm [31] (fan-beam image reconstruction formula) that is three-dimensional, because it handles integral values along oblique lines that are not orthogonal to the circular orbital plane [106].

3.5.2.1 Deep Learning Based MAR in 2D CT

Recent advances in deep learning technology are also making progress in MAR [10, 38, 47, 58, 73, 91, 126, 136]. Gjesteby et al. [38] use DL to aid the NMAR technique by providing additional correction in the projection domain to enhance image quality in critical regions. Park et al. [91] proposed a deep learning-based sinogram correction method to reduce the primary metal-induced beam-hardening factors along the metal trace in the sinogram. This method was applied to the restricted situation of patient-implant-specific model, where the mathematical beam hardening corrector [87, 88] of a given metal geometry effectively generates simulated training data. Zhang et al. [136] used DL to generate an artifact-reduced prior image, then uses the projection of the prior image to replace the metal-affected projection, and then performs the final MAR CT reconstruction. Gjesteby et al. [47] used residual learning to correct for metal streaking artifacts after the first pass by NMAR. Lin et al. [73] proposed a Dual Domain Network (DuDoNet) to restore sinogram consistency and enhance CT images simultaneously. DuDoNet pursues MAR enhancement by leveraging dual domain learning networks; sinogram and image enhancement networks. Here, the dual networks operate separately on sinogram and image domains, but are trained in end-to-end fashion. The sinogram enhancement network learns how to correct a metal-corrupted sinogram by using an inpainting loss on the metal trace and sinogram consistency loss. Since the final image is the output of the image enhancement network, the data fidelity may be compromised, resulting in anatomical structure changes. Yu et al. [126] proposed another dual domain joint learning network, that first generates a good prior image with fewer metal artifacts and then the forward projection of the prior image is utilized for sinogram enhancement. Here, the final output is the reconstructed image from the sinogram modified only in the metal trace area.

Let us discuss the MAR network mentioned above in more detail. For convenience of explanation, the following notations are used.

- **Sinogram Domain.** $\mathbf{P} \in \mathbb{R}^{n_s \times n_h}$ represents a metal-corrupted sinogram, where n_h is related to the detector size and n_s is number of projection views. The corresponding metal trace will be denoted by $\mathscr{M}_{tr} \in \{0, 1\}^{n_s \times n_h}$. $\mathbf{P}_{\text{LI}}$ represents a pre-corrected sinogram by LI (linear interpolation) on the metal trace.
- **Image Domain.** $\mu = \mathscr{R}^{\dagger}\mathbf{P}$ is the metal-induced artifacts-contaminated CT image, where $\mathscr{R}^{\dagger}$ denotes the filtered back-projection. $\mu_{\text{LI}} = \mathscr{R}^{\dagger}\mathbf{P}_{\text{LI}}$ is the CT image corresponding to $\mathbf{P}_{\text{LI}}$. A target CT image to be reconstructed can be a distribution of attenuation coefficients at a certain energy level.
- $f_{\text{DL}_{\text{sino}}}$ represents a sinogram correction network whose input can be $\mathbf{P}$, $\mathbf{P}_{\text{LI}}$, or the projection of the output of $f_{\text{DL}_{\text{img}}}$.
- $f_{\text{DL}_{\text{img}}}$ represents an image enhancement network whose input can be μ, μ_{LI}, or the filtered back-projection of the output of $f_{\text{DL}_{\text{sino}}}$.

Image domain learning approaches [38, 136] try to alleviate metal artifacts in a reconstructed image by learning $f_{\text{DL}_{\text{img}}}$ through the powerful ability of deep learning methods to capture salient image features. See Fig. 3.38 for detailed reconstruction process. These approaches heavily depend on the image quality of an input, which is contaminated by streaking and shadowing artifacts.

The sinogram domain learning approach [91] uses a deep learning method to reduce the primary metal-induced beam-hardening factors along the metal trace in the sinogram, as shown in Fig. 3.39. The learned $f_{\text{DL}_{\text{sino}}}$ is designed to work along metal traces that are contaminated by metal-related artifacts. The image reconstruction map can be expressed by $\mathscr{R}^{\dagger} \circ f_{\text{DL}_{\text{sino}}}$.

Dual domain learning approaches [73, 93, 126] employ two deep learning networks ($f_{\text{DL}_{\text{img}}}$ and $f_{\text{DL}_{\text{sino}}}$) separately working on image and sinogram domains, respectively. The sinogram domain network $f_{\text{DL}_{\text{sino}}}$ is designed to perform sinogram inpainting along metal traces on which metal artifact-inducing factors dominantly affect. These approaches learn $f_{\text{DL}_{\text{img}}}$ and $f_{\text{DL}_{\text{sino}}}$ in the end-to-end fashion by introducing the Radon forward ($\mathscr{R}$) or inversion ($\mathscr{R}^{-1}$) layer, as shown in Fig. 3.40. The image reconstruction map can be viewed as $f_{\text{DL}_{\text{img}}} \circ \mathscr{R}^{\dagger} \circ f_{\text{DL}_{\text{sino}}}$ or $\mathscr{R}^{\dagger} \circ f_{\text{DL}_{\text{sino}}} \circ \mathscr{R} \circ f_{\text{DL}_{\text{img}}} \circ \mathscr{R}^{\dagger}$.

In the supervised learning framework, a deep learning network learns a reconstruction map f_{DL} in the following sense:

$$f_{\text{DL}} = \operatorname*{argmin}_{f_{\text{DL}} \in \mathbb{DL}} \sum_{i=1}^{n_{\text{data}}} \mathbf{dist}(f_{\text{DL}}(\mathbf{x}^{(i)}), \mathbf{y}^{(i)}), \tag{3.116}$$

where $\{\mathbf{x}^{(i)}\}$ is a paired training dataset, $\mathbf{dist}$ is a metric measuring the difference between the deep learning output $f_{\text{DL}}(\mathbf{x}^{(i)})$ and label $\mathbf{y}^{(i)}$. In MAR applications, ℓ^2

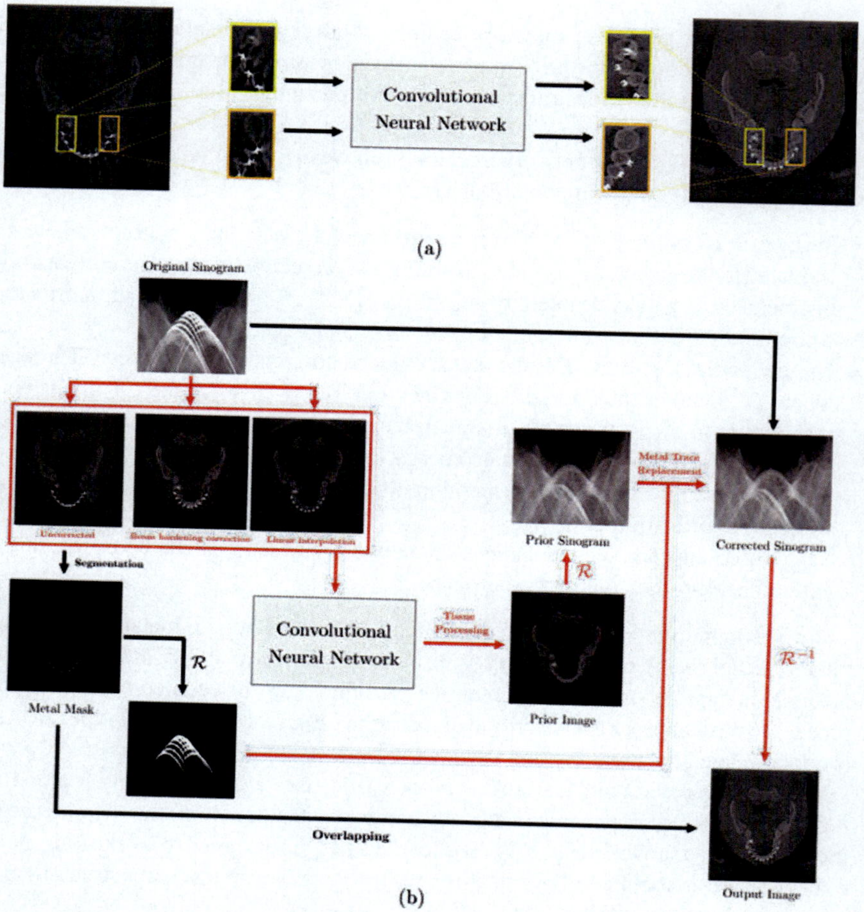

Fig. 3.38 Image domain learning method. Overall process of [38] in (a) and [136] in (b)

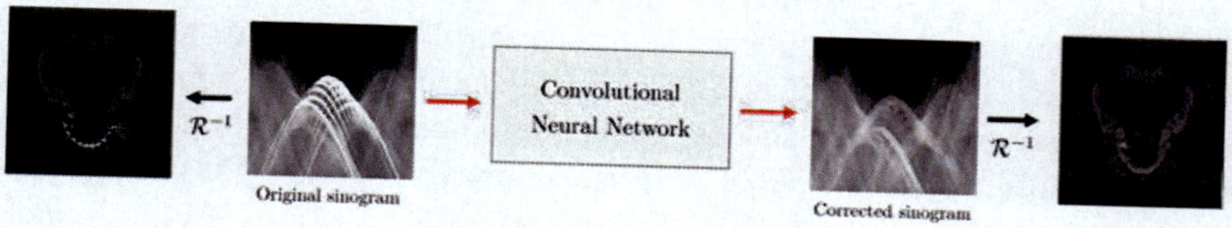

Fig. 3.39 Sinogram domain learning method [91]

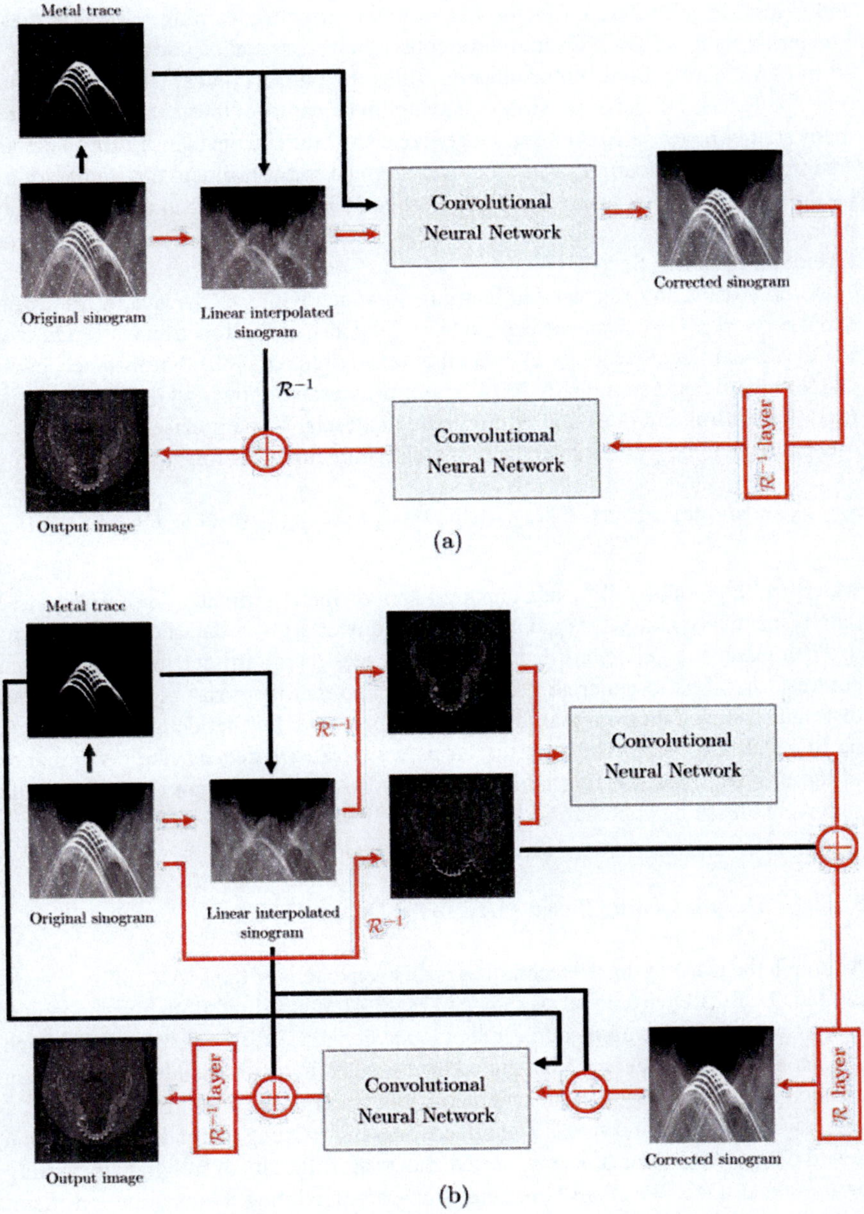

Fig. 3.40 Dual domain learning approaches. Overall process of [73] in (a) and [126] in (b)

and ℓ^1 loss are common choices for **dist**. For more effective learning, various axillary loss terms such as Radon inversion (or consistency) loss can be added to (3.116).

In MAR applications, unfortunately, collecting paired clinical CBCT data from many patients is almost impossible due to nonlinear nature of the artifacts. Therefore, many studies have used paired dataset generated by artificial metal insertion on metal-free images. Deep learning can effectively reduce metal artifacts through learning by the simulated data, whereas the gap between the simulation and actual data may cause limited reconstruction ability in real clinical situations. This issue will be further discussed in Sect. 3.6.

Alternatively, an unsupervised learning approach with various generative adversarial networks (GANs) can be used to learn a reconstruction function while addressing the fundamental difficulty in collecting real paired data [83]. For instance, cycle GAN [137], Wasserstein GAN [6, 43], and progressive growing GAN [65] can be used, whose training does not require paired datasets. To be precise, a function f_{DL} (generator) can be learned by the aid of a discriminator D as follows:

$$f_{\mathrm{DL}} = \underset{f_{\mathrm{DL}}}{\operatorname{argmin}} \ \underset{D}{\operatorname{argmax}} \ \mathbb{E}_{\mathbf{y}\sim p_{\mathrm{data}}(\mathbf{y})}(\ln D(\mathbf{y})) + \mathbb{E}_{\mathbf{x}\sim p_{\mathrm{data}}(\mathbf{x})}(\ln(1 - D(f_{\mathrm{DL}}(\mathbf{x})))), \tag{3.117}$$

where $\{\mathbf{x}^{(\mathrm{i})}\}_{\mathrm{i}=1}^{N}$ and $\{\mathbf{y}^{(\mathrm{i})}\}_{\mathrm{i}=1}^{M}$ are unpaired sets of metal-artifacted and artifact-free data respectively and p_{data} is a data distribution over a given dataset. Here, D aims to distinguish the generated data $f_{\mathrm{DL}}(\mathbf{x})$ with real metal artifact free data and, in contrast, f_{DL} aims to generate fake data $f_{\mathrm{DL}}(\mathbf{x})$ so that the discriminator D cannot distinguish such data from real metal-artifact-free data. Further details of GAN can be found in [56, 127]. However, as far as we have experienced, the performance of the unsupervised learning approach still is far from satisfactory, relative to the supervised learning approach.

3.5.2.2 Deep Learning Based MAR in 3D Dental CBCT

Although the learning-based techniques have been extended the MAR ability even in the dental CBCT environment, as shown in Fig. 3.41, many limitations and challenges have remained for the application in a practice dental CBCT environment in which the sinogram inconsistency is entangled by complex factors associated with varying geometry of metals, metal–bone and metal–teeth interaction, FOV truncation, offset detector acquisition, and so on. The effectiveness of existing methods in the clinical dental CBCT environment is very limited, due to the difficulty of sinogram inpainting on the metal trace. We should note that sinogram inpainting is much more difficult compared to image inpainting. Image inpainting is somewhat local because it uses the surrounding information of the inpainting area to fill in the missing data. On the other hand, sinogram inpainting requires sophisticated utilization of global information outside of the inpainting area to fill in missing data while maintaining sinogram consistency. In low-dose dental CBCT, sinogram data inconsistency comes from not only metal but also teeth and bones. In addition, the sinogram is contaminated under

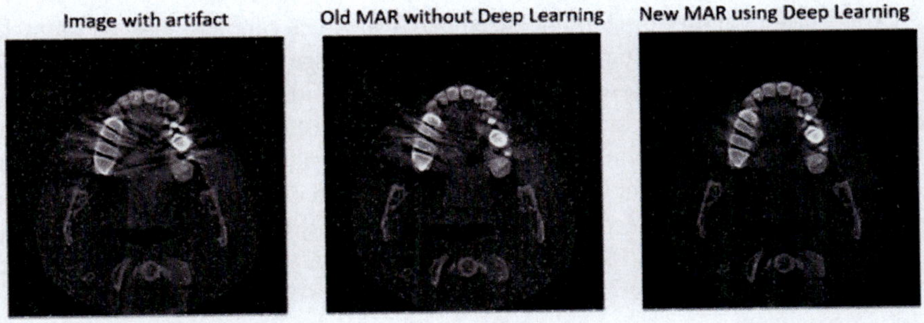

Fig. 3.41 Comparison of MAR using deep learning and MAR using a conventional method

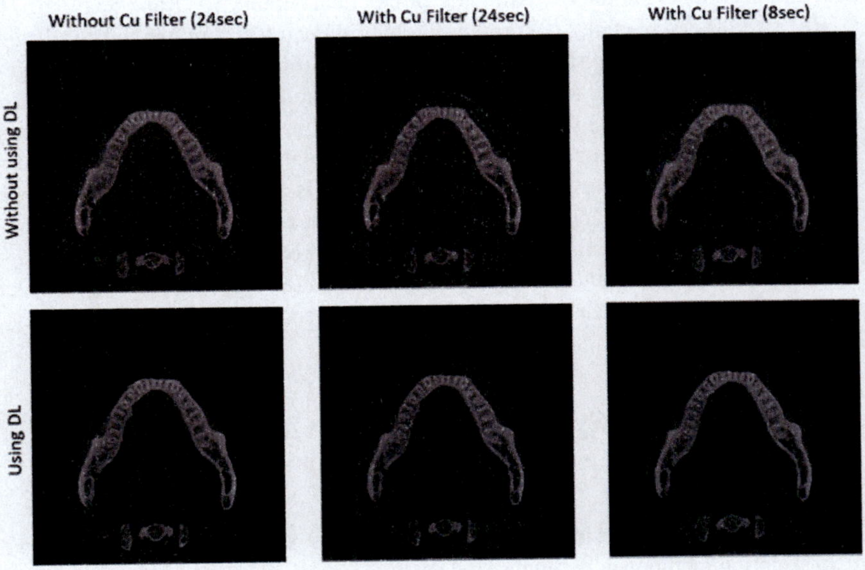

Fig. 3.42 Ablation study: Comparison of image quality with and without Cu filter (rows) and with and without deep learning (columns)

the influence of complex factors related to complex geometries of teeth and bones, FOV truncation, offset detector, scattering, etc. Without the use of a strong fidelity constraint related to the input data, using the consistency loss in these sinogram data conditions in low-dose dental CBCT can increase secondary artifacts rather than preventing them. In low-dose dental CBCT, a corrected sinogram obtained without the use of strong fidelity constraints associated with the input data tends to generate new artifacts, resulting in anatomical structure changes in the output image.

For dental MAR, there is an attempt that uses physical filtering by equipping a Cu filter (Fig. 3.34) in front of a X-ray beam generator and then applies a DL to the filtered data. See Fig. 3.42.

The learning ability of learning-based MAR methods can be improved through complex network architectures (e.g., deep layers or large feature depths) and large-scale training datasets. However, these is trade-off with total computational cost for learning that can be critical especially in practical 3D dental CBCT application dealing with high dimensional data. Even only for the simple U-net architecture, 10 days are required at least for training of 300 epochs with a dataset of 60 image voxels under our computational resource (two Intel(R) Xeon(R) CPUs E5-2630 v4, 128GB DDR4 RAM, and four NVIDIA GeForce GTX 2080ti GPUs). Here, we remark that the voxel sizes of our sinogram and CT data are $1040 \times 654 \times 658$ and $800 \times 800 \times 400$, respectively. Even though the use of sophisticated networks or large training datasets can potentially enhance the MAR capability, hurdles associated with high data dimensionality should be dealt with for practical dental CBCT application [57].

Recently, Bayaraa et al. [10] proposed a hybrid deep learning method for dental CBCT that takes advantage of analysis-based beam-hardening correction and image-to-image domain learning. As a first stage, an artifact-reduced image is reconstructed by alleviating metal-related beam hardening factors through the modified sinogram intensity adjustment method for dealing with a dental CBCT situation. In a second stage, an image-to-image deep learning network is applied to further reduce the remaining artifacts. This approach effectively alleviates metal-related beam hardening artifacts even when existing FOV truncation and offset detector configuration, whereas it does not guarantee a high quality result in the case of severe noise level or photon starvation, which commonly happens in practical dental CBCT imaging.

3.6 Generation of Synthetic Data for MAR Using Machine Learning

As mentioned in the previous sections, conventional techniques such as iterative reconstruction methods with regularized data-fitting indicates limitations in a dental CBCT environment with low X-ray dose. Lessons over four decades indicate that conventional regularization techniques have limitations in providing prior information that conveys medical image characteristics. For example, commonly used priors such as total variation (TV) regularization may result in loss of fine tooth details because the TV-norm of teeth images (i.e. solutions) is inherently high. Conventional frameworks may not be able to effectively handle nonlinear data structures of CBCT that vary with changes in metal geometry in a highly non-linear way.

Deep learning frameworks effectively leverage prior images through training data. However, in practice, it is impossible to collect many paired data (i.e., pairs of artifact-free and artifact-contaminated CBCT images) simultaneously from many patients. In order to solve the difficulty of obtaining paired learning data for MAR, it is necessary to develop a method to generate synthetic data through computer simulation.

The data generation must be done differently depending on the manufacturer and type of CBCT scanner. Since developing a generalized learning model that can be applied to all possible CBCT scanners is beyond our limits (in terms of cost, time, and manpower), it is practically appropriate to develop a machine-specific learning model for a specific company's CBCT scanner. In this section, synthetic data generation will be performed under dental CBCT hardware limitations (DENTRI, HDXWILL): tube voltage of 80–90kVp, tube current of 8–10mA, image size of 16cm × 16cm × 8cm, voxel size of 0.2mm × 0.2mm × 0.2mm, pixel number being 800 × 800 × 400 number of pixels, and a total filtration thickness of 1.6mm Al.

These synthetic data generations not only provide a low-cost source of various training data, but also systematically provide laborious and costly semantic segmentation.

3.6.1 *Semi-synthetic Data Generation*

Generating a variety of human-like head phantoms is very laborious and cumbersome. To address this problem, we use a method of semi-synthetic data generation, which uses a lot of metal artifact-free CBCT data from many patients who do not have metallic materials. For a given dental CBCT scanner, a reasonably accurate CBCT forward model can be developed using the physical formula described in Sect. 3.3.1.

For each artifact-free CBCT image, metallic materials of various shapes are inserted at various locations to generate numerous CBCT images contaminated with artifacts using the forward model. We can produce numerous combinations of CBCT images as we adjust various metal shapes and positions. For cone-beam projection and back-projection (FDK reconstruction), we used the open source TIGRE package [12] with suitable modification for dealing with an offset detector array and parameter exactly same as those of the HDXWILL machine. Figures 3.43 and 3.44 show that real patient data, simulated data and corresponding images for three dental implant models; dental crown, dental implant, and orthodontic brace.

3.6.2 *Simulated Projection Data with Simulated Dental Crowns and Implants, and Orthodontic Braces*

We generate simulated projection data by placing simulated metallic dental crown and implants, and orthodontic braces using the attenuation coefficient given in [55] and electronic noise was modeled as in [57]. We take separate strategies for dental crown, implant and orthodontic brace. For generation of a dental crown and implant, we refer to the method in [57].

The data generation method can be summarized as follows: To generate a simulated dental crown or implant, as a first step, we perform fully automated individual tooth segmentation on normal patient data by using the technique in [60] and randomly choose several tooth positions in which virtual metal implants would be

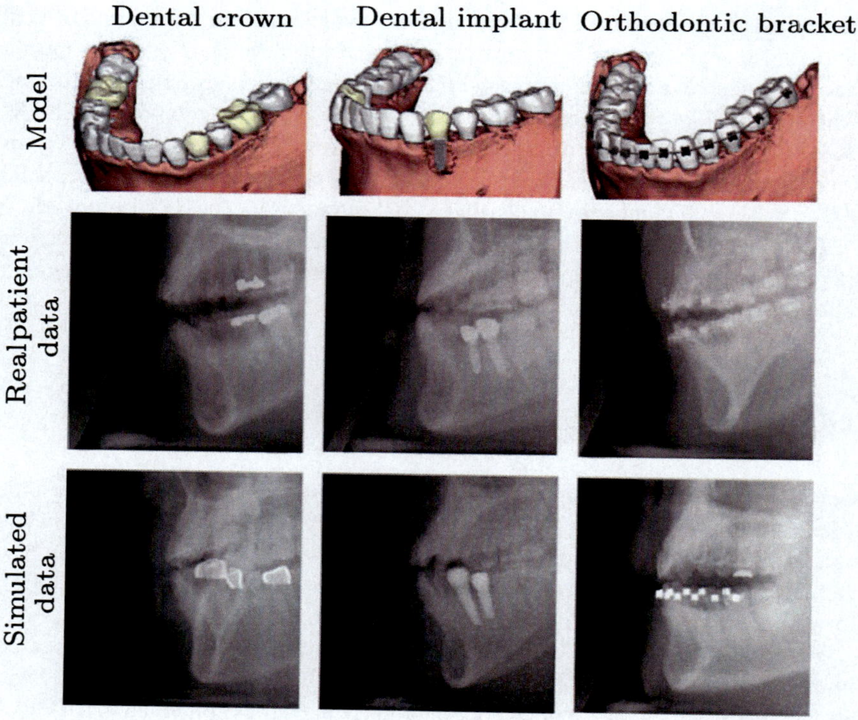

Fig. 3.43 Real patient and simulated sinogram data for three models; dental crown, dental implant, and orthodontic brace

placed. For crown case, a crown mask is constructed by cutting the root of the chosen tooth or teeth on the basis of crown height information for each tooth [85] and then by taking the erosion process. Thickness of the crown is randomly set from 0.6 to 1.4 mm. For the implant case, an additional process is applied to make an implant screw bar. For each tooth, we define the line that passes through two points of the tooth center in the lowest and middle slices except that of containing the tooth root. After then, root parts are filled with circles whose center is located at the line and the radius is empirically set. See Fig. 3.45.

We generate simulated projection data by placing simulated metallic dental crown and implants, and orthodontic braces using the attenuation coefficient given in [55] and electronic noise was modeled as in [57]. We take separate strategies for dental crown, implant and orthodontic brace. For generation of a dental crown and implant, we refer to the method in [57].

For the case of the orthodontic brace, simulated data is generated in the following procedure: By applying the individual tooth segmentation again, a binary mask of individual tooth is obtained. For each tooth, we first find z-slices in which the tooth lies and then estimate a z-slice where the center of a brace is placed by using brace

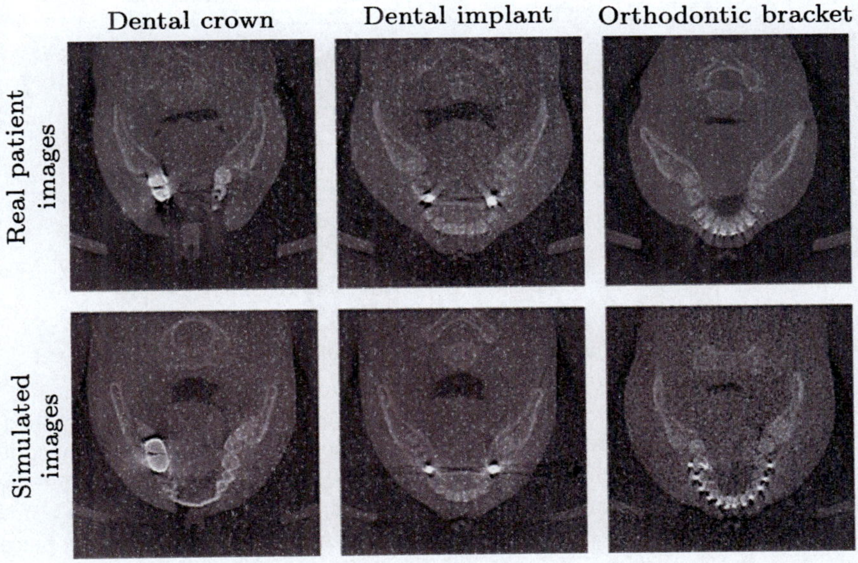

Fig. 3.44 Real patient and simulated image for three models; dental crown, dental implant, and orthodontic brace

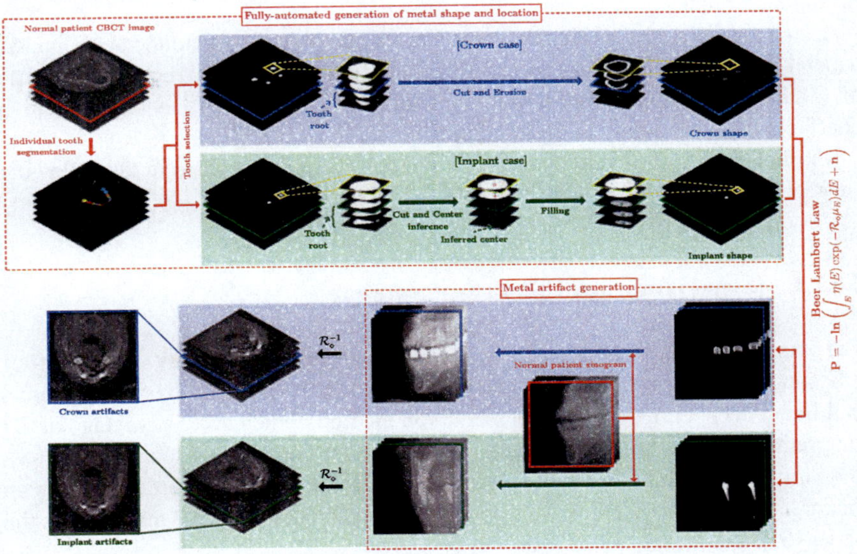

Fig. 3.45 [57] Overall process of automated data generation method for dental crown and implant. This figure was extracted from [57]

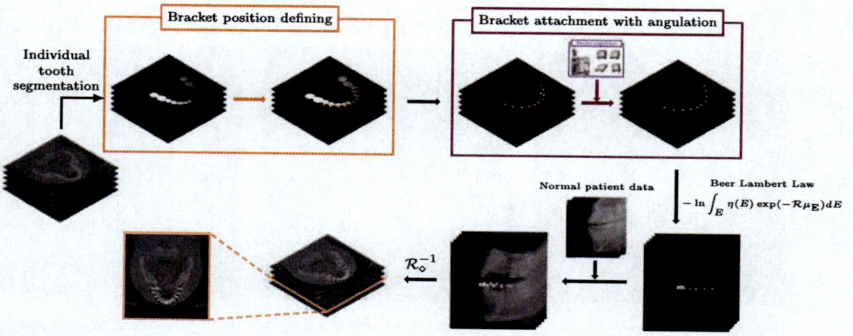

Fig. 3.46 Overall process of automated data generation method for orthodontic brace

height information depending on a tooth type and location in the skull (i.e., maxilla or mandible) in [4]. For the estimated z-slice, we define a normal vector toward the front of the tooth and find a point on the tooth surface that intersects with a line defined by the normal vector and center position. As a last step, we place a brace so that its center position is located at the point. Here, we use a brace shape given in [138] and a rotation using angulation information in [4] depending on tooth type and location in skull. Figure 3.46 shows the summary of data generation for the orthodontic brace.

Using a simulated metal insert (dental crown, implant or orthodontic brace) mask, a metal-artifacted sinogram is artificially produced via the forward model in the sense of (3.33) and combined with a normal patient sinogram. Here, a metal attenuation coefficient was randomly assigned among $\{\text{Au, Pd, Ni, Cr, Zr, Al}\}$.

Further details of the data generation method will be explained in the following subsections.

3.6.2.1 [Step 1] Individual Tooth Segmentation

The data generation is based on a 3D individual tooth mask obtained by the state-of-art fully automated segmentation method presented in [60]. The method is a hierarchical multi-step deep learning-based segmentation method, which takes advantage of 2D panoramic images of the upper and lower jaws, YOLO (you only look once) [108], and various tight and loose ROI selections on individual teeth. In order to focus on the main topic of this section, we skip the detailed explanation. One may refer to the paper [60].

Let $\mathbf{P}_{\text{normal}}$ be a normal patient's sinogram data. Using $\mathbf{P}_{\text{normal}}$, we first reconstruct a CT image μ_{normal} by the modified FDK algorithm in (3.31) and then apply the segmentation method. The final segmentation result is given by a mask $\mathbf{M}_{\text{tooth}}$, which is defined by a 3D mask where each area of an individual tooth is filled by a different value from the others. See Fig. 3.47. The mask $\mathbf{M}_{\text{tooth}}$ is the basis of the fully automated method that generates virtual but considerably realistic metal implants in normal patient data.

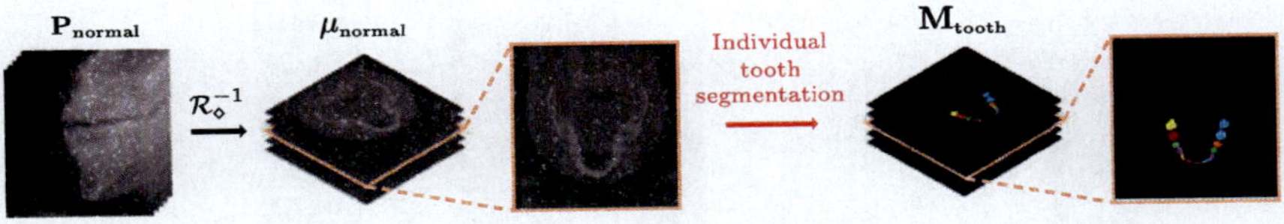

Fig. 3.47 Individual tooth segmentation. We obtain a 3D tooth segmentation mask $\mathbf{M}_{\text{tooth}}$ where each tooth is filled by a different value with the others

3.6.2.2 [Step 2] Metal Implant Shape Generation

Case 1. Dental crown
To place virtual dental crowns in normal patient data, $\mathbf{M}_{\text{tooth}}$ is utilized, as shown in Fig. 3.48. First, we select several teeth where dental crowns are desired to be placed, estimate the tooth root on the basis of the average crown height information of each tooth in [85], and fill the root part in the mask with zero. Since a dental crown covers the surface of a tooth in general cases, a region of dental crowns is defined by the result after taking an erosion process using a MATLAB built-in function `imerode` to the mask.

Case 2. Dental implant
Data generation for a dental implant follows the same process in Case 1. Generation for Dental crown involves an additional step to generate implant bars, as seen in Fig. 3.49. Let $\mathbf{M}_{\text{crown}}$ be a 3D mask where the estimated tooth root region is removed. We first find slices where the remaining tooth part is in $\mathbf{M}_{\text{crown}}$ and then estimate two center positions of teeth at the middle and bottom by taking the average at the

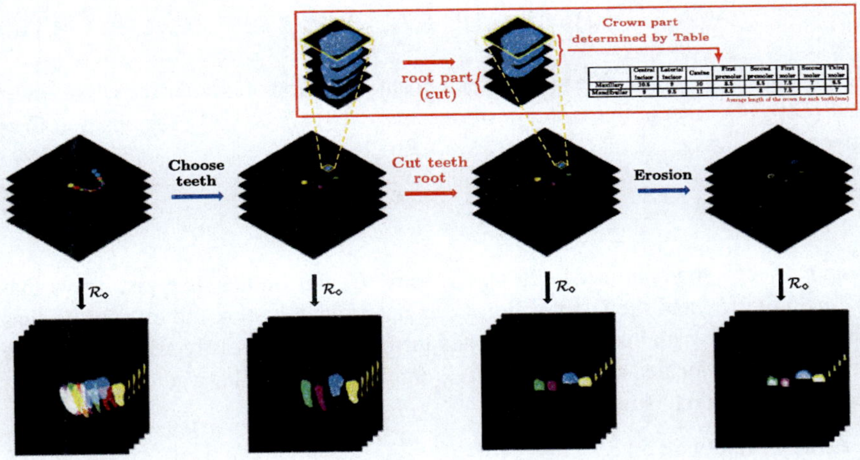

Fig. 3.48 Automated dental crown shape generation

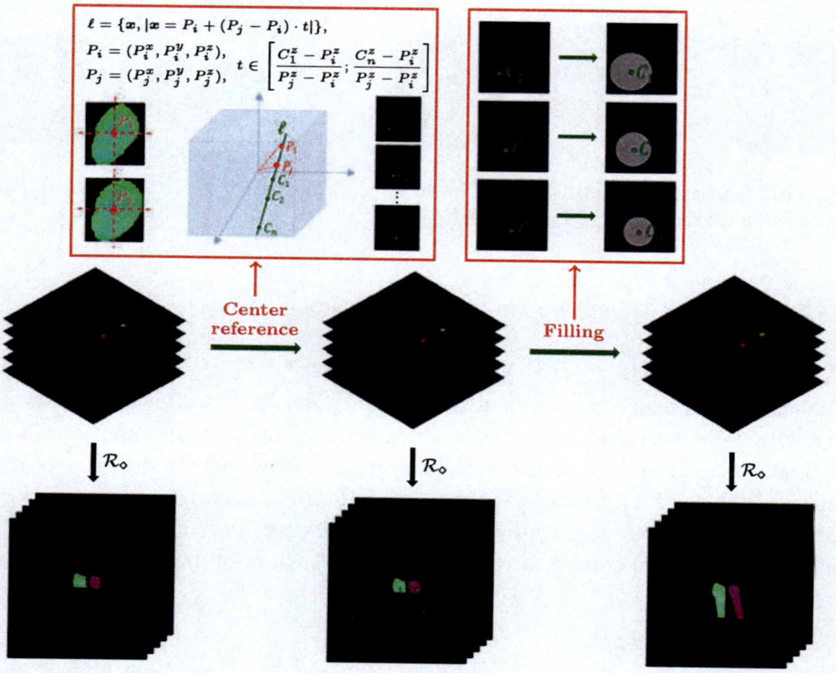

Fig. 3.49 Automated dental implant shape generation

corresponding slices in terms of position. Denoting P_i and P_j as the center positions, we define a line ℓ by

$$\ell(t) = P_i + (P_j - P_i) \cdot t, \quad \forall t \in \mathbb{R}. \tag{3.118}$$

Using this line ℓ, we are able to determine a center position of implant bars as follows: Let $\{C_i\}_{i=1}^n$ be a set of the center positions. For $i \in \{1, 2, \cdots, n\}$, The position C_i is given by

$$C_i = \ell(t_{C_i}), \quad t_{C_i} = \frac{C_i^z - P_i^z}{P_j^z - P_i^z}, \tag{3.119}$$

where C_i^z is a slice number lying on C_i. Here, C_1 lies on the slice just below that of including P_j and the last position C_n is determined so that the implant bar has the same length with that of the estimated teeth root region. Lastly, we draw a circle whose center is located at C_i. The radius is empirically set so that a generated implant bar shape looks realistic.

Case 3. Orthodontic brace

Generation of orthodontic braces requires very sophisticated understanding of tooth 3D geometry and anatomy to place braces in front of teeth and wire connecting the braces. There is a manual way using MIMICS software, which has been used

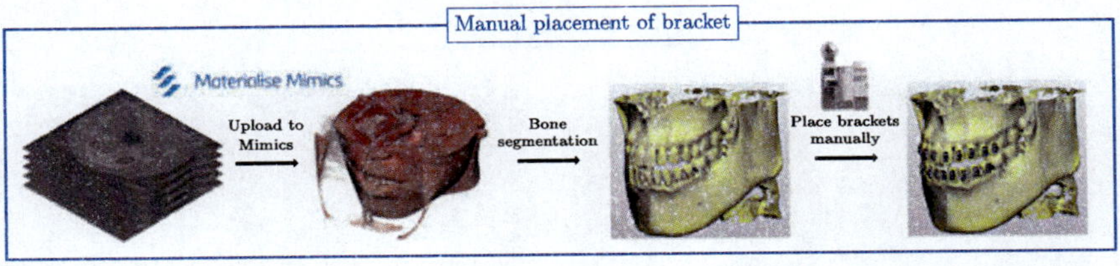

Fig. 3.50 Manual data generation of orthodontic brace with Mimics software

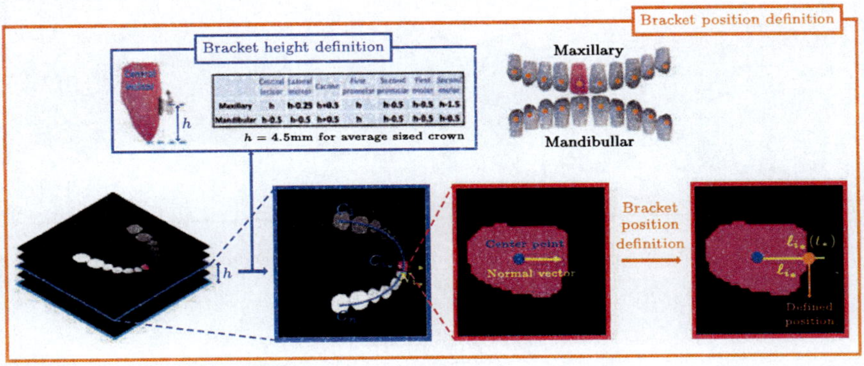

Fig. 3.51 Automated orthodontic brace position determination

for surgical planning in clinical dentistry. See Fig. 3.50. This manual procedure can provide accurate and realistic placement of orthodontic braces, but unfortunately it is very laborious and time-consuming. For this reason, we have been attempting to automate the data generation tool. This section explains our developing method, which is still not complete and far from satisfactory in realistic data generation.

In orthodontic brace placement, there are three main points; brace center position and angulation determination for an individual tooth, and wire construction.

The brace position is determined by the following process: We define a brace height for an individual tooth, which differs according to the tooth type and location in skull. This is based on the information in [4] that there are some height variations on the basis of the average height (4.5 mm), as shown in Fig. 3.51. Using this height information, for a given tooth, we are able to determine a slice where the center of the brace is placed. In the selected slice, we first find the center point of all teeth by taking the average position in an individual tooth mask on $\mathbf{M}_{\text{tooth}}$. A curve $\mathscr{C}$ is defined by

$$\mathscr{C}(t) = (C^x(t), C^y(t)) \text{ is a cubic spline and the curve } \mathscr{C} \text{ passes } \{C_i\}_{i=1}^{n_{\text{tooth}}}, \tag{3.120}$$

where $\{C_i\}_{i=1}^{n_{\text{tooth}}}$ denotes a set of all teeth center points. Let C_{i_*} be a center position of the tooth where a brace would be placed. On C_{i_*}, we define a normal vector N_{i_*} toward front of the tooth by using the Frenet frame as follows:

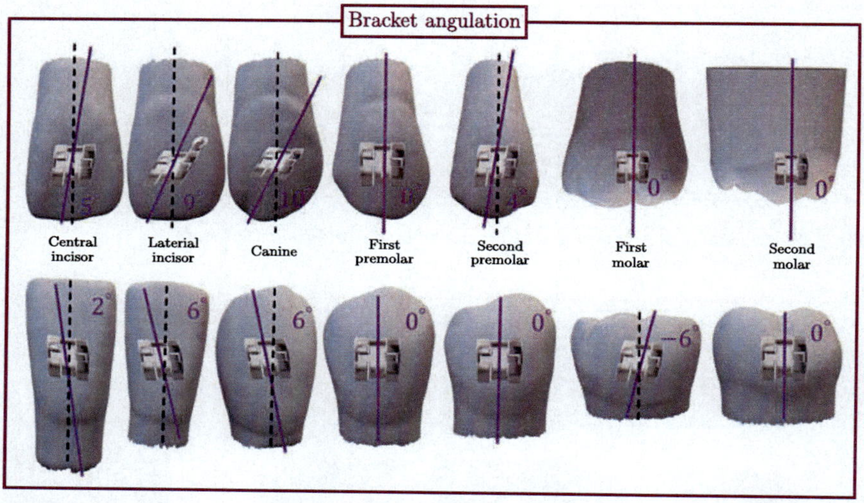

Fig. 3.52 Orthodontic brace angulation determination for each individual tooth

$$N_{i_*} = -\frac{dT/ds}{\|dT/ds\|}\bigg|_{\text{at } C_{i_*}}, \quad T = \frac{d\mathscr{C}}{ds}. \tag{3.121}$$

In our CBCT data, a dental arch is located as being convex downward. Hence, N_{i_*} is arranged toward the front of the tooth. See Fig. 3.51. Using C_{i_*} and N_{i_*}, we define a line by

$$\ell_{i_*}(t) = N_{i_*} * t + C_{i_*} \tag{3.122}$$

and find the intersection $\ell_{t_*}(t_\star)$ satisfying

$$\mathbf{M}^{z_*}_{\text{tooth}}(\ell_{i_*}(t_\star)) \neq 0 \text{ and } \mathbf{M}^{z_*}_{\text{tooth}}(\ell_{i_*}(t_\star + \epsilon)) = 0, \forall \epsilon > 0, \tag{3.123}$$

where $\mathbf{M}^{z_*}_{\text{tooth}}$ denote a 2D mask of the individual tooth at the selected slice. The point $\ell_{t_*}(t_\star)$ is set as the center position where the virtual brace is place.

We attach a brace with the typical shape given in [138] after taking a 3D rotation using angulation information in [4]. Here, the angulation is dependent on tooth type and location in the skull, as seen in Fig. 3.52.

This process is repeated to place a brace on every tooth. Lastly, in order to make a wire, the cubic spline interpolation is used again with all 3D center points of braces.

3.6.2.3 [Step 3] Metal Artifact Generation

The remaining part explains how to produce artificial metal artifacts using the generated metal implant mask. Let $\mathbf{M}_{\text{gen}}$ be a 3D binary mask of a generated metal implant.

A metal-artifacted sinogram is artificially produced via the Lambert–Beer's law in (3.33), adding electric and CT noise as follows:

$$\mathbf{P}_{\text{arti}} = -\ln\left(\int_{\mathbb{R}} \eta(E)\exp\{-\mathscr{R}\mathbf{M}_{\text{gen}} * \mu_E + \mathbf{n}\}dE\right), \tag{3.124}$$

where μ_E is the attenuation coefficient of a metallic object, which is a function of energy level, $\mathbf{n}$ is modeled by an addition of Poisson, Gaussian, and constant noise, and $\mathscr{R}$ represents a 3D dental cone-beam CBCT projection using TIGRE package [12]. Lastly, we obtain artifact data $\mathbf{P}$ by combining $\mathbf{P}_{\text{arti}}$ with a normal patient sinogram $\mathbf{P}_{\text{normal}}$ as follows:

$$\mathbf{P} = \mathbf{P}_{\text{arti}} + \mathbf{P}_{\text{normal}}. \tag{3.125}$$

The corresponding artifact-free label is given by

$$\mathbf{P}_{\star} = c\mathscr{R}\mathbf{M}_{\text{gen}} + \mathbf{P}_{\text{normal}}, \tag{3.126}$$

where c is a suitably chosen constant.

3.6.3 GAN-Based Synthetic-to-Realistic Image Refinement

This section describes a generative adversarial network (GAN)-based method that transforms the synthesized image into a more realistic CBCT image. The supervised learning process is based on the minimization of a computable empirical risk over a given training set; thereby, the size and quality of the training dataset significantly affect DL performance. Using the data generation method in the previous section, a sufficient number of paired data can be synthesized from normal patient data, whereas the synthesized data still have a lack of realism (e.g. secondary artifacts attributed to bone–metal interaction), which can cause degradation of DL performance on practical metal-artifacted CBCT image [28]. In order to compensate for the lack of realism, therefore, we attempt to use CycleGAN [137] for domain adaptation of reducing the gap between the source domain (i.e., generated metal-artifacted image) and target domain (i.e., practical metal-artifacted image). See Fig. 3.53 for a schematic explanation.

This section is based on the paper "Unpaired image-to-image translation using cycle-consistent adversarial networks" by Zhu et al.[137]. Throughout this section, the input $\mathbf{x}$ represents a synthesized image by the method in the previous section and the output $\mathbf{y}$ represents a practical CBCT image with metal artifacts. We assume that unpaired training samples $\{\mathbf{x}^{(n)}\}_{n=1}^{N}$ and $\{\mathbf{y}^{(m)}\}_{m=1}^{M}$ are drawn from unknown synthetic and real CBCT image distributions ($p_{\mathscr{X}}(\mathbf{x})$ and $p_{\mathscr{Y}}(\mathbf{y})$), respectively. The goal is

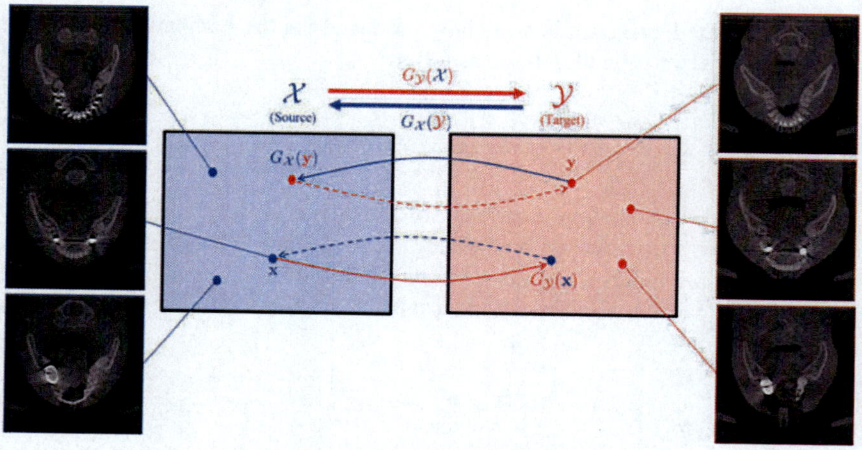

Fig. 3.53 CycleGAN-based synthetic-to-realistic image refinement

to find an optimal generator $G_{\mathscr{Y}} : \mathscr{X} \to \mathscr{Y}$ such that the output distribution over $\widehat{\mathbf{y}} = G_{\mathscr{Y}}(\mathbf{x})$ is a good approximation of $p_{\mathscr{Y}}(\mathbf{y})$ (i.e., $\widehat{\mathscr{Y}} = G_{\mathscr{Y}}(\mathscr{X})$ is distributed almost similarly to $\mathscr{Y}$).

The model of CycleGAN comprises two mappings $G_{\mathscr{Y}} : \mathscr{X} \to \mathscr{Y}$ (generating images for $\mathscr{Y}$) and $G_{\mathscr{X}} : \mathscr{Y} \to \mathscr{X}$ (generating images for $\mathscr{X}$). Similar to the standard GAN as in (3.117), this model uses the corresponding adversarial discriminators $D_{\mathscr{Y}}$ (for distinguishing between real $\mathbf{y}$ and fake $G_{\mathscr{Y}}(\mathbf{x})$) and $D_{\mathscr{X}}$ (for distinguishing between real $\mathbf{x}$ and fake $G_{\mathscr{X}}(\mathbf{y})$).

Roughly speaking, the main objective of CycleGAN is to find $(G^*_{\mathscr{Y}}, G^*_{\mathscr{X}})$ such that

$$(G^*_{\mathscr{Y}}, G^*_{\mathscr{X}}) := \underset{G_{\mathscr{Y}}, G_{\mathscr{X}}}{\text{argmin}} \ \underset{D_{\mathscr{X}}, D_{\mathscr{Y}}}{\text{argmax}} \ \mathscr{L}(G_{\mathscr{Y}}, G_{\mathscr{X}}, D_{\mathscr{Y}}, D_{\mathscr{X}}), \tag{3.127}$$

where the generator $G^*_{\mathscr{Y}}$ corresponds to our desired synthetic-to-realistic image refinement map. Here, the loss function $\mathscr{L}$ is given by

$$\mathscr{L}(G_{\mathscr{Y}}, G_{\mathscr{X}}, D_{\mathscr{X}}, D_{\mathscr{Y}}) := \mathscr{L}_{\text{GAN}}(G_{\mathscr{Y}}, D_{\mathscr{Y}}) + \mathscr{L}_{\text{GAN}}(G_{\mathscr{X}}, D_{\mathscr{X}}) + \lambda \mathscr{L}_{\text{cyc}}(G_{\mathscr{Y}}, G_{\mathscr{X}}), \tag{3.128}$$

where λ is a parameter controlling the contribution of the consistency. The first term $\mathscr{L}_{\text{GAN}}(G_{\mathscr{Y}}, D_Y)$ is for estimating optimal $G_{\mathscr{Y}}$ and $D_{\mathscr{Y}}$, given by

$$\mathscr{L}_{\text{GAN}}(G_{\mathscr{Y}}, D_{\mathscr{Y}}) = \mathbb{E}_{\mathbf{y}\sim p_{\text{data}}(\mathbf{y})}[\ln D_{\mathscr{Y}}(\mathbf{y})] + \mathbb{E}_{\mathbf{x}\sim p_{\text{data}}(\mathbf{x})}[\ln(1 - D_{\mathscr{Y}}(G_{\mathscr{Y}}(\mathbf{x}))] \tag{3.129}$$

where p_{data} denotes data distribution over a given dataset. In the same sense, the second term is for estimating $G_{\mathscr{X}}$ and $D_{\mathscr{X}}$, given by

$$\mathscr{L}_{\text{GAN}}(G_{\mathscr{X}}, D_{\mathscr{X}}) = \mathbb{E}_{\mathbf{x}\sim p_{\text{data}}(\mathbf{x})}[\ln D_{\mathscr{X}}(\mathbf{x})] + \mathbb{E}_{\mathbf{y}\sim p_{\text{data}}(\mathbf{y})}\left[\ln(1 - D_{\mathscr{X}}(G_{\mathscr{X}}(\mathbf{y})))\right]. \tag{3.130}$$

The last key term is the cycle consistency loss that attempts to impose invertability on image-to-image translation (i.e., $\mathbf{x} \approx G_{\mathscr{X}}(G_{\mathscr{Y}}(\mathbf{x}))$ and $\mathbf{y} \approx G_{\mathscr{Y}}(G_{\mathscr{X}}(\mathbf{y}))$). It is given by

$$\mathscr{L}_{\text{cyc}}(G_{\mathscr{Y}}, G_{\mathscr{X}}) = \mathbb{E}_{\mathbf{x}\sim p_{\text{data}}(\mathbf{x})}\left\|\mathbf{x} - G_{\mathscr{X}}(G_{\mathscr{Y}}(\mathbf{x}))\right\| + \mathbb{E}_{\mathbf{y}\sim p_{\text{data}}(\mathbf{y})}\left\|\mathbf{y} - G_{\mathscr{Y}}(G_{\mathscr{X}}(\mathbf{y}))\right\|, \tag{3.131}$$

where $\|\cdot\|$ is ℓ^1 or ℓ^2 norm. Here, it is desired that the generator's fidelity $\|\mathbf{x} - G_{\mathscr{X}} \circ G_{\mathscr{Y}}(\mathbf{x})\|$ is reasonably small for every $\mathbf{x} \sim p_{\mathscr{X}}(\mathbf{x})$.

The additional training of $G^*_{\mathscr{X}}$ $(\approx (G^*_{\mathscr{Y}})^{-1}|_{\text{on } \mathscr{Y}})$ can be viewed as a means of preventing loss of information on original data $\mathbf{x}$ during the image-to-image conversion via $G^*_{\mathscr{Y}}$. In our case, especially, it may produce an effect that the primary metal artifacts in $\mathbf{x}$ (generated by the CT forward model) are not considerably affected while making up for the lack of realism via $G^*_{\mathscr{Y}}$.

In practice, the adversarial loss in (3.129) and (3.130) can be replaced by various kinds of GAN loss function (e.g., least squares loss [82] and Wasserstein loss [43]) in order to get high performance or training stability.

3.7 Discussion and Conclusion

In CT, the presence of highly attenuating materials such as metal, concentrated iodinated contrast media or bone complicates reconstruction [8] by violating an assumption of the forward model: that sinogram data are the Radon transform of an image. The increasing use of metallic implants in a generally aging population makes metal-induced artifacts a major impediment to CT diagnosis.

In the field of dental and medical radiography, there is an increasing demand for effective metal artifact reduction (MAR) in CT, since the metallic objects-related artifacts seriously degrade the image of CT, resulting in loss of information about teeth and/or other biological structures. The presence of metallic objects such as hip replacements, dental fillings, surgical clips, and pacemaker wires causes streak artifacts in the field of view of CT. The artifacts appear as dark and bright streaks, and are caused by physical effects such as beam hardening, scattered radiation, nonlinear partial volume effect (NLPV), and noise. MAR is becoming increasingly crucial as the number of the aged with artificial prostheses and metallic implants is rapidly increasing with a rapid aging population.

The metal artifacts are caused mainly from the mismatch of the forward model of the filtered back-projection (FBP); the presence of metallic subjects in an imaging subject violates the model's assumption of the CT sinogram data being the Radon transform of an image. The mismatched projection data due to these effects cause severe streaking and shading artifacts in the reconstructed CT images. Metal artifacts are caused by the effects of beam-hardening of polychromatic X-ray photon beams and the various complicated metal-tissue interactions such as scattering, nonlinear partial volume effects and electric noise [8, 53]. Although extensive research efforts have sought to improve CT reconstruction methods [24, 61, 79, 128], tackling metal-related artifacts is a very challenging problem because the inconsistent data induced by metal depend non-linearly on the geometries and placements of the metal objects.

The mismatched projection data due to these effects cause severe streaking and shading artifacts in the reconstructed CT images, as shown in Fig. 3.21. Metal artifacts are caused by the effects of beam-hardening of polychromatic X-ray photon beams and the various complicated metal–tissue interactions such as scattering, nonlinear partial volume effects and electric noise [8, 53]. Although extensive research efforts have sought to improve CT reconstruction methods [24, 61, 79, 128], tackling metal-related artifacts is a very challenging problem because the inconsistent data induced by metal depend non-linearly on the geometries and placements of the metal objects.

Accurate and robust identification of cephalometric landmarks for three-dimensional (3D) CT images is an important task for the diagnosis, surgical planning, growth analysis, and treatment evaluation. Recent advances in imaging technology have led to the transition from two-dimensional (2D) cephalometry to 3D using CT scan images. 3D cephalometry holds several advantages, including the accurate identification of anatomical structures, avoidance of geometric distortion of the image, and ability to evaluate complicated facial structure. However, 3D cephalometry being a manual operation, it requires time-consuming and labor-intensive work.

Cephalometric analysis is conducted basically through cephalometric annotation, i.e., landmark detection for meaningful anatomical structures. It requires a high level of expertise, experience, and time. These difficulties are aggravated as 2D cephalometric analysis gives way to 3D because of remarkable increase in data volume and geometric complexity. Several different approaches have been proposed to address these limitations using automatic 3D annotation systems [18, 41, 42, 76, 81, 111]. However, because our search of the literature found no trials of machine learning-based automatic 3D cephalometric analysis, we wanted to try it.

Various MAR methods have been suggested in the last few decades. Most MAR methods fall into one of three categories: inpainting-based methods (e.g. interpolation, wavelet, heat diffusion, Euler's Elastica, TV-H^{-1} flow), iterative reconstruction methods (e.g. iterative FBP, weighted least squares methods), and hybrid methods that combine the first two methods. However, existing MAR methods do not reduce the metal artifacts effectively, and they may introduce new streaking artifacts which do not exist before the application of the methods.

Recently, our team established a rigorous characterization on the metal streaking artifacts using the notion of the wavefront set from microlocal analysis for the first time. We found that the metal streaking artifacts arise mainly from the boundary geometry of the metal region. The metal streaking artifacts are produced only when the wavefront set of the Radon transform of the characteristic function of a metal region does not contain the wavefront set of the square of the Radon transform. We also developed a new MAR method, making it possible to reveal background projection data hidden in metal regions.

Based on prior knowledge of the structure of metal artifacts, it is possible to remove the metal streaks efficiently to provide usefulness in the field of diagnosis, preoperative and presurgical assessment, and surgical navigation and the workup for rapid prototyping.

Despite various effort seeking to reduce metal artifacts, metal streaking artifacts continue to pose difficulties, and the development of suitable reduction methods remains challenging. As mentioned in Sect. 3.3.2, the streaking artifacts arise mainly from the geometry of the boundary of the metallic objects using the concept of micro-local analysis. From these theoretical studies, the structure of streaking artifacts can be extracted from the reconstructed CT image provided that the geometry information of the teeth and mandible is given.

Acknowledgements This research were supported by Samsung Science & Technology Foundation (No. SRFC-IT1902-09). Seo was supported by a grant of the Korea Health Technology R &D Project through the Korea Health Industry Development Institute (KHIDI), funded by the Ministry of Health & Welfare, Republic of Korea (grant number : HI20C0127). We are deeply grateful to HDXWILL for their help and collaboration.

References

1. Abdoli, M., Ay, M.R., Ahmadian, A., Dierckx, R.A., Zaidi, H.: Reduction of dental filling metallic artifacts in CT-based attenuation correction of PET data using weighted virtual sinograms optimized by a genetic algorithm. Med. Phys. **37**, 6166–6177 (2010)
2. Abdurahman, S., Frysch, R., Bismark, R., Melnik, S., Beuing, O., Rose, G.: Beam hardening correction using cone beam consistency conditions. IEEE Trans. Med. Imag. **37**(10), 2266–2277 (2018)
3. Aichert, A., Berger, M., Wang, J., Maass, N., Doerfler, A., Hornegger, J., Maier, A.K.: Epipolar consistency in transmission imaging. IEEE Trans. Med. Imag. **34**(11), 2205–2219 (2015)
4. Alexander, R.G.: The principles of the Alexander discipline. Semin. Orthodont. **7**(2), 62–66 (2001)
5. Alvarez, R.E., Macovski, A.: Energy-selective reconstructions in X-ray computerized tomography. Phys. Med. Biol. **21**, 733 (1976)
6. Arjovsky, M., Chintala, S., Bottou, L.: Wasserstein GAN (2017)
7. Bal, M., Spies, L.: Metal artifact reduction in CT using tissue-class modeling and adaptive prefiltering. Med. Phys **33**, 2852–2859 (2006)
8. Barrett, J.F., Keat, N.: Artifacts in CT: recognition and avoidance. Radiographics **24**, 1679–1691 (2004)
9. Bazalova, M., Beaulieu, L., Palefsky, S., Verhaegen, F.: Correction of CT artifacts and its influence on Monte Carlo dose calculations. Med. Phys. **34**, 2119–2132 (2007)
10. Bayaraa, T., Hyun, C.M., Jang, T.J., Lee, S.M., Seo, J.K.: A two-stage approach for beam hardening artifact reduction in low-dose dental CBCT. IEEE Access **8**, 225981–225994 (2020)
11. Beer "Bestimmung der Absorption des rothen Lichts in farbigen Flussigkeiten". Annalen der Physik und Chemie vol. 86, pp. 78–88 (1852)
12. Biguri, A., Dosanjh, M., Hancock, S., Soleimani, M.: TIGRE: a MATLAB-GPU toolbox for CBCT image reconstruction. Biomed. Phys. Eng. Express. **2**(5), 055010 (2016)
13. Van de Casteele, E.: Model-based approach for beam hardening correction and resolution measurements in microtomography. Ph.D. dissertation, Dept. Natuurkunde., Antwerpen Univ., Antwerp, Belgium (2004)

14. Chang, W., Loncaric, S., Huang, G., Sanpitak, P.: Asymmetric fan transmission CT on SPECT systems. Phys. Med. Biol. **40**, 913 (1995)
15. Cho, P.S., Johnson, R.H., Griffin, T.W.: Cone-beam CT for radiotherapy applications. Phys. Med. Biol. **40**, 1863 (1995)
16. Choi, J., Kim, M.W., Seong, W., Ye, J.C.: Compressed sensing metal artifact removal in dental CT. In: Proceedings of the IEEE International Symposium Biomedical Imaging, pp. 334–337 (2009)
17. Choi, J., Kim, K.S., Kim, M.W., Seong, W., Ye, J.C.: Sparsity driven metal part reconstruction for artifact removal in dental ct. J. Xray. Sci. Technol. **19**(4), 457–475 (2011)
18. Codari, M., Caffini, M., Tartaglia, G.M., Sforza, C., Baselli, G.: Computer-aided cephalometric landmark annotation for CBCT data. Int. J. Comput. Assist. Radiol. Surg. **12**(1), 113–121 (2017)
19. Cormack, A.M.: Representation of a function by its line integrals, with some radiological applications. J. Appl. Phys. **34**, 2722–2727 (1963)
20. Clackdoyle, R.: Necessary and sufficient consistency conditions for fanbeam projections along a line. IEEE Trans. Nucl. Sci. **60**(3) (2013)
21. Clackdoyle, R., Desbat, L., Lesaint, J., Rit, S.: Data consistency conditions for cone-beam projections on a circular trajectory. IEEE Signal Process. Lett. **23**(12), 1746–1750 (2016)
22. Duistermaat, J.J., Hörmander, L.: Fourier integral operators. II. Acta Mathematica **128**(1), 183–269 (1972)
23. De Man, B., Nuyts, J., Dupont, P., Marchal, G., Suetens, P.: Metal streak artifacts in X-ray computed tomography: a simulation study. In: 1998 IEEE Nuclear Science Symposium Conference Record. In 1998 IEEE Nuclear Science Symposium and Medical Imaging Conference (Cat. No. 98CH36255), vol. 3, pp. 1860–1865 (1998)
24. De Man, B., Nuyts, J., Dupont, P., Marchal, G., Suetens, P.: Metal streak artifacts in X-ray computed tomography?: a simulation study. IEEE Trans. Nucl. Sci. **46**, 691–696 (1999)
25. De Man, B., Nuyts, J., Dupont, P., Marchal, G., Suetens, P.: An iterative maximum-likelihood polychromatic algorithm for CT. IEEE Trans. Med. Imag. **20**, 999–1008 (2001)
26. De Hoop, M.V., Smith, H., Uhlmann, G., Van der Hilst, R.D.: Seismic imaging with the generalized Radon transform: a curvelet transform perspective. Inverse Probl. **25**, 025005 (2009)
27. Draenert, F., Coppenrath, E., Herzog, P., Muller, S., Mueller-Lisse, U.: Beam hardening artefacts occur in dental implant scans with the NewTom R cone beam CT but not with the dental 4-row multidetector CT. Dentomaxillofac Radiol. **36**, 198–203 (2007)
28. Du, M., Gao, H., Liang, K., Liu, Y., Xing, Y.: Unsupervised domain adaptation for practical metal artefact reduction in X-ray CT. In: The 6th International Conference on Image Formation in X-Ray Computed Tomography (2020)
29. Duan, X., Zhang, L., Xiao, Y., Cheng, J., Jianping, C., Chen, Z., Xing, Y.: Metal artifact reduction in CT images by sinogram TV inpainting. In: 2008 IEEE Nuclear Science Symposium Conference Record (2008)
30. Elbakri, I.A., Fessler, J.A.: Statistical image reconstruction for polyenergetic X-ray computed tomography. IEEE Trans. Med. Imag. **21**, 89–99 (2002)
31. Feldkamp, L.A., Davis, L.C., Kress, J.W.: Practical cone-beam algorithm. J. Opt. Soc. Am. A **1**(6), 612–619 (1984)
32. Finch, D., Lan, I.R., Uhlmann, G.: Microlocal analysis of the X-ray transform with sources on a curve. In: Inside Out, Inverse Problems and Applications, vol. 47 (2003)
33. Frikel, J., Quinto, E.T.: Characterization and reduction of artifacts in limited angle tomography. Inverse Probl. **29**(12), 125007 (2013)
34. Finch, D.V., Solmon, D.C.: A characterization of the range of the divergent beam X-ray transform. SIAM J. Math. Anal. **14**, 767–771 (1983)
35. Greenleaf, A., Uhlmann, G.: Nonlocal inversion formulas for the X-ray transform. Duke Math. J. **58**, 205–240 (1989)
36. Grangeat, P.: Mathematical framework of cone beam 3D reconstruction via the first derivative of the radon transform. In: Mathematical Methods in Tomography. Lecture Notes in Mathematics, vol. 1497 (1991)

37. Gjesteby, L., De Man, B., Jin, Y., Paganetti, H., Verburg, J., Giantsoudi, D., Wang, G.: Metal artifact reduction in CT: where are we after four decades? IEEE Access **4**, 5826–5849 (2016)
38. Gjesteby, L., Yang, Q., Xi, Y., Shan, H., Claus, B., Jin, Y., Wang, G.: Deep learning methods for CT image-domain metal artifact reduction. In Developments in X-Ray Tomography XI. Proceedings of the SPIE International Society for Optics and Engineerings, vol. 10391, p. 103910W (2017)
39. Grangeat, P.: Mathematical framework of cone beam 3D reconstruction via the first derivative of the Radon transform. In: Mathematical Methods in Tomography, pp. 66–97. Springer, Berlin (1991)
40. Gupta, J., Ali, S.P.: Cone beam computed tomography in oral implants. Natl. J. Maxillofac. Surg. **4**, 2 (2013)
41. Gupta, A., Kharbanda, O.P., Sardana, V., Balachandran, R., Sardana, H.K.: A knowledge-based algorithm for automatic detection of cephalometric landmarks on CBCT images. Int. J. Comput. Assist. Radiol. Surg. **10**(11), 1737–1752 (2015)
42. Gupta, A., Kharbanda, O.P., Sardana, V., Balachandran, R., Sardana, H.K.: Accuracy of 3D cephalometric measurements based on an automatic knowledge-based landmark detection algorithm. Int. J. Comput. Assist. Radiol. Surg. **11**(7), 1297–1309 (2016)
43. Gulrajani, I., Ahmed, F., Arjovsky, M., Dumoulin, V., Courville, A.: Improved training of wasserstein gans (2017). arXiv:1704.00028
44. GE Healthcare, Smart Metal Artifact Reduction (MAR) (2013)
45. Goodfellow, I., Pouget-Abadie, J., Mirza, M., Xu, B., Warde-Farley, D., Ozair, S., Courville, A., Bengio, Y.: Generative adversarial nets. In: Advances in Neural Information Processing Systems (NIPS2014) (2014)
46. Gu, J., Zhang, L., Yu, G., Xing, Y., Chen, Z.: Metal artifacts reduction in CT images through Euler's elastica and curvature based sinogram inpainting. In: Medical Imaging 2006: Image Processing. International Society for Optics and Photonics, vol. 6144, p. 614465 (2006)
47. Gjesteby, L., Shan, H., Yang, Q., Xi, Y., Claus, B., Jin, Y., De Man, B., Wang, G.: A dual-stream deep convolutional network for reducing metal streak artifacts in CT images. Phys. Med, Biol (2019)
48. Helgason, S.: The Radon Transform. Birkhauser, Boston (1980)
49. Herman, G.T., Trivedi, S.S.: A comparative study of two postreconstruction beam-hardening correction methods. IEEE Trans. Med. Imag. **2**, 128–135 (1983)
50. Hörmander, L.: Fourier integral operators. Acta Mathematica **127**(1), 79 (1971)
51. Hörmander, L.: The Analysis of Linear Partial Differential Operators. III, volume 274 of Grundlehren der Mathematischen Wissenschaften [Fundamental Principles of Mathematical Sciences]. Springer, Berlin (1983)
52. Hounsfield, G.N.: Computerized transverse axial scanning (tomography): I. Description of system Br. J. Radiol. **46**, 1016–22 (1973)
53. Hsieh, J.: Computed Tomography Principle, Design, Artefacts, and Recent Advances. SPIE, Belingham WA (2003)
54. Hsieh, J., Nett, B., Yu, Z., Sauer, K., Thibault, J.B., Bouman, C.A.: Recent advances in CT image reconstruction. Curr. Radiol. Rep. **1**, 39–51 (2013)
55. Hubbell, J.H., Seltzer, S.M.: Tables of X-ray mass attenuation coefficients and mass energy-absorption coefficients from 1 keV to 20 MeV for elements Z = 1 to 92 and 48 additional substances of dosimetric interest. NIST (1996)
56. Hyun, C.M., Baek, S.H., Lee, M., Lee, S.M., Seo, J.K.: Deep learning-based solvability of underdetermined inverse problems in medical imaging. Med. Image Anal. **69**, 101967 (2021)
57. Hyun, C.M., Bayaraa, T., Yun, H.S., Jang, T.J., Park, H.S., Seo, J.K.: Deep learning method for reducing metal artifacts in dental cone-beam CT using supplementary information from intra-oral scan. Physics in Medicine and Biology **67**(17), 175007 (2022)
58. Huang, X., Wang, J., Tang, F., Zhong, T., Zhang, Y.: Metal artifact reduction on cervical ct images by deep residual learning. Biomed. Eng. Online **17** (2018)
59. Rehani, M.M., Gupta, R., Bartling, S., Sharp, G.C., Pauwels, R., Berris, T., Boone, J.M.: Radiological Protection in Cone Beam Computed Tomography (CBCT), vol. 129. ICRP Publication (2015)

60. Jang, T.J., Kim, K.C., Cho, H.C., Seo, J.K.: A fully automated method for 3D individual tooth identification and segmentation in dental CBCT. IEEE PAMI (2021)
61. Kalender, W.A., Hebel, R., Ebersberger, J.: Reduction of ct artifacts caused by metallic implants. Radiology **164**(2), 576–577 (1987)
62. Katsevich, A.: Cone beam local tomography. SIAM J. Appl. Math. **59**, 2224–2246 (1999)
63. Katsevich, A.: Improved cone beam local tomography. Inverse Probl. **22**, 627 (2006)
64. Kachelriess, M., Krauss, A.: Iterative Metal Artifact Reduction (iMAR): Technical Principles and Clinical Results in Radiation Therapy. In: Siemens Healthcare (2016)
65. Karras, T., Aila, T., Laine, S., Lehtinen, J.: Progressive growing of gans for improved quality, stability, and variation. ICLR (2018)
66. Klein, O., Nishina, Y.: Über die Streuung von Strahlung durch freie Elektronen nach der neuen relativistischen Quantendynamik von Dirac. Z. Phys. **52**(11–12), 853–868 (1929)
67. Lambert, J.H., Anding, E.: Lamberts Photometrie: Photometria, Sive De Mensura et Gradibus Luminis, Colorum et Umbrae. W. Engelmann, pp. 1728–1777 (1892)
68. Lewitt, R.M., Bates, R.H.T.: Image reconstruction from projections: Iv: projection completion methods (computational examples). Optik **50**, 269–278 (1978)
69. Lehmann, L., Alvarez, R., Macovski, A., Brody, W., Pelc, N., Riederer, S.J., Hall, A.: Generalized image combinations in dual KVP digital radiography. Med. Phys. **8**, 659–667 (1981)
70. Lee, S.M., Bayaraa, T., Jeong, H., Hyun, C.M., Seo, J.K.: A direct sinogram correction method to reduce metal-related beam-hardening in computed tomography. IEEE Access **7**, 128828–128836 (2019)
71. Lemmens, C., Faul, D., Nuyts, J.: Suppression of metal artifacts in CT using a reconstruction procedure that combines MAP and projection completion. IEEE Trans. Med. Imag. **28**, 256–260 (2008)
72. Lesaint, J.: Data consistency conditions in X-ray transmission imaging and their application to the self-calibration problem. Doctoral dissertation, Université Grenoble Alpes (2019)
73. Lin, W.A., Liao, H., Peng, C., Sun, X., Zhang, J., Luo, J., Chellappa, R., Zhou, S.K.: Dudonet: Dual domain network for ct metal artifact reduction. In: Proceedings of the IEEE Conference on Computer Vision and Pattern Recognition, pp. 10512–10521 (2019)
74. Ludwig, D.: The radon transform on Euclidean space. Commun. Pure Appl. Math. **19**, 49–81 (1966)
75. Lyu, Y., Fu, J., Peng, C., Zhou, S.K.: U-DuDoNet: Unpaired dual-domain network for CT metal artifact reduction. arXiv preprint arXiv:2103.04552 (2021)
76. Makram, M., Kamel, H.: Reeb graph for automatic 3D cephalometry. IJIP **8**(2), 17–29 (2014)
77. Miracle, A., Mukherji, S.: Conebeam CT of the head and neck, part 2: clinical applications. Am. J. Neuroradiol. **30**, 1285–1292 (2009)
78. Mehranian, A., Ay, M.R., Rahmim, A., Zaidi, H.: X-ray CT metal artifact reduction using wavelet domain sparse regularization. IEEE Trans. Med. Imag. **32**, 1702–1722 (2013)
79. Meyer, E., Raupach, R., Lell, M., Schmidt, B., Kachelrie, M.: Normalized metal artifact reduction (NMAR) in computed tomography. Med. Phys. **37**, 5482–5493 (2010)
80. Menvielle, N., Goussard, Y., Orban, D., Soulez, G.: Reduction of beam-hardening artifacts in X-ray CT. In: 2005 IEEE Engineering in Medicine and Biology 27th Annual Conference, pp. 1865–1868 (2006)
81. Montùfar, J., Romero, M., Scougall-Vilchis, R.J.: Automatic 3-dimensional cephalometric landmarking based on active shape models in related projections. Am. J. Orthod. Dentofacial. Orthop. **153**(3), 449–458 (2018)
82. Mao, X., Li, Q., Xie, H., Lau, R.Y., Wang, Z., Smolley, S.P.: Least squares generative adversarial networks. In: CVPR. IEEE (2017)
83. Nakao, M., Imanishi, K., Ueda, N., Imai, Y., Kirita, T., Matsuda, T.: Regularized three-dimensional generative adversarial nets for unsupervised metal artifact reduction in head and neck CT images. IEEE Access **8**, 109453–109465 (2020)
84. Natterer, F.: The Mathematics of Computerized Tomography. SIAM (1986)
85. Nelson, S.J.: Wheeler's Dental Anatomy. Physiology and Occlusion-e-Book, Elsevier Health Sciences (2014)

86. Philips Healthcare "Metal artifact reduction for orthopedic implants (O-MAR)," White Paper, Philips CT Clinical Science. Andover, Massachusetts (2012)
87. Park, H.S., Choi, J.K., Seo, J.K.: Characterization of metal artifacts in X-ray computed tomography. Commun. Pure Appl. Math. (2017). https://doi.org/10.1002/cpa.21680
88. Park, H.S., Hwang, D., Seo, J.K.: Metal artifact reduction for polychromatic X-ray CT based on a beam hardening corrector. IEEE Tran. Med. Imag. **35**, 480–487 (2016)
89. Park, H.S., Chung, Y.E., Seo, J.K.: Computed tomographic beam hardening artefacts: mathematical characterization and analysis. Philos. Trans. R. Soc. A (2015). https://doi.org/10.1098/rsta.2014.0388
90. Park, H.S., Choi, J.K., Park, K.R., Kim, K.S., Lee, S.H., Ye, J.C., Seo, J.K.: Metal artifact reduction in CT by identifying missing data hidden in metals. J. X-ray Sci. Technol. **21**, 357–72 (2013)
91. Park, H.S., Lee, S.M., Kim, H.P., Seo, J.K., Chung, Y.E.: CT sinogram-consistency learning for metal-induced beam hardening correction. Med. Phy. **45**, 5376–5384 (2018)
92. Pauwels, R., Jacobs, R., Bosmans, H., Schulze, R.: Future prospects for dental cone beam CT imaging. Imaging Med. **4**(5), 551–563 (2012)
93. Peng, C., Li, B., Li, M., Wang, H., Zhao, Z., Qiu, B., Chen, D.Z.: An irregular metal trace inpainting network for x-ray CT metal artifact reduction. Med. Phys. **47**, 4087–4100 (2020)
94. Petersen, B.E.: Introduction to the Fourier Transform and Pseudo-Differential Operators. Pitman Advanced Publishing Program (1983)
95. Poludniowski, G., Landry, G., DeBlois, F., Evans, P.M., Verhaegen, F.: Spekcalc: a program to calculate photon spectra from tungsten anode x-ray tubes. Phys. Med. Biol. **54**, N433 (2009)
96. Patch, S.: Consistency conditions upon 3D CT data and the wave equation. Phys. Med. Biol. **47**, 2637–2650 (2002)
97. Quinto, E.T.: The dependence of the generalized Radon transform on defining measures. Trans. Amer. Math. Soc. **257**, 331–346 (1980)
98. Quinto, E.T.: Singularities of the X-ray transform and limited data tomography in r^2 and r^3. SIAM J. Math. Anal. **24**, 1215–1225 (1993)
99. Quinto, E.T.: An introduction to X-ray tomography and Radon transforms. In: Proceedings of Symposia in Applied Mathematics, vol. 63 (2006)
100. Quinto, E.T.: Local algorithms in exterior tomography. J. Comput. Appl. Math. **199**, 141–148 (2007)
101. Quinto, E.T., Rullgaard, H.: Local singularity reconstruction from integrals over curves in r^3. Inverse Probl. Imag. **7**, 585–609 (2013)
102. Radon, J.H.: Uber die Bestimmung von Funktionen durch ihre Integralwerte längs gewisser Mannigfaltig. Ber. vor Sächs. Akad. Wiss. **69**, 262–77 (1917)
103. Radon, J.: 1.1 über die bestimmung von funktionen durch ihre integralwerte längs gewisser mannigfaltigkeiten. Classic Papers Modern Diagnost. Radiol. **5**, 21 (2005)
104. Ramm, A.G., Katsevich, A.I.: The Radon Transform and Local Tomography. Taylor & Francis (1996)
105. Ramm, A.G., Zaslavsky, A.I.: Reconstructing singularities of a function from its Radon transform. Math. Comput. Model. **18**, 109–138 (1993)
106. Rodet, T., Noo, F., Defrise, M.: The cone-beam algorithm of Feldkamp, Davis, and Kress preserves oblique line integrals. Med. Phys. **31**(7), 1972–1975 (2004)
107. Razavi, T., Palmer, R.M., Davies, J., Wilson, R., Palmer, P.J.: Accuracy of measuring the cortical bone thickness adjacent to dental implants using cone beam computed tomography. Clin. Oral Implants Res. **21**, 718–725 (2010)
108. Redmon, J., Divvala, S., Girshick, R., Farhadi, A.: You only look once: Unified, real-time object detection. In: Proceedings of the IEEE Conference on Computer Vision and Pattern Recognition, pp. 779–788 (2016)
109. Roeske, J.C., Lund, C., Pelizzari, C.A., Pan, X., Mundt, A.J.: Reduction of computed tomography metal artifacts due to the Fletcher-Suit applicator in gynecology patients receiving intracavitary brachytherapy. Brachytherapy **2**, 207–214 (2003)

110. Ronneberger, O., Fischer, P., Brox, T.: U-net: Convolutional networks for biomedical image segmentation. In: Proceedings of the Medical Image Computing and Computer Assisted Intervention, pp. 234–241 (2015)
111. Shahidi, S., Oshagh, M., Gozin, F., Salehi, P., Danaei, S.M.: Accuracy of computerized automatic identification of cephalometric landmarks by a designed software. Dentomaxillofac. Radiol. **42**(1), 20110187–20110187 (2013)
112. Sanders, M., Hoyjberg, C., Chu, C., Leggitt, V., Kim, J.: Common orthodontic appliances cause artifacts that degrade the diagnostic quality of CBCT images. J. Calif. Dent. Assoc. **35**, 850–857 (2007)
113. Sittig, D.F., Ash, J.S., Ledley, R.S.: The story behind the development of the first whole-body computerized tomography scanner as told by Robert S. Ledley. J. Am. Med. Inform. Assoc. **13**(5), 465–469 (2006)
114. Sukovic, P.: Cone Beam Computed Tomography in craniofacial imaging. Orthod. Craniofac. Res. **6**, 31–36 (2003)
115. Scarfe, W., Azevedo, B., Toghyani, S., Farman, A.: Cone beam computed tomographic imaging in orthodontics. Aust. Dent. J. 33–50 (2017)
116. Schulze, R.K.W., Berndt, D., D'Hoedt, B.: On cone-beam computed tomography artifacts induced by titanium implants. Clin. Oral Implants Res. **21**, 100–107 (2010)
117. Schulze, R., Heil, U., Grob, D., Bruellmann, D., Dranischnikow, E., Schwanecke, U., Schoemer, E.: Artefacts in CBCT: a review. Dentomaxillofac Radiol. **40**, 265–273 (2011)
118. Shen, J., Chan, T.F.: Mathematical models for local nontexture inpaintings. SIAM J. Appl. Math. **62**, 1019–1043 (2002)
119. O'Sullivan, J.A., Benac, J.: Alternating minimization algorithms for transmission tomography. IEEE Trans. Med. Imag. **26**, 283–297 (2007)
120. Stonestrom, J.P., Alvarez, R.E., Macovski, A.: A framework for spectral artifact corrections in X-ray CT. IEEE Trans. Biomed. Eng. **28**(2), 128–141 (1981)
121. Sharma, K.S., Gong, H., Ghasemalizadeh, O., Yu, H., Wang, G., Cao, G.: Interior micro?CT with an offset detector. Med. Phys. **41**, 061915 (2014)
122. Trèves, F.: Introduction to Pseudodifferential and Fourier Integral Operators Volume 2: Fourier Integral Operators, vol. 2. Springer Science & Business Media (1980)
123. Tuy, H.K.: An inversion formula for cone-beam reconstruction. SIAM J. Appl. Math. **43**(3), 546–552 (1983)
124. Tisson, G.: Reconstruction from transversely truncated cone beam projections in micro-tomography. Universiteit Antwerpen, Faculteit Wetenschappen, Departement Fysica (2006)
125. Yu, L., Leng, S., McCollough, C.H.: Dual-energy CT based monochromatic imaging. AJR Am. J. Roentgenol. **199**, S9–S15 (2012)
126. Yu, L., Zhang, Z., Li, X., Xing, L.: Deep sinogram completion with image prior for metal artifact reduction in CT images. IEEE Trans. Med. Imag. **40**(1), 228–238 (2020)
127. Yi, X., Walia, E., Babyn, P.: Generative adversarial network in medical imaging: a review. Med. Image Anal. **58**, 101552 (2019)
128. Wang, G., Snyder, D.L., O'Sullivan, J.A., Vannier, M.W.: Iterative deblurring for CT metal artifact reduction. IEEE Trans. Med. Imag. **15**, 657–64 (1996)
129. Williamson, J.F., Whiting, B.R., Benac, J., Murphy, R.J., Blaine, G.J., O'Sullivan, J.A., Politte, D.G., Snyder, D.L.: Prospects for quantitative computed tomography imaging in the presence of foreign metal bodies using statistical image reconstruction. Med. Phys. **29**, 2404–2418 (2002)
130. Wei, Y., Yu, H., Wang, G.: Integral invariants for computed tomography. IEEE Signal Process. Let. **13**(9), 549–552 (2006)
131. Wurfl, T., Maab, N., Dennerlein, F., Huang, X., Maier, A.K.: Epipolar consistency guided beam hardening reduction-ECC2. In: Proceedings of the 14th International Meeting Fully Three-Dimensional Image Reconstruction Radiology and Nuclear Medicine, pp. 181–185 (2017)
132. Zhao, S., Robeltson, D.D., Wang, G., Whiting, B., Bae, K.T.: X-ray CT metal artifact reduction using wavelets: an application for imaging total hip prostheses. IEEE Trans. Med. Imag. **19**, 1238–1247 (2000)

133. Zhao, S., Bae, K.T., Whiting, B., Wang, G.: A wavelet method for metal artifact reduction with multiple metallic objects in the field of view. J. Xray Sci. Technol. **10**, 67–76 (2002)
134. Zhang, Y., Yan, H., Jia, X., Yang, J., Jiang, S.B., Mou, X.: A hybrid metal artifact reduction algorithm for x-ray CT. Med. Phys. **40**, 041910 (2013)
135. Zhang, D.: Single energy metal artifact reduction a reliable metal management tool in CT, White Paper (2017)
136. Zhang, Y., Yu, H.: Convolutional neural network based metal artifact reduction in X-ray computed tomography. IEEE Trans. Med. Imag. **37**, 1370–1381 (2018)
137. Zhu, J.Y., Park, T., Isola, P., Efros, A.A.: Unpaired image-to-image translation using cycle-consistent adversarial networks. In: Proceedings of the IEEE International Conference on Computer Vision, pp. 2223–2232 (2017)
138. Thingiverse "Orthodontic bracket." https://www.thingiverse.com/thing:2044227

Chapter 4
Artificial Intelligence for Digital Dentistry

Tae Jun Jang, Sang-Hwy Lee, Hye Sun Yun, and Jin Keun Seo

Abstract Digital dentistry is evolving rapidly with the rapid innovation of artificial intelligence (AI) and the advancement of an AI-based digital platform that integrates 3D jaw–teeth–face data from various imaging devices such as cone-beam computerized tomography (CBCT), oral scanner, face scanner, 3D tracking devices, and others. Digital dentistry equipped with the AI-based integrated platform enables dentists to provide accurate diagnoses and treatment while saving time and cost, significantly improving digital workflow. Additionally, digital dentistry also improves patient satisfaction by increasing the patient's comfort and decreasing the chance of revisiting the dental clinic thanks to the enhanced accuracy level of dental treatment on the teeth, gingiva and occlusion. In digital dentistry, the 3D digital composite model obtained from image data of CBCT, and oral and facial scan will be an essential tool for almost all processes, including virtual treatment planning and on-screen simulation of surgical or dental treatment. Noting that the dental regions of 3D CT data do not have the level of resolution to be used directly for treatment, the jaw–tooth composite model, which accurately fuses the individual tooth geometry obtained from the dental impression model or oral scan with the jaw bone obtained from CT, is important for planning and performing dental treatment and predicting treatment outcome.

T. J. Jang · H. S. Yun · J. K. Seo (✉)
School of Mathematics and Computing (Computational Science and Engineering), Yonsei University, Seoul, Korea
e-mail: seoj@yonsei.ac.kr

T. J. Jang
e-mail: taejunjang@yonsei.ac.kr

S.-H. Lee
Department of Oral and Maxillofacial Surgery, Oral Science Research Center, College of Dentistry, Yonsei University, Seoul, Korea
e-mail: sanghwy@yonsei.ac.kr

J. K. Seo (ed.), *Deep Learning and Medical Applications*, Mathematics in Industry 40,
https://doi.org/10.1007/978-981-99-1839-3_4

4.1 Introduction

Digital dentistry refers to any dental technology that effectively combines computer-controlled components with traditional dental technology to enhance dental treatment and digital workflow such as virtual surgical planning. Detailed 3D images of patient's teeth, jaw bone and dental anatomy are important for advanced dental care such as dental implant planning and placement, oral and maxillofacial surgery, orthodontic treatment, cephalometric analysis, and fabricating the dental crown and bridge. With recent advances in dental imaging technologies such as cone-beam computerized tomography (CBCT), intraoral scanner (IOS), and 3D facial scanner [53], dentists can digitally manage the entire process including scanning, planning, design and production in dental practices [35].

Yet, moving from one module to another, comprehensively merging different technologies, and conveniently managing the entire process is still inconvenient for most dentists [20]. Most dentists do not have the time or energy to dedicate themselves to processes that require significant proficiency. Hence, there exists a great need for the development of a smart dental platform that can dramatically improve clinical workflow for both dentists and patients by reducing manual mistakes and treatment time. If digital workflows are well built for accuracy and user convenience, dentists can provide fast, accurate and convenient treatment, shorten treatment duration, and predict treatment outcomes [44].

Digital dentistry is experiencing drastic changes through the rapid enhancement of artificial intelligence (AI) [6]. Digital dentistry will be applied to most dental fields such as implants, prosthodontics, orthognathic surgery, and orthodontic treatment [12, 16, 25, 60]. For the example of dental implant therapy, a digital dentistry-based digital implant planning system integrates CBCT data with intraoral scans to provide computer-aided design and manufacturing of the implant-retained prosthesis and drill guides [27]. In the future, it could be used to remote monitoring and interim follow-up process through an interconnected platform between patient, dentist, and surgeon [10].

This chapter mainly focuses on explaining a digital platform for a 3D jaw–teeth–face model with a variety of applications, including 3D digital treatment planning and orthognathic surgery. The digital platform for the 3D jaw–teeth–face model aims to integrate 3D multi-modal data of CBCT, intraoral scans, and face scans, in order to create an integrated treatment plan for the patient as a single digital anatomic model, including bone, teeth, gingiva, and face [10]. Note that dental CBCT alone may not provide accurate detailed information on dental geometry and occlusal relationships. In dental CBCT, metal-induced artifacts are becoming an increasingly frequent problem, as the number of aged people with metallic dental prostheses and implants is rapidly increasing with the aging population [52]. Metallic objects present in the CBCT field of view produce streaking artifacts that highly degrade the reconstructed CT images [4, 11, 48], resulting in a loss of information on the teeth and other anatomical structures. Because dental CBCT is designed to use a much lower radiation dose than the conventional multidetector CT (MDCT), it tends to produce more artifacts [51]. The field of view (FOV) size in dental CBCT is usually small [49] compared to the size of a patient's head because a small detector is employed to reduce system costs. Moreover, the contrast for soft tissue or mass is critically low and its radiodensity (i.e., Hounsfield unit value) is not standardized at all. Due to these limitations of dental CBCT, it is difficult to describe the crown morphology and occlusal relationship using only CBCT images. When composite models for accurate radiological diagnosis and surgical management of pathologies are required, MDCT is recommended instead of CBCT, even at the expense of increased radiation exposure and increased costs. Intraoral scans make up for the aforementioned weaknesses of dental CBCT wonderfully. The intraoral scanner can acquire accurate 3D images of the teeth surfaces and gingiva in high resolution [28, 62], and its accuracy somewhat is approaching the level of the clinical application for digital impressions and occlusal analysis [33, 37, 61]. The intraoral scans can be merged into CBCT scans through a surface matching method (or a rigid registration) to build a jaw–teeth model for a realistic digital simulation. This digital jaw–teeth model facilitates the process of virtual surgical planning, treatment simulation, and design and delivery of orthodontic and surgical treatment [2, 10, 32].

The 3D jaw–teeth–face model can be built by integrating facial surface scans into the aforementioned jaw–teeth model (based on CBCT and intraoral scans) [24, 57]. It allows dentists to provide a prediction of surgical outcome in patients with facial deformities and to achieve better outcomes. The 3D jaw–teeth–face model can be used to create simulations of various osteotomies with an idea of the expected esthetic changes. Here, the main issues are (1) how to merge the face scan and CBCT; (2) how to get the change of soft tissue associated with the skeletal changes by surgery or other treatments.

We hope that this chapter helps in providing future research directions for an all–in–one solution from CT scan to clinical treatment for full digital dentistry and education for future generations.

4.2 Development of AI-Based Data Integration Platform for Digital Dentistry

This section explains necessity and usefulness of the 3D jawbone–tooth–face composite model. The 3D jaw–tooth–face composite model is not only necessary in the entire process of basic dental treatment, but also is a basic tool in almost all studies and treatments related to teeth and jawbones such as 3D morphological analysis of jaw or tooth structure, kinematic modeling of the jaw, and virtual surgery. The jaw–tooth–face composite model is required for CAD/CAM device manufacturing and is useful for reproducing personalized jaw movement patterns and confirming occlusion, including a virtual articulator.

4.2.1 Traditional Versus Digital Dentistry

Traditional dentistry uses plain radiographs, dental plaster models, and articulator mountings for analysis, diagnosis, and treatment planing. See Fig. 4.1. Based on the analysis, the required guide device or prosthesis is manufactured by the dental laboratories. However, with a planar radiograph, a 3D analysis is not possible, and it is difficult to reproduce a tooth model. In the process of making a prosthesis, only limited numbers of occlusal spots that record the relationship between the maxilla and the mandible can be checked. Due to these limitations, the treatment plan is highly dependent on the operator's experience, and it is difficult to cope with unexpected problems. Moreover, in the process of final finishing of the prosthesis in the oral cavity, the chair time is very long for the repeated checkups and corrections of the prosthesis.

With recent advances in medical technology and computer technology, we are experiencing a paradigm shift in dental care. The advancement of medical imaging technology (e.g. CBCT, intraoral scanner) and the convergence of computer applications (e.g., 3D imaging simulation program, digital dental lab using CAD/CAM, navigation surgery) have been introduced into the dental field, which has dramatically improved treatment results. See Fig. 4.2.

Digital dentistry will be innovatively applied to all dental fields such as implants/prosthodonitcs, oral and maxillofacial surgery, and orthodontic treatment. The reason is that dentists can perform a digital dentistry-based digital surgical simulation to provide an elaborate surgical guide and skip the process of repeated dental impressions that make the patient uncomfortable. More importantly, the dentist has the advantage of being able to completely control the final result while performing elaborate treatment, which can dramatically improve the satisfaction of both the dentist and the patient. It is useful for reproducing personalized jaw movement patterns and confirming occlusion.

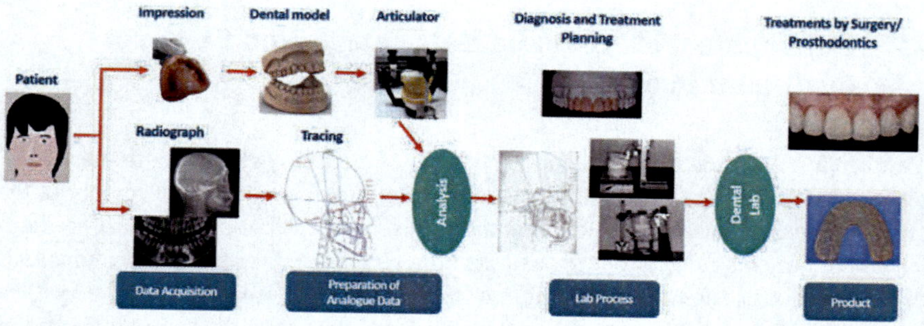

Fig. 4.1 Traditional dental practice methods. Analysis, diagnosis, and treatment plan are established using planar radiographs, dental plaster models, articulators, and mountings. The necessary guide devices or prostheses are manufactured by a separate dental laboratory

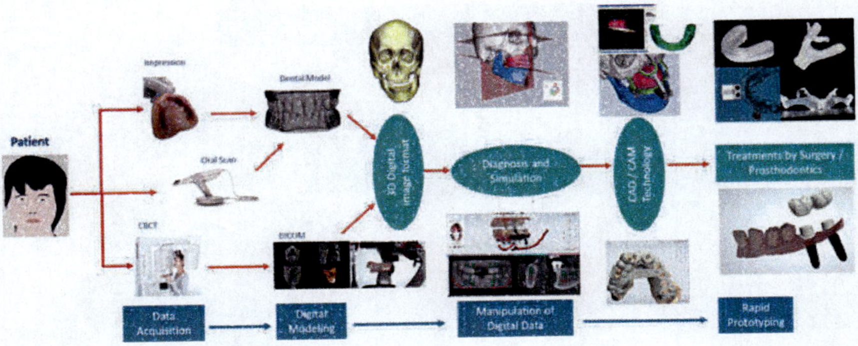

Fig. 4.2 Paradigm-shift in dental practice: digital dentistry

4.2.2 *Necessity and Usefulness of AI-Based Digital Platform Integrating 3D Jaw–Teeth–Face Data*

In order to reinforce the efficiency and connectivity of dental treatment, it would be desirable to form multi-modal image data by collecting various digital data produced in dentistry in one place. See Fig. 4.3. However, there are many obstacles to creating such an integrated platform, including the quality of data (e.g., metal-induced artifacts in CT, stitching error in IOS), the accuracy of the overlapping/replacement process of two sets of data, and the time and skill required to create the model. The most promising way to solve this is to create a composite model by combining the teeth using artificial intelligence. See Fig. 4.4. By introducing an AI-based data integration algorithm for multi-mode image data configuration and building a smart

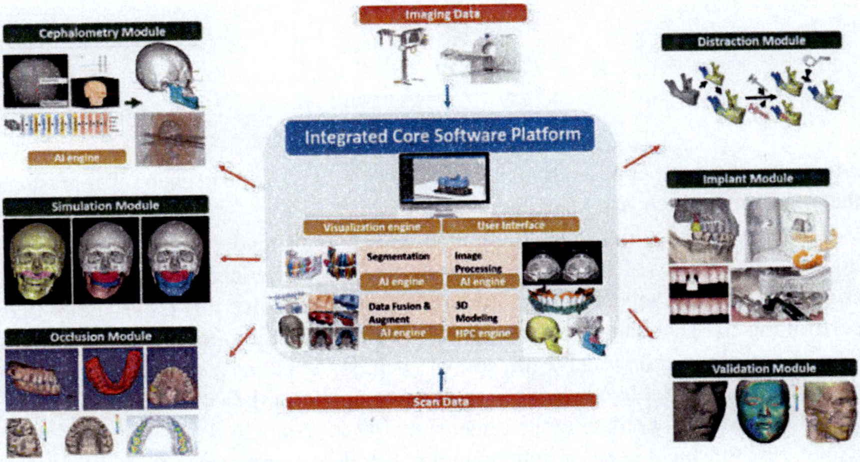

Fig. 4.3 AI-based digital platform integrating 3D jaw–teeth–face data

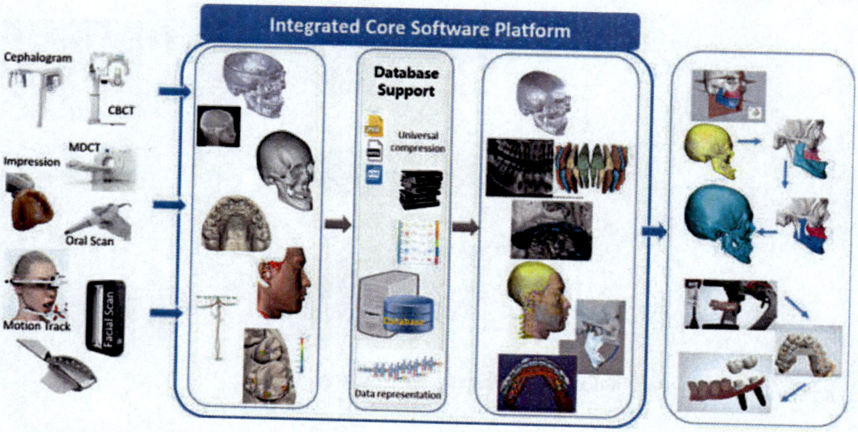

Fig. 4.4 Development of a jawbone–teeth–face complex model platform and construction of a dentist–patient-friendly digital workflow through smart three-dimensional 'teeth segmentation' and 'multi-modal data fusion'

platform, dentists can quickly and conveniently improve the accuracy of cephalometric analysis, patient-customized prosthodontic CAD, computer surgery simulation, and digital implant surgery.

4.3 Individual Tooth Segmentation in IOS

Let X denote a point cloud of a mandibular complete-arch scan by IOS. The goal of the segmentation is to decompose X into individual teeth ($X_1^\diamond \cup X_2^\diamond \cup \cdots X_J^\diamond$) and the rest including gingiva (X_{gingiva}):

$$X = \underbrace{X_1^\diamond \cup X_2^\diamond \cup \cdots X_J^\diamond}_{X_{\text{teeth}}} \cup X_{\text{gingiva}}, \tag{4.1}$$

where J is the number of teeth in the mandible (i.e. $J \leq 16$) and each $X_j^\diamond$ represents the j-th tooth, which will be explained later.

There exist difficulties in handling 3D IOS data X using deep learning due to the high dimensionality of the input data and the limited availability of training data. High accuracy and robustness are important here. To resolve this issue, rather than performing 3D segmentation directly, first, we use multiple bounding boxes (less than 16) to detect individual teeth, so that each box contains a single tooth that fits tightly. Then, the problem is reduced to segmenting the tooth from the bounding box, thereby significantly reducing the dimensionality of the input data.

For the robustness issue of tight bounding box detection, instead of using 3D data X, we generate 2D images that highlight the features of the teeth and then detect bounding boxes using 2D images. Experiments show that using this 2D decomposition as prior information for 3D decomposition (4.39) makes the result more robust and accurate.

4.3.1 Tooth Feature-Highlighted 2D Image Generation

Let Ω_X be a point-set surface determined by the point cloud data X. From the point-set surface Ω_X, we can generate tooth feature-highlighted 2D images; a 2D rendered image with lighting effect (denoted by $\mathscr{I}_r$) and a depth image (denoted by $\mathscr{I}_d$). These two images will be used for individual tooth detection and identification.

In order to generate $\mathscr{I}_r$ and $\mathscr{I}_d$, we first need to align X in a new coordinate system with three axes $\mathbf{u}_1, \mathbf{u}_2, \mathbf{u}_3$ in such a way that $\mathbf{u}_1, \mathbf{u}_2, \mathbf{u}_3$ are roughly horizontal, sagittal, and vertical directions, respectively. Now, we will explain how to select the coordinate system. The origin of the coordinate system is selected as $\bar{\mathbf{x}} = \frac{1}{N}\sum_{i=1}^{N} \mathbf{x}_i$. We apply principal component analysis (PCA) to obtain three principal bases $\{\mathbf{pc}_1, \mathbf{pc}_2, \mathbf{pc}_3\}$ for X. Then, the three coordinate directions $\{\mathbf{u}_1, \mathbf{u}_2, \mathbf{u}_3\}$ are chosen by

$$\mathbf{u}_2 = \begin{cases} \mathbf{pc}_2 & \text{if } \langle \mathbf{pc}_2, \sum_{\mathbf{x}\in X} \frac{\mathbf{x}-\bar{X}}{\|\mathbf{x}-\bar{X}\|}\rangle \geq 0 \\ -\mathbf{pc}_2 & \text{otherwise} \end{cases}, \tag{4.2}$$

$$\mathbf{u}_3 = \begin{cases} \mathbf{pc}_3 & \text{if } \langle \mathbf{pc}_3, \sum_{\mathbf{x}\in X} \mathbf{n}_\mathbf{x}\rangle \geq 0 \\ -\mathbf{pc}_3 & \text{otherwise} \end{cases}, \tag{4.3}$$

$$\mathbf{u}_1 = \mathbf{u}_2 \times \mathbf{u}_3. \tag{4.4}$$

Now, we are ready to explain how to generate $\mathscr{I}_r$ and $\mathscr{I}_d$. Without loss of generality, we may assume $\bar{\mathbf{x}} = 0$ and $\mathbf{u}_1 = (1, 0, 0)$, $\mathbf{u}_2 = (0, 1, 0)$, $\mathbf{u}_3 = (0, 0, 1)$. The rendered 2D image $\mathscr{I}_r$ is defined on the occlusal plane Π:

$$\Pi = \left\{ \frac{1}{s}((u, -v, 0) + \mathbf{a}) : i = 1, \ldots, N_1,\ j = 1, \ldots, N_2 \right\} \tag{4.5}$$

with s being a pixel spacing and $\mathbf{a} = (-\frac{N_1+1}{2}, \frac{N_2+1}{2}, \max\{\|\mathbf{p}\| : \mathbf{p} \in \Omega_X\})$ being a translation vector, as shown in Fig. 4.5a. Here, s and $N_1 \times N_2$ (e.g., $N_1 \times N_2 = 400 \times 400$) are related to image resolution and field of view, respectively. To be precise, $\mathscr{I}_r$ is given by

$$\mathscr{I}_r(u, v) = \begin{cases} \max\{\langle \mathbf{n}_{\mathbf{p}^*_{u,v}}, \mathbf{u}_3\rangle, 0\} & \text{if } \ell_{u,v} \cap \Omega_X \neq \emptyset \\ 0 & \text{otherwise} \end{cases}, \tag{4.6}$$

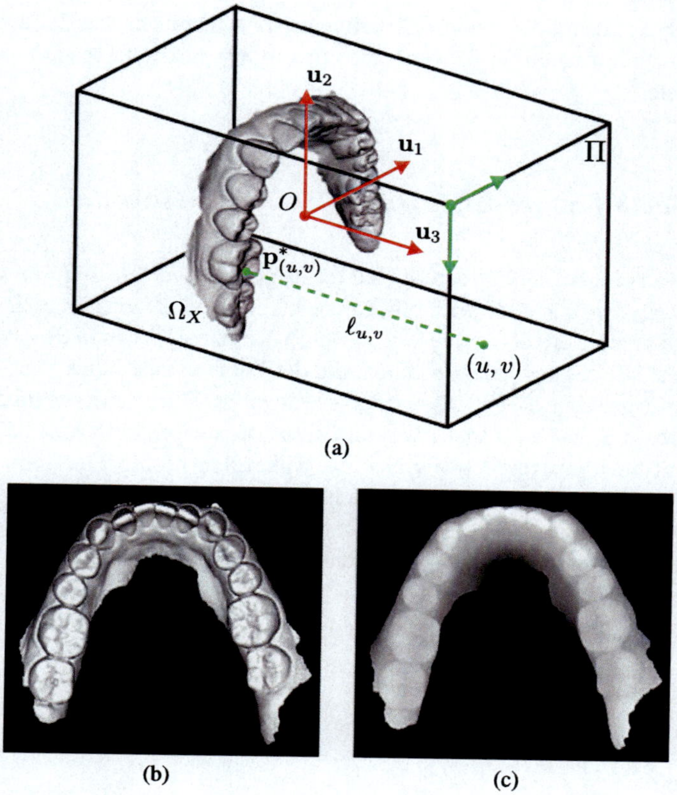

Fig. 4.5 Generation of 2D rendered image $\mathscr{I}_r$ with lighting effect and depth image $\mathscr{I}_d$ from X. **a** Alignment of X in a new coordinate system with three axes $\mathbf{u}_1, \mathbf{u}_2, \mathbf{u}_3$ in such a way that $\mathbf{u}_1, \mathbf{u}_2, \mathbf{u}_3$ are roughly horizontal, sagittal, and vertical directions, respectively, **b** 2D image of $\mathscr{I}_r$, and **c** 2D image of $\mathscr{I}_d$ where the grayscale stands for the depth information

where $\ell_{u,v}$ is the line passing through $\frac{1}{s}((u, -v, 0) + \mathbf{a})$ with the direction $\mathbf{u}_3$, and $\mathbf{p}^*_{u,v}$ is a point lying on the tooth surface Ω_X given by

$$\mathbf{p}^*_{u,v} = \text{argmax}\{\langle \mathbf{p}, \mathbf{u}_3 \rangle : \mathbf{p} \in \ell_{u,v} \cap \Omega_X\}, \tag{4.7}$$

and $\mathbf{n}_\mathbf{p}$ is a unit normal vector at $\mathbf{p}$.

The depth image $\mathscr{I}_d$ is given by

$$\mathscr{I}_d(u, v) = \begin{cases} 1 - \frac{\langle \mathbf{p}^*_{u,v}, \mathbf{u}_3 \rangle - z^*_{\text{min}}}{z^*_{\text{max}} - z^*_{\text{min}}} & \text{if } \ell_{u,v} \cap \Omega_X \neq \emptyset \\ 0 & \text{otherwise} \end{cases}, \tag{4.8}$$

where $z^*_{\min} = \min\{\langle \mathbf{p}^*_{u,v}, \mathbf{u}_3 \rangle : 1 \le u \le N_1,\ \ 1 \le v \le N_2\}$ and $z^*_{\max} = \max \{\langle \mathbf{p}^*_{i,j}, \mathbf{u}_3 \rangle : 1 \le u \le N_1,\ 1 \le v \le N_2\}$.

As shown in Fig. 4.5b, c, the depth values of the tooth crowns are distinct because the tooth positions protrude forward from the gingiva and other tissues. While the rendered image contains clear geometric features by lighting and shading the surface, the depth image provides the tooth reliability by expressing the relative distance.

4.3.2 Tooth Bounding Box Detection and 3D Tooth ROI Extraction Using Generated 2D Images

We use the deep learning method [43] to find a bounding box detection map

$$f_{det} : (I_r, I_d) \to \{\mathbf{b}_1, \ldots, \mathbf{b}_J\}, \tag{4.9}$$

where $\{\mathbf{b}_1, \ldots, \mathbf{b}_J\}$ denote a set of vectors associated with 2D bounding boxes, corresponding to individual teeth $X_1^\diamond, \ldots, X_J^\diamond$ in (4.39). Here, each bounding box $\mathbf{b}_j$ should contain a single tooth and can be uniquely determined by bounding box coordinates with center position, width, and height.

For bounding box detection, we divide the images I_r and I_d into uniform squares of size $k \times k$ (*e.g.* $k = 20$). Let G_{ij} be an (i, j) square where $(i, j) \in \{(i, j) : i = 1, 2, \ldots, \frac{N_1}{k},\ j = 1, 2, \ldots, \frac{N_2}{k}\}$.

As shown in Fig. 4.6,

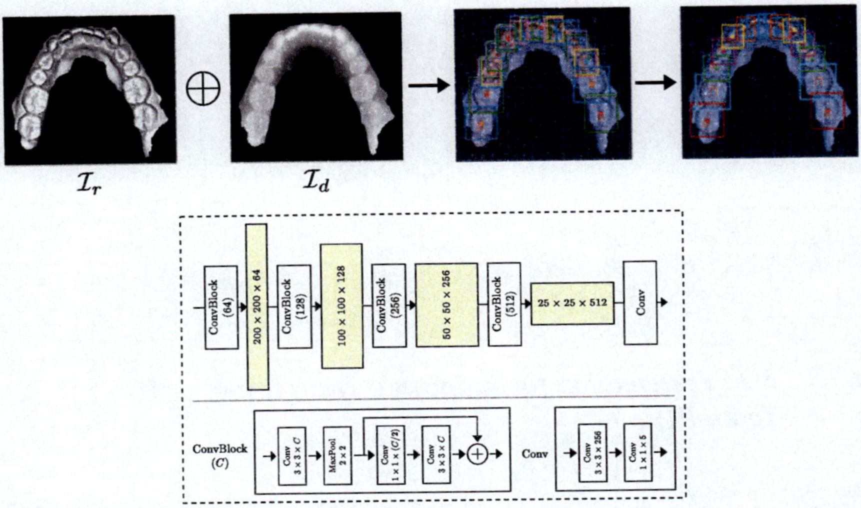

Fig. 4.6 The architecture of the bounding box detection map f_{det}. The input of f_{det} is two 2D images of $\mathcal{I}_r$ and $\mathcal{I}_d$. The output is the teeth detection using bounding boxes. The map f_{det} is based on YOLO

The map f_{det} is composed of two parts:

$$f_{det} = f_{NMS} \circ f^1_{det}, \tag{4.10}$$

where f^1_{det} is the network for predicting the bounding box and estimating a confidence score for each grid cell $G_{i,j}$, and f_{NMS} is a non-maximal suppression (NMS) process for filtering overlapping boxes in such a way that only one bounding box is left per tooth. The output of f^1_{det} is given by

$$f^1_{det}(I_r, I_d) = \begin{pmatrix} \mathcal{O}_{1,1} & \mathcal{O}_{1,2} & \cdots & \mathcal{O}_{1,\frac{N_1}{k}} \\ \mathcal{O}_{2,1} & \mathcal{O}_{2,2} & \cdots & \mathcal{O}_{2,\frac{N_1}{k}} \\ \vdots & \vdots & \ddots & \vdots \\ \mathcal{O}_{\frac{N_2}{k},1} & \mathcal{O}_{\frac{N_2}{k},2} & \cdots & \mathcal{O}_{\frac{N_2}{k},\frac{N_1}{k}} \end{pmatrix}, \tag{4.11}$$

where $\mathcal{O}_{ij} = (c_{ij}, \mathbf{b}_{ij})$ predicts a confidence score $c_{ij} \in [0, 1]$ representing belief of existence of the tooth center in G_{ij} and a bounding box component $\mathbf{b}_{ij} = (u_{ij}, v_{ij}, w_{ij}, h_{ij})$ consisting of the center position, width and height of a box, respectively. For detailed explanation on c_{ij} and $\mathbf{b}_{ij}$, see Sect. 2.4.5.

Next, the network f_{NMS} in (4.10) is to filter overlapping boxes using the non-maximal suppression technique [1]. The f_{NMS} eliminates overlapped bounding boxes $\mathbf{b}_{ij}$ of the same tooth by selecting the box with the highest score.

Using a labeled dataset $\{\mathcal{I}_r^{(n)}, \mathcal{I}_d^{(n)}), \mathcal{O}^{*(n)}\}_{n=1}^N$ where $\mathcal{O}^*$ is the ground-truth, a map f^1_{det} is trained by minimizing the loss between the output $\mathcal{O} = f^1_{det}(\mathcal{I}_r, \mathcal{I}_d)$ and the ground-truth $\mathcal{O}^*$ as follows:

$$\begin{aligned} \mathcal{L}_{det} = \sum_{n=1}^{N} \Bigg[& \lambda_0 \sum_{(i,j)\in\Omega_0^{(n)}} (0 - c_{ij}^{(n)})^2 + \sum_{(i,j)\in\Omega_1^{(n)}} (1 - c_{ij}^{(n)})^2 \\ & + \lambda_1 \sum_{(i,j)\in\Omega_1^{(n)}} \left\| \mathbf{b}_{ij}^{*(n)} - \mathbf{b}_{ij}^{(n)} \right\|^2 \Bigg], \end{aligned} \tag{4.12}$$

where $\Omega_0^{(n)} = \{(i, j) : c_{ij}^{*(n)} = 0\}$, $\Omega_1^{(n)} = \{(i, j) : c_{ij}^{*(n)} = 1\}$, $\lambda_0 = 0.1$, and $\lambda_1 = 5$.

4.3.3 3D Segmentation for Individual Teeth from the 3D Tooth ROIs

We perform individual tooth segmentation and identification using individual tooth ROIs obtained by the detected bounding boxes. The individual tooth ROIs $\{X_{\mathbf{b}_1}, \ldots, X_{\mathbf{b}_J}\}$ are determined by the detected bounding boxes $\{\mathbf{b}_1, \ldots, \mathbf{b}_J\}$:

$$X_{\mathbf{b}} = X \cap \mathrm{BOX}_{\mathbf{b}}, \tag{4.13}$$

where $\mathrm{BOX}_{\mathbf{b}}$ is the bounding box corresponding to the component $\mathbf{b} = (u, v, w, h)$, i.e.,

$$\mathrm{BOX}_{\mathbf{b}} = \left[su - \frac{sw}{2}, su + \frac{sw}{2}\right) \times \left[-sv - \frac{sh}{2}, -sv + \frac{h}{2}\right) \times \mathbb{R}. \tag{4.14}$$

Given $X_{\mathbf{b}}$, we apply a deep learning-based segmentation network $f_{seg} : X_{\mathbf{b}} \mapsto \mathscr{S}_{\mathbf{b}}$ to get a tooth segmentation $\mathscr{S}_{\mathbf{b}}$ which is selected from the point cloud $X_{\mathbf{b}}$. Figure 4.7 shows the architecture of the segmentation network f_{seg}, which is based on PointNet [40] and the EdgeConv [56]. The f_{seg} is designed to estimate the probability that the points belong to the target tooth in $X_{\mathbf{b}}$, using a k-nearest neighbor (k-NN) graph and multi-layer perceptron (MLP).

For the sake of clarity, we will describe the process of f_{seg} using a concrete example shown in Fig. 4.7. In the first layer, the output $\mathbf{h}_j^{\ell_1}$ corresponding to the input $\mathbf{x}_j \in X_{\mathbf{b}}$ is computed as follows:

$$\mathbf{h}_j^{\ell_1} = \max_{\mathbf{x} \in NN_k^{\ell_1}(\mathbf{x}_j)} ReLU\left[\Theta^{\ell_1} \cdot \mathbf{x}_j + \Psi^{\ell_1} \cdot (\mathbf{x} - \mathbf{x}_j))\right], \tag{4.15}$$

where Θ^{ℓ_1} and Ψ^{ℓ_1} are the learnable parameters in $\mathbb{R}^{3 \times d_1}$ and $NN_k^{\ell_1}(\mathbf{x}_j)$ is the set of k_{ℓ_1} points in $X_{\mathbf{b}}$ closest to $\mathbf{x}_j$. Here, the superscript ℓ_j stands for j-th layer. If $d_1 = 64$, the output of the first layer has the dimension $d_1 = 64$, i.e., $\mathbf{h}_j^{\ell_1} \in \mathbb{R}^{64}$. The k-nearest neighborhood $NN_k^{\ell_1}(\mathbf{x}_j)$ of $\mathbf{x}_j$ is used to capture local feature through the relations between each point and its neighbors. Similarly, in the second layer, the output at the j-th point is computed by

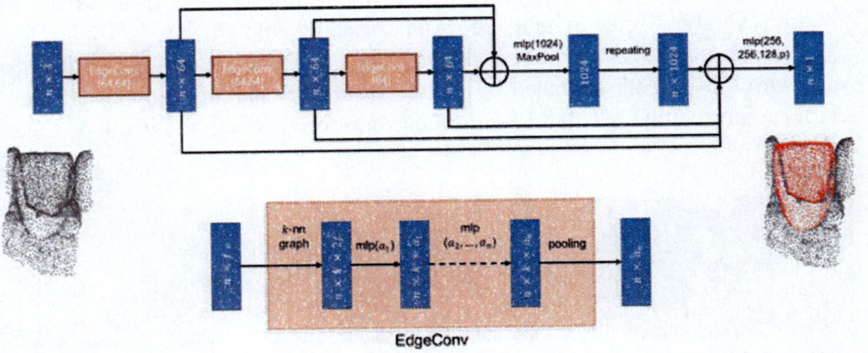

Fig. 4.7 Framework of the individual tooth segmentation network f_{seg}. Its input is the point cloud $X_{\mathbf{b}} = X \cap \mathrm{BOX}_{\mathbf{b}}$, as shown in the left image. The output is a segmented tooth, as shown in the red part of the right image

$$\mathbf{h}_j^{\ell_2} = \max_{\mathbf{h} \in NN_k^{\ell_2}(\mathbf{h}_j^{\ell_1})} ReLU\left[\Theta^{\ell_2} \cdot \mathbf{h}_j^{\ell_1} + \Psi^{\ell_2} \cdot (\mathbf{h} - \mathbf{h}_j^{\ell_1}))\right], \tag{4.16}$$

where Θ^{ℓ_2} and Ψ^{ℓ_2} are the learnable parameters in $\mathbb{R}^{d_1 \times d_2}$ and $NN_k^{\ell_2}(\mathbf{h}_j^{\ell_1})$ is the set of k_{ℓ_2} points closest to $\mathbf{h}_j$. We continue this process until the last layer. In the last layer, the network estimates whether $\mathbf{x}_j$ belong to the target tooth.

Using a labeled data $\{X_{\mathbf{b}}^{(n)}, \mathscr{S}^{(n)}\}_{n=1}^N$, a map f_{seg} is learned by minimizing the loss as follows:

$$\mathscr{L}_{seg} = -\sum_{n=1}^{N} \mathscr{S}^{(n)} \log\left[f_{seg}(X_{\mathbf{b}}^{(n)})\right]. \tag{4.17}$$

4.4 A Fully Automated Method for 3D Individual Tooth Identification and Segmentation in Dental CBCT

3D segmentations of teeth, jaws, and skulls from CBCT images are important components for future digital dentistry [29, 34, 41]. Accurate digital models of individual tooth geometry and jaws facilitate surgical planning of impacted teeth, orthodontic planning, computer-aided digital implant surgical guides, bite irregularity prediction, cephalometric analysis, and so on [3, 7, 19, 50, 54]. Recently, with the widespread use of dental CBCT equipment, it is rapidly establishing itself as a standard imaging equipment in dentistry [18]. Most dental CBCTs have significantly lower radiation doses and relatively low prices and maintenance costs, compared to MDCTs [26].

Automatic individual tooth segmentation from dental CBCT image is a challenging task because dental CBCT images are often affected by the metal-related artifacts, which are associated with beam hardening, scattering, partial volume effects, etc. [49]. Figure 4.8 shows CBCT images before and after applying metal artifact reduction (MAR). The CBCT image after applying the state-of-the-art MAR still remains severely affected by metal artifacts caused by gold crowns.

For tooth segmentation to have practical value in a clinical setting, it must also work with CBCT data degraded by metal artifacts. Hence, developing 3D tooth segmentation method seems to be difficult, due to occluded or overlapped teeth

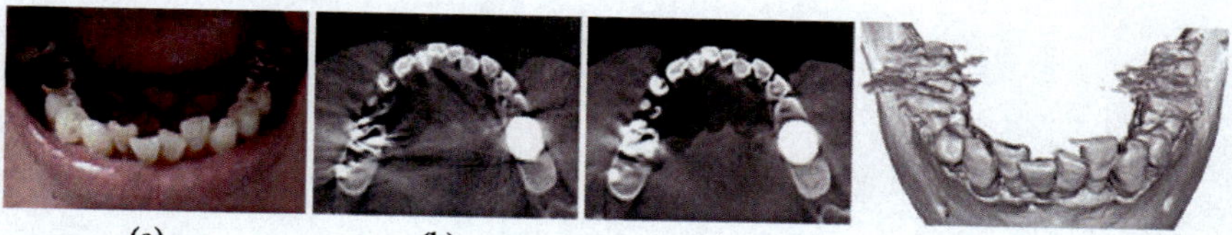

(a) (b) (c) (d)

Fig. 4.8 Metal artifacts in dental CBCT. **a** A patient having multiple gold dental prostheses; **b** CBCT image reconstructed by the standard reconstruction algorithm; **c** CBCT image after applying the state-of-the-art metal artifact reduction; **d** bone segmentation from CBCT image (**c**)

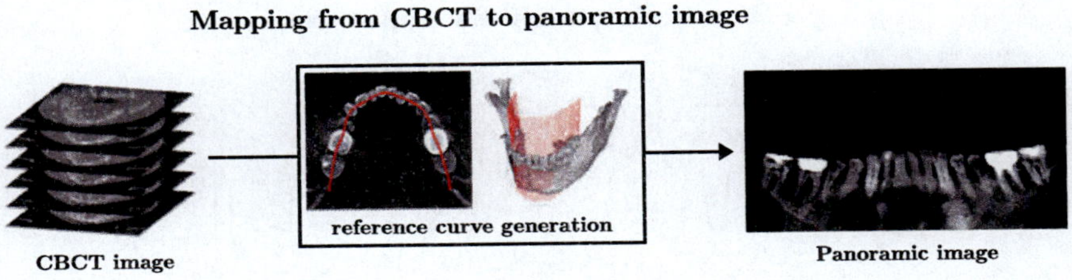

Fig. 4.9 Generating a panoramic image from metal artifact-contaminated CBCT image

images by metal-related artifacts. Several deep learning methods [5, 8, 30, 42, 58] have been developed for direct 3D tooth segmentation, but their performance is limited in noisy environments where metal artifacts are very common in dental CBCT.

We explain a deep learning model to circumvent the aforementioned limitations associated with low-dose CBCT images. It is crucial to observe that panoramic images generated from CBCT images are not significantly affected by metal-related artifacts. Figure 4.9 shows the panoramic image created from CBCT data degraded by metal artifacts. This occurs because the cone beam projection configuration is advantageous in composing panoramic image reconstructions. We take advantage of these panoramic images to accurately perform 3D tooth segmentation and identification.

The process of this deep learning method is as follows:

1. From a CBCT image I, we generate upper-jaw panoramic image $P_{\text{upper-jaw}}$ and lower-jaw panoramic image $P_{\text{lower-jaw}}$. The reason for separating the upper and lower jaws is to reduce the overlap between adjacent teeth in the generated panoramic images.
2. The next step is to identify individual teeth by numbers according to FDI dental notation, as shown in Fig. 4.10. The developed tooth detection method is to localize bounding boxes that enclose each tooth and classifies them into four types (incisors, canines, premolars, and molars) according to tooth morphology. Then, the tooth detection results are used to identify individual teeth. Finally, we perform 2D segmentation on individual teeth.
3. The final step is to provide the 3D tooth segmentation, utilizing the 2D bounding boxes and segmentations of the individual teeth from the previous step.

A schematic diagram of our method is described in Fig. 4.11.

Let I denote a 3D CT image with voxel grid $\Omega := \{(x, y, z) \in \mathbb{N}^3 : 1 \le x \le N_x, 1 \le y \le N_y, 1 \le z \le N_z\}$, where N_x, N_y and N_z are the voxel sizes in directions x (sagittal axis), y (frontal axis) and z (longitudinal axis), respectively. The value $I(x, y, z)$ at the voxel position (x, y, z) is represented as the attenuation coefficient.

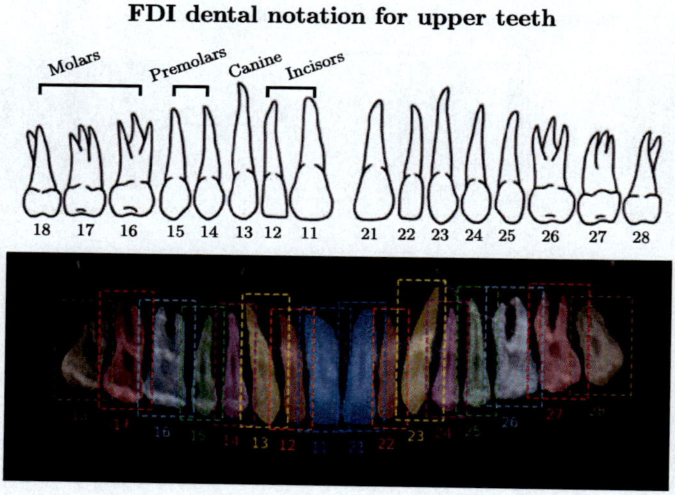

Fig. 4.10 Tooth identification process according to FDI dental notation system

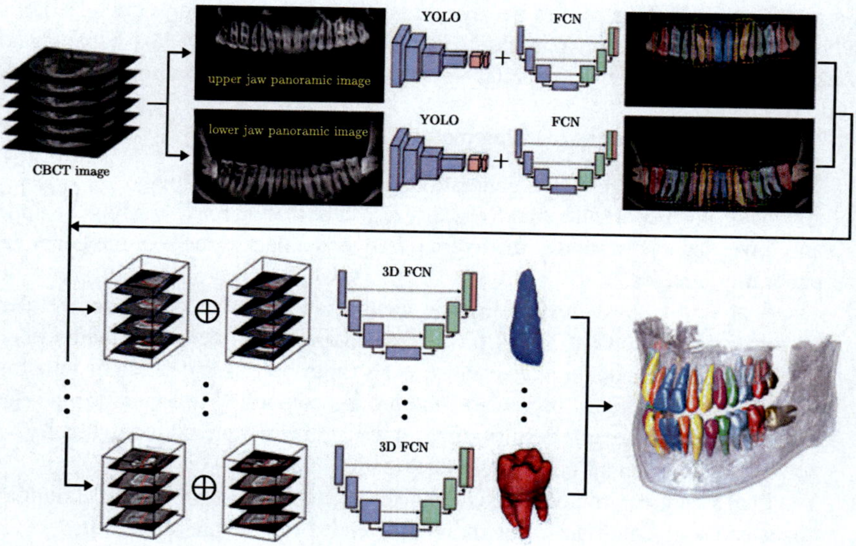

Fig. 4.11 Schematic diagram of the proposed method, which consists of three steps: (1) panoramic image reconstruction of the upper and lower jaws from a 3D CBCT image; (2) tooth identification, 2D bounding box detection, and segmentation of individual teeth in the panoramic images; (3) 3D segmentation for individual teeth from the 3D tooth ROIs that are obtained from the bounding boxes and 2D segmentations

4.4.1 Generation of Panoramic Images of the Upper and Lower Jaws from a 3D CBCT Image

This step describes the automatic reconstruction of the upper- and lower-jaw panoramic images from a 3D CBCT image I. Figure 4.12 illustrates this workflow.

[Step 1–1] Given I, we use Otsu's thresholding technique [39] to get a binary image $\tilde{I}(x, y, z)$ that can be viewed as a rough segmentation of upper and lower jaws.

[Step 1–2] Given the binary image $\tilde{I}$, the connected-component labeling (CCL) [47] is used to extract the upper-jaw part $I_{\text{upper-jaw}}$ and the lower-jaw part $I_{\text{lower-jaw}}$. The CCL method generates all the connected components in a binary image. The lower-jaw part is the largest connected component in $\tilde{I}$, and the upper-jaw part is the second largest connected component.

[Step 1–3] We apply maximum intensity projection (MIP) in the z direction to $I_{\text{upper-jaw}} := I \odot I_{\text{upper-jaw}}$, in order to generate the following 2D image $\mho_{\text{upper-jaw}}$ displaying the upper dental arch:

$$\mho_{\text{upper-jaw}}(x, y) = \max_{z} I_{\text{upper-jaw}}(x, y, z). \tag{4.18}$$

Similarly, we obtain $I_{\text{lower-jaw}}$ and $\mho_{\text{lower-jaw}}$ for the lower jaw.

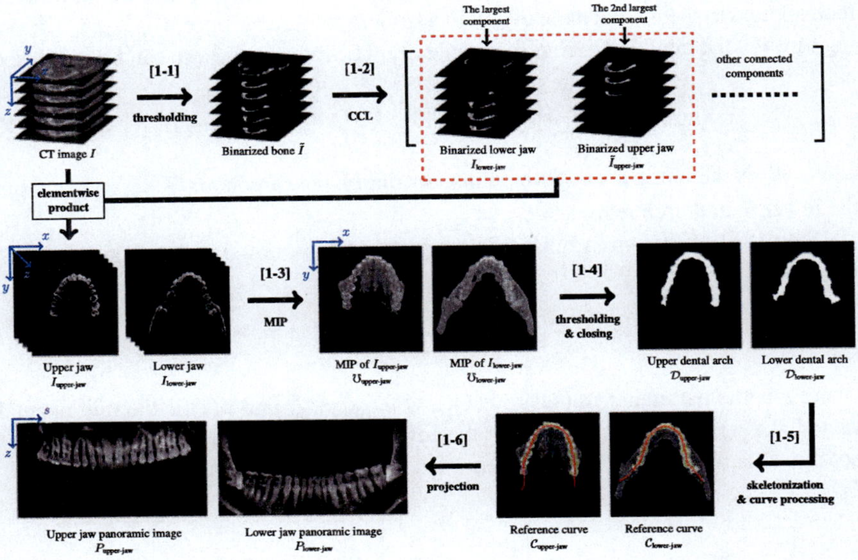

Fig. 4.12 Generation of panoramic images from a CBCT image. This figure shows the process of reconstruction of upper-jaw panoramic image $P_{\text{upper-jaw}}$ and lower-jaw panoramic images $P_{\text{lower-jaw}}$ from a 3D CT image I

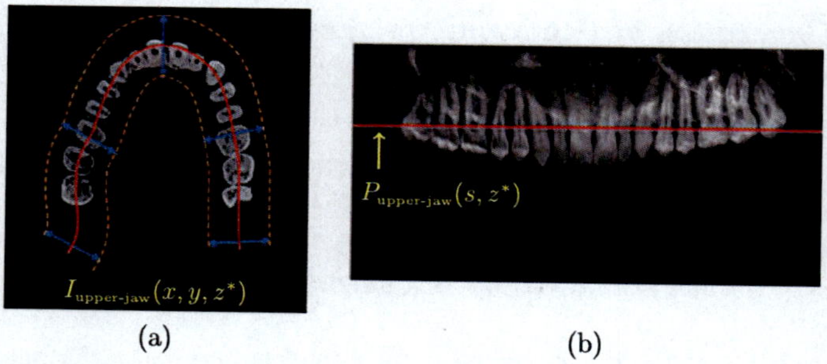

Fig. 4.13 Illustration of Step 1–6. **a** The z^*th slice of the upper–jaw image $I_{\text{upper-jaw}}$: the reference curve (red line), normal directions of the curve (blue arrow lines) and domain of projection (region inside orange dotted line). **b** Reconstructed panoramic image: the z^*th horizontal line of panoramic image $P_{\text{upper-jaw}}$ corresponds to the z^*th slice

[Step 1–4] Next, binary dental arch regions $\mathscr{D}_{\text{upper-jaw}}$ and $\mathscr{D}_{\text{lower-jaw}}$ are obtained by applying Otsu's method [39] and the morphological closing [15] to the MIP images $\mho_{\text{upper-jaw}}$ and $\mho_{\text{lower-jaw}}$, respectively. Here, Otsu thresholding is adopted to get rough dental arch regions and the morphological closing is used to smoothen out the rough regions.

[Step 1–5] Given the upper dental arch region $\mathscr{D}_{\text{upper-jaw}}$ in the previous step, we employ the morphological skeletonization [31] to extract a medial axis of the dental arch region. Cubic spline curve fitting, interpolation and extrapolation techniques are then applied to the medial axis, to obtain a smooth reference curve $\mathscr{C}_{\text{upper-jaw}}$ passing through the dental arch region completely. The reference curve can be expressed as

$$\mathscr{C}_{\text{upper-jaw}} = \{\mathbf{r}(s) = (x(s), y(s)) : s \in 1, 2, \ldots, N_s\}, \tag{4.19}$$

where N_s is the number of curve points. Similarly, we can obtain $\mathscr{C}_{\text{lower-jaw}}$ from the lower dental arch region $\mathscr{D}_{\text{lower-jaw}}$.

[Step 1–6] As shown in Fig. 4.13, an upper-jaw panoramic image is given by

$$P_{\text{upper-jaw}}(s, z) = \int_{-\alpha}^{\alpha} I_{\text{upper-jaw}}(\mathbf{r}(s) + t\mathbf{n}(s), z)\, dt, \tag{4.20}$$

where s is the parameter in (4.19), $\mathbf{r}(s) \in \mathscr{C}_{\text{upper-jaw}}$, and $\mathbf{n}(s)$ is the unit normal vector at $\mathbf{r}(s)$. Similarly, we obtain the lower panoramic image $P_{\text{lower-jaw}}$. For notational simplicity, we refer to both $P_{\text{upper-jaw}}$ and $P_{\text{lower-jaw}}$ as P.

4.4.2 Individual Tooth Detection, Identification and Segmentation in the 2D Reconstructed Panoramic Images

This step aims to identify and segment individual teeth in the reconstructed panoramic images. To achieve the goal, we first perform individual tooth detection. Here, the teeth are classified as incisor (class 1), canine (class 2), premolar (class 3), and molar (class 4).

We use YOLO [43] to find a bounding box map:

$$f_{det} : P \to \{\mathbf{b}_1, \ldots, \mathbf{b}_J\}. \tag{4.21}$$

This step is exactly the same as that in Sect. 4.3.2. For each tooth in the detected bounding box, a number is assigned to identify the unique tooth according to the FDI system. Figure 4.14 shows that the detected bounding boxes are listed in ascending order of s coordinates of the box center. The upper right and left quadrants are divided from the middle of four sequential incisor boxes. For the two right incisors and the two left incisors, numbers 1 and 2 are assigned from the inside to the outside, respectively. Number 3 is assigned to the canines since there is only one in each quadrant. On each side, premolars are assigned numbers 4 and 5 from the inside to the outside. Likewise, molars are assigned numbers 6, 7 and 8 (if a wisdom tooth exists).

Next, we use the U-shaped FCN [45] to get 2D tooth segmentation, taking advantage of bounding box knowledge obtained from previous step.

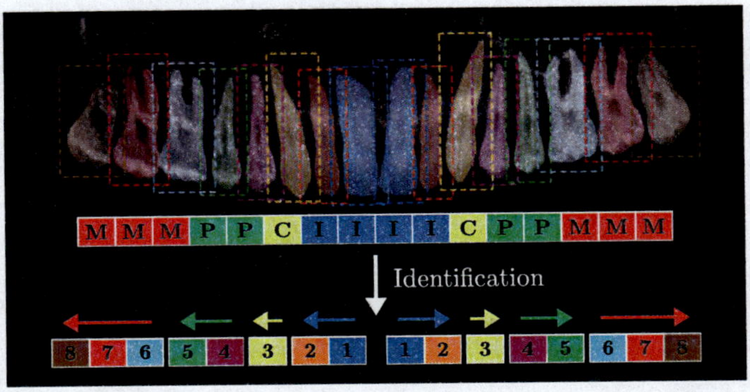

Fig. 4.14 Tooth identification process using the classification results in Step 2–1. The capital letters represent the first letters of the tooth type and the numbers are tooth codes

4.4.3 3D Segmentation for Individual Teeth from the 3D Tooth ROIs

In this final step, 3D individual tooth segmentation is performed by applying the 3D version of a U-shaped FCN [45], as shown in Fig. 4.15. Here, the input of this network is the pair of the loose ROI ($\mathscr{ROI}_{loose,j}$) and tight ROI ($\mathscr{ROI}_{tight,j}$), where j is the number representing j-th teeth in the CBCT image I. Figure 4.16 shows the process of determining the loose and tight ROIs from detected bounding boxes and 2D segmented regions in the previous step. Note that the tight ROI is crucial for improving the segmentation accuracy at the attached boundaries between a target tooth and its neighboring teeth.

The input of the network is $I_{\text{roi3}} = \mathscr{ROI}_{loose} \oplus \mathscr{ROI}_{tight}$, which represents the concatenating vector of two ROIs. Let Y_{roi3} denote a binary vector representing 3D tooth segmentation corresponding to I_{roi3}. Using a training dataset $\{I^{(n)}_{\text{roi3}}, Y^{(n)}_{\text{roi3}}\}_{n=1}^{N}$, we learn a 3D segmentation map $f_{seg3} : I_{\text{roi3}} \mapsto Y_{\text{roi3}}$ by minimizing the following loss:

$$\mathscr{L}_{seg3} = \frac{1}{N}\sum_{n=1}^{N}\left[-\frac{1}{V}\sum_{\mathbf{v}} Y^{(n)}_{\text{roi3}}(\mathbf{v}) \log\left[f_{seg3}\left(I^{(n)}_{\text{roi3}}\right)(\mathbf{v})\right]\right], \tag{4.22}$$

where $\mathbf{v}$ is a voxel position and V is the number of voxels of Y_{roi3}.

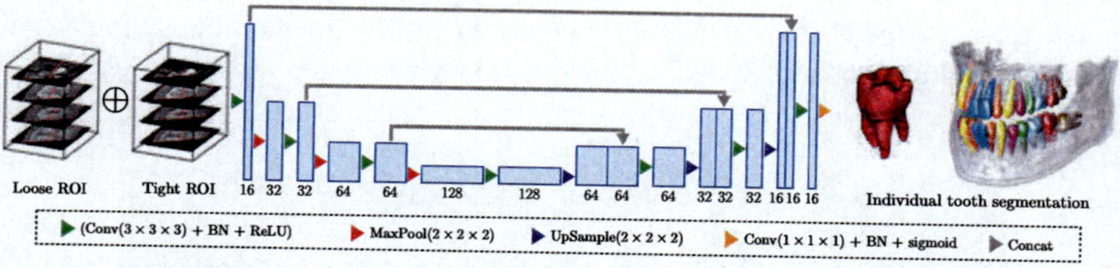

Fig. 4.15 Architecture of the proposed network in Step 4, which is a 3D version of U-net [45]. Individual tooth segmentation is performed using loose and tight 3D ROIs of a target tooth obtained in Step 3

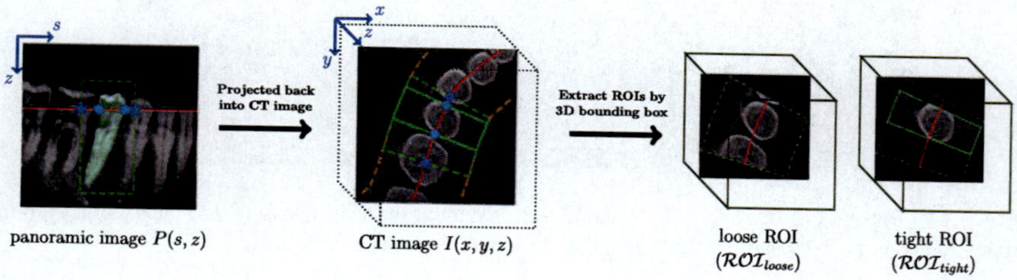

Fig. 4.16 Extraction of loose and tight 3D tooth ROIs from a detected bounding box and a segmented tooth region

4.5 Accurate Digital Impression Extraction Method of the Entire 3D Tooth Using CBCT and Intraoral Scanner

This section describes a novel image registration technique that aligns two different 3D medical images of CBCT and IOS into one coordinate system. The goal of the registration is to integrate two images from different imaging modalities in one scene by complementing each image's weaknesses.

To illustrate the registration method in a simple and practical way, this section only focuses on aligning the patient's IOS image (source data) with the same patient's CBCT image (target data). This registration is intended to provide accurate details of the occlusal relationship to aid in building a jaw–tooth model for realistic digital simulation. This digital simulation can eliminate the hassle of traditional dental prosthodontic treatments that are labor intensive, costly, require at least two individual visits, and require temporary prosthodontics to be worn until the final crown is in place. Moreover, if the final crown made in the dental laboratory does not fit properly at the second visit, the patient and dentist will have to repeat the previous operation, and the laboratory may have to redesign the restoration prosthesis.

The reason why CBCT and IOS data should be integrated into one scene by compensating for the weaknesses of each image is as follows. In dental CBCT images, the gingival structure cannot be seen, and the tooth surface is often degraded by the metal-related artifacts, which are associated with beam hardening, scattering, partial volume effects, etc. As shown in Fig. 4.17, intraoral scanning provides precise tooth surface images, whereas dental CBCT images can be affected by the metal-related artifacts. IOS compensates for these shortcomings of dental CBCT. On the other hand, IOS is very accurate for narrow area scans, but produces cumulative stitching errors for full arch scans [9, 36, 37]. Reducing stitching errors to within 0.2mm is essential for clinical practice. Using CBCT data can reduce these stitching errors in IOS.

Rigid registration can be used by taking advantage of the property that the upper and lower jawbones are rigid. The teeth in the upper and lower jaw bones are a partially overlapping area between the IOS and CBCT data. Hence, an important

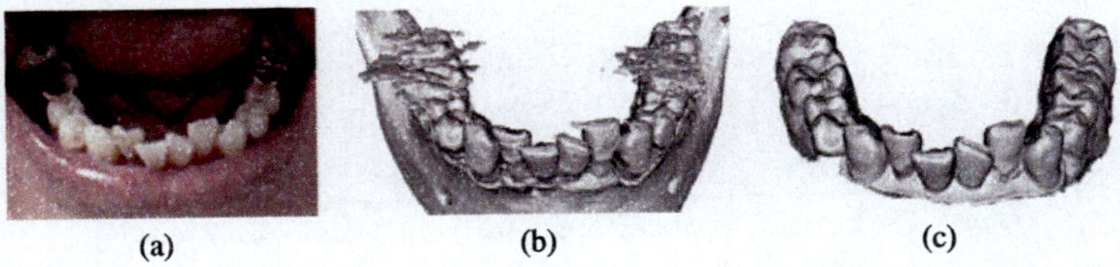

Fig. 4.17 Accurate registration of IOS and CBCT with discrepancy up to 0.2 mm is a challenging problem due to metal artifacts in CBCT and stitching error in IOS. **a** photo of a patient with gold crown, **b** maxillofacial image obtained from a CBCT image of the patient, and **c** IOS model of the same patient

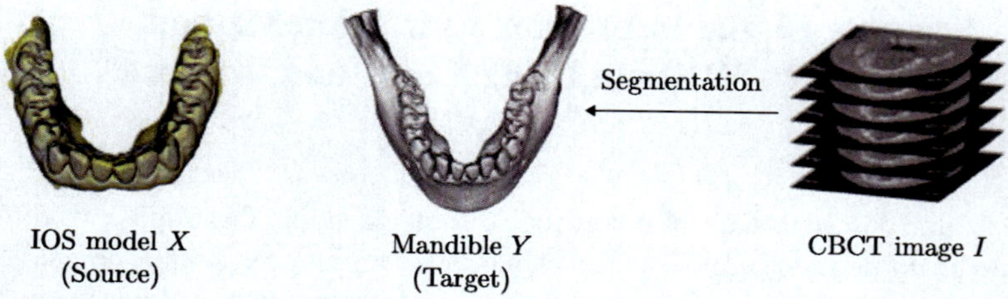

Fig. 4.18 IOS image X (left), CBCT image I of the same patient (left), and mandible Y segmented from I (middle)

pre-processing task for registration is to automatically segment teeth from CBCT data as well as from IOS data and semantically label each individual tooth. However, in this section, we will not discuss individual tooth segmentation and identification methods, in order to focus on the registration techniques.

For ease of explanation, we only describe the registration method for the mandibular part (lower jaw). The goal is to find a rigid transformation between the source point set from IOS (denoted by X) and the target point set from CBCT (denoted by Y). See Fig. 4.18 for X and Y. Now, let us clearly define the mathematical symbols X and Y.

- I represents a CBCT image.
- $Y = \{\mathbf{y}_1, \mathbf{y}_2, \ldots, \mathbf{y}_M\}$ is a set of points referring to a mandible segmented from the CBCT image I.
- $X = \{\mathbf{x}_1, \mathbf{x}_2, \ldots, \mathbf{x}_N\}$ is a point set of the mandibular part of IOS data.

4.5.1 Rigid Transformation

The rigid transformation (that preserves relative distances) is modeled with 6 degrees of freedom: 3×3 rotation matrix (denoted by $\mathscr{R}$) determined by three angles and translation vector (denoted by $\mathbf{t}$) in $\mathbb{R}^3$. To be precise, the rotation matrix $\mathscr{R}$ can be expressed in terms of three rotation angles ϕ, θ, ψ about the three axes (x_1, x_2, x_3):

$$\mathscr{R} = \underbrace{\begin{pmatrix} 1 & 0 & 0 \\ 0 & \cos\phi & -\sin\phi \\ 0 & \sin\phi & \cos\phi \end{pmatrix}}_{\text{rotation about } x_1\text{-axis}} \underbrace{\begin{pmatrix} \cos\theta & 0 & -\sin\theta \\ 0 & 1 & 0 \\ \sin\theta & 0 & \cos\theta \end{pmatrix}}_{\text{rotation about } x_2\text{-axis}} \underbrace{\begin{pmatrix} \cos\psi & -\sin\psi & 0 \\ \sin\psi & \cos\psi & 0 \\ 0 & 0 & 1 \end{pmatrix}}_{\text{rotation about } x_3\text{-axis}}. \tag{4.23}$$

Note that the rotation $\mathscr{R}$ can be represented by using unit quaternions [17].

The rigid transformation, denoted by $\mathscr{T} = [\mathscr{R}, \mathbf{t}]$, is defined by

$$\mathscr{T}\mathbf{x} := \mathscr{R}\mathbf{x} + \mathbf{t} \quad \text{for } \mathbf{x} \in X. \tag{4.24}$$

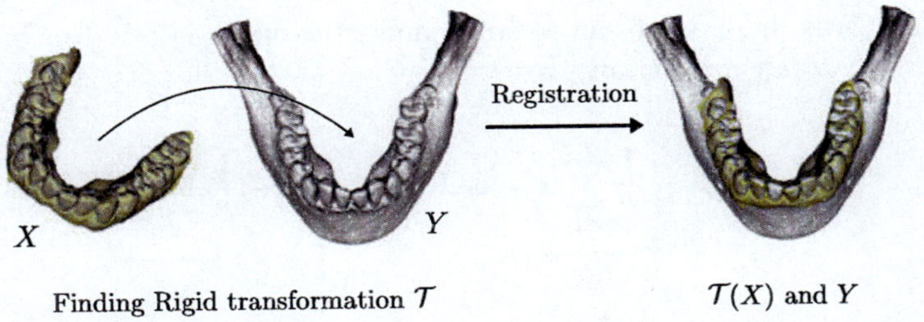

Fig. 4.19 Point cloud registration between X (left) and Y (middle). It is the process of finding a rigid transformation $\mathscr{T}$ that aligns X and Y into one coordinate system

Hence, the goal is to find a suitable rigid transformation $\mathscr{T}$ such that the transformed point cloud $\{\mathbf{y} = \mathscr{T}\mathbf{x} : \mathbf{x} \in X\}$ best aligns with the reference Y (i.e., registration of X and Y using partially overlapping tooth surfaces), as shown in Fig. 4.19.

4.5.2 Paired Point Methods

In paired point methods, operators manually or automatically select three or more fiducial points in both X and Y that are considered to be correspondences between two data. See Fig. 4.20 for fiducial points on teeth. The paired points are used to determine a rigid registration of X and Y. Let $C_P := \{(\mathbf{x}_{p_j}, \mathbf{y}_{p_j}) : j = 1, \ldots, P\}$ be a set of fiducial points in $X \times Y$. The registration can be achieved with the following mean square error minimization to best match the pairs of fiducial points:

$$\mathscr{T}^* = \underset{\mathscr{T}}{\operatorname{argmin}} \sum_{j=1}^{P} \left\| \mathscr{T}\mathbf{x}_{p_j} - \mathbf{y}_{p_j} \right\|^2 \quad \text{subject to } \mathscr{R} \in \mathrm{SO}(3), \tag{4.25}$$

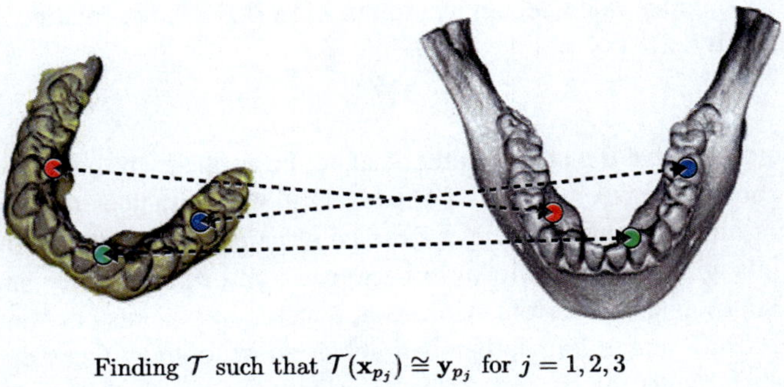

Fig. 4.20 Paired point method using three fiducial points on teeth

where SO(3) is the group of all rotations about the origin in $\mathbb{R}^3$. To explain the matching procedure more clearly, write

$$\tilde{\mathbf{x}}_{p_j} := \mathbf{x}_{p_j} - \underbrace{\frac{1}{P}\sum_{j=1}^{P}\mathbf{x}_{p_j}}_{\mu_{X_P}} \quad \text{and} \quad \tilde{\mathbf{y}}_{p_j} := \mathbf{y}_{p_j} - \underbrace{\frac{1}{P}\sum_{j=1}^{P}\mathbf{y}_{p_j}}_{\mu_{Y_P}}. \tag{4.26}$$

Then, setting $\mathbf{t}^* = \mu_{Y_P} - \mathscr{R}\mu_{X_P}$, the problem of finding $\mathscr{T}^* = [\mathscr{R}^*, \mathbf{t}^*]$ in (4.25) is reduced to finding $\mathscr{R}^*$:

$$\mathscr{R}^* = \operatorname*{argmin}_{\mathscr{R}} \sum_{j=1}^{P} \left\| \mathscr{R}\tilde{\mathbf{x}}_{p_j} - \tilde{\mathbf{y}}_{p_j} \right\|^2. \tag{4.27}$$

Note that

$$\sum_{j=1}^{P} \left\| \mathscr{R}\tilde{\mathbf{x}}_{p_j} - \tilde{\mathbf{y}}_{p_j} \right\|^2 = \sum_{j=1}^{P} \left(\|\tilde{\mathbf{x}}_{p_j}\|^2 + \|\tilde{\mathbf{y}}_{p_j}\|^2 \right) - 2\operatorname{tr}(\mathscr{R}G), \tag{4.28}$$

where G is the following matrix related to the covariance:

$$G = \sum_{j=1}^{P} \tilde{\mathbf{x}}_{p_j} \tilde{\mathbf{y}}_{p_j}^T. \tag{4.29}$$

Here, the superscript T stands for the transpose and the operator tr is the trace of matrix. Hence, (4.25) is reduced to

$$\mathscr{R}^* = \operatorname*{argmax}_{\mathscr{R}} \operatorname{tr}(\mathscr{R}G). \tag{4.30}$$

Hence, if the singular value decomposition of G is $U\Sigma V^T$, the rotation matrix can be determined by

$$\mathscr{R}^* = VU^T. \tag{4.31}$$

We should note that the above method cannot be applied without knowing more than three pairings between X and Y. The registration accuracy using this approach is dependent on the accurate finding of the point pairs. Even if manually marking fiducial points with discernible features between X and Y, precise registration may fail due to slight mismatch errors. Moreover, in cases of patients containing metallic objects, CBCT image fails in detailed depiction of tooth surfaces due to metal artifacts, which hinder the accurate selection of reference points.

In the following, we will explain well-known registration methods in situations where the pairings between X and Y are not provided in advance.

4.5.2.1 Principal Component Analysis (PCA) Based Registration

Principal component analysis (PCA)-based methods are a simple way to achieve registration between X and Y by aligning the three eigenvectors of the three largest variances of X and Y. To be precise, we first compute the centroids of X and Y that are given by

$$\mu_X = \frac{1}{N}\sum_{j=1}^{N} \mathbf{x}_j \quad \text{and} \quad \mu_Y = \frac{1}{M}\sum_{j=1}^{M} \mathbf{y}_j. \tag{4.32}$$

We set the translation vector to $\mathbf{t} = \mu_Y - \mu_X$ so that the centroid of the translated set $X + \mathbf{t} := \{\mathbf{x}_j + \mathbf{t} : j = 1, \ldots, N\}$ is aligned with μ_Y. Now, it remains to find $\mathscr{R}$ in such a way that the three principal axes of $X - \mu_X$ are aligned with those of $Y - \mu_Y$. To find the rotation $\mathscr{R}$, set

$$\mathfrak{X} = \begin{pmatrix} \mathbf{x}_1 - \mu_X \\ \vdots \\ \mathbf{x}_N - \mu_X \end{pmatrix}. \tag{4.33}$$

The first principal component is the vector $\mathbf{v}_1 \in \mathbb{R}^3$ which maximizes the variance:

$$\mathbf{v}_1 = \underset{\|\mathbf{v}\|=1}{\operatorname{argmax}}[\mathbf{v}^T \mathfrak{X}^T \mathfrak{X} \mathbf{v}]. \tag{4.34}$$

A simple computation shows that $\mathbf{v}_1$ is the unit eigenvector corresponding to the biggest eigenvalue (λ_1) of $X^T X$:

$$\mathfrak{X}^T \mathfrak{X} \mathbf{v}_1 = \lambda_1 \mathbf{v}_1. \tag{4.35}$$

In this way, we can find three unit eigenvectors $\{\mathbf{v}_1, \mathbf{v}_2, \mathbf{v}_3\}$ of the covariance matrix $\mathfrak{X}^T \mathfrak{X}$ and $\{\mathbf{w}_1, \mathbf{w}_2, \mathbf{w}_3\}$ of the covariance matrix corresponding to Y. Then, $\mathscr{R}$ is determined by aligning PCA axes:

$$\begin{pmatrix} \mathbf{w}_1 \\ \mathbf{w}_2 \\ \mathbf{w}_3 \end{pmatrix} = \mathscr{R} \begin{pmatrix} \mathbf{v}_1 \\ \mathbf{v}_2 \\ \mathbf{v}_3 \end{pmatrix}. \tag{4.36}$$

Limitations of this approach are the following:

- Principal axes and centroids are very sensitive to noise and outliers
- If the eigenvalues are similar (*e.g.*, $\lambda_1 \approx \lambda_2$), then the registration becomes unreliable.

Therefore, this approach is mostly used for providing an initial guess for the registration (Fig. 4.21).

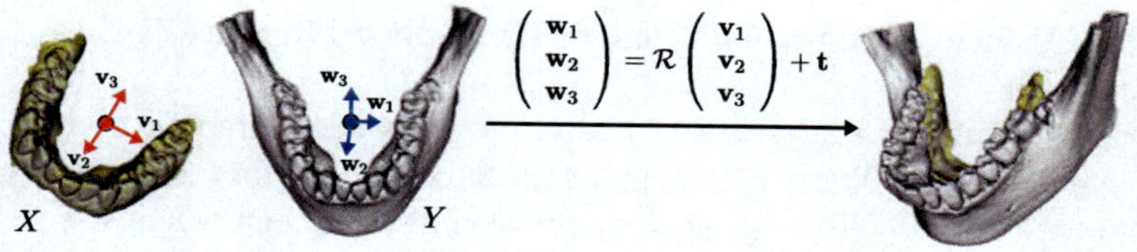

Fig. 4.21 PCA-based alignment. PCA was used to find the three principal eigenvectors $[\mathbf{v}_1, \mathbf{v}_2, \mathbf{v}_3]$ and $[\mathbf{w}_1, \mathbf{w}_2, \mathbf{w}_3]$ of X and Y, respectively. For the initialization, we compute the rotation matrix $\mathscr{R}$ and translation vector $\mathbf{t}$ so that $[\mathbf{w}_1, \mathbf{w}_2, \mathbf{w}_3]^T = \mathscr{R}[\mathbf{v}_1, \mathbf{v}_2, \mathbf{v}_3]^T + \mathbf{t}$. It would be desirable to eliminate outliers (non-common portions of two data) to obtain a better initialization

4.5.2.2 Iterative Closest Point (ICP) Method

ICP is an iterative method to gradually improve the correspondences by searching for the nearest neighbor target point $\mathbf{y}_i$ of $\mathbf{x}_j$ in source X. The rough steps of ICP are as follows:

Step 1 Get a good initial guess $\mathscr{T}^{(0)} = [\mathscr{R}^{(0)}, \mathbf{t}^{(0)}]$. The initial guess can be obtained in various ways, including PCA and paired point method.
Step 2 Set $\hat{X}^{(0)} = \{\hat{\mathbf{x}}_j^{(0)} = \mathscr{T}^{(0)}\mathbf{x}_j : j = 1, \ldots, N\}$.
Step 3 For each $\hat{\mathbf{x}}_j^{(0)}$, find the closest point $\hat{\mathbf{y}}_j^{(0)} \in Y$:

$$\hat{\mathbf{y}}_j^{(0)} = \underset{\mathbf{y}\in Y}{\text{argmin}} \|\mathbf{y} - \hat{\mathbf{x}}_j^{(0)}\|^2. \tag{4.37}$$

Step 4 Get $\mathscr{T}^{(1)}$ by

$$\mathscr{T}^{(1)} = \underset{\mathscr{T}}{\text{argmin}} \sum_{j=1}^{P} \left\| \mathscr{T}\hat{\mathbf{x}}_j^{(0)} - \hat{\mathbf{y}}_j^{(0)} \right\|^2 \tag{4.38}$$

Step 5 Set $\hat{X}^{(1)} = \{\hat{\mathbf{x}}_j^{(1)} = \mathscr{T}^{(1)}\hat{\mathbf{x}}_j^{(0)} : j = 1, \ldots, N\}$. This step is similar to Step 2.
Step 6 Repeat Step 3-Step 4 to get $\mathscr{T}^{(2)}$.
Step 7 ICP iteratively get $\mathscr{T}^{(3)}, \mathscr{T}^{(4)}, \ldots$ by repeating Step 3–Step 4.
Stop * Stopping criterion: Stop the iteration if $\mathscr{T}^{(n+1)} \approx \mathscr{T}^{(n)}$ (unchanged).

In our registration model, without a very good initial guess $\mathscr{T}^{(0)}$, the above ICP method has a serious flaw in that the results depend heavily on the quality of the initial guess and are prone to getting stuck in wrong local optima. This wrong local optima problem may be caused from a large non-overlapping sets. See Fig. 4.22.

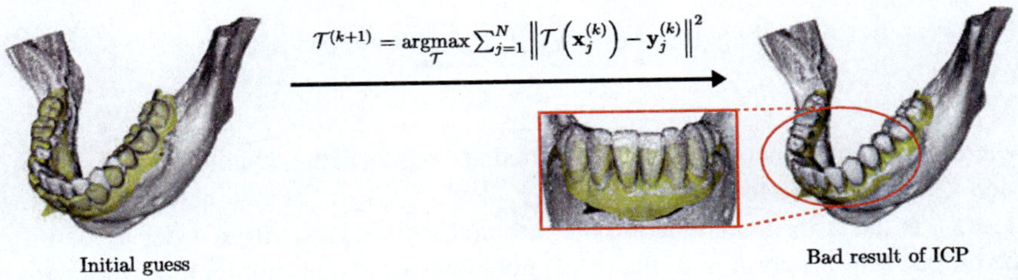

Fig. 4.22 The local minimum problem of ICP registration method. The ICP method depends heavily on the quality of the initial guess (left), and are prone to getting stuck in wrong local optima (right)

4.5.3 *Removing Non-overlapping Parts Between X and Y that Can Affect ICP Registration*

In registration processes using ICP, the non-overlapping part of the two sets of data can become mismatched. Hence, it would be desirable to minimize the non-overlapping point sets in X and Y that can affect ICP registration. To achieve the goal, the authors [21, 22] developed an automated methods of segmentation and identification of individual teeth from X and Y. These segmentations allow one to trim off non-overlapping points X and Y, which are the major hurdle in the registration. With conventional segmentation techniques (without using deep learning techniques), the above mentioned individual tooth segmentations may be a very challenging task. However, with careful use of deep learning techniques, it is possible to achieve individual tooth segmentation and identification from X and Y. See Fig. 4.23.

The semantic segmentation of individual teeth from IOS data is described in Sect. 4.3, where X is decomposed into

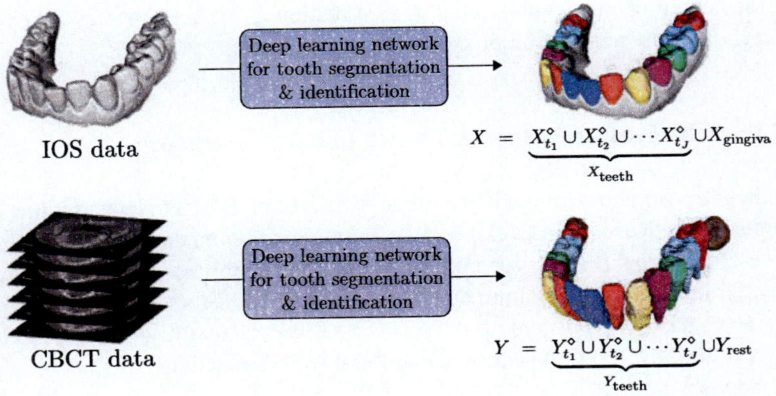

Fig. 4.23 Individual tooth segmentation and identification of IOS and CBCT images. X and Y are decomposed into individual teeth and the rest

$$X = \underbrace{X_{t_1}^{\diamond} \cup X_{t_2}^{\diamond} \cup \cdots X_{t_J}^{\diamond}}_{X_{\text{teeth}}} \cup, X_{\text{gingiva}} \tag{4.39}$$

where J is the number of teeth exposed outside the gum in the mandible (i.e. $J \leq 16$), each $X_{t_j}^{\diamond}$ represents a tooth with the code c_j, and X_{gingiva} is the rest including gingiva. Here, t_j is the number assigned to the detected individual tooth in order to identify the unique tooth according to the (FDI) notation. X_{teeth} does not include unexposed teeth in X (e.g., impacted wisdom teeth).

Section 4.4 deals with CBCT data with the semantic segmentation of individual tooth, where several complex procedures were used to deal with metal-induced artifacts. As (4.39) for IOS, Y is decomposed into

$$Y = \underbrace{Y_{t_1}^{\diamond} \cup Y_{t_2}^{\diamond} \cup \cdots Y_{t_J}^{\diamond}}_{Y_{\text{teeth}}} \cup Y_{\text{rest}}, \tag{4.40}$$

where each $Y_j^{\diamond}$ represents the j-th tooth and Y_{rest} is the rest that may include unexposed wisdom teeth.

For the registration, we utilize the semantically segmented data $\{X_{t_j}^{\diamond} : j = 1, \ldots, J\}$ and $\{Y_{t_j}^{\diamond} : j = 1, \ldots, J\}$. The goal is to find an optimal transformation $\mathscr{T}^*$ such that $\mathscr{T}(X_{t_j}^{\diamond})$ is aligned with $Y_{t_j}^{\diamond}$ for $j = 1, \ldots, J$:

$$\mathscr{T}^* = \operatorname*{argmin}_{\mathscr{T}} \sum_{j=1}^{J} \text{dist}(\mathscr{T}(X_j^{\diamond}) , Y_j^{\diamond}). \tag{4.41}$$

This registration approach aims to eliminate the correspondence between different teeth, while maintaining the individual tooth correspondence, which makes optimization for transformation more efficient. The ICP method, which allows for high-accuracy alignment, will be used as the final registration, and the key issue in this case is finding a very good initial transformation $\mathscr{T}^{(0)}$. See Fig. 4.24 for overall architecture of the registration process [22].

4.5.3.1 Point Feature Histograms-Based Initial Registration

To address the limitation of the ICP method, which often falls into local minima during registration, Rusu [46] developed a global registration method called Fast Point Feature Histograms (FPFH) for initial matching between two point clouds X_{teeth} and Y_{teeth}. In FPFH, we compute the two sets of FPFH vectors; $FPFH(X_{teeth}) = \{FPFH(\mathbf{x}) : \mathbf{x} \in X_{teeth}\}$ and $FPFH(Y_{teeth}) = \{FPFH(\mathbf{y}) : \mathbf{y} \in Y_{teeth}\}$. Here, $FPFH(\mathbf{x})$ (or $FPFH(\mathbf{x})$) is designed to reliably find matching pairs between two point clouds.

Now, we explain the definition of $FPFH(\mathbf{x})$, which is a bit complicated. The $FPFH(\mathbf{x})$ uses the geometric features of the normal vector and the curvature on

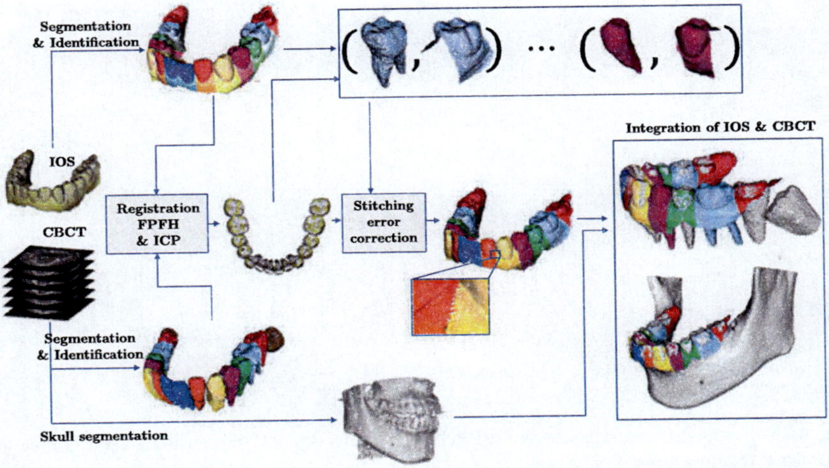

Fig. 4.24 Architecture of the registration process. The semantic segmentations of individual tooth from IOS and CBCT play an important role for the registration

the surface X_{teeth}. However, such simple feature information alone cannot perform a one-to-one matching between X_{teeth} and Y_{teeth}, because there are too many points with similar geometric features in the clouds. The $FPFH(\mathbf{x})$ represents not only the geometric information of the normal vector and curvature of the surface X_{teeth} at $\mathbf{x}$, but also the relevant information from its surrounding neighborhood points over X_{teeth}. Furthermore, $FPFH$ is defined in a discriminatory manner to reliably discriminate regional geometric features.

Let $NN_k(\mathbf{x})$ denote the k-nearest neighborhood of $\mathbf{x}$ over the point cloud X_{teeth} (i.e., the set of k points such that $\max\{\|\mathbf{x}' - \mathbf{x}\| : \mathbf{x}' \in NN_k(\mathbf{x})\} \leq \|\tilde{\mathbf{x}} - \mathbf{x}\|$ for all $\tilde{\mathbf{x}} \in X_{teeth} \backslash NN_k(\mathbf{x})$). Let $\mathbf{n_x}$ denote the unit normal vector of the surface X_{teeth} at $\mathbf{x}$. The $FPFH(\mathbf{x})$ is based on angular variations of the surface normals on $NN_k(\mathbf{x})$. To be precise, for $\mathbf{x}' \in NN_k(\mathbf{x}) \backslash \{\mathbf{x}\}$, we compute the following three angles $\{\rho(\mathbf{x}, \mathbf{x}'), \phi(\mathbf{x}, \mathbf{x}'), \theta(\mathbf{x}, \mathbf{x}')\}$ that are designed to be symmetric:

$$\rho(\mathbf{x}, \mathbf{x}') = \cos^{-1}\left(\langle \mathbf{u}, \frac{\mathbf{x}_t - \mathbf{x}_s}{\|\mathbf{x}_t - \mathbf{x}_s\|}\rangle\right), \tag{4.42}$$

$$\phi(\mathbf{x}, \mathbf{x}') = \cos^{-1}\left(\langle \mathbf{v}, \mathbf{n}_{\mathbf{x}_t}\rangle\right), \tag{4.43}$$

$$\theta(\mathbf{x}, \mathbf{x}') = \begin{cases} \cos^{-1}\left(\langle \mathbf{u}, \mathbf{n}_{\mathbf{x}_t}\rangle\right) & \text{if } \langle \mathbf{w}, \mathbf{n}_{\mathbf{x}_t}\rangle \geq 0 \\ \cos^{-1}\left(\langle \mathbf{u}, \mathbf{n}_{\mathbf{x}_t}\rangle\right) + \pi & \text{otherwise} \end{cases}, \tag{4.44}$$

where $(\mathbf{x}_s, \mathbf{x}_t)$ is either $(\mathbf{x}, \mathbf{x}')$ or $(\mathbf{x}', \mathbf{x})$ that is determined by

$$(\mathbf{x}_s, \mathbf{x}_t) = \begin{cases} (\mathbf{x}, \mathbf{x}') & \text{if } \langle \mathbf{n_x}, \mathbf{x}' - \mathbf{x}\rangle \leq \langle \mathbf{n_{x'}}, \mathbf{x} - \mathbf{x}'\rangle \\ (\mathbf{x}', \mathbf{x}) & \text{otherwise} \end{cases}, \tag{4.45}$$

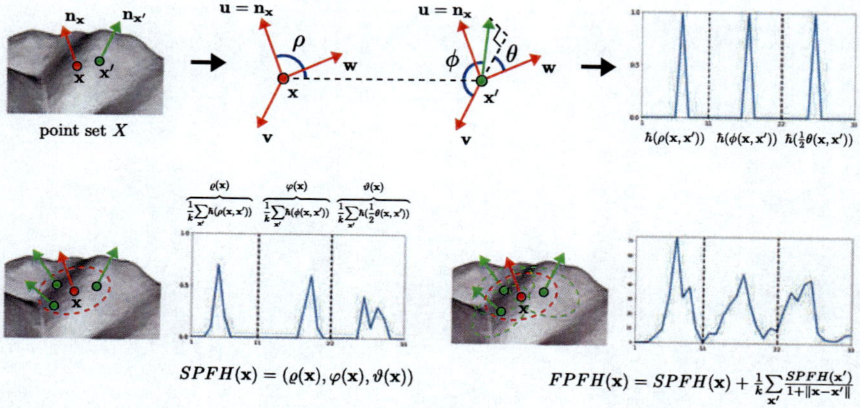

Fig. 4.25 Fast point feature histogram (FPFH). FPFH encodes a 33-dimensional vector reflecting geometric features around each point in X. The top image shows the feature vector (ρ, ϕ, θ) of $FPFH(\mathbf{x})$ that represents intrinsic geometrical relation. The bottom left image shows $SPFH(\mathbf{x})$, which is the sum of the feature vectors between $\mathbf{x}$ and points in $NN_k(\mathbf{x})$. The bottom right image shows $FPFH(\mathbf{x})$ that is the weighted sum of $SPFH(\mathbf{x}')$ for $\mathbf{x}' \in NN_k(\mathbf{x})$

and the triple $\mathbf{u}$, $\mathbf{v}$, $\mathbf{w}$ is the Darboux frame defined by

$$\mathbf{u} = \mathbf{n}_{\mathbf{x}_s}, \quad \mathbf{v} = \frac{\mathbf{x}_t - \mathbf{x}_s}{\|\mathbf{x}_t - \mathbf{x}_s\|} \times \mathbf{u}, \quad \mathbf{w} = \mathbf{u} \times \mathbf{v}.$$

Note that the use of $(\mathbf{x}_s, \mathbf{x}_t)$ in (4.45) is needed in order that the three angles ρ, ϕ, and θ have the symmetric property. These three angles provide consistent geometric features representing the difference between $\mathbf{n_x}$ and $\mathbf{n_{x'}}$.

Now, we are ready to define $FPFH(\mathbf{x})$, which is visually illustrated in Fig. 4.25. It uses a Simplified Point Feature Histogram (SPFH) given by

$$SPFH(\mathbf{x}) = (\varrho(\mathbf{x}), \varphi(\mathbf{x}), \vartheta(\mathbf{x})) \in \mathbb{R}^{11} \times \mathbb{R}^{11} \times \mathbb{R}^{11} \tag{4.46}$$

and

$$\varrho(\mathbf{x}) = \frac{1}{k} \sum_{\mathbf{x}' \in NN_k(\mathbf{x}) \setminus \{\mathbf{x}\}} \hbar(\rho(\mathbf{x}, \mathbf{x}')), \tag{4.47}$$

$$\varphi(\mathbf{x}) = \frac{1}{k} \sum_{\mathbf{x}' \in NN_k(\mathbf{x}) \setminus \{\mathbf{x}\}} \hbar(\phi(\mathbf{x}, \mathbf{x}')), \tag{4.48}$$

$$\vartheta(\mathbf{x}) = \frac{1}{k} \sum_{\mathbf{x}' \in NN_k(\mathbf{x}) \setminus \{\mathbf{x}\}} \hbar(\frac{1}{2}\theta(\mathbf{x}, \mathbf{x}')), \tag{4.49}$$

where $\hbar : [0, \pi) \mapsto \mathbb{R}^{11}$ is the map defined by

$$\hbar(s) = (\hbar_1(s), \ldots, \hbar_{11}(s)), \quad \hbar_j(s) = \begin{cases} 1 & \text{if } s \in [\frac{j-1}{11}\pi, \frac{j}{11}\pi) \\ 0 & \text{otherwise} \end{cases}. \tag{4.50}$$

Here, the vector-valued function $\hbar$ is used to increase the ability to determine differences between different local geometric features. Then, $FPFH(\mathbf{x})$ is the following weighted sum of $SPFH(\mathbf{x}')$ over $NN_k(\mathbf{x})$:

$$FPFH(\mathbf{x}) = SPFH(\mathbf{x}) + \frac{1}{k} \sum_{\mathbf{x}' \in NN_k(\mathbf{x}) \setminus \{\mathbf{x}\}} \frac{SPFH(\mathbf{x}')}{1 + \|\mathbf{x} - \mathbf{x}'\|}. \tag{4.51}$$

Here, the weight $1/(1 + \|\mathbf{x} - \mathbf{x}'\|)$ depends on the center point $\mathbf{x}$ and its neighbor $\mathbf{x}' \in NN_k(\mathbf{x}) \setminus \{\mathbf{x}\}$.

Similarly, we compute $FPFH(Y_{teeth}) = \{FPFH(\mathbf{y}) : \mathbf{y} \in Y_{teeth}\}$. Next, we find matching pairs between $FPFH(X_{teeth})$ and $FPFH(Y_{teeth})$. For each $\mathbf{x} \in X_{teeth}$, we select $\mathbf{y} \in FPFH(Y_{teeth})$, denoted by $Corr_{Y_{teeth}}(\mathbf{x})$, whose $FPFH$ is most similar to $FPFH(\mathbf{y})$:

$$Corr_{Y_{teeth}}(\mathbf{x}) = \operatorname*{argmin}_{\mathbf{y} \in Y} \|FPFH(\mathbf{x}) - FPFH(\mathbf{y})\|. \tag{4.52}$$

Similarly, we compute $Corr_{X_{teeth}}(\mathbf{y})$ for all $\mathbf{y} \in Y_{teeth}$. Then, we get the correspondence pair set by

$$Corr = \{(\mathbf{x}, Corr_{Y_{teeth}}(\mathbf{x})) : \mathbf{x} \in X_{teeth}\} \cap \{(Corr_{X_{teeth}}(\mathbf{y}), \mathbf{y}) : \mathbf{y} \in Y_{teeth}\}. \tag{4.53}$$

See Fig. 4.26.

To filter out inaccurate pairs from the set $Corr$, we randomly sample three matches $(\mathbf{x}_1, \mathbf{y}_1), (\mathbf{x}_2, \mathbf{y}_2), (\mathbf{x}_3, \mathbf{y}_3) \in Corr$ and select them if the following conditions are met, and drop them otherwise:

$$\tau < \frac{\|\mathbf{x}_i - \mathbf{x}_j\|}{\|\mathbf{y}_i - \mathbf{y}_j\|} < \frac{1}{\tau}, \quad \text{for } 1 \le i < j \le 3, \tag{4.54}$$

where τ is a number close to 1. We denote this filtered subset by $Corr^*$. These selected points remain less than about 2% of the total points. Then, an initial transformation is determined by

$$\mathscr{T}^{(0)} = \operatorname*{argmin}_{\mathscr{T}} \sum_{(\mathbf{x},\mathbf{y}) \in Corr^*} \|\mathbf{y} - \mathscr{T}(\mathbf{x})\|_2^2. \tag{4.55}$$

We denote X_{teeth} transformed by the previously obtained $\mathscr{T}^{(0)}$ as $X_{\text{teeth}}^{(0)} = X_{t_1}^{(0)} \cup \cdots \cup X_{t_n}^{(0)}$, where $X_{t_j}^{(0)} = \mathscr{T}^{(0)}(X_{t_j})$ for $j = 1, \ldots, n$. Then $X_{\text{teeth}}^{(0)}$ and Y_{teeth} are roughly aligned, but fine-tuning is needed for accurate registration.

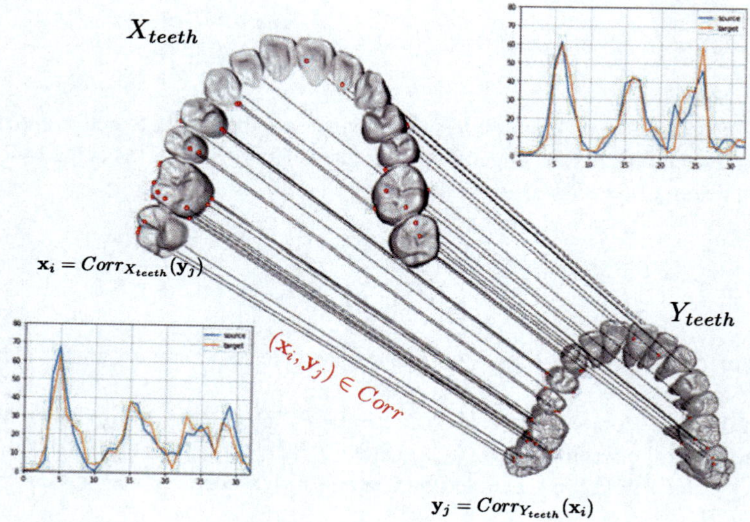

Fig. 4.26 Representation of the set of correspondences: $Corr = \{(\mathbf{x}, Corr_{Y_{teeth}}(\mathbf{x})) : \mathbf{x} \in X_{teeth}\} \cap \{(Corr_{X_{teeth}}(\mathbf{y}), \mathbf{y}) : \mathbf{y} \in Y_{teeth}\}$. The number of the set $Corr$ is around 1% of X_{teeth}

A fine rigid transformation is obtained through iterative process, which gradually improves correspondence finding. We improve the ICP (T-ICP) by using the individual tooth segmentation and identification results. For $k \geq 1$, denote $X^{(k)}_{\text{teeth}} = \mathscr{T}^{(k)}(X^{(k-1)}_{\text{teeth}})$. Here, the k-th rigid transformation $\mathscr{T}^{(k)}$ is determined by

$$\mathscr{T}^{(k)} = \underset{\mathscr{T} \in SE(3)}{\operatorname{argmin}} \sum_{(\mathbf{x},\mathbf{y}) \in Corr^{(k)}} \|\mathbf{y} - \mathscr{T}(\mathbf{x})\|^2, \tag{4.56}$$

where

$$\begin{aligned} Corr^{(k)} = & \left\{(\mathbf{x}, m(\mathbf{x}; Y_{\text{teeth}})) : \mathbf{x} \in X^{(k-1)}_{\text{teeth}}\right\} \\ & \cap \bigcup_{j=1}^{n} \left\{(\mathbf{x}, \mathbf{y}) \in X^{(k-1)}_{t_j} \times Y_{t_j}\right\} \end{aligned} \tag{4.57}$$

and

$$m(\mathbf{x}; Y_{\text{teeth}}) = \underset{\mathbf{y} \in Y_{\text{teeth}}}{\operatorname{argmin}} \|\mathbf{x} - \mathbf{y}\|. \tag{4.58}$$

The k-th correspondence set $Corr^{(k)}$ contains pairs of $\mathbf{x} \in X^{(k-1)}_{t_j}$ and $m(\mathbf{x}; Y_{\text{teeth}}) \in Y_{t_j}$ closest to $\mathbf{x}$ for $j = 1, \ldots, n$. Using the segmented teeth prevents undesired correspondences between two teeth with different codes. See Fig. 4.27. Note that it is the vanilla ICP when not using the set of tooth pairs. The final rigid transformation $\mathscr{T}^*$ is obtained by the following composition of transformations:

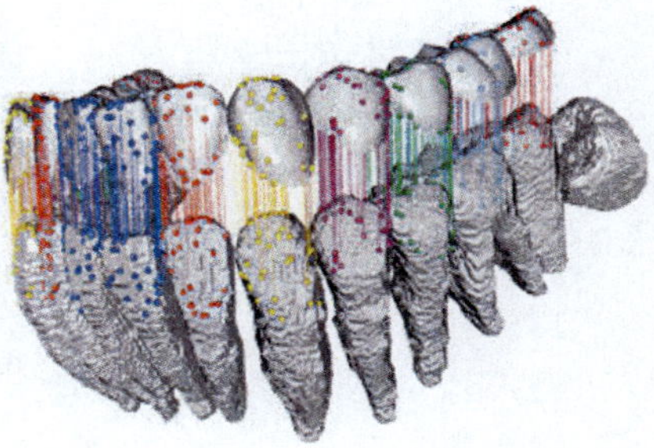

Fig. 4.27 Correspondences obtained by individual tooth segmentation and identification

$\mathscr{T}^* = \mathscr{T}^{(K)} \circ \cdots \circ \mathscr{T}^{(0)}$, where K is the number of iterations until the stopping criterion is satisfied for a given $\varepsilon > 0$:

$$\sum_{(\mathbf{x},\mathbf{y})\in Corr^{(K)}} \|\mathscr{T}^{(K)} \circ \cdots \circ \mathscr{T}^{(0)}(\mathbf{x}) - \mathbf{y}\| < \varepsilon. \tag{4.59}$$

4.5.3.2 Stitching Error Correction in IOS

Next, we edit the IOS models with stitching errors by referring to the CBCT images. As shown in Fig. 4.28a, a 3D point set X of the IOS model can be decomposed into

$$X = \underbrace{X_{t_1} \cup \cdots \cup X_{t_J}}_{X_{\text{teeth}}} \cup \underbrace{X_{g_1} \cup \cdots \cup X_{g_J}}_{X_{\text{gingiva}}}, \tag{4.60}$$

where

$$X_{g_j} = \left\{ \mathbf{x} \in X_{\text{gingiva}} : \underset{\mathbf{x}'\in X_{\text{teeth}}}{\text{argmin}} \|\mathbf{x} - \mathbf{x}'\| \in X_{t_j} \right\}. \tag{4.61}$$

Here, t_j is the number between 1 and 32 that is assigned to an individual tooth to identify the unique tooth according to the universal notation system [38].

We denote $X^*_{t_j} = \mathscr{T}^*(X_{t_j})$ and $X^*_{g_j} = \mathscr{T}^*(X_{g_j})$ for $j = 1, \ldots, J$. Each tooth $X^*_{t_j}$ is transformed by a corrective rigid transformation $\mathscr{T}^{**}_j$, which is obtained by applying the vanilla ICP to sets $X^*_{t_j-1} \cup X^*_{t_j} \cup X^*_{t_j+1}$ and $Y^*_{t_j-1} \cup Y^*_{t_j} \cup Y^*_{t_j+1}$ as the source and target. Here, $X^*_{t_j-1}$ (or $X^*_{t_j+1}$) is an empty set if $t_j - 1$ (or $t_j + 1$) is not equal to $t_{j'}$ for every $j' = 1, \ldots, J$. Using the individual corrective transformations, IOS stitching errors are corrected separately by $X^{**}_{t_j} = \mathscr{T}^{**}_j(X^*_{t_j})$ for $j = 1, \ldots, J$. In this procedure, we use one tooth and two adjacent teeth on both sides for reliable correction. It takes advantage of the fact that narrow digital scanning is accurate. Now it remains to fix the gingiva area whose boundary shares the boundaries with the teeth. To fit the boundaries between the gingiva and individually transformed teeth,

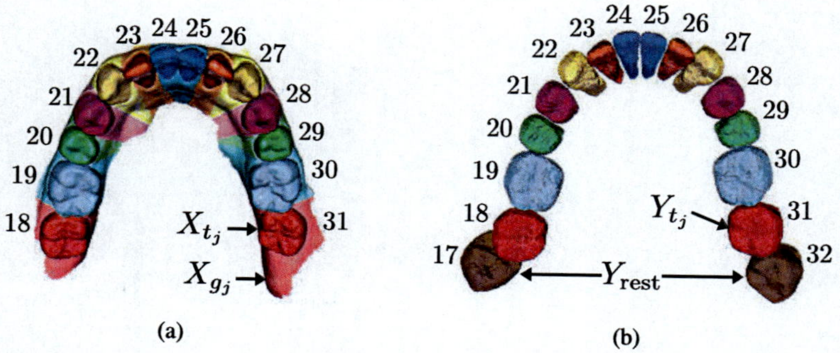

Fig. 4.28 Results of TSIM-IOS and -CBCT, respectively. The indicated numbers represent mandibular teeth by the universal notation. **a** Individual IOS teeth and their split gingiva parts, and **b** CBCT teeth containing unexposed wisdom teeth

the gingival surface is divided according to the areas in contact with the individual teeth by Eq. (4.60). Therefore, the rectified gingiva is obtained by $X^{**}_{g_j} = \mathscr{T}^{**}_j(X^*_{g_j})$ for $j = 1, \ldots, J$.

4.6 Discussion and Future Research Direction

Digital dentistry is undergoing a paradigm shift with dazzling advances in AI tools, CBCT, IOS, 3D printing, and 3D CAD/CAM. Digital dentistry will be applied to all dental fields such as implants/prosthodontics, oral and maxillofacial surgery, and orthodontic treatment. In the near future, dentists will use digital dentistry to perform virtual surgery simulations and fully control the final result while providing precise treatment, such as a sophisticated surgical guide, omitting the classical impression process that is uncomfortable for patients, and consistent denture production using a 3D printer and CAD/CAM.

The success of digital dentistry depends on marketable innovation. Digital dentistry can learn a lot from Kodak's downfall because the reasons for its downfall are very subtle. It seems that Kodak has not neglected innovation. In the 1980s, Kodak invested in digital technology and developed the first digital camera. Kodak predicted that cameras would be digitized and photos would be shared online. Their subtle mistake was not to foresee that online photo sharing will evolve into a whole new form of business (such as Facebook). Kodak made the mistake of viewing this digital transformation as an extension of their existing printing business. This is the reason that technological innovation alone is not enough.

Digital dentistry must move toward increasing the satisfaction of both dentists and patients while generating revenue. Digital dentistry should allow dentists to use a variety of software in an integrated way, even if they do not have software expertise. Therefore, there is a great demand for a smart platform that allows doctors to easily transform and integrate various modality images. The recent development of DL tools has made it possible to develop such a smart platform, and it is expected that economic value will be created as the burden of medical expenses increases due to aging population.

Recently, as the use of CBCT equipment has become common in dentistry, it has established itself as a standard imaging equipment in dentistry. The reason is that CBCT has a significantly lower radiation dose and is relatively inexpensive in price and maintenance. With the development of CBCT technology with built-in artificial intelligence, it provides high resolution comparable to existing CT images while significantly lowering radiation dose exposure. The presence of highly attenuating materials such as metallic objects complicates reconstruction techniques by violating the forward model assumption of the sinogram data being equal to the Radon transform of an image. As the number of aged patients with metallic implants is increasing, metal-induced artifacts are becoming a major factor degrading the CBCT diagnostic performance. Additionally, in the era of low-dose CBCT protocols requiring the reduction of either the tube voltage or tube current or both, artifacts related to these high-attenuation materials could be accentuated. This occurs because the reduced X-ray tube voltage or current causes more severe beam hardening, scattering, photon starvation, and photon noise. Due to these effects, the mismatched projection data cause severe streaking and shading artifacts in the reconstructed CBCT images. Recently, deep learning-based research to improve CBCT reconstruction methods to minimize artifacts caused by high-attenuation materials has been actively conducted and significant progress is being made.

The development of artificial intelligence is expected to automate the convergence of CBCT, oral scanners, and facial scanners, which will be very useful elements for both patients and doctors to manage dental care and dental health. The integration of CBCT and IOS can provide highly accurate digital impressions by compensating for the shortcomings of CBCT's metal artifacts and IOS's stitching errors. Traditional impression-making methods have a number of factors that limit accuracy, such as patient movement, tearing and deformation of the impression during removal, and soft tissue contraction. Therefore, this fusion can eliminate the cumbersome procedure of traditional impressions for both the dentist and the patient, and significantly shorten the treatment time.

In the field of medical image segmentation, segmentation performance is significantly different after DL (2014–) and before DL (1970–2012). Before 2012, there were several attempts to develop 3D tooth segmentation methods, most of which are based on level set methods [13, 14, 23, 55, 59]. Unfortunately, level set-based methods have fundamental limitations in achieving fully automated segmentation. This difficulty arises from the dependence of such methods on the initialization of level set, and the automatic initialization is hindered by the complex image structure associated with adjacent teeth, the jaw, the alveolar bone, etc. Hence, user intervention through manual initialization is inevitable in this approach. From 2014, these difficult segmentation problems began to be solved with the rapid development of deep learning methods. This is due to DL's remarkable ability to capture the spatial relationships between pixels to figure out local and global interconnections.

Digital dentistry will advance in the direction of overcoming the limitations of existing dental diagnosis/treatment methods. Conventional methods used planar radiographs, dental plaster casts, and articulator mounting to perform analysis and diagnosis, treatment planning, and fabrication of necessary guide devices and prostheses through dental laboratory procedures. Because 3D analysis is not possible in planar radiographs, it is difficult to model and reproduce teeth. Therefore, it is difficult to accurately understand the relationship between the maxilla and the mandible during the prosthodontic manufacturing process. Because accurate analysis and diagnosis are not possible due to the above limitations, the establishment of a treatment plan depends on the experience of the operator, and unexpected problems may occur. Therefore, even in the final finishing process of the oral prosthodontics, the chair time is very long and correction of the prosthodontics is required. Recently, innovations in dental imaging equipment and rapidly improved deep learning technologies have been introduced into the dental field, showing remarkable improvement in treatment results. AI-based digital dentistry is expected to occupy an important position in dentistry in the near future due to its more precise, accurate, convenient and effective compared to existing dental tools.

Acknowledgements This research was supported by a grant of the Korea Health Technology R & D Project through the Korea Health Industry Development Institute (KHIDI), funded by the Ministry of Health & Welfare, Republic of Korea (grant number : HI20C0127). We are deeply grateful to HDXWILL for their help and collaboration.

References

1. Alexe, B., Deselaers, T., Ferrari, V.: Measuring the objectness of image windows. IEEE Trans. Pattern Anal. Mach. Intell. **34**(11), 2189–2202 (2012)
2. Baan, F., Bruggink, R., Nijsink, J., Maal, T.J.J., Ongkosuwito, E.M.: Fusion of intra-oral scans in cone-beam computed tomography scans. Clin. Oral Investig. **25**(1), 77–85 (2021)
3. Baan, F., de Waard, O., Bruggink, R., Xi, T., Ongkosuwito, E.M., Maal, T.J.J.: Virtual setup in orthodontics: planning and evaluation. Clin. Oral Investig. **24**(7), 2385–2393 (2020)
4. Barrett, J.F., Keat, N.: Artifacts in CT: recognition and avoidance. Radiographics **24**(6), 1679–1691 (2004)
5. Chen, Y., Haiyan, D., Yun, Z., Yang, S., Dai, Z., Zhong, L., Feng, Q., Yang, W.: Automatic segmentation of individual tooth in dental CBCT images from tooth surface map by a multi-task FCN. IEEE Access **8**, 97296–97309 (2020)
6. Chen, Y., Stanley, K., Att, W.: Artificial intelligence in dentistry: current applications and future perspectives. Quintessence Int. **51**(3), 248–57 (2020)
7. Chin, S.-J., Wilde, F., Neuhaus, M., Schramm, A., Gellrich, N.-C., Rana, M.: Accuracy of virtual surgical planning of orthognathic surgery with aid of cad/cam fabricated surgical splinta novel 3d analyzing algorithm. J. Cranio-Maxillofac. Surg. **45**(12), 1962–1970 (2017)
8. Cui, Z., Li, C., Wang, W.: Toothnet: automatic tooth instance segmentation and identification from cone beam CT images. In: Proceedings of the IEEE Conference on Computer Vision and Pattern Recognition, pp. 6368–6377 (2019)
9. Diker, B., Tak, O.: Comparing the accuracy of six intraoral scanners on prepared teeth and effect of scanning sequence. J. Adv. Prosthodont. **12**(5), 299 (2020)
10. Elnagar, M.H., Aronovich, S., Kusnoto, B.: Digital workflow for combined orthodontics and orthognathic surgery. Oral Maxillofac. Surg. Clin. North Am. **32**(1), 1–14 (2020)

11. Esmaeili, F., Johari, M., Haddadi, P.: Beam hardening artifacts by dental implants: comparison of cone-beam and 64-slice computed tomography scanners. Dental Res. J. **10**(3), 376 (2013)
12. Farronato, M., Maspero, C., Lanteri, V., Fama, A., Ferrati, F., Pettenuzzo, A., Farronato, D.: Current state of the art in the use of augmented reality in dentistry: a systematic review of the literature. BMC Oral Health **19**(1), 1–15 (2019)
13. Gan, Y., Xia, Z., Xiong, J., Zhao, Q., Ying, H., Zhang, J.: Toward accurate tooth segmentation from computed tomography images using a hybrid level set model. Med. Phys. **42**(1), 14–27 (2015)
14. Gao, H., Chae, O.: Individual tooth segmentation from CT images using level set method with shape and intensity prior. Pattern Recogn. **43**(7), 2406–2417 (2010)
15. Haralick, R.M., Sternberg, S.R., Zhuang, X.: Image analysis using mathematical morphology. IEEE Trans Pattern Anal Mach. Intell. (4), 532–550 (1987)
16. Ho, C.-T., Lai, H.-C., Lin, H.-H., Denadai, R., Lo, L.-J.: Outcome of full digital workflow for orthognathic surgery planning in the treatment of asymmetric skeletal class iii deformity. J. Formosan Med. Assoc. (2021)
17. Horn, B.K.P.: Closed-form solution of absolute orientation using unit quaternions. J. Opt. Soc. Am. A **4**(4), 629–642 (1987)
18. Horner, K., Jacobs, R., Schulze, R.: Dental CBCT equipment and performance issues. Radiat. Protect. Dosim. **153**(2), 212–218 (2013)
19. Jacobs, R., Salmon, B., Codari, Hassan, B., Bornstein, M.M.: Cone beam computed tomography in implant dentistry: recommendations for clinical use. BMC Oral Health **18**(1), 1–16 (2018)
20. Jahangiri, L., Akiva, G., Lakhia, S., Turkyilmaz, I.: Understanding the complexities of digital dentistry integration in high-volume dental institutions. British Dental J. **229**(3), 166–168 (2020)
21. Jang, T.J., Kim, K.C., Cho, H.C., Seo, J.K.: A fully automated method for 3d individual tooth identification and segmentation in dental CBCT. IEEE Trans. Pattern Anal. Mach. Intell. (2021)
22. Jang, T.J., Yun, H.S., Hyun, C.M., Kim, J.-E., Lee, S.-H., Seo, J.K.: Fully automatic integration of dental cbct images and full-arch intraoral impressions with stitching error correction via individual tooth segmentation and identification (2021). arXiv:2112.01784
23. Ji, D.X., Ong, S.H., Foong, K.W.C.: A level-set based approach for anterior teeth segmentation in cone beam computed tomography images. Comput. Biol. Med. **50**, 116–128 (2014)
24. Joda, T., Gallucci, G.O.: The virtual patient in dental medicine. Clin. Oral Implant Res **26**(6), 725–726 (2015)
25. Joda, T., Zarone, F., Ferrari, M.: The complete digital workflow in fixed prosthodontics: a systematic review. BMC Oral Health **17**(1), 1–9 (2017)
26. Kaasalainen, T., Ekholm, M., Siiskonen, T., Kortesniemi, M.: Dental cone beam CT: an updated review. Physica Medica **88**, 193–217 (2021)
27. Kernen, F., Kramer, J., Wanner, L., Wismeijer, D., Nelson, K., Flugge, T.: A review of virtual planning software for guided implant surgery-data import and visualization, drill guide design and manufacturing. BMC Oral Health **20**(1), 1–10 (2020)
28. Kravitz, N.D., Groth, C., Jones, P.E., Graham, J.W., Redmond, W.R.: Intraoral digital scanners. J. Clin. Orthod **48**(6), 337–347 (2014)
29. Lahoud, P., EzEldeen, M., Beznik, T., Willems, H., Leite, A., Van Gerven, A., Jacobs, R.: Artificial intelligence for fast and accurate 3-dimensional tooth segmentation on cone-beam computed tomography. J. Endodont. **47**(5), 827–835 (2021)
30. Lee, S., Woo, S., Yu, J., Seo, J., Lee, J., Lee, C.: Automated CNN-based tooth segmentation in cone-beam CT for dental implant planning. IEEE Access **8**, 50507–50518 (2020)
31. Lee, T.-C., Kashyap, R.L., Chu, C.-N.: Building skeleton models via 3-d medial surface axis thinning algorithms. CVGIP: Graph. Models Image Proc. **56**(6), 462–478 (1994)
32. Li, J., Sommer, C., Wang, H.-L., Lepidi, L., Joda, T., Mendonca, G.: Creating a virtual patient for completely edentulous computeraided implant surgery: a dental technique. J. Prosth. Dent. **125**(4), 564–568 (2021)
33. Lim, J.-H., Park, J.-M., Kim, M., Heo, S.-J., Myung, J.-Y.: Comparison of digital intraoral scanner reproducibility and image trueness considering repetitive experience. J. Prosthe. Dent. **119**(2), 225–232 (2018)

34. Linares, O.C., Bianchi, J., Raveli, D., Neto, J.B., Hamann, B.: Mandible and skull segmentation in cone beam computed tomography using super-voxels and graph clustering. Vis. Comput. **35**(10), 1461–1474 (2019)
35. Marin, I., Goga, N., Vasilateanu, A., Pavaloiu, I.-B.: Integrated platform for extracting dental knowledge imaging and 3d modeling. In: 2017 E-Health and Bioengineering Conference (EHB), pp. 157–160. IEEE (2017)
36. Moon, Y.-G., Lee, K.-M.: Comparison of the accuracy of intraoral scans between complete-arch scan and quadrant scan. Prog. Orthod. **21**(1), 1–6 (2020)
37. Nagy, Z., Simon, B., Mennito, A., Evans, Z., Renne, W., Vag, J.: Comparing the trueness of seven intraoral scanners and a physical impression on dentate human maxilla by a novel method. BMC Oral Health **20**(1), 1–10 (2020)
38. Nelson, S.J.: Wheeler's dental anatomy, physiology and occlusion-e-book. Elsevier Health Sciences (2014)
39. Otsu, N.: A threshold selection method from gray-level histograms. IEEE Trans. Syst. Man Cybern. **9**(1), 62–66 (1979)
40. Qi, C.R., Su, H., Mo, K., Guibas, L.J.: Pointnet: Deep learning on point sets for 3d classification and segmentation. In: Proceedings of the IEEE Conference on Computer Vision and Pattern Recognition, pp. 652–660 (2017)
41. Qiu, B., Guo, J., Kraeima, J., Glas, H.H., Borra, R.J.H., Witjes, M.J.H., van Ooijen, P.M.A.: Automatic segmentation of the mandible from computed tomography scans for 3d virtual surgical planning using the convolutional neural network. Phys. Med. Biol. **64**(17), 175020 (2019)
42. Rao, Y., Wang, Y., Meng, F., Jiansu, P., Sun, J., Wang, Q.: A symmetric fully convolutional residual network with DCRF for accurate tooth segmentation. IEEE Access **8**, 92028–92038 (2020)
43. Redmon, J., Divvala, S., Girshick, R., Farhadi, A.: You only look once: Unified, real-time object detection. In Proceedings of the IEEE Conference on Computer Vision and Pattern Recognition, pp. 779–788 (2016)
44. Rekow, E.D.: Digital dentistry: the new state of the artis it disruptive or destructive? Dent. Mater. **36**(1), 9–24 (2020)
45. Ronneberger, O., Fischer, P., Brox, T.: U-net: convolutional networks for biomedical image segmentation. In: International Connference on Medical Image Computing and Computer-Assisted Intervention, pp. 234–241. Springer (2015)
46. Rusu, R.B., Blodow, N., Beetz, M.: Fast point feature histograms (FPFH) for 3d registration. In: 2009 IEEE International Conference on Robotics and Automation, pp. 3212–3217. IEEE (2009)
47. Samet, H., Tamminen, M.: Efficient component labeling of images of arbitrary dimension represented by linear bintrees. IEEE Trans. Pattern Anal. Mach. Intell. **10**(4), 579–586 (1988)
48. Schulze, R., Heil, U., Groβ, Bruellmann, D.D., Dranischnikow, E., Schwanecke, U., Schoemer, E.: Artefacts in CBCT: a review. Dentomaxillofacial Radiol. **40**(5), 265–273 (2011)
49. Schulze, R.K.W., Berndt, D., d'Hoedt, B.: On cone-beam computed tomography artifacts induced by titanium implants. Clin. Oral Implants Res. **21**(1), 100–107 (2010)
50. Scott, J., Stagnell, S., Downie, I.: The use of 3d model planning in the management of impacted teeth. Oral Surg. **11**(2), 125–130 (2018)
51. Tang, X., Krupinski, E.A., Xie, H., Stillman, A.E.: On the data acquisition, image reconstruction, cone beam artifacts, and their suppression in axial MDCT and CBCT-a review. Med. Phys. **45**(9), e761–e782 (2018)
52. William Murray Thomson and Sunyoung Ma: An ageing population poses dental challenges. Singapore Dent. J. **35**, 3–8 (2014)
53. Vandenberghe, B.: The crucial role of imaging in digital dentistry. Dent. Mater. **36**(5), 581–591 (2020)
54. Wang, R.H., Ho, C.-T., Lin, H.-H., Lo, L.-J.: Threedimensional cephalometry for orthognathic planning: Normative data and analyses. J. Formosan Med. Assoc. **119**(1), 191–203 (2020)

55. Wang, Y., Liu, S., Wang, G., Liu, Y.: Accurate tooth segmentation with improved hybrid active contour model. Phys. Med. Biol. **64**(1), 015012 (2018)
56. Wang, Y., Sun, Y., Liu, Z., Sarma, S.E., Bronstein, M.M., Solomon, J.M.: Dynamic graph CNN for learning on point clouds. ACM Trans. Graph. (TOG) **38**(5), 1–12 (2019)
57. Yamamoto, S., Miyachi, H., Fujii, H., Ochiai, S., Watanabe, S., Shimozato, K.: Intuitive facial imaging method for evaluation of postoperative swelling: a combination of 3-dimensional computed tomography and laser surface scanning in orthognathic surgery. J. Oral Maxillofac. Surg. **74**(12), 2506-e1 (2016)
58. Yang, Y., Xie, R., Jia, W., Chen, Z., Yang, Y., Xie, L., Jiang, B.X.: Accurate and automatic tooth image segmentation model with deep convolutional neural networks and level set method. Neurocomputing **419**, 108–125 (2021)
59. Yau, H.-T., Yang, T.-J., Chen, Y.-C.: Tooth model reconstruction based upon data fusion for orthodontic treatment simulation. Comput. Biol. Med. **48**, 8–16 (2014)
60. Zaharia, C., Gabor, A.-G., Gavrilovici, A., Stan, A.T., Idorasi, L., Sinescu, C., Negruiu, M.-L.: Digital dentistry-3d printing applications. J. Interdiscip. Med. **2**(1), 50–53 (2017)
61. Zarone, F., Ruggiero, G., Ferrari, M., Mangano, F., Joda, T., Sorrentino, R.: Accuracy of a chairside intraoral scanner compared with a laboratory scanner for the completely edentulous maxilla: an in vitro 3-dimensional comparative analysis. J. Prosth. Dent. **124**(6), 761-e1 (2020)
62. Zimmermann, M., Mehl, A., Mormann, W.H., Reich, S.: Intraoral scanning systems-a current overview. International Journal of Computerized Dentistry **18**(2), 101–129 (2015)

Chapter 5
Artificial Intelligence for Fetal Ultrasound

Hyun Cheol Cho, Siyu Sun, Sung Wook Park, Ja-Young Kwon, and Jin Keun Seo

Abstract Diagnostic ultrasound is the most commonly used imaging method in the field of obstetrics and gynecology to estimate various biometrics related to fetal development, fetal well-being, and perinatal prognosis. Until now, ultrasound measurements of fetal health parameters (i.e., amniotic fluid volume, biparietal diameter, head circumference, abdominal circumference, and others) have been made through a cumbersome and time-consuming manual process, and their accuracy depends heavily on the operator's skill and experience. Therefore, there has been a high demand for an easy-to-use interface for collecting biometrics from fetal ultrasound images to improve clinician workflow efficiency. Traditional methods have fundamental limitations in automating biometric measurements from noisy ultrasound images that are often degraded by signal dropouts, reverberation artifacts, missing boundaries, attenuation, shadows, speckles, and so on. Medical imaging is experiencing a paradigm shift due to the remarkable and rapid advancement of deep learning technology, and ultrasound companies, including Samsung Medison, are making every effort to develop a new AI-based system for automated fetal ultrasound diagnosis. The reason for these efforts of ultrasound companies is that AI technology is expected to become

H. C. Cho · S. W. Park
AI Vision Group, Samsung Medison, Seoul, Korea
e-mail: hc5310.jo@samsungmedison.com

S. W. Park
e-mail: sw724.park@samsungmedison.com

S. Sun · J. K. Seo (✉)
School of Mathematics and Computing (Computational Science and Engineering), Yonsei University, Seoul, Korea
e-mail: seoj@yonsei.ac.kr

S. Sun
e-mail: 2018323306@yonsei.ac.kr

J.-Y. Kwon
Department of Obstetrics and Gynecology, Institute of Women's Life Medical Science, Yonsei University College of Medicine, Seoul, Korea
e-mail: jaykwon@yuhs.ac

J. K. Seo (ed.), *Deep Learning and Medical Applications*, Mathematics in Industry 40,
https://doi.org/10.1007/978-981-99-1839-3_5

a turning point in diagnostic ultrasound. This chapter focuses on fetal ultrasound, explains deep learning-based medical imaging technology, and hopes to help readers discover new possibilities and to provide future directions.

5.1 Introduction

Ultrasound (US) is routinely used in obstetrics and gynecology to evaluate fetal development and well-being. A fetal ultrasound imaging system uses a transducer (probe) to transmit sound waves in the frequency range of 1–5 MHz (higher than human audible frequency), hitting an object to be imaged and generating reflected echo waves to be measured using the same probe. The measured echo signal is used to produce gray-scale images of organs and a color Doppler imaging of blood movement. US imaging provides gynecologists with a lot of information about fetal development and well-being, but obstetric diagnosis is a difficult task that requires integrated and balanced knowledge, as there are many types of diagnoses and various disorders.

In early pregnancy, transabdominal US is usually used to confirm pregnancy and to determine the gestational age and the estimated due data (EDD), where the date of the last menstrual period may be also taken into account. In the case when a transabdominal US does not provide enough information, a transvaginal US may be done. During the first trimester, the anatomical features of the fetus (i.e. eyes, mouth, brain, head, arms, legs, figures, teeth, etc.) are developed and are fully formed at the end of the third month. The first-trimester fetal US scan measures the crown-rump length (CRL) for estimating gestational age, nuchal translucency (NT) thickness for chromosomal defects, and other factors related to fetal malformations. It is important to detect serious structural fetal abnormalities as early as possible through a first-trimester fetal US scan, because this helps reduce the serious emotional and economic costs of parents by offering them the opportunity to choose termination of pregnancy.

During the second trimester, anatomical details can be viewed via US, allowing clinicians to monitor the anatomical and functional surveillance of fetuses. At this point, the gynecologist looks at the defined anatomical structures, including the maternal structure (amniotic fluid, umbilical cord, uterine arteries, etc.) and the fetal anatomical structure (heart, head, brain, chest, abdomen, limbs, spine, etc.). For quantitative analysis of US images to predict intrauterine growth restriction and fetal maturity for obstetric diagnosis and for estimating gestational age [11], gynecologists use the measurements of fetal biometric parameters such as estimation of amniotic fluid volume (AFV), biparietal diameter (BPD), head circumference (HC), abdominal circumference (AC), femur length (FL), humeral length (HL), fetal pyelectasis (FP), trans-cerebellar diameter (TCD), occupito-frontal diameter (OFD), iterocular distance (IOD), binocular distance (BOD), and ratios of fetal cardio-thoracic, thoracic-abdominal, and thoracic-head circumference [25].

Fetal ultrasound examination for evaluation of fetal structure involves the steps of finding a reference plane to be checked while scanning, checking the presence or absence of a structure in the acquired standard plane, and estimating whether the shape and size of a specific structure are within the normal range. Based on this estimation, dysfunction/malformation/probable disease of fetal organs is inferred.

In fetal ultrasound examination, the stage where the examiner's subjectivity is involved in the stage of obtaining a standard plane and the stage of placing caliper during measurement, because anatomical knowledge of the constituent fetal structure is involved. More specifically, manually measuring the biometric parameters mentioned above is cumbersome and requires time-consuming steps that involve multiple keystrokes and probe motion. Clinicians or sonographers perform cumbersome tasks of continuously moving the transducer based on their anatomical knowledge until they find a proper plane for each measurement. Hence, the accuracy of manual measurements depends highly on the operator's experiences [12, 70].

To address this operator dependence, there has been a great demand for automated fetal ultrasound measurements. This automation not only reduces operator dependence, but also shortens US examination time, reducing the workload of the ultrasound technician and providing patient convenience. Therefore, automation not only increases the reliability of diagnosis by reducing the subjective judgment error of the examiner, but also provides convenience to the patient.

Before the advent of recent deep learning, these fetal biometric measurements were beyond the reach of existing technologies (e.g. energy-based segmentation methods using active contour or level set), due to difficulties in handling noisy ultrasound images affected by signal dropouts, artifacts, missing boundaries, attenuation, shadows, and speckling [83]. The typical energy-based segmentation techniques involving shape-based models start with a good initial contour and use iterative contour evolution through the energy minimization process. The reason why the automation of this approach is difficult is because the performance depends on the selection of the initial contour, and it is difficult to properly set the criteria for terminating the iterative process at the target boundary, taking into account both local image patterns and the global image structure simultaneously. US images are often degraded by various confusing factors. Most existing methods use image-intensity information to probe the boundaries of target anatomies [11, 37, 74, 99, 100], but have difficulties in handling target anatomies having low contrast against surroundings, non-uniform contrast, and irregular shape in US images. In the noisy fetal US environment, it is very difficult to find an energy functional that effectively handles such factors.

Automation of fetal biometric measurement may require methods equipped with techniques that mimic clinicians' measurement procedures: Considering the prior knowledge of the anatomical structure and ultrasound imaging characteristics, the target object is searched based on the global image structure as well as its local pattern. This is because, without using anatomical knowledge while looking at the entire image, it is often difficult even for doctors to recognize where a small patch image is on the body.

Recently, deep learning techniques have been successfully applied for automatic fetal biometry measurements given the assumption that a standard plane is selected [15, 36, 44, 45] and have reached the stage of commercial use in the clinical field.

Automation of fetal biometric measurement may require methods equipped with techniques that mimic clinicians' measurement procedures: Considering the prior knowledge of the anatomical structure and ultrasound imaging characteristics, the target object is searched based on the global image structure as well as its local pattern. The success of deep learning techniques lies in their ability to capture the spatial relationships between pixels to figure out local and global interconnections while taking prior knowledge about anatomical structures and ultrasound image characteristics into account. The automation that has been successful so far relates to anatomical recognition, automatic caliper placement, structural localization and classification under conditions given a standard plane.

Let us briefly explain U-net, a network frequently used in deep learning-based fetal ultrasound automation. Its encoding path is based on a series of convolutions followed by pooling to reliably recognize image features so that the resulting output is fairly resilient to variations in position and scale of a target structure. It also utilizes concatenative skip connections between the same-scale feature maps obtained from the encoder and decoder subnetworks, which play an important role in restoring spatial information lost during downsampling. U-nets have shown remarkable overall performance in medical image segmentations by simultaneously making use of global features and local spatial information [45, 57, 58, 82]. The optimal network depth may depend on a number of factors, including the size of the input image, the amount of training data, the variance in the size of the target features in the image, and the difficulty of the task. Instead of increasing the network depth and the filter size, spatially-adaptive filters such as atrous convolution can be used to increase the receptive field size. This technique has been used to properly handle large differences in a target's size. Various modifications or auxiliary means of U-net have been developed to supplement the limitations of the classical U-net architecture. These include Attention U-net [87], M-net [22], U-net++ [104], and MultiResUnet [31].

Despite the remarkable success of the DL approach in fetal US, there are several important issues where automation has not yet reached the level of clinical application. The most difficult and important issue to solve is the automatic selection of the standard axial planes (SAPs). In practice, selection of proper standard planes for fetal structural screening involves coordinating and scanning through the fetal body in search of the landmark anatomy that defines certain standard planes. This process is extremely time-consuming and most dependent on the level of experience of the operator. Thus, without anatomical and spatio-temporal knowledge, the operator will not be able to localize the scan position and produce the proper standard plane.

There have been many efforts towards the automatic SAPs selection [8, 12, 14, 45, 56, 93]. Let us briefly explain the existing results for automatically finding the SAPs for estimating central nervous system (CNS) malformations. The SAPs include the transthalamic plane, transventricular plane, and transcerebellar plane. Kim et al. [45] developed an automated plane acceptance check method to determine whether the input image is acceptable for the standard plane. It uses geometric placement of three feature points (the 'box-like' cavum septum pellucidum, the 'V-shaped' ambient cistern, and cerebellum) for the plane acceptance check. Lin et al. [56] proposes a new multi-task learning framework using a faster regional convolutional neural network (MF R-CNN) architecture for standard plane detection and quality assessment. It uses scoring protocol for standard plane detection based on six predefined key anatomical structures. An automatic framework based on deep learning was proposed to detect standard planes from fetal US videos, where the output of the network is simply whether each plane is standard or not [12, 14]. Cai et al. [8] used visual attention of sonographers to guide standard plane detection in consecutive US video frames. Existing deep learning-based SAPs selection methods focus on intrinsic features detection to evaluate the presence or absence of specific anatomical structures (e.g. cranium, falx, cavum septum pellucidum, lateral ventricle, thalamus, crus cerebri, ambient cistern, and cerebellum). Unfortunately, experiments using such deep learning methods have shown that such total scoring approaches of the key anatomical structures may be inappropriate or insufficient to reach reliable decisions for SAPs selection at a clinically applicable level. Even among experts, evaluation of the suitability of SAPs with only a single frame are often inconsistent.

We observed that clinicians typically choose the optimal SAPs from a given video stream rather than rigorously evaluating whether each frame meets all of the standard requirements for SAPs. Hence, it seems fundamentally vague to distinguish between SAP and non-SAP by looking at only one plane. This vagueness makes it difficult to get high-quality training data for supervised learning. This shows that a single frame alone may not be sufficient to support judgment when evaluating SAPs. Anatomical structures are three-dimensional with irregular shape, and spatial relationships between organs are unique to the region of the body. When the operator scans through the fetal body, unique shape and signal change sequences according to their spatial relationship are produced by the movement of the transducer. The operator recognizes such change sequences across the series of image frames and uses the knowledge to determine the transducer's movement and level where the transducer is scanning. This is referred to as contextual knowledge or clues. The operator pays attention to such contextual clues during scanning to determine whether the probe is approaching, reaching, or passing the standard plane.

Years of experience have shown that in deep learning, it is still difficult for a machine to learn the subtle context of an ultrasound image on its own. Achieving a level of automation for clinical use requires a careful strategy of deep learning models to provide a sufficiently machine-learning-friendly environment. The structure of the loss function associated with the labeling strategy in the network has a significant impact on training neural networks and affects the reliability of the automation. In particular, just inserting a little supplemental information into the network can significantly improve network performance. Deep neural networks are vulnerable to adversarial attacks, which can provide erroneous outputs with small perturbations of their inputs, so a lot of effort is needed to handle adversarial attacks and add contextual information appropriately to improve reliability.

This chapter discusses deep learning techniques for automated or semi-automated fetal biometric systems for fetal US image diagnosis.

5.2 Basics of Fetal Ultrasound

5.2.1 Sound Wave

Ultrasound in medical imaging is a longitudinal wave of mechanical energy with alternating compression and rarefaction of a physical medium (tissue, air, liquid), as shown in Fig. 5.1a. The speed of a sound wave, denoted as c, is determined by the properties of the medium it is propagating through, and is expressed as

$$c = \sqrt{\frac{B}{\rho}}, \tag{5.1}$$

where B (Pa) is the bulk modulus reflecting the stiffness of the medium, and ρ $(\mathrm{kg} \cdot \mathrm{m}^{-3})$ is the density of the medium. Hence, low stiffness and high density result in low speed of sound, while high stiffness and low density lead to high speed of sound. Figure 5.1b shows the speed of sound in various media [20, 29]. It can be seen that the speed of sound is low in air but much higher in bone. In addition, a sound of frequency f, propagating in a specific medium, has a wavelength λ, where $\lambda = c/f$. The wavelength of a sound is related to an imaging factor—axial resolution [69], which would be explained by the transabdominal US and transvaginal US.

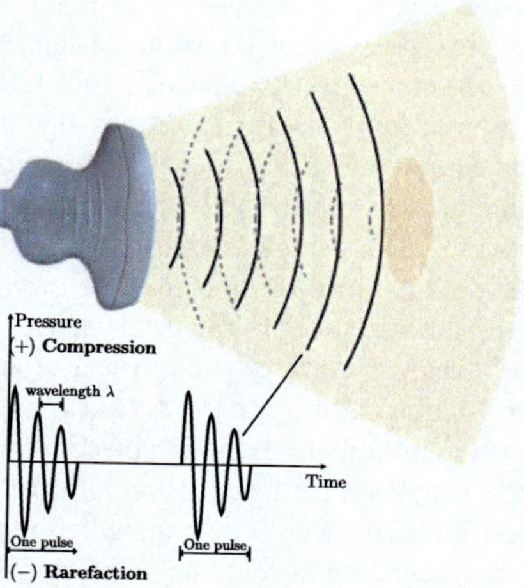

(a)

Medium	$c(\mathrm{m} \cdot \mathrm{s}^{-1})$
Air	333
Fat	1430
Water	1480
Blood	1560
Muscle	1600
Bone	2000-4080

(b)

Fig. 5.1 Illustration of US waves. **a** Description of US wave penetrating inside human body. **b** The speed of sound in some medium

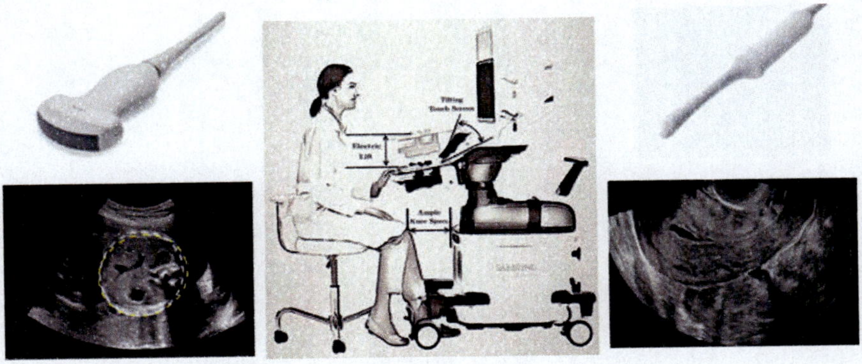

Fig. 5.2 Transabdominal (left) and transvaginal (right) ultrasound. The center figure is modified from the production manual of HERA W9 (SAMSUNG MEDISON CO.)

5.2.2 *Transabdominal and Transvaginal Ultrasound*

Figure 5.2 shows the transabdominal and transvaginal US, where the transabdominal scan is performed with a curvilinear transducer and the transvaginal scan is done with a wandlike transducer. Transabdominal US uses a low frequency in the range 3–5MHz to visualize deep regions including fetus, whereas transvaginal US uses a higher frequency in the range 5–8MHz to visualize fetus and organs in the pelvic region including cervix, uterus, and placenta near the probe in higher axial resolution.

Axial resolution in US, as illustrated in Fig. 5.3, is defined as the minimum reflector separation to distinguish reflectors along the axis of the US beam. It is one half of the spatial pulse length (SPL), which is the product of the wavelength and the number of cycles within a pulse (n) that is expressed as

$$\text{Axial resolution} = \frac{\text{SPL}}{2} = \frac{n\lambda}{2}. \tag{5.2}$$

Hence, the higher the frequency, the smaller the wavelength, and the better the axial resolution. However, the depth penetration for imaging can be reduced due to higher frequency. This is because US energy of high frequencies is attenuated more rapidly than that of lower frequencies [65]. Therefore, transvaginal US has a higher axial resolution but a smaller field of view, compared to transabdominal US.

5.2.3 *Principle of Two-Dimensional (2D) B-mode Ultrasound Imaging*

Various imaging techniques including 2D B-mode imaging, real-time three-dimensional (3D) imaging, and color Doppler imaging are used in transabdominal US

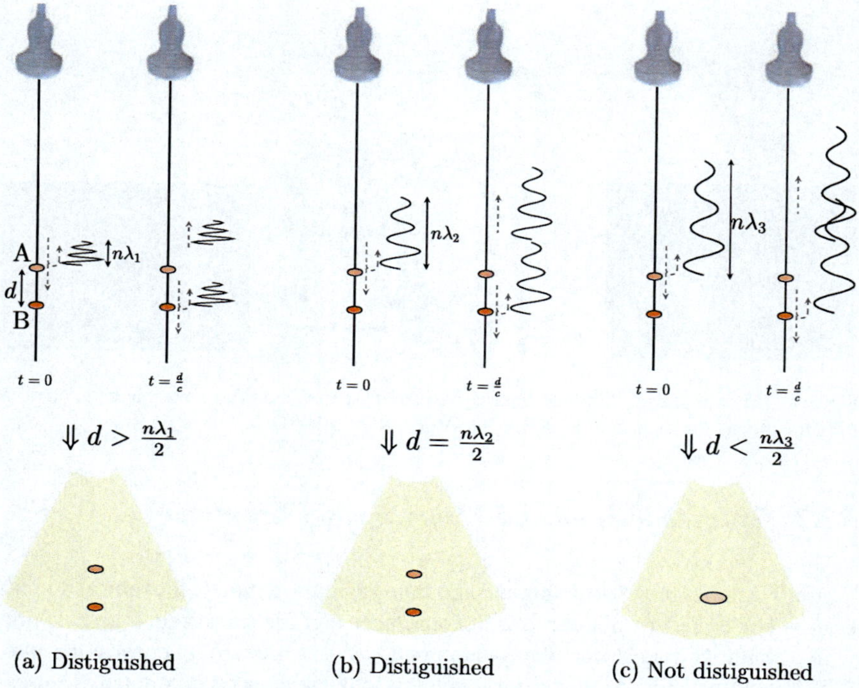

Fig. 5.3 Illustration of axial resolution of two imaging objects (denoted as A and B with distance d along US beam axis). In **b**, objects A and B are just distinguished, and the axial resolution is same as d. Comparing **a–c**, where $\lambda_1 < \lambda_2 < \lambda_3$, and $n = 3$, it can be seen that the smaller the wavelength, the better the axial resolution

and transvaginal US. In this section, we describe the basic principle of US imaging by taking 2D B-mode imaging as an example, as shown in Fig. 5.4.

Beam-forming and time gain compensation: A 2D B-mode US image is a gray-scale cross-sectional image showing tissues and organs within the body. Each image consists of multiple scanlines (about 128–512 scanlines) that are obtained through the transmission and reception beam-forming with transducer arrays [35].

When the transducer works as a transmitter, an alternating current passes through a set of closely spaced transducer elements attached to piezoelectrical crystals. The current causes crystals to vibrate, and the vibrating crystals produce US waves in the form of short pulses. To achieve the optimum lateral resolution at a focal point, denoted as $\boldsymbol{r}_f$, on each scanline, transmission beam-forming is used to control time delays electronically to focus pulses at $\boldsymbol{r}_f$, as shown in Fig. 5.5. Let $\boldsymbol{r}_i$ denote the position of the i-th element for $i = 1, \ldots, N$, where N is the number of elements. Assuming the pulse response is given by the Dirac delta function [35], the pulse at $\boldsymbol{r}_f$ and time t is represented as

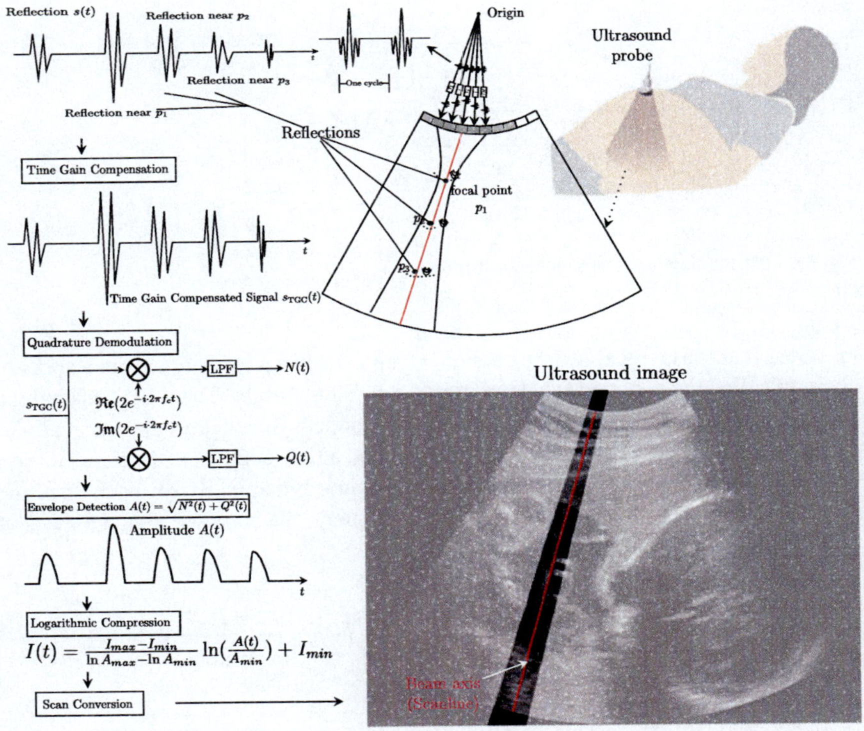

Fig. 5.4 Schematic description of 2D B-mode US imaging mechanism

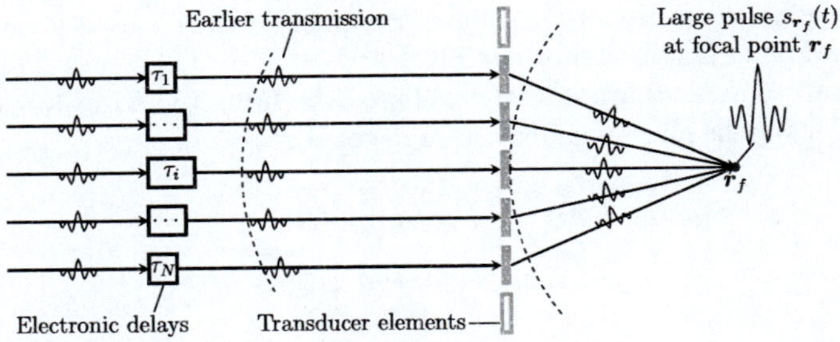

Fig. 5.5 The diagram of transmission beam-forming

$$s_{r_f}(t) = \sum_{i=1}^{N} s_i(t - \tau_i) = K \sum_{i=1}^{N} \delta(t - \frac{|r_i - r_f|}{c}), \tag{5.3}$$

where τ_i is the delayed time, $c = 1540$m/s the constant speed of sound in human tissue assumed by US image processing, and K a constant that depends on depth.

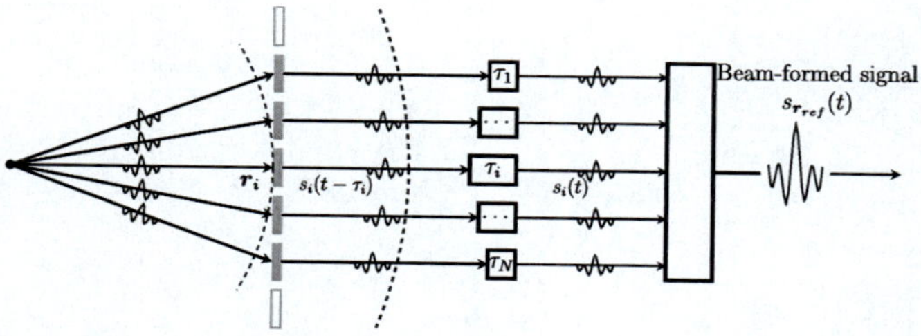

Fig. 5.6 The diagram of reception beam-forming

When it acts as a receiver, the crystals receive the reflected echos from a specific position called the received focal point $\boldsymbol{r}_{ref}$, and then vibrate. The vibrations induce electrical signals. The reception beam-forming controls time delays electronically to make the signals phase summed, as depicted in Fig. 5.6. In contrast to the transmission beam-forming, during reception, focusing is done dynamically on each scanline [101]. Let τ_i be the delayed time of the i-th element, the summed signal $s_{r_{ref}}(t)$ at time t is denoted as

$$s_{r_{ref}}(t) = \sum_{i=1}^{N} s_i(t-\tau_i) = K\sum_{i=1}^{N} \delta(t - \frac{|\boldsymbol{r}_i - \boldsymbol{r}_{ref}|}{c}) \tag{5.4}$$

When sound propagates through tissues, attenuation occurs, which is the reduction of amplitude due to absorption, reflection, and scattering. The attenuation, mainly due to absorption, is the limiting factor for the depth penetration of US beam. To compensate the attenuation, time gain compensation (TGC) is used to amplify the signal based on its depth, as shown in Fig. 5.4.

Quadrature demodulation and envelope detection: The beam-formed signal after TGC along the given scanline can be denoted as

$$s_{\mathrm{TGC}}(t) = A(t)\cos(2\pi f_c t + \phi(t)), \tag{5.5}$$

$$= \frac{A(t)}{2}(e^{i(2\pi f_c t+\phi(t))} + e^{-i(2\pi f_c t+\phi(t))}), \tag{5.6}$$

where $A(t)$, $\phi(t)$, and f_c represent the amplitude, phase, and center frequency of the signal, respectively. To remove the high-frequency signals and preserve the amplitude information with low-frequency, quadrature demodulation is performed. Specifically, by multiplying $s_{\mathrm{TGC}}(t)$ with $2e^{-i\cdot 2\pi f_c t}$ and applying a low-pass filter (LPF), a complex signal $A(t)e^{i\phi(t)}$ is acquired. Its real and imaginary parts are named in-phase and quadrature signals, denoted by $N(t)$ and $Q(t)$, respectively:

$$N(t) = \Re\mathrm{e}(A(t)e^{i\phi(t)}), \tag{5.7}$$

$$Q(t) = \Im\mathrm{m}(A(t)e^{i\phi(t)}). \tag{5.8}$$

Hence, the envelope detection can be achieved by

$$A(t) = \sqrt{N^2(t) + Q^2(t)}. \tag{5.9}$$

Logarithmic compression: The variation of signal amplitude is usually high and most clinically significant signals appear with small values. To display these meaningful signals in the 0–255 gray-scale image, logarithmic compression is employed. The image intensity, denoted as $I(t)$, is represented as

$$I(t) = \frac{I_{max} - I_{min}}{\ln A_{max} - \ln A_{min}} \ln(\frac{A(t)}{A_{min}}) + I_{min}, \tag{5.10}$$

where I_{min} and I_{max} are the minimum and maximum gray-scale intensities, respectively, and A_{min} and A_{max} the minimum and maximum amplitudes of the signal, respectively.

Scan conversion: To convert input data captured in different coordinates into Cartesian coordinates that are more suitable for display, scan conversion in US imaging is applied. The commonly used interpolation method is bilinear interpolation [54].

5.3 Ultrasound Artifacts

5.3.1 *Interactions of Ultrasound with Tissues and Imaging Artifact*

US is prone to various artifacts that are commonly encountered in obstetrics and gynecology. These artifacts may interfere with image interpretation, but can be recognized and interpreted with a basic appreciation of the US-tissue interactions and the physical properties of the US beam [20]. See Fig. 5.7.

5.3.2 *Reflection and Reverberation Artifact*

Reverberation artifacts occur when the US beam encounters strong reflectors of the fat–muscle interfaces in the abdominal wall and the transducer–skin interface that are approximately orthogonal to the beam propagation direction.

The difference in acoustic impedance between the fat and fascia makes the incident US beam reflect back and forth several times before returning to the transducer. Reverberation is then generated due to the assumption that US beam returns to the transducer after a single reflection.

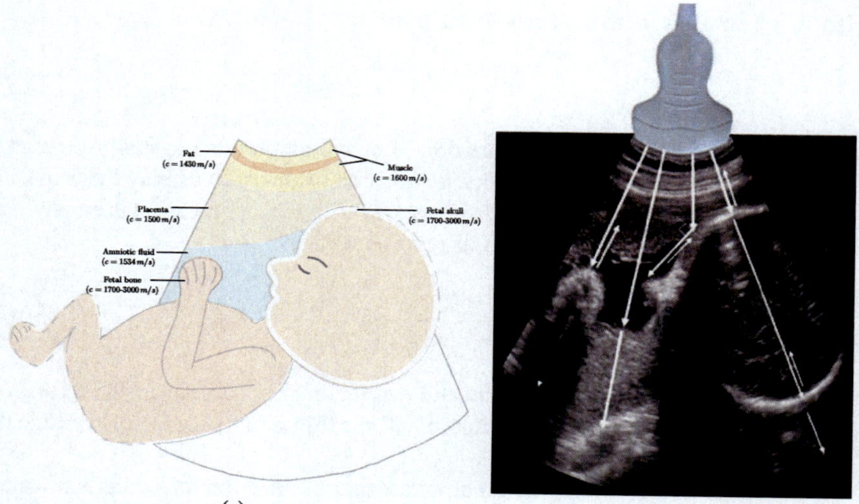

(a) (b)

Fig. 5.7 Ultrasound image artifacts due to the assumption of constant propagation speed $c = 1540$ m/s, constant attenuation coefficient $\alpha = 0.6$ dB/cm, and a single reflection. **a** US cartoon. **b** Real US image

As shown in Fig. 5.8, the image characteristics (e.g., pattern, intensity, and shape) of severe reverberation artifacts are similar to that of fat–muscle interfaces in the abdominal wall. These artifacts may obscure the amniotic fluid situated anteriorly or can be mistaken for the anterior uterine wall, thus causing inaccurate amniotic fluid measurement [85]. Hence, it is necessary to recognize these artifacts by using the physical correlation among US, reverberation artifacts, and abdominal wall [91].

5.3.3 *Refraction and Edge Shadowing Artifact*

Edge shadowing artifacts are commonly seen as a hypoechoic area posterior to the edge of the fetal skull, as depicted in Fig. 5.9b. It is caused by the deflection of US and curvature of the skull. Specifically, according to Snell's law [49], the US beam is refracted when it meets the curved skull due to the speed difference between the skull bone and the tissues it penetrates through. Edge shadowing artifacts then occur if the US signal is deflected and unfocused when the transmitted angle equals 90°C, as shown in Fig. 5.9a.

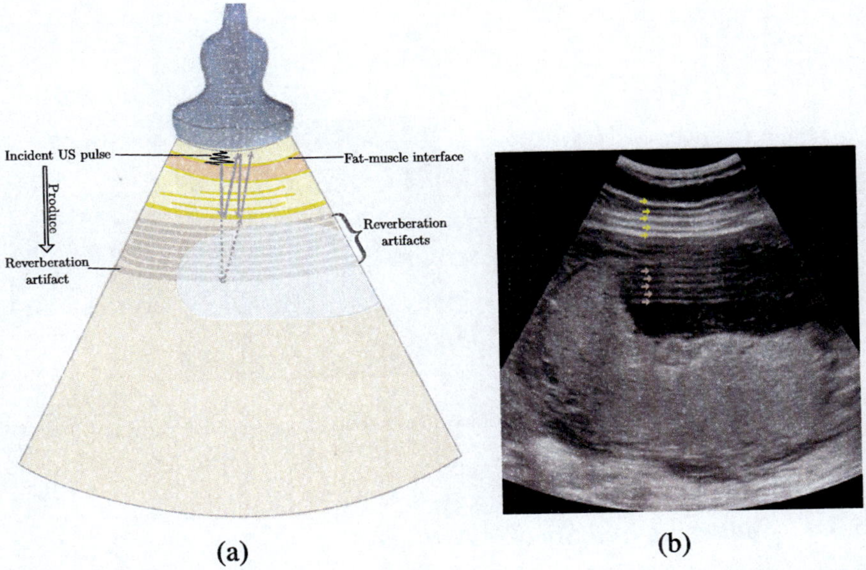

Fig. 5.8 The cause of reverberation artifact in US imaging. **a** Mechanism of reverberation artifacts. **b** Real US image. The yellow and light gray arrows indicate the fat–muscle interfaces and reverberation artifacts, respectively

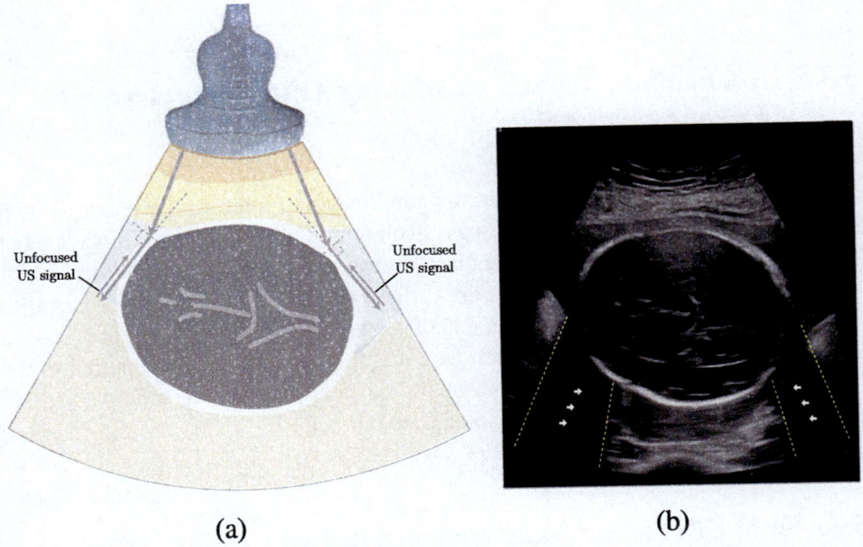

Fig. 5.9 The cause of edge shadowing artifacts in US imaging. **a** Mechanism of edge shadowing artifact. **b** Real US images. The white arrows indicate edge shadowing artifacts

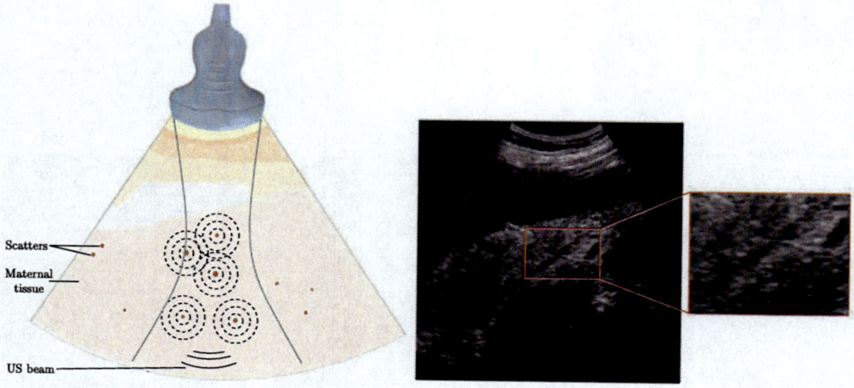

Fig. 5.10 The mechanism of speckle artifacts in ultrasound imaging. **a** US cartoon. **b** Real US image

5.3.4 Scattering and Speckle Artifact

Speckle artifacts in ultrasound result from the constructive and destructive interference of the US beam with multiple small-scale scatter in tissue. See Fig. 5.10a. This acoustic interference effect gives it a granular appearance, as shown in Fig. 5.10b.

5.3.5 Attenuation, Acoustic Shadowing Artifact and Acoustic Enhancement Artifacts

As US penetrates through the tissues, its amplitude and intensity decreases, which is called attenuation. It is the result of reflection and absorption of US waves, but mainly due to absorption, that is, the direct conversion of sound energy into heat.

Acoustic shadowing, an attenuation artifact, appears as a dark region distal to a high-attenuation object such as fetal skull bone, ribs, spine, and femur. Acoustic enhancement, on the contrary, is seen as a bright area since TGC is applied to the anterior structure of low attenuation, including fetal gallbladder, urinary bladder, gestation sac of amniotic fluid, and so on. See Fig. 5.11.

5.4 Fetal Ultrasound Measurements in List

The International Society of Ultrasound in Obstetrics and Gynecology (ISUOG) [86] provided ISUOG practice guidelines, which aim to describe the appropriate assessment of fetal biometry and diagnosis of fetal growth disorders. These disorders consist

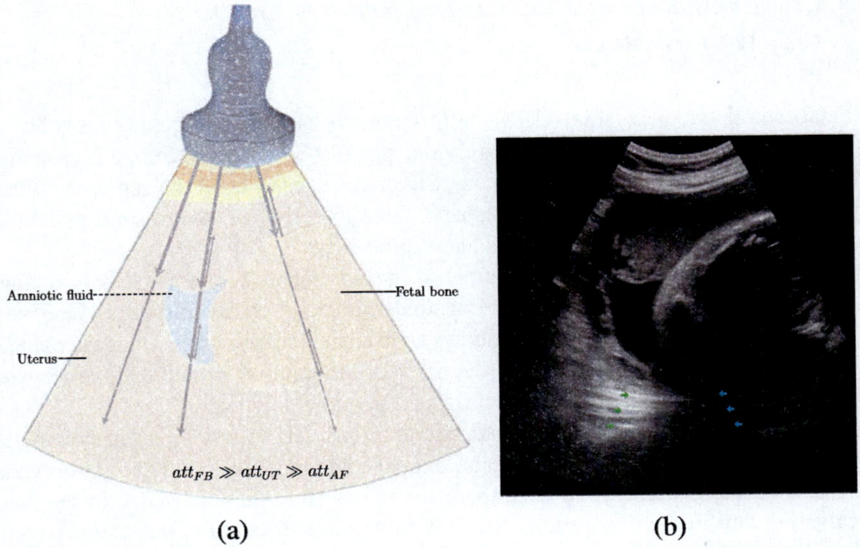

Fig. 5.11 The cause of acoustic shadowing artifacts and acoustic enhancement artifacts in US imaging. **a** US cartoon. "att" denotes the attenuation coefficient, "FB", "UT" and "AF" represent fetal bone, uterus and amniotic fluid, respectively. **b** US image with acoustic shadowing artifacts (cyan arrows) and enhancement artifacts (green arrows)

mainly of fetal growth restriction, and ultrasound screening is the most commonly used technique for observing various factors related to fetal growth abnormalities.

Fetal biometric measurements such as the fetal biparietal diameter (BPD), head circumference (HC), and abdominal circumference (AC) are useful for predicting intrauterine growth restriction and fetal maturity as well as for estimating gestational age [11]. For fetal ultrasound examination for evaluation of fetal structure, the clinician continuously moves the transducer to find several standard planes containing the key structures to be examined. This process is very cumbersome because fetal movements, breathing, and fetal posture interfere with acquisition. When a standard plane is found, the clinician then checks for the presence of structures and estimates whether the shape and size of a particular structure are within normal range. Based on this estimation, dysfunction/malformation/probable disease of fetal organs is inferred.

Until now, fetal biometric measurements have been performed manually, and the accuracy of the measurements depends on the skill level of clinicians, where the factors that influence the accuracy are (i) whether a proper plane has been obtained and (ii) whether the calipers are properly positioned. To improve workflow and reduce user variability in data collection, ultrasound companies are rushing to develop automatic systems for estimating fetal biometric parameters.

5.4.1 *Evaluation of Fetal Central Nervous System Malformations*

Fetal central nervous system (CNS) malformations constitute a frequent cause of congenital abnormalities, affecting roughly one per 100 births. Therefore, it is essential to detect fetal CNS abnormalities early. Structures usually noted on screening ultrasound CNS examination are: head shape, lateral ventricles, cavum septi pellucidi, thalami, cerebellum, cisterna magna, and spine [63].

The most important step of biometric measurements for fetal CNS is to find three standard axial planes (SAPs): transthalamic plane, transventricular plane, and transcerebellar plane. Figure 5.13a shows a visual description of the standard planes. Three SAPs allow visualization of the relevant intracranial structures to assess the anatomic integrity of the brain and provide a possible diagnosis.

On the transthalamic plane, head circumference (HC) and biparietal diameter (BPD) are measured to evaluate gestational age and diagnose fetal central nervous system (CNS) pathology. BPD calipers are placed from the outer edge of the near calvarial wall to the inner edge of the far calvarial wall, and the line joining them is orthogonal to the central axis of the head. HC is estimated by calculating the boundary of an ellipse drawn around the outside of the calvarium.

US-based biometric measurements of the fetal CNS have so far been performed manually, and the accuracy of the measurements depends on whether acceptable SAPs have been obtained and whether the calipers are properly positioned. Obtaining the three SAPs in clinical practice requires the operator's extensive fetal anatomy knowledge, manipulation of the ultrasound probe, and considerable clinical experience. Hence, SAPs selection is highly operator-dependent and time consuming, involving multiple keystrokes and probe movements. Therefore, there has been a high demand for automation of SAPs selection to reduce operator workload and variability [19, 47].

5.4.1.1 Criteria of Standard Planes

The definitions and criteria of standard planes are as follows.

- **Transthalamic plane** is the transverse plane containing the thalamus as a cross-section measuring the size of the fetus's head. This plane shows the "box-like" cavum septum pellucidum (CSP) and the "V-shaped" ambient cistern (AC), but should not show the entire cerebellum (Cbll). The criteria of the standard plane are:

1. Both craniums and hemispheres should be observed uniformly and symmetrically.
2. In the center of the skull, the third ventricle, both sides of the thalami, and the midlind falx should all be clearly observed.

3. Regardless of whether cavum septum pellucidum is visible or not, if the above two conditions (1)–(2) are satisfied, it is judged as an appropriate image.
4. If the intracranial structure is not clearly visible, the two craniums are asymmetric, or the cerebellum is visible, it is considered inappropriate image.

- **Transventricular plane** is the cross-section using ventricles and the choroid plexus as anatomical indicators. The criteria of the standard plane are:

1. Both craniums and hemispheres should be observed uniformly and symmetrically.
2. The border of the lateral ventricle and the midline falx inside the skull should be clearly observed.
3. Regardless of whether the cavum septum pellucidum is visible or not, if the conditions of (1)–(2) are satisfied, it is judged as an appropriate image.
4. If the intracranial structure is not clearly visible, the two craniums are asymmetric, or the cerebellum is visible, it is considered an inappropriate image.

- **Transcerebellar plane** is the cross section using cerebellar hemispheres and cisterna magna as anatomical indicators. The criteria of standard planes are:

1. Both craniums and hemispheres should be observed uniformly and symmetrically.
2. Midline falx should be observed, and the boundary of the cerebellum and cisterna magna should be clearly observed.
3. Regardless of whether the cavum septum pellucidum is visible or not, if the conditions of (1)–(2) are satisfied, it is judged as an appropriate image.
4. If the intracranial structure is not clearly visible, the two craniums are asymmetric, or the cerebellum is visible, it is considered an inappropriate image.

Finding these planes using an US probe is tricky, even for expert sonographers, because the position or orientation of the fetal head is not fixed. Sonographers have to move the US probe back and forth to find anatomical cues that match the criteria of each standard plane, but ultrasound images are often degraded by various artifacts, making it difficult to find these anatomical clues.

5.4.2 *Abdominal Circumference Measurement*

Among biometric measurements, abdominal circumference (AC) is most predictive of fetal weight thus, a variation in AC measurement leads to inaccurate fetal weight estimation [9]. AC is useful for predicting intrauterine growth restriction and fetal maturity [11]. To ensure a precise AC plane that is perpendicular to the true fetal longitudinal axis, the clinician has to continuously move the transducer to find a plane consisting of accurate landmarks. This process, firstly, is cumbersome as fetal movement, breathing movement, and fetal position hinder prompt acquisition of the plane; and secondly, may lead to inaccurate measurement as inexperienced operators often fail to adhere to multiple landmarks of correct the AC plane [19].

5.4.2.1 Criteria of Standard Plane

In order to measure AC, it is prerequisite to determine the standard plane from the US image. The standard plane must include the fetal stomach bubble (SB), spine and the hockey-stick-shaped proximal portion consists of the umbilical vein (UV) and the right portal vein (RV) connection.

- **Axial plane of the abdomen** is a cross-section in which the abdomen of a fetus is measured in a circular shape, and the stomach and umbilical vein are used as anatomical indicators. The criteria of the standard plane are:

1. The whole oval-shaped abdominal skin should be clearly observed.
2. A low-shaded circular stomach should be observed inside the abdomen.
3. The umbilical vein and the right portal vein are connected and observed like a hockey stick, and the umbilical vein on the side not connected to the right portal vein is separated from the skin surface of the abdomen and should be observed briefly.
4. The upper and right portal veins are located opposite to each other based on the center of the abdomen.
5. It is ideal if both ribs are observed symmetrically, but if the conditions of (1)–(4) are satisfied regardless of the symmetry of the ribs, it is judged as an appropriate image.

5.4.3 *Assessment of Amniotic Fluid Index*

This section is based on the papers [15, 91]. Amniotic fluid (AF) is a complex substance essential to fetal development, as AF is required for promotion of maturation of the lungs, gastrointestinal development, and musculoskeletal system development [64]. The AF volume (AFV) is an important indicator for reflecting pregnancy progress and fetal development [15, 18, 64]. Therefore, its assessment is indispensable during an antenatal ultrasound (US) [16], and AFV is commonly estimated by measuring the AF index (AFI) [75], as shown in Fig. 5.12.

AFV is usually measured by evaluating the four-quadrant amniotic fluid index (AFI) or the single deep vertical pocket (SDP) technique [43, 62, 75]. To measure the AFI or SDP, sonographers manually follow the following time-consuming steps, which involve multiple keystrokes and probe motions [15]: (i) The clinician determines a proper AF pocket and then (ii) measure the depth of the AF by estimating the suitable point. Although AFI and SDP are known to be reproducible and semi-quantitative, the manual AFI measurement is highly dependent on the sonographer's experience in practice.

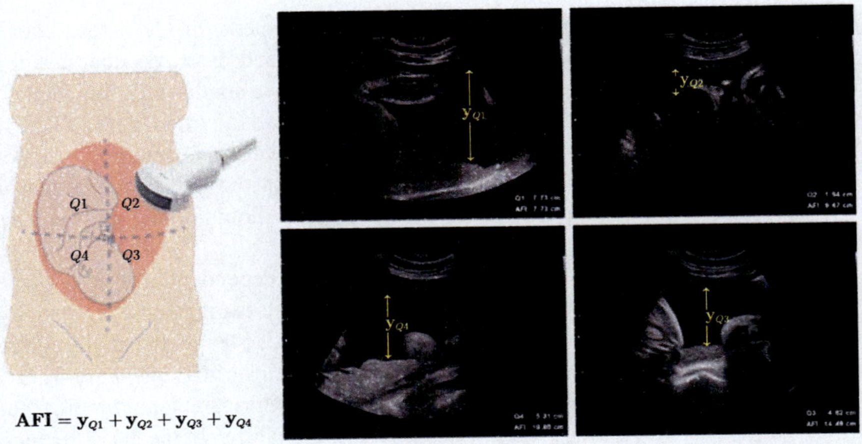

Fig. 5.12 AFI measurement. AFI is the sum of deepest depth of amniotic fluid in each quadrant of ultrasound images. Here, Q1, Q2, Q3, and Q4 indicate each quadrant and y_{Q1}, y_{Q2}, y_{Q3}, and y_{Q4} indicate the deepest depth of amniotic fluid in each quadrant, respectively. This figure is extracted from [15]

5.4.4 Measurement of Cervical Length

Preterm birth (PTB) is the delivery of a neonate prior to 37 weeks of gestation and occurs in 5–18% of all pregnancies [4, 7, 79]. PTB is the leading cause of perinatal morbidity and mortality and long-term disabilities [66, 102]. Despite numerous studies on prediction of pregnancies at high-risk of spontaneous PTB, reliable and reproducible strategies to predict PTB remain a long-standing challenge in obstetrical practice. Numerous studies have shown that a short cervix is more prone to dilation and tends to increase the risk of PTB [6, 34, 88]. Therefore, cervical length (CL) measurement may provide an opportunity for obstetricians and gynecologists to offer interventions to reduce the risks of PTB. Transvaginal ultrasound is used for accurate CL measurement, where the entire CL can be imaged by inserting a transducer probe into the vagina.

The Uterine cervix is a firm cylindrical structure located at the lower part of the uterine corpus with a size of about 3.0 cm or longer, which is very important in maintaining pregnancy because it supports the weight of the conceptus and acts as a barrier against the rise of vaginal microorganisms throughout gestation. As structural modification of the cervix precedes labor and delivery, the potential roles of CL measurement as a clinical parameter in prediction of spontaneous preterm delivery, labor induction success, and differential for false labor have been studied.

As incorrect CL measurement can negatively impact decision-making leading to unnecessary admission or interventions, CL measurement should be performed strictly adhering to the standardized criteria. The key anatomical landmarks for CL measurements are the cervical canal, internal cervical os, external cervical os, vagina, bladder, and presenting part of the fetus. Caliper placement for CL measurement must be preceded by correct identification of internal and external ostia, and the endocervical canal in between. CL is measured either by a single- or 2-line method depending upon the shape of cervical canal to characterize the true length [42].

To avoid inaccurate CL measurement, the patient should have an empty bladder and be placed in the dorsal lithotomy position, to ensure that the anterior and posterior lips of the cervix are of similar thickness, and the probe should not be pressed hard against the cervix [92].

The quality of CL measurement is highly operator dependent, errors in plane acquisition or caliper placement leading to poor quality screening and consequently having a detrimental effect on patient care. Kuusela et al. [48] previously reported that the difference in transvaginal ultrasound measurements of CL may vary from 5 to 10 mm between examiners. Thus, strict adherence to criteria for sonographic measurement of CL is mandatory to ensure agreement and reliability of the measurement. Detection of ostia where the anterior and posterior walls of the cervix meet can be confusing due to speckle noise and echo from abutting structures such as cervical mucus or the posterior vaginal wall. Furthermore, diverse echogenicity found in the internal cervical canal and surrounding mucosal layer may contribute to inaccurate tracing of the cervical canal. These issues may culminate in inaccurate caliper placement or measurement technique. Yost et al. [97] reported the anatomic and technical difficulties encountered with transvaginal ultrasound imaging of the cervix in a consecutive series of women at risk for preterm delivery. In real clinical settings, the process of acquisition and selection of a standard plane is time-consuming, cumbersome and operator dependent. Scanning time or accuracy of the standard plane produced will depend greatly on the operator's skill and knowledge, and pregnant women undergoing the exam must endure many minutes of discomfort.

Hence, there has been a high demand for the development of an automatic CL measurement method to improve workflow and shorten the scanning time to lessen patient discomfort while reducing user variability. Automating CL measurement is a very challenging task. Delineation of the external os, where the external surfaces of the anterior and posterior lips come in contact oftentimes is difficult due to abutting posterior vaginal wall echo or image noise caused by bowel content and peristalsis. Moreover, the shape of the cervix and the amount of mucus in the cervix vary greatly between individuals, and the echogenicity of the cervical canal is very heterogeneous according to the gestational age at the time of the scan. In the presence of myometrial contraction at the isthmic region of the uterus, differentiating the true cervix requires identification of the cervical mucosal layer surrounding the true canal; however, echo of the mucosal layer is inconsistent.

To overcome the difficulties mentioned above, we need a good strategy that puts the concept of attention mechanism into a deep neural network, selectively focusing only on the necessary area, and ignoring the rest.

5.4.5 *Other Fetal Examinations*

Congenital heart defects (CHDs), which represent structural heart malformations, are a leading cause of infant mortality. Because B-mode alone may not be sufficient to diagnose the fetal heart, color Doppler is used supplementarily to view semiquantitative overall blood flow. Detecting CHDs in fetal US is sometimes missed, so efforts are being made to improve it using deep learning techniques.

3D fetal US provides a 3D view of internal structures, and although it has a lower resolution than 2D US, many clinicians are actually benefiting from 3D US. In fetal spine examination, 3D US is advantageous for diagnosis of various neural tube defects such as spina bifida, abnormal curvature of the spine, and hemivertebra. 3D US examination of the fetal profile can be performed to obtain diagnostic information about malformations.

3D US can cover the entire embryo and fetus in the first and second trimesters of pregnancy. 3D US has the advantage of being able to provide whole fetal structural visualization and fetal volume measurement, and may facilitate examination by less trained ultrasound technicians. In addition, 3D US may be advantageous over 2D US in terms of developing a fully automated fetal biometric system, since finding standard planes from 2D US is difficult. The problem of automatic segmentation of fetal envelopes in 3D US data is a challenging task as it is difficult to separate fetuses from maternal tissues due to the factors such as poor image quality with artifacts and noise, and missing boundaries and limbs owing to dropouts of the US signals. 3D segmentation of the fetus should include limbs that occupy a significant portion of the fetal volume.

Estimation of placental volume with 3D US in the first trimester can be used as a screening test to predict a potentially "at-risk" pregnancy [60]. Manual segmentation of the placenta is difficult and time consuming, so there is a need to develop an automated method.

5.4.6 *Remarks*

US imaging is widely used in obstetrics and gynecology due to its non-invasive, real-time monitoring, and relatively low cost, and there is currently no imaging technology to replace it. CT is of very limited use to evaluate the fetus due to the risk of radiation exposure, and MRI (having long scan acquisition time) is also used in a very limited way for maternal indications during the first trimester of pregnancy.

Weaknesses of US imaging, compared to other imaging modalities, are that it varies from patient to patient, operator to operator, and machine to machine, and acquired images may be distorted or incomplete depending on the orientation of the transducer and the skill of the operator.

The reason why the demand for automatic estimation of ultrasound images is high is to shorten the examination time and reduce factors due to patient-specific and operator dependence. Ultrasound companies (e.g., Samsung, GE, Philips, Siemens Healthcare) are focusing their efforts on developing technologies that leverage deep learning technology to reduce keystrokes, improve workflows and reduce operator dependency. Perhaps this time is the turning point in ultrasound imaging technology with the rapid development of deep learning.

5.5 Deep Learning Methods for Fetal Ultrasound Measurement

DL approaches can play an important role in automation of fetal ultrasound image systems, as well as improving the efficiency of ultrasound examination and women's health. Until recent advances in deep learning techniques [46, 51, 59, 80–82, 84], automation was far beyond the reach of conventional technologies [71, 83], due to difficulties in handling noisy ultrasound images affected by signal dropouts, artifacts, missing boundaries, attenuation, shadows, and speckling [83]. Most traditional fetal organ segmentation is based on image intensity-based or gradient-based approaches that are preferred to extract the boundaries of target anatomies [21, 61, 76, 90, 99]. However, these traditional approaches have fundamental difficulties in reaching full automation at a clinically applicable level, because rule-based models have limitations in comprehensively handling the complex, diverse and uncertain intrinsic characteristics of the fetal US images mentioned above.

Recently, with a marked and rapid advance in deep learning techniques, automation of fetal US systems is in fact experiencing a paradigm shift, enabling automated measurements of some fetal biometrics through supervised learning [15, 36, 44, 45]. This success is due to advances in convolutional neural networks (CNNs) used as core DL techniques and groundbreaking improvements in computational costs (e.g., fast implementations on GPUs). A rudimentary CNN model for pattern recognition mechanisms (having convolutional layers and downsampling layers) was proposed by Fukushima [23] in 1980. In 1989, LeCun et al. [52] used back-propagation to learn the convolution kernel coefficients for hand-written digit recognition, and succeeded in its practical application [53]. However, until 2012, the use of CNNs did not receive much attention due to serious difficulties in efficient training of deep neural networks. The turning point came when a GPU-based CNN named AlexNet won the ImageNet challenge in December 2012 [46]. Since then, DL technologies have advanced rapidly and have revolutionized the automation of medical image analysis. In 2015, Ronneberger et al. [82] proposed a convolutional network, called U-net, for medical image segmentation, which is now widely used as a basic tool in various medical image analysis methods. The fundamental concept of U-net comes from the fully convolutional network (FCN) [59].

CNNs are composed of convolution layers, ReLU layers, pooling layers, fully connected layers, and loss layers. In CNN, convolution filters are applied along the input image of each layer, and the extracted features are hierarchically more and more abstract as the layers get deeper. Training is the process of optimizing convolution filters or learning hierarchies of features, which is performed to minimize a loss function (e.g., the difference between the output of the model and the ground-truth label) through backpropagation and stochastic gradient descent.

One of the main issues in neural network research is a generalization gap, which is the difference between the training error and the test error. It is well known that over-parameterized deep neural networks (i.e. deep neural networks which use significantly more parameters than the number of samples in the training data) can learn an arbitrary dataset with 100% training accuracy [103]. However, an overparameterized network has a risk of overfitting, and therefore, in the field of medical imaging where training data is often scarce, it would be desirable to avoid increasing the number of nodes at risk of overfitting to improve trainability. Experiences have shown that even with nearly similar deep learning methods, learning ability is dramatically improved when contextual information about data is added to the network.

The success of DL in automated fetal biometrics is attributed to its superior ability to extract local and global interconnections by capturing spatial relationships between pixels from training data. However, while convolutional networks are good at integrating spatial contexts, they lack the ability to distinguish between two images that are very similar in overall structure. This shortcoming has been a major impediment to automatic navigation to standard planes, which is essential for accurate biometric measurements. For each plane, it is ambiguous whether this is a standard plane or not, and each clinician may have slightly different judgment criteria. This vagueness makes it difficult to get high quality training data for supervised learning. In practice, clinicians typically choose the optimal standard planes for a given video stream, taking into account the overall anatomical cues, rather than rigorously evaluating whether each frame meets all of the standard requirements for standard planes.

Currently, the most difficult challenge in fetal US automation is achieving standard planes selection (SPS) mentioned above. Although many efforts have been made to automate SPS using DL [8, 45, 56, 78], it has not yet reached the level of clinical application. What makes fetal US much more difficult to deal with DL compared to other imaging modalities such as CT or MRI is the complex diversity of fetal US images that are contaminated by several uncertain factors (e.g., various artifacts depending on fetal position, scan direction, and transducer position). Therefore, in order to develop a reliable DL model, it is necessary that DL architecture comprehensively reflects uncertain factors such as anatomical structure, the doctor's decision-making process, and fetal US image characteristics degraded by artifacts. Numerous failures of SPS have shown the importance of including subtle anatomical contextual information in DL architectures to address the difficulties mentioned above. Schlemper et al. [87] developed an attention-gated network to capture small differences in local anatomical structures along with capturing the global context. Cai et al. [8] used visual attention of sonographers to guide SPS in consecutive US video frames. Pu et al. [78] used a deep learning network consisting of CNN and a recurrent neural network (RNN) to learn the spatial and temporal features of the US video stream. However, our experiences have shown that it is still very difficult to extract subtle anatomical contextual information (i.e., ambiguous anatomical clues of local structures) with machine learning without human intervention.

This section intends to discuss these difficult topics with DL.

5.5.1 DL-Based Automatic US Examination of the Fetal Central Nervous System

Fetal US examination for evaluation of central nervous system (CNS) malformations involves the steps of finding three standard axial planes (SAPs) to be checked while scanning, checking the presence or absence of a structure in the acquired standard planes, and estimating whether the shape and size of a specific structure are within the normal range [63]. Figure 5.13 shows the three SAPs, which are the transventricular (TV) plane, transthalamic (TT) plane, and transcerebellar (TC) plane. On the TV plane, the anterior and posterior portions of the lateral ventricles are visible, and the anterior horns of the right and left ventricles are separated by the cavum septi pellucidi (CSP). On the TV plane, we measure the atrial width of the lateral ventricles. On the TC plane, the frontal horns of the lateral ventricles, CSP, thalami, cerebellum and cisterna magna are visible. On the TT plane, The transthalamic plane, head circumference (HC) and biparietal diameter (BPD) are measured to evaluate gestational age and diagnose CNS pathology. On the TC plane, we measure the transverse cerebellar diameter and cisterna magna depth.

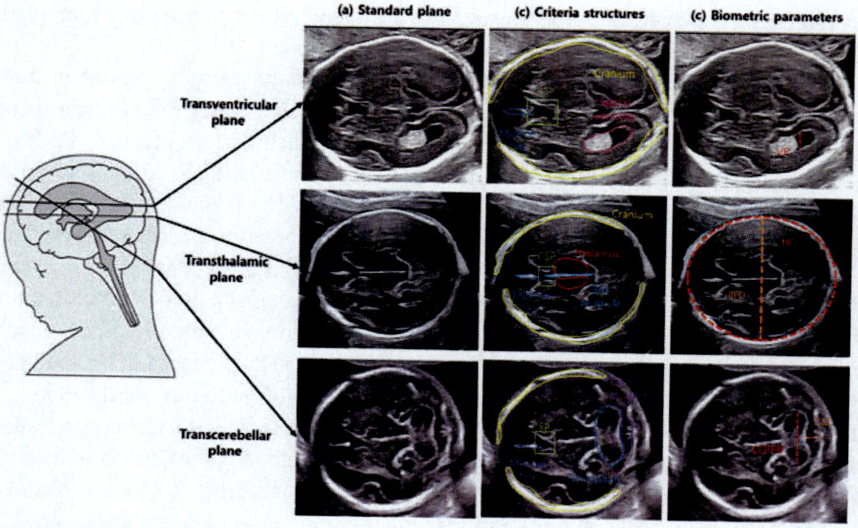

Fig. 5.13 Three three standard axial planes for the fetal central nervous system

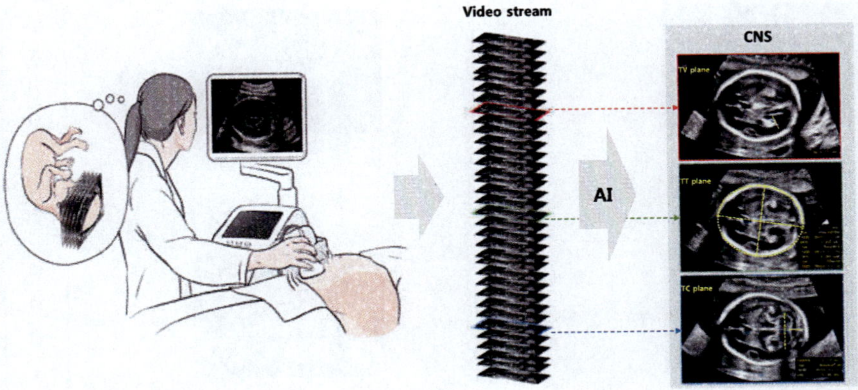

Fig. 5.14 Finding standard planes using a US probe is tricky, even for expert sonographers, because the position or orientation of the fetal head is not fixed. Sonographers have to move the US probe back and forth to find anatomical cues that match the criteria of each standard plane, but ultrasound images are often degraded by various artifacts, making it difficult to find these anatomical clues. This figure was extracted from [15]

Assessment of the fetal brain structural integrity on the three SAPs plays an important role in inferring dysfunction/malformation/probable disease. The pivotal step in the process of sonographic assessment of the fetal brain is acquiring the right plane, which, despite numerous efforts, has been very difficult to achieve with DLs so far.

The goal of selecting the three SAPs is to find a function

$$f_{SAP} : \mathbb{I} \mapsto \mathbb{S}, \tag{5.11}$$

where $\mathbb{I} = \{I_t : t = 1, \ldots, T\}$ is US video frames and $\mathbb{S} = (I_{t_1}, I_{t_2}, I_{t_3})$ is a triple of TV plane, TT plane, and TC plane in that order. Here, each I_t is in $\{0, 1, \ldots 255\}^{H_I \times W_I}$ that represents 256-level gray-scale US images ranging from 0 to 255 with the height H_I and width W_I. See Fig. 5.14.

In order to develop a robust automated system for SAPs selection and fetal biometric measurements, it is necessary to understand the sonographer's procedure. When developing deep learning methods, the network structure and its labeled data must take into account the ultrasound examiner's procedures. We begin with reviewing DL methods for automation of SAPs selection.

5.5.1.1 CNN Model for SAPs Selection

The simplest and most naive DL model is the following CNN model that targets classification for each image $I \in \mathbb{I}$:

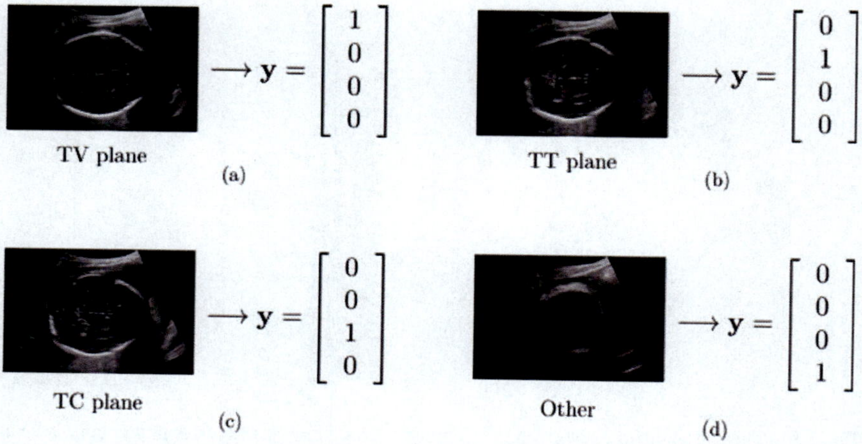

Fig. 5.15 **a**, **b**, **c**, and **d** indicate a class vectorization of TV plane, TT plane, TC plane, and other, respectively. The goal is to learn $f_{cnn} : \mathbf{I} \mapsto \mathbf{y}$ by using the labeled training data

$$f_{cnn} : \mathbf{I} \mapsto \mathbf{y}, \tag{5.12}$$

where $\mathbf{y} = (y_1, y_2, y_3, y_4) \in [0, 1]^4$ is a vector representing the TV plane, TT plane, TC plane, and other in that order, as seen in Fig. 5.15.

Given a labeled training data $\{(I^{(j)}, \mathbf{y}^{*(j)}) : j = 1, \ldots, N\}$, the CNN model f_{cnn} can be obtained by

$$f_{cnn} = \text{argmin} - \frac{1}{N} \sum_{j=1}^{N} \mathbf{y}^{*(j)} \odot \log f_{cnn}(I^{(j)}), \tag{5.13}$$

where $\odot$ stands for the sum of elementary-wise products and "argmin" represents the operation to find the parameters (i.e., weights and bias) of the neural network f_{cnn} that gives the minimum in the average cross-entropy loss function. A detailed description of the f_{cnn} structure and its parameters will be given below.

The f_{cnn} mainly consists of a (i) convolutional layer, (ii) pooling layer, and (iii) fully-connected layer, as seen in Fig. 5.16. At the first convolution layer, 16 feature maps of size 224 × 224 (denoted by $\mathbf{h}_1$) are computed using 16 convolution filters of size 3 × 3 and 16 biases of size 1 × 1 (denoted by $\mathbf{W}_1$ and $\mathbf{b}_1$) as follows:

$$\mathbf{h}_1 = \sigma(\mathbf{W}_1 \circledast_1 I + \mathbf{b}_1) = [\sigma(\mathbf{w}_1^1 \circledast_1 I + b_1^1), \ldots, \sigma(\mathbf{w}_1^{16} \circledast_1 I + b_1^{16})], \tag{5.14}$$

where $\circledast_s$ stands for convolution with stride s, $\sigma(\mathbf{h}) = max(0, \mathbf{h})$ is the rectified linear unit (ReLU) and

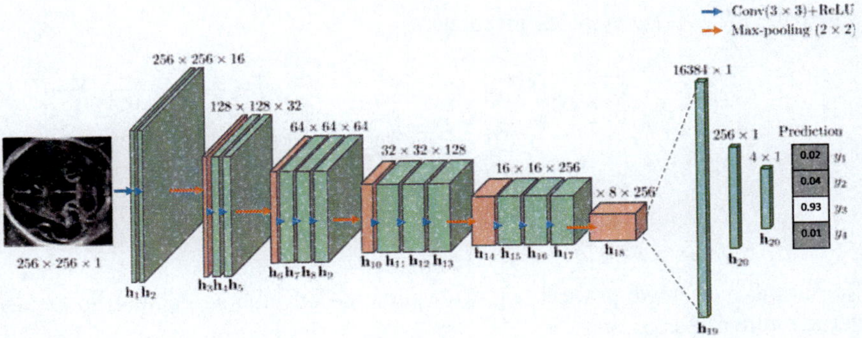

Fig. 5.16 Simple architecture of the CNN model

$$\mathbf{w}_1^i = \begin{bmatrix} w_{1,(1,1)}^i & w_{1,(1,2)}^i & w_{1,(1,3)}^i \\ w_{1,(2,1)}^i & w_{1,(2,2)}^i & w_{1,(2,3)}^i \\ w_{1,(3,1)}^i & w_{1,(3,2)}^i & w_{1,(3,3)}^i \end{bmatrix}, \ i = 1, \ldots, 16. \tag{5.15}$$

Then, the $\sigma(\mathbf{w}_1^i \circledast_1 I + b_1^i)$ produces a feature map of dimension 256 × 256 that is given by

$$\sigma(\mathbf{w}_1^i \circledast_1 I + b_1^i) = \begin{bmatrix} \sigma(\sum_{j,k=0}^{3} w_{1,(j,k)}^i I_{j,k} + b_1^i) & \cdots & \sigma(\sum_{j,k=0}^{3} w_{1,(j,k)}^i I_{j,k+223} + b_1^i) \\ \vdots & \ddots & \vdots \\ \sigma(\sum_{j,k=0}^{3} w_{1,(j,k)}^i I_{j+223,k} + b_1^i) & \cdots & \sigma(\sum_{j,k=0}^{3} w_{1,(j,k)}^i I_{j+223,k+223} + b_1^i) \end{bmatrix}. \tag{5.16}$$

Here, the zero-padding is used to the image I to control spatial size of the output feature map and to preserve the information of the borders. Throughout this chapter, we denote $\circledast_1 = \circledast$. Similarly, another convolution layer can follow the first convolution layer to obtain the second feature map of $\mathbf{h}_2$ given by

$$\mathbf{h}_2 = ReLU(\mathbf{W}_2 \circledast \mathbf{h}_1 + \mathbf{b}_2). \tag{5.17}$$

After two convolution layers mentioned above, max pooling operation of size 2 × 2 with stride 2 is applied to $\mathbf{h}_2$ of size 256 × 256 × 16 in (5.17) for dimensionality reduction as follows:

$$\mathbf{h}_3 = Max - Pool_2(\mathbf{h}_2), \tag{5.18}$$

$$Max - Pool_2(\mathbf{h}_2^1, \ldots, \mathbf{h}_2^{16}) = [(Max - Pool_2(\mathbf{h}_2^1), \ldots, Max - Pool_2(\mathbf{h}_2^{16}))], \tag{5.19}$$

where the $Max - Pool_2(\mathbf{h}_2^i)$ is given by

$$Max - Pool_2(\mathbf{h}_2^i) = \begin{bmatrix} max\begin{pmatrix} h^i_{2,(1,1)} & h^i_{2,(1,2)} \\ h^i_{2,(2,1)} & h^i_{2,(2,2)} \end{pmatrix} & \cdots & max\begin{pmatrix} h^i_{2,(1,223)} & h^i_{2,(1,224)} \\ h^i_{2,(2,223)} & h^i_{2,(2,224)} \end{pmatrix} \\ \vdots & \ddots & \vdots \\ max\begin{pmatrix} h^i_{2,(223,1)} & h^i_{2,(223,2)} \\ h^i_{2,(224,1)} & h^i_{2,(224,2)} \end{pmatrix} & \cdots & max\begin{pmatrix} h^i_{2,(223,223)} & h^i_{2,(223,224)} \\ h^i_{2,(224,223)} & h^i_{2,(224,224)} \end{pmatrix} \end{bmatrix}. \tag{5.20}$$

Repeat this successive process (e.g. Convolution–Convolution–Pooling) to extract deep feature maps.

Lastly, for classification purposes, the output of the last layer ($\mathbf{h}_{18}$) of the above-mentioned successive process is flattened to create a single long feature vector. This feature vector $\mathbf{h}_{18}$ is the input of a fully connected layer, and the final output for the classification prediction $\mathbf{y}$ is obtained by

$$\mathbf{h}_{19} = \sigma(\mathbf{W}_{19}\mathbf{h}_{18} + \mathbf{b}_{19}), \tag{5.21}$$

$$\mathbf{h}_{20} = \mathbf{W}_{20}\mathbf{h}_{19} + \mathbf{b}_{20}, \tag{5.22}$$

$$\mathbf{y} = \sigma_s(\mathbf{h}_{19}), \tag{5.23}$$

where $\mathbf{W}_{19}, \mathbf{W}_{20}$ are $256 \times 16384, 4 \times 256$ matrices and $\mathbf{b}_{19}, \mathbf{b}_{20}$ are $256 \times 1, 4 \times 1$ dimensional vectors and σ_s is a softmax activation function. The i-th class probability $y_i = \sigma_s(\mathbf{h}_{19})_i$ can be obtained by

$$y_i = \sigma_s(\mathbf{h}_{19})_i = \frac{\exp(h^i_{19})}{\sum_{k=1}^{4} \exp(h^k_{19})}. \tag{5.24}$$

Chen et al. [12], for the first time, attempted to use this type of CNN architecture to locate the fetal abdominal standard plane in US videos. Wu et al. [93] presented two CNN models, which are designed to find the region of interest (ROI) of the fetal abdominal region and to evaluate the image quality by assessing the goodness of depiction for the key anantomical structures, respectively. Baumgartner et al. [3] adopted a CNN architecture for real-time plane detection and weakly supervised target anatomy localization. In our experience, these CNN-based methods have difficulties in distinguishing subtle differences of key anatomical structures between frames, making it difficult to reach a clinically applicable level.

5.5.1.2 Attention-CNN Model for SAPs Selection

The failure experiences of the previous CNN-based SAP selection methods suggest that expert knowledge (anatomical features, decision-making process) and the characteristics of fetal ultrasound should be well reflected in deep learning. This is

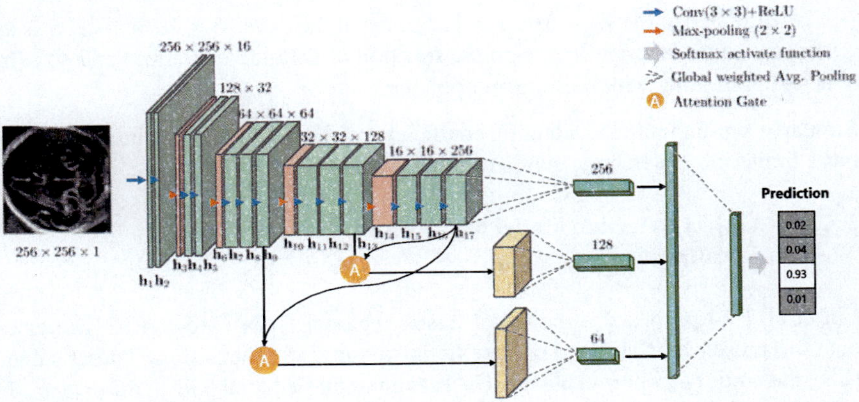

Fig. 5.17 Simple architecture of the attention-CNN model

because fetal ultrasound images contain various artifacts and noise, which are complexly related to the maternal, fetal position, scan direction, and transducer position.

Schlemper et al. [87] developed an attention gate (AG) in the CNN structure to suppress the irrelevant region of the input image while highlighting salient features for classification tasks.

The AG is a mechanism that can be incorperated into the CNN architecture described in Sect. 5.5.1.1. As shown in Fig. 5.17, the last convolutional feature map $\mathbf{h}_{17}$ is downsampled four times to capture a reasonably large receptive field. Hence, $\mathbf{h}_{17}$ can be viewed as a global feature grid vector that is used to focus on target-relevant regions and suppress irrelevant feature contents.

In Fig. 5.17, attention gates are added to $\mathbf{h}_9$ and $\mathbf{h}_{13}$, where the size of $\mathbf{h}_9 = (\mathbf{h}_9^1, \ldots, \mathbf{h}_9^{64})$ is $64 \times 64 \times 64$ and the size of $\mathbf{h}_{13} = (\mathbf{h}_{13}^1, \ldots, \mathbf{h}_{13}^{128})$ is $32 \times 32 \times 128$. The AG computes a vector $\alpha^9 \in [0, 1]^{64\times 64}$ and $\alpha^{13} \in [0, 1]^{32\times 32}$ and their outputs are

$$\hat{\mathbf{h}}_9 = (\alpha^9 \odot h_{9,1}, \ldots, \alpha^9 \odot h_{9,64}), \tag{5.25}$$

$$\hat{\mathbf{h}}_{13} = (\alpha^{13} \odot h_{13,1}, \ldots, \alpha^{13} \odot h_{13,128}). \tag{5.26}$$

The attention coefficient α^{13} is computed as follows:

- Compute $\mathbf{q}_{att} = ReLu(\mathbf{h}_{17} \circledast W_{att,17} + \downarrow_{dwon} (\mathbf{h}_{13} \circledast W_{att,13}) + \mathbf{b_q})$, where the size of $W_{att,17}$ is $1 \times 1 \times 256 \times F_{int}$, the size of $W_{att,13}$ is $1 \times 1 \times 128 \times F_{int}$, the size of the $\mathbf{b_q}$ is $16 \times 16 \times F_{int}$ and $\downarrow_{dwon}$ **(h)** is a downsampling with bilinear interpolation.
- Compute

$$\alpha^{13} = \uparrow_{up} (\hat{\sigma} (\mathbf{q}_{att} \circledast W_{\mathbf{q}} + \mathbf{b}_{\alpha})), \tag{5.27}$$

where the size of the $W_{\mathbf{q}}$ is $1 \times 1 \times 1$, the size of the $\mathbf{b}_\alpha$ is $16 \times 16 \times 1$ and $\hat{\sigma}$ is an normalization function (e.g. sigmoid function or softmax operation) and $\uparrow_{up}$ (**h**) is an upsampling with bilinear interpolation.

Similarly, we compute the attention coefficient α^9. However, this attention gate also has a limitation in detecting subtle differences between frames.

5.5.1.3 Object Detection Model for Key Anatomical Structures Localization

Lin et al. [56] proposed a multi-task faster regional CNN (MF R-CNN) to evaluate the presence or absence of specific anatomical structures (e.g., lateral sulcus (LS), thalamus (T), choroid plexus (CP), cavum septi pellucidi (CSP), third ventricle (TV), brain midline (BM)), where presence scores of the key structures are used for determination of the SAPs acceptance.

For the ease of explaination, we consider the following simplified detection model of the TT plane using YOLO [80]:

$$f_{det} : I \to \mathbf{O} = \{(c_1, \mathbf{b}_1), \ldots, (c_5, \mathbf{b}_5)\}, \tag{5.28}$$

where the output $\mathbf{O}$ represents a sequence of bounding boxes associated with the five anatomical structures: (1) falx, (2) CSP, (3) lateral ventricle, (4) AC, and (5) cerebellum. Here, $\mathbf{b}_k$ stands for a bounding box related to k-th anatomcial structures and c_k refers to its predicted confidence score. The $\mathbf{b}_k$ is a 4-dimensional vector representing bounding box coordinates with center position, width, and height. If a k-th anatomcial structure is not detected, we set $c_k = 0$ and $\mathbf{b}_k = (0, 0, 0, 0)$. An evaluation score of TT plane acceptance can be a weighted sum of $c_1, \ldots, c_5$. Please refer to [45] for TT plane acceptance check.

Now, we explain the detailed architecture of f_{det}. The map f_{det} is composed of three parts:

$$f_{det} = f_{post} \circ f_{NMS} \circ f_{grid}, \tag{5.29}$$

where f_{grid} is the network for predicting bounding boxes and estimating a confidence score for each grid cell, f_{NMS} is a non-maximal suppression (NMS) process for filtering overlapping boxes, and f_{post} is a post-processing process to get the final output.

We begin with explaining f_{grid}. We divide the image $I \in \{0, \ldots, 255\}^{W_I \times H_I}$ into uniform squares $\{G_{ij} : i = 1, \ldots, \frac{W_I}{k},\ j = 1, \ldots, \frac{H_I}{k}\}$, where G_{ij} is a grid cell of size $k \times k$. For each grid cell G_{ij}, the f_{grid} predicts a bounding box $\mathbf{b}_{ij}$, confidence score c_{ij} , and its class probability $\mathbf{y}_{ij}$:

$$f^1_{det}(I) = \begin{pmatrix} \mathcal{O}_{1,1} & \mathcal{O}_{1,2} & \cdots & \mathcal{O}_{1,\frac{W_I}{k}} \\ \mathcal{O}_{2,1} & \mathcal{O}_{2,2} & \cdots & \mathcal{O}_{2,\frac{W_I}{k}} \\ \vdots & \vdots & \ddots & \vdots \\ \mathcal{O}_{\frac{H_I}{k},1} & \mathcal{O}_{\frac{H_I}{k},2} & \cdots & \mathcal{O}_{\frac{H_I}{k},\frac{W_I}{k}} \end{pmatrix}, \tag{5.30}$$

where $\mathcal{O}_{ij} = (c_{ij}, \mathbf{b}_{ij}, \mathbf{y}_{ij})$. Here, $c_{ij} \in [0, 1]$ represents belief of existence of the target center in G_{ij}, $\mathbf{b}_{ij}$ is a bounding box component and $\mathbf{y}_{ij} = (y_1, y_2, y_3, y_4, y_5) \in [0, 1]^5$ is a class probability in G_{ij}. For details please refer to [80].

The map f_{grid} is trained by using a labeled dataset $\{I^{(n)}, \mathcal{O}^{*(n)}\}_{n=1}^N$. $\mathcal{O}^{*(n)}$ represents the ground-truth corresponding to $I^{(n)}$. The loss function of f_{grid} is

$$\mathcal{L}_{grid} = \sum_{n=1}^{N} \left[\lambda_0 \sum_{(i,j)\in\Omega_0^{(n)}} (0 - c_{ij}^{(n)})^2 + \sum_{(i,j)\in\Omega_1^{(n)}} (1 - c_{ij}^{(n)})^2 \right. \\ \left. + \lambda_1 \sum_{(i,j)\in\Omega_1^{(n)}} \left\| \mathbf{b}_{ij}^{*(n)} - \mathbf{b}_{ij}^{(n)} \right\|^2 \right], \tag{5.31}$$

where $\Omega_0^{(n)} = \{(i, j) : c_{ij}^{*(n)} = 0\}$, $\Omega_1^{(n)} = \{(i, j) : c_{ij}^{*(n)} = 1\}$, $\lambda_0 = 0.1$, and $\lambda_1 = 5$.

Next, the network f_{NMS} in (5.29) is to filter overlapping boxes using the non-maximal suppression technique [1]. The f_{NMS} eliminates overlapped bounding boxes $\mathbf{b}_{ij}$ of the same structure by selecting a box with the highest score.

Finally, f_{post} selects a maximum value of the confidence score of each class. If k-th anatomcial structure is not detected, we set $c_k = 0$ and $\mathbf{b}_k = (0, 0, 0, 0)$.

Experiments have shown that the results of evaluating the suitability of SAPs with only a single frame are often inconsistent, even among experts. In practice, clinicians typically choose the optimal SAPs for a given video stream rather than rigorously evaluating whether each frame meets all of the standard requirements for SAPs. Hence, it seems fundamentally vague to distinguish between SAPs and non-SAPs by looking at only one plane. This vagueness makes it difficult to get high quality training data for supervised learning.

5.5.1.4 CNN-RNN Model

Cai et al. [8] and Pu et al. [78] proposed a CNN-RNN model to deal with the difficulties of above-mentioned model by adding a recurrent neural network (RNN) structure to the CNN. In [78], the CNN component is designed to identify fetal key anatomical structures from each video frame and recognize SAPs, and the RNN component is designed to obtain the temporal information between adjacent planes in order to realize precise locations and track fetal organs accross the frame. These CNN-RNN models are based on the need to utilize the context information of the ultrasound video stream to overcome the limitations of SAP evaluation with only a single frame.

The CNN-RNN model is used to recover frame-level classifications:

$$f_{cnn-rnn} : \mathbf{I} \mapsto \mathbf{Y}, \tag{5.32}$$

where $\mathbf{I} = \{I_1, \ldots, I_5\} \subset \mathbb{I}$ is a set of five consecutive frames and $\mathbf{Y} = \{\mathbf{y}_1, \ldots, \mathbf{y}_5\}$ is the corresponding class probability.

We denote by f_{cnn^*} a part of the network f_{cnn} in (5.12) excluding deeper layers including the last two fully connected layers. For example, the output of f_{cnn^*} can be $\mathbf{h}_9$ in Fig. 5.16. Let us write $\mathbf{z}_t = f_{cnn^*}(I_t),\ t \in \{1, \ldots, 5\}$. We use a convolutional long short-term memory (ConvLSTM) [94] that is the following map:

$$LSTM_t^{Conv} : (\mathbf{z}_t, \mathfrak{h}_{t-1}) \to (o_t, \mathfrak{h}_t), \tag{5.33}$$

where

$$\begin{pmatrix} i_t \\ f_t \\ o_t \\ g_t \end{pmatrix} = \begin{pmatrix} \sigma \\ \sigma \\ \sigma \\ tanh \end{pmatrix} \left[\begin{pmatrix} W_{zi} & W_{hi} \\ W_{zf} & W_{hf} \\ W_{zy} & W_{hy} \\ W_{zc} & W_{hc} \end{pmatrix} \circledast \begin{pmatrix} \mathbf{z}_t \\ \mathfrak{h}_{t-1} \end{pmatrix} + \begin{pmatrix} W_{ci} \\ W_{cf} \\ W_{cy} \\ 0 \end{pmatrix} \odot C_{t-1} + B \right] \tag{5.34}$$

and

$$\begin{cases} C_t = f_t \odot C_{t-1} + i_t \odot g_t, \\ \mathfrak{h}_t = o_t \odot tanh(C_t). \end{cases} \tag{5.35}$$

The expressions of (5.34) and (5.35) are equivalent to

$$\begin{cases} i_t = \sigma(W_{zi} \circledast \mathbf{z}_t + W_{hi} \circledast \mathfrak{h}_{t-1} + W_{ci} \odot C_{t-1} + b_i), \\ f_t = \sigma(W_{zf} \circledast \mathbf{z}_t + W_{hf} \circledast \mathfrak{h}_{t-1} + W_{cf} \odot C_{t-1} + b_f), \\ C_t = f_t \odot C_{t-1} + i_t \odot tanh(W_{zc} \circledast \mathbf{z}_t + W_{hc} \circledast \mathfrak{h}_{t-1} + b_c), \\ o_t = \sigma(W_{zy} \circledast \mathbf{z}_t + W_{hy} \circledast \mathfrak{h}_{t-1} + W_{cy} \circledast C_t + b_y), \\ \mathfrak{h}_t = o_t \odot tanh(C_t). \end{cases} \tag{5.36}$$

Here, i_t is the *input gate*, f_t is the *forget gate* (whether to erase cell), C_t is the *cell state*, and o_t is the *output gate*.

For classification of the input image I_t, the output o_t of the $LSTM_t^{Conv}$ is flattened to create a single long feature vector, denoted by $\tilde{o}_t$. This feature vector $\tilde{o}_t$ takes matrix multiplication, and the final output for the classification prediction is obtained by

$$\mathbf{y}_t = \sigma(\mathbf{W}_o \tilde{o}_t + \mathbf{b}_o), \tag{5.37}$$

where $\mathbf{W}_o$ is a $\sharp(\tilde{o}) \times 4$ matrix, and $\mathbf{b}_o$ is a 4-dimensional vector.

Given a labeled training data $\{(\mathbf{I}^{(j)}, \mathbf{Y}^{*(j)}) : j = 1, \ldots, N\}$, the CNN-RNN model $f_{cnn-rnn}$ can be obtained by

$$f^*_{cnn-rnn} = \text{argmin} \frac{1}{N} \sum_{j=1}^{N} \mathbf{Y}^{*(j)} \odot \log f_{cnn-rnn}(\mathbf{I}^{(j)}), \tag{5.38}$$

where $f_{cnn-rnn}(\mathbf{I}^{(j)}) = (\mathbf{y}_1^{(j)}, \ldots, \mathbf{y}_5^{(j)})$ and

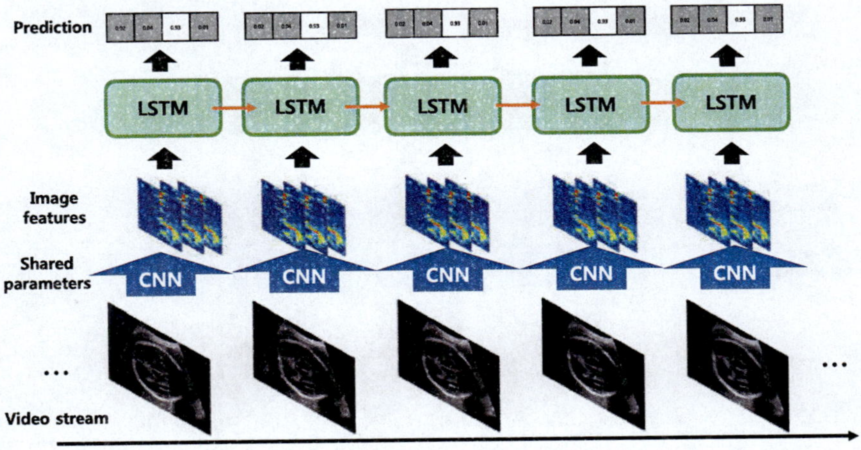

Fig. 5.18 Simple architecture of the CNN-RNN model

$$\mathbf{Y}^{*(j)} \odot \log f_{cnn-rnn}(\mathbf{I}^{(j)}) = \frac{1}{5}\sum_{t=1}^{5} \mathbf{y}_t^{*(j)} \odot \log \mathbf{y}_t^{(j)}. \tag{5.39}$$

Chen et al. [14] applied this CNN-RNN method for detecting the following three standard planes: (1) fetal abdominal standard plane; (2) fetal face axial standard plane; and (3) fetal four-chamber view standard plane of the heart. The network architecture of their method is a bit different from that of $f_{cnn-rnn}$: The number of $\mathbf{I}$ is 30; the $\mathbf{z}_t$ is $f_{cnn}(I_t) \in [0, 1]^2$ described in (5.12); and the standard LSTM is used instead of ConvLSTM. Here, each evaluation class was divided separately whether the input I_t was a standard plane or not. The result of their method shows poor performance in terms of individual output. On the other hand, their result shows that one correct standard plane can be successfully selected from a video which encloses at least one standard plane.

Cai et al. [8] proposed a bi-directional RNN to use the contextual information of the back and forth frames as in Fig. 5.19. Their network achitecture is as follows: The number of $\mathbf{I}$ is 3; and the network $f_{cnn-rnn}$ determines $\mathbf{y}_t$, which gives whether the middle plane I_t is acceptable or not. In order to utilize its neighbor plane I_{t-1}, I_{t+1} when determining $\mathbf{y}_t$, the positive order of output o_{t-1} and the reverse order of output o_{t+1} are concantenated and fed into a classification layer, where the equation (5.37) is replaced by

$$\mathbf{y}_t = \sigma(\mathbf{W}_o[\tilde{o}_{t-1}, \tilde{o}_t, \tilde{o}_{t+1}] + \mathbf{b}_o). \tag{5.40}$$

Moreover, their method exploits a visual navigation process of sonographers by learning to generate visual attention maps of ultrasound images around standard biometry planes of the fetal abdomen, head (trans-ventricular plane) and femur.

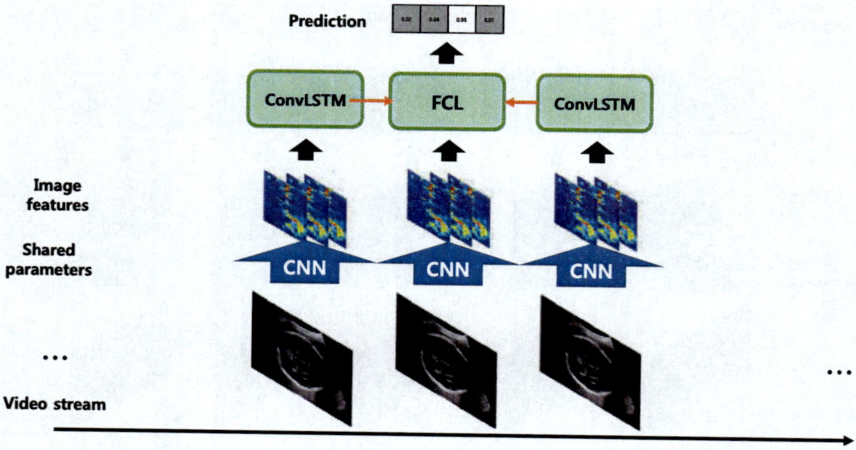

Fig. 5.19 Simple architecture of the bi-directional RNN model

This model $f_{cnn-rnn}$ is based on the paper by Pu et al. [78], where the CNN component identifies fetal key anatomical structures from each video frame while the RNN component obtains the temporal information between adjacent frames.

Although the RNN structures are used to exploit the context of the US video, these deep learning structures alone seem to have limitations in adequately detecting subtle transitions in key anatomical structures between frames.

5.5.1.5 Hierarchical DL Models with Anatomical Contexts Embedding and Attention-Guiding Mechanisms

Our experience examining the previous methods has shown that extracting subtle anatomical contextual information (i.e., ambiguous anatomical clues of local structures) is essential for developing a robust automated SAPs selection system. Unfortunately, learning with machine learning to extract useful context for choosing SAPs seems difficult at its current level. Therefore, in order to provide a good machine learning friendly environment, it is recommended to use the classical methods to highlight useful contexts while ignoring unnecessary information.

Specifically, we need to provide an environment for deep learning to focus on anatomical cues of SAPs based on criteria of standard planes in Fig. 5.13: TV plane cues include midline falx, CSP, 3rd ventricle, and thalamus; TT plane cues include midline falx, CSP, lateral ventricles, and choroid plexus; and TC plane cues include midline falx, CSP, cerebellum, and cisterna magna. See Fig. 5.20.

Anatomical contextual information can be extracted from US probe motion-induced differences between successive frames. The underlying idea behind this approach is based on the empirical observation of sonographers' process of finding

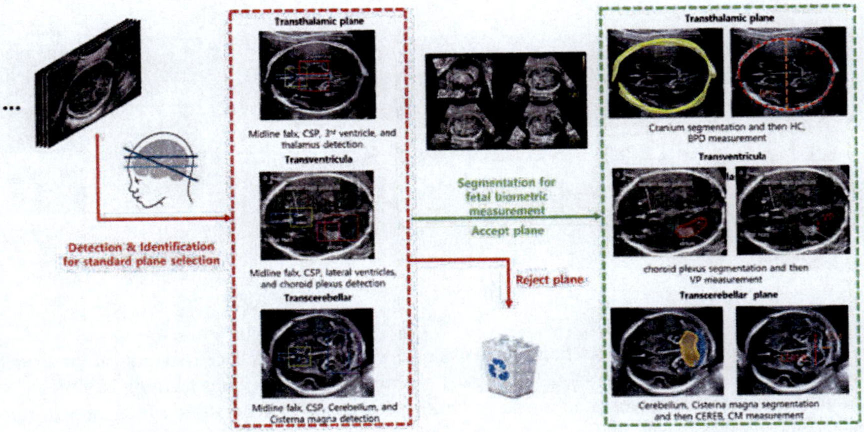

Fig. 5.20 Key anatomical cues for selecting SAPs. The development of semi-automated labeling tools is important because deep learning models require huge amounts of annotated data to train the system

SAPs; sonographers often move the US probe across the mother's body surface to obtain complementary information of contextual cues along successive frames for SAPs, rather than relying solely on instant-level recognition of anatomical structures in each single frame.

The proposed method mainly divides into two parts: (i) zone-classification network f_{zone}; and (ii) SAPs selection network $f_{selection}$. Zone-classification network f_{zone} classifies the input frames into predefined 7 zones (A—G) and others and generates the anatomical attention map utilizing contextual information of between input frames. Next, SAPs selection network $f_{selection}$ selects the best three SAPs from the three SAPs zones (TT,TV, and TC zone) classified in the previous step using three specialized CNNs for each plane.

Firstly, we consider the following zone-classification network that consists of roughly three sub-networks:

$$f_{zone} = f_{zn} \circ (f_{ctx} \circ f_{ROI}; f_{ROI}), \tag{5.41}$$

where the above networks are roughly described in Fig. 5.21. The detailed roles of the above subnetworks are as follows:

- f_{ROI} is a function that maps from $\mathbb{I} = \{I_t : 1, \ldots, T\}$ to the corresponding bounding boxes $\mathbb{I}_{ROI} = \{I_{t,ROI} : 1, \ldots, T\}$ containing the region of interest (ROI) of the fetal head. To be precise, the head region in I_t is segmented through U-net [82], and the divided image is rotated around the center so that the BPD is aligned in the vertical direction. The detected bounding boxes can be represented by storing the coordinates of the corners. Since the bounding boxes have different sizes, each image is padded with zeros to get a rectangular region of fixed width and height.

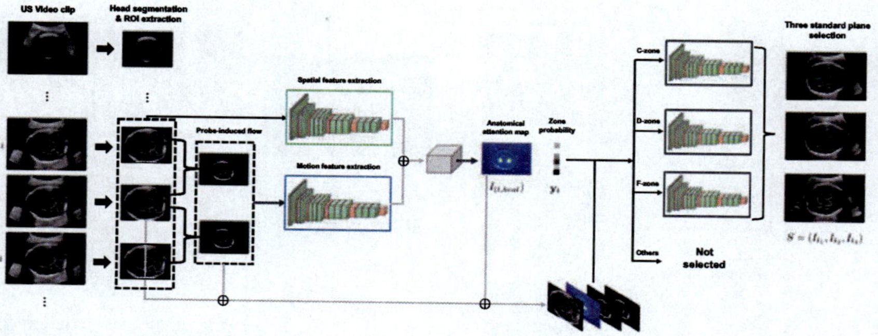

Fig. 5.21 Automated SAPs selection system using US probe motion-induced anatomical contextual information. Since extracting subtle contextual information through deep learning is difficult, it would be desirable to develop an image processing technique to generate images that help find nuanced context between frames

- f_{ctx} is a function that maps from $\mathbb{I}_{ROI}$ to $\mathbb{I}_{ctx} = \{\hat{I}_t : t = 2, \ldots, T-1\}$ where $\hat{I}_t$ is a contextual image between $I_{t-1:ROI}$ and $I_{t:ROI}$. The method of generating $\hat{I}_t$ will be explained later.
- f_{zn} is a function whose output is $\mathbb{Y}_{zone} = \{(\mathbf{y}_t, I_{t,heat}) : t = 2, \ldots, T-1\}$ where $\mathbf{y}_t \in [0, 1]^8$ is a vector giving the probability that each frame I_t belongs to one of 7 zones (A–G) and others based on change of the anatomical cues and $I_{t:heat}$ represents heat maps of key anatomical structures. The t-th component $(\mathbf{y}_t, I_{t,heat})$ is determined by $I_{t-1:ROI}, I_{t:ROI}, I_{t+1:ROI}, \hat{I}_{t-1}, \hat{I}_t$.

The map f_{zone} is trained by using a labeled dataset $\{I_t, S_t^*, \mathbf{y}_t^*, I_{t,heat}^*\}_{t=1}^N$, where $S_t^* \in \mathbb{R}^{H_I \times W_I}$ denotes a ground-truth segmentation image for the fetal head, $\mathbf{y}_t^* \in [0, 1]^8$ is a ground-truth zone classification labeling, and $I_{t,heat}^* \in \mathbb{R}^{H_I \times W_I}$ represents a ground-truth anatomical structures attention map.

For segmentation of the fetal head, U-net $f_{ROI,U-net}$ is learned by

$$f_{ROI,U-net} = \text{argmin} \frac{1}{N} \sum_{t=1}^{N} [\frac{1}{M} \sum_{\mathbf{X}} S_t^*(\mathbf{X}) \odot \log f_{ROI,U-net}(I_t(\mathbf{X}))], \tag{5.42}$$

where $\mathbf{X}$ is a pixel position and M is the number of pixels of $\mathbf{Y}_{\Omega_{head}}$. Then, fetal head ROI $I_{t,ROI}$ is extracted by using the segmentation result S_t and resizes into a fixed size 256×256.

Next, for zone classification and generating an anatomical structures attention map, f_{zn} is trained by minimizing the loss between output $(\mathbf{y}_t, I_{t,heat}) = f_{zn}(I_{t-1}, I_t, I_{t+1})$ and ground-truth $(\mathbf{y}_t^*, I_{t,heat}^*)$ as follows:

$$\mathcal{L}_{zn} = -\frac{1}{N} \sum_{t=1}^{N} [\mathbf{y}_t^* \odot \log \mathbf{y}_t + \sum_{\mathbf{X}} (I_{t,heat}(\mathbf{X}) - I_{t,heat}^*(\mathbf{X}))^2]. \tag{5.43}$$

Secondly, we need a sorting network, denoted by f_{sort}, that maps from $\mathbb{Y}_{zone}$ to $\mathbb{Z}_{sort} = \{\mathbf{z}_t : t = 2, \ldots, T-1\}$, where $\mathbf{z}_t \in \{(1,0,0,0), (0,1,0,0), (0,0,1,0), (0,0,0,1)\}$. Here, $(1,0,0,0)$ indicates that I_t lies in the C zone; $(0,1,0,0)$ D zone; $(0,0,1,0)$ F zone; and $(0,0,0,1)$ otherwise. Hence, each $\mathbf{z}_t$ can be regarded as a zone-class prediction corresponding to the image frame I_t.

The final network is a selective network consisting of three CNN-type networks $f_{sel} = (f_{TV-sel}, f_{TT-sel}, f_{TC-sel})$:

- the input of f_{TV-sel} is $\{(I_{t,ROI}, I_{t,heat}, \hat{I}_{t-1}, \hat{I}_t) : \mathbf{z}_t = (1,0,0,0)\}$ and its output is a selected TV plane I_{t_1};
- the input of f_{TT-sel} is $\{(I_{t,ROI}, I_{t,heat}, \hat{I}_{t-1}, \hat{I}_t) : \mathbf{z}_t = (0,1,0,0)\}$ and its output is a selected TT plane I_{t_2};
- the input of f_{TC-sel} is $\{(I_{t,ROI}, I_{t,heat}, \hat{I}_{t-1}, \hat{I}_t) : \mathbf{z}_t = (0,0,1,0)\}$ and its output is a selected TC plane I_{t_3}.

Now, we will explain how to generate $\hat{I}_t$ (i.e., the output of the network f_{ctx}) that is generated from the difference between I_{t-1} and I_t. Since the difference $I_t - I_{t-1}$ is an image caused by the motion of the probe, it contains a lot of noise to the extent that it is almost useless by itself. Therefore, careful image processing is required to suppress extraneous background artifacts to highlight key anatomical contexts between successive frames. As shown in Fig. 5.22, the $\hat{I}_t$ is generated through the following steps:

1. Produce the difference image $I_{t:ROI} - I_{t:ROI}$.
2. Apply Gaussian filtering, Otsu thresholding, and histogram equalization sequentially to the difference image to remove noise and highlight the key anatomical cues.
3. Apply the Farneback algorithm to the image in the previous step to compute the motion vector flow of the key anatomical structures.

Figure 5.22 shows probe motion-induced flow during the scanning process.

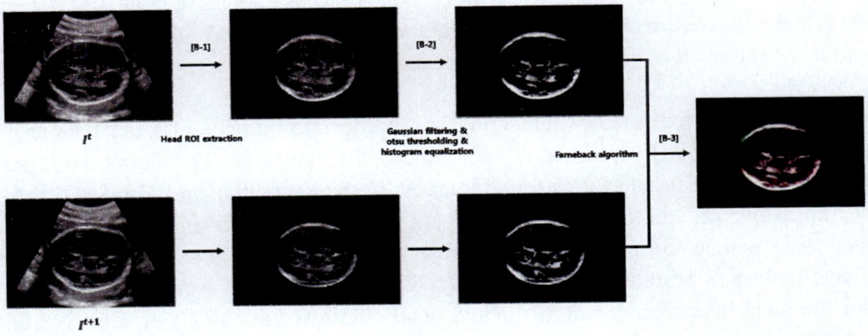

Fig. 5.22 Workflow of the probe motion-induced flow

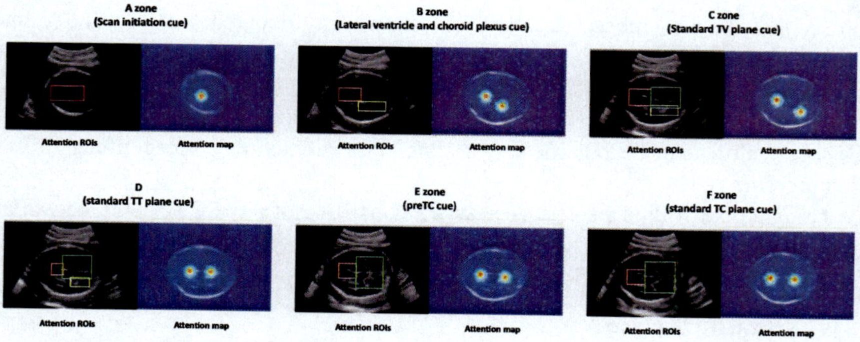

Fig. 5.23 Attention ROIs and map

Next, we will explain how to generate $I_{t,heat}$ (i.e., the second output of the network f_{zn}) using the predefined key anatomical structures ROIs from each zone. It is generated using the distance between the center position of key anatomical structures ROIs and every pixel position of the image. To be precise, let $\{\mathbf{C}_t^k\}_{k=1}^K$ be a set of center positions of key anatomical structures and the number of K depends on which zone I_t belongs to, then the heat map can be computed as

$$I_{t,heat}^k(\mathbf{p}) = \exp\left(-\frac{\|\mathbf{p} - \mathbf{C}_t^k\|^2}{\sigma_k^2},\right) \tag{5.44}$$

$$I_{t,heat}^*(\mathbf{p}) = max\{I_{t,heat}^1(\mathbf{p}), \ldots, I_{t,heat}^K(\mathbf{p})\}, \tag{5.45}$$

where $\mathbf{p}$ is a pixel position of the image and σ_k is a parameter controlling the spread of the peak. See Fig. 5.23.

5.5.1.6 Head Boundary Detection

This section describes the f_{ROI} network in (5.41), where its main goal is to detect the fetal head. We use an ellipse fitting method to detect the head contour of the fetus and it is conducted by placing an ellipse around the outside of the skull bone echoes [63, 73].

Ellipse fitting is based on detecting an appropriate number of head boundary pixels [5, 67, 77, 95]. However, it is difficult to distinguish head borders from other border-like pixels in noisy ultrasound images. To overcome this problem, we utilize our knowledge of the ultrasound propagation direction and image features of the maternal tissue near the probe. In the case of the parent tissue close to the probe, the image pattern is oriented perpendicular to the direction of ultrasound propagation, but the head boundary is not. To detect head circumference (HC) effectively and robustly while preventing misjudgement of boundary-like patterns, we detect three main features: maternal tissues of the US probe's directional pattern, upper head boundary of the concave arc pattern, and lower head boundary of the convex arc pattern.

Then, the goal is to learn a map $f_{head} : I \to U$ such that

$$f_{head}\underbrace{(\text{input US image})}_{I} = \underbrace{\text{images of three features}}_{U}. \tag{5.46}$$

Here, the output $U = (U_1, U_2, U_3)$ is the vector of the following three feature images:

- U_1 is an image of maternal tissues of the US probe's directional pattern.
- U_2 is an image of upper head boundary of the concave arc pattern.
- U_3 is an image of lower head boundary of the convex arc pattern.

Let $\{(I^{(i)}, U^{(i)}) : i = 1, \ldots, N\}$ be the labelled training data. The network f_{head} is learned from

$$f_{head} = \underset{f \in \mathbb{NN}}{\text{argmin}} \frac{1}{N} \sum_{n=1}^{N} \text{dist}(f(I^{(n)}), U^{(n)}), \tag{5.47}$$

where $\mathbb{NN}$ denotes a set of functions described in a given form of neural network and $\text{dist}(f(I^{(n)}), U^{(n)})$ is a distance between $f(I^{(n)})$ and $U^{(n)}$.

We select $\mathbb{NN}$ as U-net [82], which is one of the most popular networks for medical image segmentation. The U-net $f_{head} \in \mathbb{NN}$ consists of encoding and decoding paths:

$$f_{head} = f_{\text{decoding}} \circ_{\text{skip}} f_{\text{encoding}}, \tag{5.48}$$

where $\circ^{\text{skip}}$ is the composition operator with skip connections which will be explained later.

The encoding path f_{encoding} is based on a series of convolutions followed by pooling to reliably recognize image features so that the resulting output is fairly resilient to variations in position and scale on a target structure. The basic component of the encoding path f_{encoding} is the following operation consisting of "convolution-Relu-covolution-Relu" followed by max pooling: Given input $\mathbf{h}$, the output $\mathbf{h}'$ is

$$\mathbf{h}' = \mathscr{H}_{W',W}(\mathbf{h}) := \mathscr{P}_{ooling}(\text{conv}_{W'} \circ \text{conv}_{W}(\mathbf{h})). \tag{5.49}$$

Here, conv_W and $\mathscr{P}_{ooling}$ are defined as follows:

- W is a trainable parameter consisting of convolution filters $[\mathbf{w}_1, \ldots, \mathbf{w}_k]$ of size $d \times d$ matrix and biases $[\mathbf{b}_1, \ldots, \mathbf{b}_k]$. For example, if $d = 3$,

$$\mathbf{w}_j = \begin{pmatrix} w_{j,(1,1)} & w_{j,(1,2)} & w_{j,(1,3)} \\ w_{j,(2,1)} & w_{j,(2,2)} & w_{j,(1,3)} \\ w_{j,(3,1)} & w_{j,(3,2)} & w_{j,(3,3)} \end{pmatrix} \quad \text{and} \quad \mathbf{b}_j = \begin{pmatrix} b_{1;1} \\ b_{1;2} \\ b_{1;3} \end{pmatrix}. \tag{5.50}$$

Note that the number of parameters $W = [\mathbf{w}_1, \ldots, \mathbf{w}_k, \mathbf{b}_1, \ldots, \mathbf{b}_k]$ is $(d \times d + d) \times k$.

- conv_W denotes a convolution operation with Relu. For example, if W has k convolution filters, then

$$\text{conv}_W(\mathbf{h}) := [ReLU(\mathbf{w}_1 \circledast_s \mathbf{h} + \mathbf{b}_1), \ldots, ReLU(\mathbf{w}_k \circledast_s \mathbf{h} + \mathbf{b}_k)], \qquad (5.51)$$

where $ReLU(h) = max\{h, 0\}$ is the rectified linear unit and $\circledast_s$ stands for standard convolution with stride s. For example, if $s = 1$ and $\mathbf{w}_j$ is $d \times d$ matrix, then

$$(a, b)\text{-component of } \mathbf{w}_j \circledast_1 \mathbf{h} := \sum_{m,n=1}^{d} w_{j,(m,n)} h_{(a+m,b+n)}, \qquad (5.52)$$

which is almost the same as the standard convolution.
- $\mathscr{P}_{ooling}$ is the max pooling operation that is applied to reduce the dimensions of the feature maps.

The above consecutive operations are repeated up to the end of the encoding path to extract feature maps:

$$f_{\text{encoding}} = \mathscr{H}_{W^{(8)},W^{7)}} \circ H_{W^{(6)},W^{(5)}} \circ H_{W^{(4)},W^{(3)}} \circ H_{W^{(2)},W^{(1)}}, \qquad (5.53)$$

where $\circ$ is the standard composition operator.

The decoding path f_{decoding} is used to produce a segmented output from the semantic information extracted by the encoder. It is a reverse process of the encoding path, replacing the pooling operator by the average unpooling to restore the size of the output. Also, each unpooled output is concatenated with the corresponding feature from the encoding path. The basic component of the decoding path f_{decoding} is the operation consisting of "convolution–Relu–covolution–Relu"followed by the average unpooling: Given unpooled input $\mathbf{h}$, the output $\mathbf{h}'$ is

$$\mathbf{h}' = \mathscr{H}^*_{W',W}(\mathbf{h}) := \mathscr{U}_{npooling}(\text{conv}_{W'} \circ \text{conv}_W(\mathbf{h}, \mathbf{h}_{encoding})), \qquad (5.54)$$

where $\mathbf{h}_{encoding}$ is the feature from the encoding path corresponding to $\mathbf{h}$ and $\mathscr{U}_{npooling}$ denotes the average unpooling. Here, the superscript $*$ means dual operation, because the notation $\mathscr{H}^*_{W',W}$ is somewhat similar to a dual of $\mathscr{H}_{W',W}$.

At the last layer, we take the pixel-wise softmax activation function after applying a 1×1 convolution.

From the head boundary points, it is easy to obtain HC and BPD measurements. These measurements can be estimated by placing an ellipse around the outside of the skull. We use a least-squares-based geometric ellipse fitting method to obtain the five ellipse parameters $\Theta = (a, b, \theta_c, x_c, y_c)$, which provide the following ellipse representation:

$$\alpha(x - x_c)^2 + \beta(y - y_c)^2 + \gamma(x - x_c)(y - y_c) = a^2 b^2, \qquad (5.55)$$

where

$$\alpha = a^2 \sin^2 \theta_c + b^2 \cos^2 \theta_c,$$
$$\beta = a^2 \cos^2 \theta_c + b^2 \sin^2 \theta_c,$$
$$\gamma = (a^2 - b^2) \sin 2\theta_c.$$

The parameter Θ is obtained by solving

$$\underset{\Theta}{\operatorname{argmin}} \sum_{(p_1,p_2)\in P} \frac{|p_2 - m(\Theta; p_1, p_2)p_1 - c(\Theta; p_1, p_2)|^2}{1 + m(\Theta; p_1, p_2)^2}, \tag{5.56}$$

where

$$m(\Theta; p_1, p_2) = -\frac{2\alpha(p_1 - x_c) + \gamma(p_2 - y_c)}{2\beta(p_2 - y_c) + \gamma(p_1 - x_c)}, \tag{5.57}$$

$$c(\Theta; p_1, p_2) = \frac{2\beta(p_2 - y_c)p_2 + 2\alpha(p_1 - x_c)p_x - \gamma(p_2 x_c + p_1 y_c - 2p_1 p_2)}{2\beta(p_2 - y_c) + \gamma(p_1 - x_c)}. \tag{5.58}$$

For the details of this method, please refer to [77]. Once $\Theta = (a, b, \theta_c, x_c, y_c)$ is determined by 5.56, BPD and HC, respectively, are given by

$$\text{BPD} = 2b \tag{5.59}$$

and

$$\text{HC} = \pi \left[3(a + b) - \sqrt{(3a + b)(a + 3b)} \right]. \tag{5.60}$$

5.5.2 Discussion

For reliable SAP selection, automated methods must reflect the decision-making process of ultrasound engineers. In particular, it is important to reflect knowledge of the geometrical arrangement of anatomical landmarks (e.g., "box-like" cavum septum pellucidum and the "V-shaped" ambient cistern, and cerebellum). To do this, w must crop the image into an ellipse around the fetal head and rotate it in such a way that the image is aligned in terms of the direction of the fetus' midline and the landmarks in the rotated image are positioned almost uniformly on the ellipse. Using this normalized US image, the search area can be significantly narrowed and the 3D position of the US image can be reliably predicted.

Since it is difficult to obtain reliable results by directly applying the SAP selection network $f_{selection}$, we first run the stable zone classification network f_{zone} to filter out irrelevant frames. Next, the SAP selection network $f_{selection}$ is applied to the set of frames associated with the three SAP zones (TT, TV, TC zone) obtained through the previous network f_{zone}. In order to ensure reliable SAP selection, a deep neural solution must ensure a visual interpretation of the anatomical context for details related to the decision-making process. The black box-type decision-making process makes it difficult to understand the limitations of deep learning, thereby hindering network performance improvement.

5.6 Transvaginal Ultrasound: Automatic Measurement of Cervical Length

This section discusses deep learning-based automatic measurement of cervical length (CL) from a transvaginal ultrasound image. CL indicates the length of the lower end of the uterus, and CL measurement is performed on the cervical canal (CC) visible throughout after checking the internal os and external os. This section is based on the paper [50].

Acquisition of the standard midsagittal plane of the cervix and recognition of cervical shape and key features required for CL measurement is crucial in a clinical setting as funneling, abrupt shortening of the cervical length, or cervix length of less than 2.5 cm are associated with preterm birth [4]. As fetal scanning, examination of the cervix using ultrasound is knowledge-intensive and time-consuming and highly operator dependent. Nonetheless, the automation of cervix examination has been underexplored.

The challenges of developing a DL algorithm for the detection of cervical length can be elaborated as follows. A transvaginal image consisting of the cervix is a constellation of soft tissue echogenicity. The anterior and posterior lips of the cervix and its abutting structures that include the bladder, low uterine wall, anterior and posterior vaginal walls, and bowels, depict similar echogenicity. This low image contrast between the cervix and adjacent structures makes pattern recognition training difficult. Furthermore, feature detection of the true CC is a very challenging issue due to three-dimensional anatomical geometry and unique features of the cervix. The cervical canal, which is the key structure to detect in the midsagittal plane of the cervix for cervical length measurement, is extremely thin, having an inner width of 2–4 mm [2], is not always linear, and is irregularly filled with mucin. As a consequence, CC is rarely presented as a continuous curve having homogenous echogenicity but more usually as diverse patterns having heterogeneous echogenicity.

In real clinical settings, the process of acquisition and selection of the standard plane is even more time-consuming, cumbersome, and operator dependent. To remove operator dependency in scanning while reducing time and labor, the workflow of automated CL measurement is composed into: (i) standard plane selection from the input video frames, and (ii) CL measurement from the selected plane.

The above-mentioned CL measurement model can be expressed as a function

$$f : \mathbb{I} \mapsto Y, \tag{5.61}$$

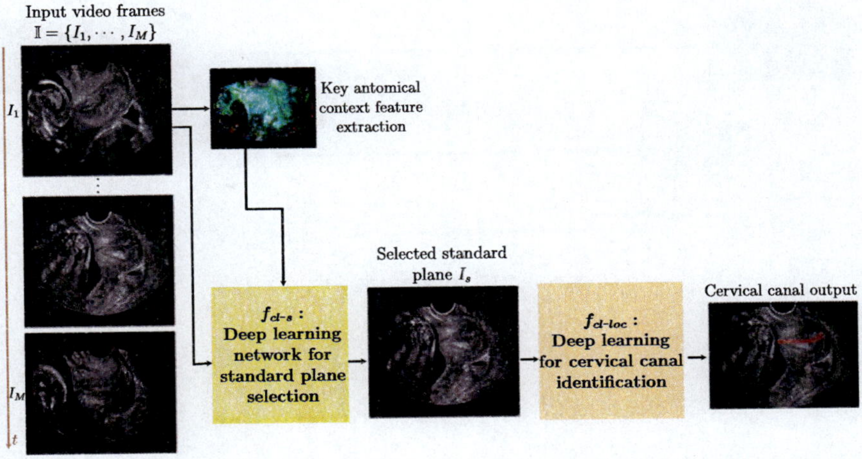

Fig. 5.24 Schematic diagram of automated CL measurement using a transvaginal US video

where $\mathbb{I} = \{I_1, \ldots, I_M\}$ is the US video of M frames and Y is a binary image of CC for CL measurement, as shown in Fig. 5.24. The function f is composed into a standard plane selection network ($f_{cl\text{-}s}$) and a CL detection network ($f_{cl\text{-}loc}$):

$$f = f_{cl\text{-}loc} \circ f_{cl\text{-}s}. \tag{5.62}$$

Denoting by I_s the selected standard plane, we need to find $f_{cl\text{-}s} : \mathbb{I} \mapsto I_s$ and $f_{cl\text{-}loc} : I_s \mapsto Y$. Finding $f_{cl\text{-}s}$ is a very challenging problem that will not be discussed in this section. Instead, we only discuss finding $f_{cl\text{-}loc}$ under the assumption that I_s is successfully selected.

5.6.1 U-Net Based Models Without the Aid of Supplementary Learning of CL-related Features

We first explore the feasibility and effectiveness of CC segmentation using U-Net [82], which is the most popular segmentation network in medical imaging. Next, we will also apply various modified U-Nets, such as Attention U-Net [87], UNet++ [105], ResUNet++ [39], and TransUNet [13].

The function $f_{cl\text{-}loc}$ is to be learned using labeled training data $\{(I_s^{(j)}, Y^{(j)}) : j = 1, \ldots, N\}$ and the loss function is

$$\mathcal{L}(\Theta) = -\frac{1}{N}\sum_{j=1}^{N} Y^{(j)} \odot_{ave} \log(f_{cl\text{-}loc}(I_s^{(j)})), \tag{5.63}$$

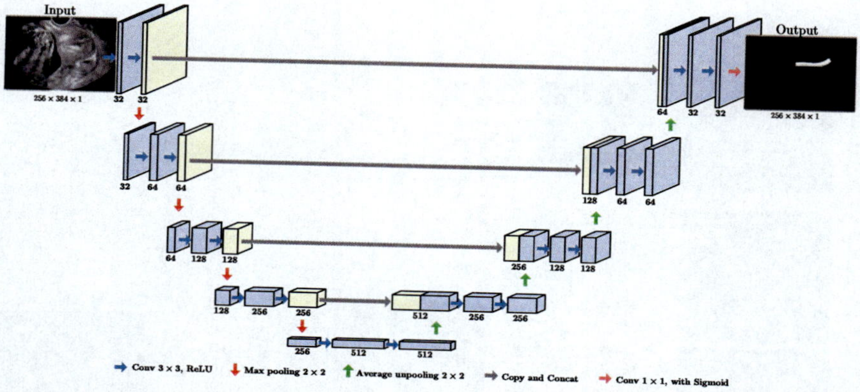

Fig. 5.25 U-Net architecture for CC segmentation

where Θ is a set of parameters of the neural network $f_{cl\text{-}loc}$, $\odot_{ave}$ is the average of element-wise multiplication, and $Y^{(j)}$ is the ground truth CC segmentation of the image $I_s^{(j)}$.

5.6.1.1 U-Net Model

U-Net, as shown in Fig. 5.25, adopts an encoder–decoder structure with skip connections. The encoding path consists of several blocks, which are the consecutive operation of two 3 × 3 convolutions followed by a ReLU activation and a 2 × 2 max-pooling layer, to extract abstract features in the input image. At the deeper stage of the encoding path, the network achieves rich context feature representations. The decoding path adopts 2 × 2 up-convolution for upsampling, followed by two successive 3 × 3 convolutions and ReLU activation. Skip connections are used to concatenate the low-level feature maps in the encoding path and high-level features in the decoding path to help restore the lost spatial information [17].

Unfortunately, experiments in Fig. 5.28 showed that the simple U-Net structure performed poorly in reliably capturing the CC. The reasons for the low performance appear to be low image contrast and heterogeneous echogenicity between the CC and abutting structures (anterior and posterior lips, bladder, low uterine wall, anterior and posterior vaginal walls, and so on). Moreover, the CC is very thin and so occupies only a tiny part of US images. This poor image environment makes it difficult for the loss function to selectively sensitize the CC, and consequently pattern recognition training for the network through gradient-based back propagation is difficult. Even an expert may find it difficult to localize the CC by looking at only patch images without viewing the entire US image. Therefore, for robust CL measurement, the loss function (or convolution filters) must be sensitive to the global anatomical structures that are closely linked to the CC.

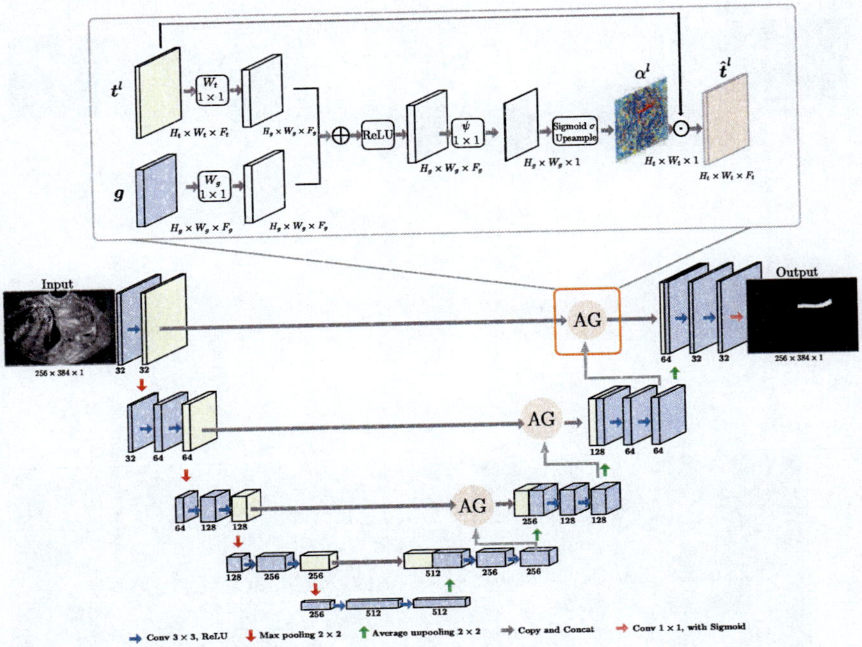

Fig. 5.26 Attention U-Net architecture

5.6.1.2 Attention U-Net

We apply the Attention U-Net to pay more attention to the CC. As shown in Fig. 5.26, the Attention U-Net is a standard U-Net structure integrated with an attention gate (AG) module [87], which is designed to increase attention to salient features passing through the skip connections while discarding irrelevance information. U-Net uses skip connections that combines the spatial information of the downsampling pathes and the upsampling pathes, which results in a lot of redundant low-level feature extraction due to poor feature representation in the initial layers. In the Attention U-Net, context features extracted from the decoding path are utilized to actively suppress activations in irrelevant regions in feature maps of the encoding path.

As shown at the top of Fig. 5.26, the AG takes in two inputs: the low-level feature maps $\boldsymbol{t}^l$ in the encoding subnetwork and the global features $\boldsymbol{g}$ at a coarser scale in the decoding path. The two feature maps $\boldsymbol{t}^l$ and $\boldsymbol{g}$ are subjected to 1×1 convolution with stride 2 and 1×1 convolution with stride 1, respectively. Then the resulting feature maps are summed element-wise and we next apply ReLU activation. Then the resulting feature maps with the size $H_g \times W_g \times F_g$ are reduced to the dimensions $H_g \times W_g \times 1$ through a 1×1 convolution. Then, we obtain an attention coefficient $\alpha^l \in [0, 1]^{H_g \times W_g}$ by applying a sigmoid. The attention coefficient α^l is upsampled to the size $H_t \times W_t$, which is multiplied element-wise by $\boldsymbol{t}^l$.

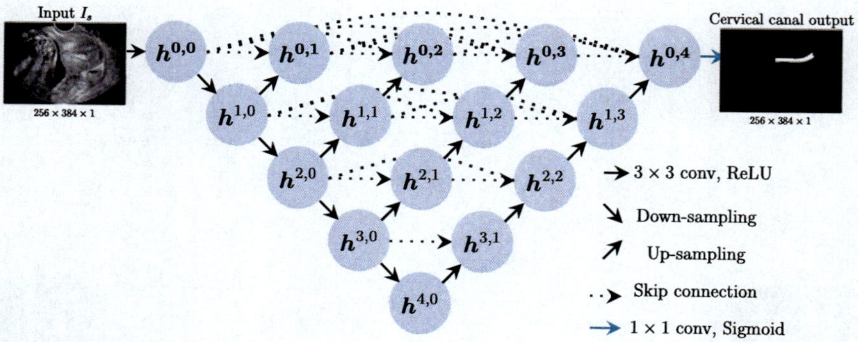

Fig. 5.27 UNet++ structure

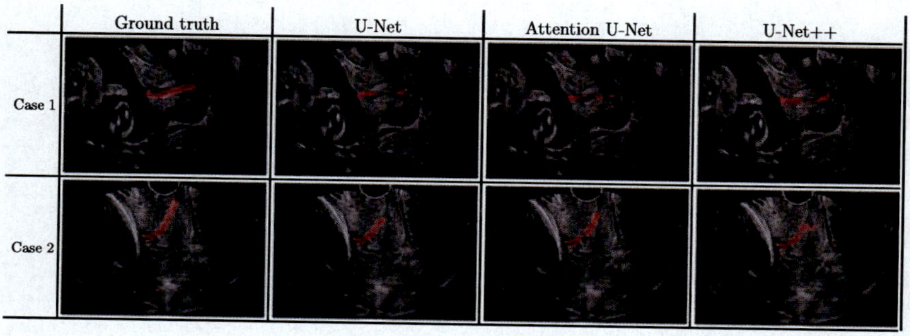

Fig. 5.28 An example of qualitative comparison of CC segmentation between U-Net, Attention U-Net, and UNet++

Unfortunately, Attention U-Net also performed poorly for the same reasons described for U-Net. See Fig. 5.28.

5.6.1.3 UNet++

UNet++ [105] was developed for more accurate segmentation in medical images. UNet++ uses redesigned skip connections, where the encoder and decoder subnetworks are connected through a series of nested, dense skip pathways. This architecture of the re-designed skip connections is intended to aggregate features of different semantic scales to provide a more flexible fusion scheme than U-Net. As depicted in Fig. 5.27, each node of the decoder combines all the multi-scale features from its previous nodes, resulting in the densely connected skip connections.

According to the experiments shown in Fig. 5.28, for reaching the level of clinical application, CL measurement using the UNet++ model also seems to be difficult.

5.6.2 *Deep Learning Models with the Aid of Supplementary Learning of CL-Related Features*

A lesson learned from failures of various U-Nets described before is that it is difficult to succeed with the loss function focused only on the CC. The loss function should include auxiliary forces of capturing the surrounding anatomical structures (e.g., the anterior and posterior lips and internal and external os) that help to accurately and rigidly capture the CC.

A strategy should be devised to force the network parameters to learn the interconnections of the CC and the surrounding anatomical structures in the process of optimizing the loss function via backpropagation. Additionally, it is desirable to exclude areas that are not relevant to achieving the goal, so that the network can focus more on the goal and learn local and global spatial relationships between pixels.

Recently, Sun et al. [50] developed CL-Net, that reflects the above-mentioned strategy. It consists of two neural networks: (1) a semantic image decomposition network and (2) a CL measurement network including CC-related anatomical structures. The first neural network divides the input image into three regions: pre-cervix, cervix, and post-cervix. The pre-cervix region includes bladder, anterior vaginal wall, and uterine cavity. The cervix region is a region of interest (ROI) that includes anterior lip, posterior lip, CC, and posterior vaginal wall. The post-cervix region includes cul-de-sac and rectum. The second network focuses its attention on the ROI and is conducted to identify the CC. The second network is designed to supplementarily extract key features (e.g., anterior and posterior lips, mucosa layer, and both ends of CC), which are the important factors to help identify the thin CC.

The network $f_{cl\text{-}loc}$ is composed of two steps:

$$f_{cl\text{-}loc} = f_{cl\text{-}loc_2} \circ f_{cl\text{-}loc_1}. \tag{5.64}$$

All input images are resized as 256 × 384 pixels and image intensities are normalized to zero mean and unit variance. The $f_{cl\text{-}loc_1}$ is the semantic image decomposition network that is the map:

$$f_{cl\text{-}loc_1} : I_s \mapsto \mathbf{S}, \tag{5.65}$$

where I_s is a transvaginal US image, $\mathbf{S} = \{\mathcal{S}_1, \mathcal{S}_2, \mathcal{S}_3\}$ is the set of the three binary images of pre-cervix, cervix, and post-cervix regions. The $f_{cl\text{-}loc_2}$ is a CL measurement network that is the map:

$$f_{cl\text{-}loc_2} : (I_s, \mathcal{S}_2) \mapsto Y. \tag{5.66}$$

5.6.2.1 Semantic Image Decomposition Network

For study purposes, let us try the Fully Convolutional dense Dilated Net (FCdDN) [72] for the semantic image decomposition into pre-cervix, cervix, and post-cervix regions.

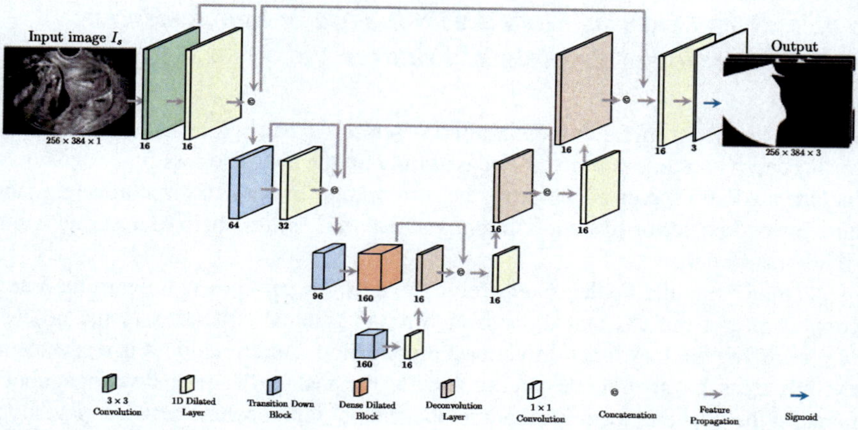

Fig. 5.29 Architecture of FCdDN

Fully Convolutional dense Dilated Net (FCdDN) The FCdDN was proposed for real-time medical image segmentation recently. The architecture, as shown in Fig. 5.29, is similar to the U-Net structure. A novel layer called the "1D dilated layer" [72] was developed as the basic layer. In addition, it also consists of various layers, including the convolution layer, transition down block [38], dense dilated layer [38], and deconvolution layer.

Convolution layer: At the first layer in Fig. 5.30, a set of feature maps (denoted as $\mathbf{h}^1$) is computed by convolving fiters (denoted as $\boldsymbol{W}^1$) across the input. The feature map h^1 is given by

$$\mathbf{h}^1 = I_s \circledast_{3\times3} \boldsymbol{W}^1, \tag{5.67}$$

where $\circledast_{3\times3}$ denotes a 3×3 standard convolution with stride 1. Figure 5.30b shows the filters $\boldsymbol{W}^1 = \{\boldsymbol{w}_i^1\}_{i=1}^{48}$ and each $\boldsymbol{w}_i^1$ is represented as

$$\boldsymbol{w}_i^1 = \begin{bmatrix} w_{i,(1,1)}^1 & \cdots & w_{i,(1,3)}^1 \\ \vdots & \ddots & \vdots \\ w_{i,(3,1)}^1 & \cdots & w_{i,(3,3)}^1 \end{bmatrix}. \tag{5.68}$$

The generated feature map $\mathbf{h}^1$, as shown in Fig. 5.30c, is a channel-wise concatenation of the convolved feature $I_s \circledast_{3\times3} \boldsymbol{w}_i^1$, for $i = 1, \ldots, 48$.

1D dilated layer: The second layer takes advantages of dense connectivity [30], dilated convolutions [98], and factorized filters [40] to improve the segmentation efficiency while retaining high accuracy, as illustrated in Fig. 5.31. Dense connectivity in DenseNet [30] is used to connect each layer to its previous layers directly. To be precise, the i-th layer uses all feature maps in its front layers to generate the features $\mathbf{h}^i$, which is given by:

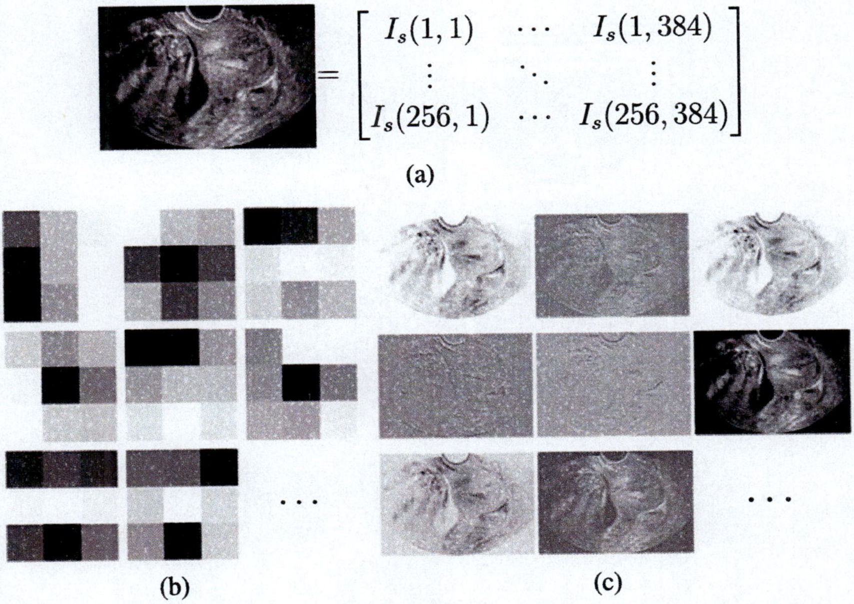

Fig. 5.30 Convolution layer: **a** Input image I_s, **b** convolution filters W^1, and **c** the generated feature map $\mathbf{h}^1$

$$\mathbf{h}^i = H_i([\mathbf{h}^0, \mathbf{h}^1, \ldots, \mathbf{h}^{i-1}]), \tag{5.69}$$

where $H_i(\cdot)$ is a consecutive operation of batch normalization (BN) [32], RuLU, convolution, and dropout rate 0.2 [89]. Here, $[\cdot, \cdot]$ denotes the concatenation. This dense connectivity encourages feature reuse, strengthens feature propagation, and is parameter efficient that may reduces over-fitting on tasks with small training set [30].

For a real-time network, this 1D dilated layer uses factorized filters to reduce parameters: 2D 3×3 convolution filters are decomposed into two 1D convolution filters of 3×1 and 1×3. To achieve a larger receptive field without losing the spatial information of the image, atrous convolution is adopted, which allows aggregation of more global contextual information.

In Fig. 5.31, $\mathbf{h}^{1,1} = ReLU(BN(\mathbf{h}^1)) = \max\{BN(\mathbf{h}^1), 0\}$. As illustrated in Fig. 5.32, two atrous convolutions using 1D factorized filters W^2 and W^3 are computed as

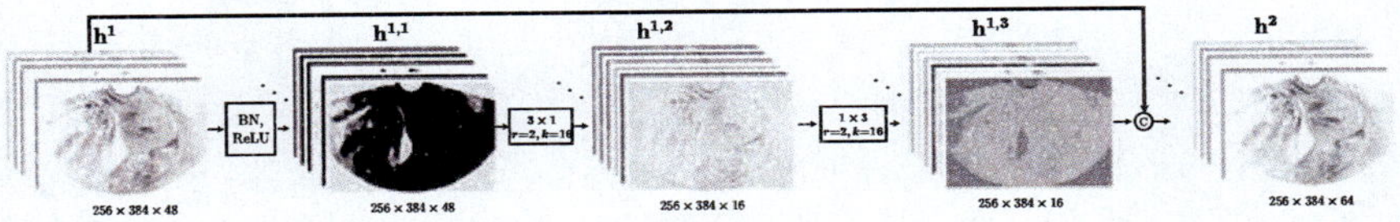

Fig. 5.31 Architecture of the 1D dilated layer, where r is the dilation rate of atrous convolution and k is the channel number of output

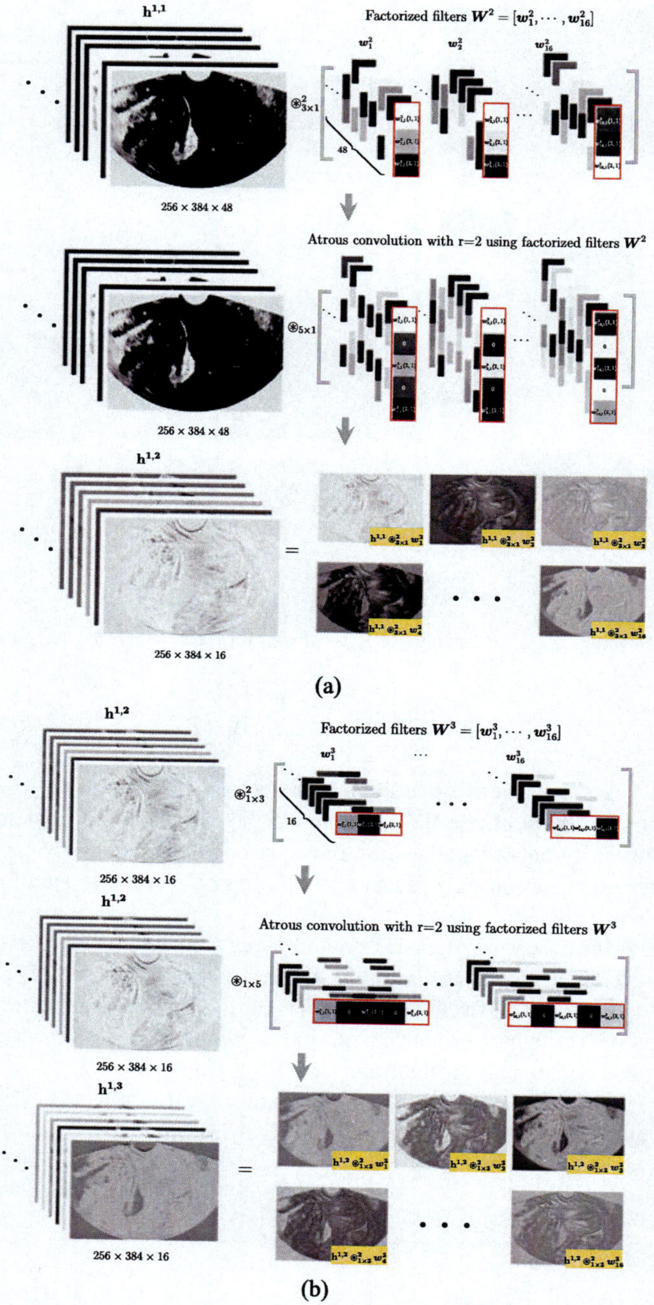

Fig. 5.32 **a** Computing feature maps $\mathbf{h}^{1,2}$ by convolving a 3 × 1 factorized filter set W^2 across $\mathbf{h}^{1,1}$ using atrous convolution of rate 2. **b** Computing feature maps $\mathbf{h}^{1,3}$ by convolving a 1 × 3 factorized filter set W^3 across $\mathbf{h}^{1,2}$ using atrous convolution of rate 2

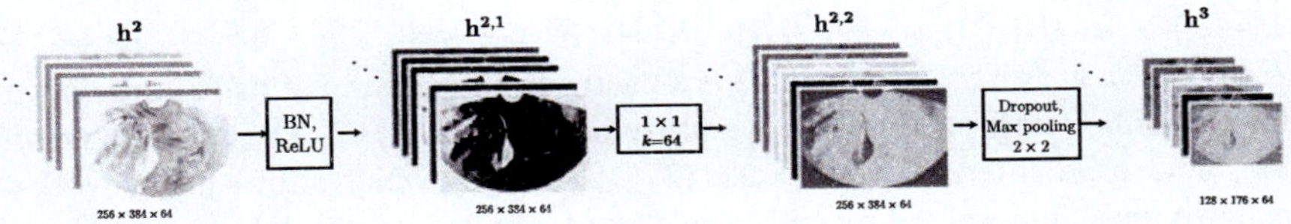

Fig. 5.33 Architecture of the transition down block

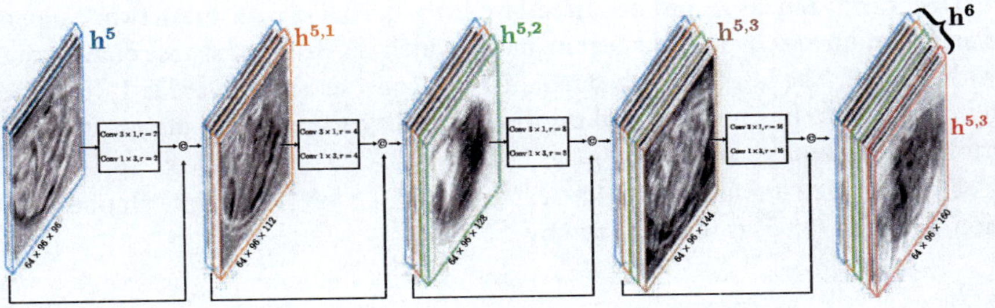

Fig. 5.34 Diagram of dense dilated block

$$\mathbf{h}^{1,2} = \mathbf{h}^{1,1} \circledast^2_{3\times1} \boldsymbol{W}^2 = [\mathbf{h}^{1,1} \circledast^2_{3\times1} \boldsymbol{w}^2_1, \ldots, \mathbf{h}^{1,1} \circledast^2_{3\times1} \boldsymbol{w}^2_{16}], \tag{5.70}$$

$$\mathbf{h}^{1,3} = \mathbf{h}^{1,2} \circledast^2_{1\times3} \boldsymbol{W}^3 = [\mathbf{h}^{1,2} \circledast^2_{1\times3} \boldsymbol{w}^3_1, \ldots, \mathbf{h}^{1,2} \circledast^2_{1\times3} \boldsymbol{w}^3_{16}], \tag{5.71}$$

where $\circledast^2_{3\times1}$ and $\circledast^2_{1\times3}$ denote atrous convolutions of rate 2 with filter size 3×1 and 1×3, respectively. See Fig. 5.32 for images of $\mathbf{h}^{1,2}$ and $\mathbf{h}^{1,3}$. Afterwards, dropout is used. Finally, the output of this layer, denoted as $\mathbf{h}^2$, is given by the concatenation of feature maps after dropout and $\mathbf{h}^1$, as shown in Fig. 5.31.

Transition down block: At the third layer, a transition down block is applied to down-sample feature maps. It consists of BN, ReLU, 1×1 convolution, dropout, and 2×2 max pooling. See Fig. 5.33. In this down-sampling block, BN and ReLU are first performed to obtain feature maps $\mathbf{h}^{2,1}$, where $\mathbf{h}^{2,1} = ReLU(BN(\mathbf{h}^2))$. Then a 1×1 convolution (with parameters $\boldsymbol{W}^4$) and dropout are used to extract feature maps $\mathbf{h}^{2,2}$. Finally, a 2×2 max pooling down-samples $\mathbf{h}^{2,2}$ to $\mathbf{h}^3$:

$$\mathbf{h}^3 = [\mathbf{h}^3_1, \ldots, \mathbf{h}^3_{64}] = \text{max-pooling}([\mathbf{h}^{2,2}_1, \ldots, \mathbf{h}^{2,2}_{64}]). \tag{5.72}$$

Dense dilated block: The fourth layer and fifth layer are a 1D dilated layer and transition down block, respectively. Motivated by the dense block in FC-DenseNet [38], a dense dilated block, denoted as $\mathscr{D}(\cdot)$, was developed and used as the sixth layer of FCdDN. This block is also a combination of dense connectivity, dilated convolution, and factorized filters. See Fig. 5.34.

The output of this layer $\mathbf{h}^6$ is a dense connectivity of all the preceding feature maps in this block, which is given by

$$\mathbf{h}^6 = \mathscr{D}(\mathbf{h}^5) = [\mathbf{h}^5, \mathbf{h}^{5,1}, \mathbf{h}^{5,2}, \mathbf{h}^{5,3}, \mathbf{h}^{5,4}], \tag{5.73}$$

where $\mathbf{h}^{5,1} = H([\mathbf{h}^5])$, $\mathbf{h}^{5,2} = H([\mathbf{h}^5, \mathbf{h}^{5,1}])$, $\mathbf{h}^{5,3} = H([\mathbf{h}^5, \mathbf{h}^{5,1}, \mathbf{h}^{5,2}])$, and $\mathbf{h}^{5,4} = H([\mathbf{h}^5, \mathbf{h}^{5,1}, \mathbf{h}^{5,2}, \mathbf{h}^{5,3}])$. Here, $H(\cdot)$ is a consecutive operation of two atrous convolutions with factorized filter 3×1 and factorized filter 1×3. The convolutions in this block have different dilation rates (2, 4, 8, and 16), which allows the network to achieve multiple levels of receptive field sizes and thus multi-scale feature fusion. Such fusion helps deal with the large variations in sizes and locations of pre-cervix, cervix, and post-cervix.

Deconvolution layer and loss function: In the decoding path, the network adopts deconvolution to upsample the feature maps, which is a 3×3 transpose convolution with stride 2. The last layer is a standard 1×1 convolution with stride 1, followed by the softmax function, to produce the probability maps of the uterus, cervical, irrelevant remaining, and background regions.

Using labeled training data $\{(I_s^{(j)}, S^{(j)}) : j = 1, \ldots, N\}$, the segmentation function $f_{cl\text{-}loc_1}$ in (5.65) is to be learned by

$$f_{cl\text{-}loc_1} = \underset{f_{cl\text{-}loc_1} \in \mathbb{NN}}{\operatorname{argmin}} \; -\frac{1}{N}\sum_{j=1}^{N}\sum_{i=1}^{3} \mathscr{S}_i^{(j)} \odot_{ave} \log(f_{1,i}(I_s^{(j)})) + \frac{1}{N}\sum_{j=1}^{N}\left(1 - \frac{1}{3}\sum_{i=1}^{3} \frac{2\tau \mathscr{S}_i^{(j)} \odot_{ave} f_{1,i}(I_s^{(j)})}{\|\mathscr{S}_i^{(j)}\|^2 + \|f_{1,i}^2(I_s^{(j)})\|^2 + \epsilon}\right), \tag{5.74}$$

where $\odot_{ave}$ is the average of element-wise multiplication and τ is the number of pixels in the domain of input image I_s, $f_{1,i}$ is an i-th class output of $f_{cl\text{-}loc_1}$, and ϵ is set 10^{-7} to ensure the loss stability.

Cycle-Consistent Adversarial Networks (CycleGAN) For study purposes, let's try the semantic decomposition using CycleGAN [106], which is proposed for unpaired image-to-image translation. For ease of explanation, we denote $\boldsymbol{X}$ as a set of transvaginal US images $\{\boldsymbol{x}^{(j)}\}_{j=1}^N$, where $\boldsymbol{x}$ temporarily represents US image I_s. The set of semantic labels is denoted as $\boldsymbol{Y}$, where $\boldsymbol{Y} = \{\boldsymbol{y}^{(j)}\}_{j=1}^N$ and $\boldsymbol{y}$ represents S in (5.65).

This approach uses adversarial training to learn an image-to-semantic label translation $f_{cl\text{-}loc_1} : \boldsymbol{X} \mapsto \boldsymbol{Y}$ such that the output $\hat{\boldsymbol{y}} = f_{cl\text{-}loc_1}(\boldsymbol{x})$ approximates to the corresponding label $\boldsymbol{y}$. For convenience, we denote $f_{cl\text{-}loc_1}$ by G from now on. A cycle-consistent architecture [106] having two mapping functions $G : \boldsymbol{X} \mapsto \boldsymbol{Y}$ and $F : \boldsymbol{Y} \mapsto \boldsymbol{X}$ is used to guarantee that the learned function G could map an individual $\boldsymbol{x}$ to a desired output $\boldsymbol{y}$. This network trains G, F, the adversarial discriminator D_X for $\boldsymbol{X}$, and the adversarial discriminator D_Y for $\boldsymbol{Y}$ simultaneously, as illustrated in Fig. 5.35. Here, the discriminator D_Y aims to differentiate between real labels $\boldsymbol{y}$ and generated labels $G(\boldsymbol{x})$, and likewise, the discriminator D_X aims to differentiate between real US images $\boldsymbol{x}$ and generated images $F(\boldsymbol{y})$. The full objective is composed of two adversarial losses and two cycle consistency losses.

Loss function: The loss function for G and D_Y is expressed as

$$\mathfrak{L}_{\text{GAN}}(G, D_Y, \boldsymbol{X}, \boldsymbol{Y}) = \mathbb{E}_{\boldsymbol{y} \sim p_{\text{data}}(\boldsymbol{y})}[\log D_Y(\boldsymbol{y})] + \mathbb{E}_{\boldsymbol{x} \sim p_{data}(\boldsymbol{x})}[\log(1 - D_Y(G(\boldsymbol{x})))]. \tag{5.75}$$

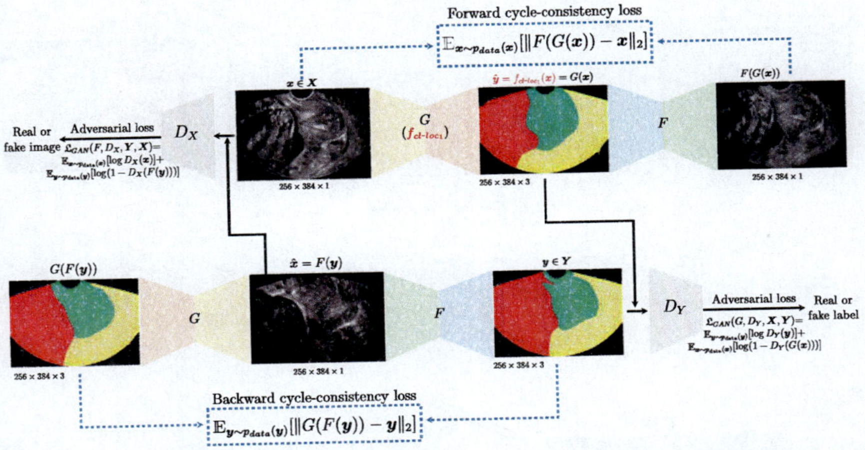

Fig. 5.35 CycleGAN architecture. We aim to achieve desirable $G(\boldsymbol{x})$ from cycleGAN for segmenting the pre-cervix (red), cervix (green), post-cervix (yellow) regions from an US image $\boldsymbol{x}$

The generator G is roughly related to

$$G \approx \min_{G} \max_{D_Y} \mathfrak{L}_{\text{GAN}}(G, D_Y, \boldsymbol{X}, \boldsymbol{Y}). \tag{5.76}$$

Similarly, the backward generator F is roughly related to

$$F \approx \min_{F} \max_{D_X} \mathfrak{L}_{\text{GAN}}(F, D_X, \boldsymbol{Y}, \boldsymbol{X}), \tag{5.77}$$

where

$$\mathfrak{L}_{\text{GAN}}(F, D_X, \boldsymbol{Y}, \boldsymbol{X}) = \mathbb{E}_{x\sim p_{\text{data}}(\boldsymbol{x})}[\log D_X(\boldsymbol{x})] + \mathbb{E}_{\boldsymbol{y}\sim p_{data}(\boldsymbol{y})}[\log(1 - D_X(F(\boldsymbol{y})))]. \tag{5.78}$$

To regularize the mapping functions, two cycle consistency losses are adopted to encourage $F(G(\boldsymbol{x})) \approx \boldsymbol{x}$ and $G(F(\boldsymbol{y})) \approx \boldsymbol{y}$. These two cycle consistency are called (i) forward cycle-consistency: $\boldsymbol{x} \mapsto G(\boldsymbol{x}) \mapsto F(G(\boldsymbol{x})) \approx \boldsymbol{x}$, and (ii) backward cycle-consistency: $\boldsymbol{y} \mapsto F(\boldsymbol{y}) \mapsto G(F(\boldsymbol{y})) \approx \boldsymbol{y}$. The cycle consistencies loss is given by

$$\mathfrak{L}_{cyc}(G, F) = \mathbb{E}_{\boldsymbol{x}\sim p_{data}(\boldsymbol{x})}[\|F(G(\boldsymbol{x})) - \boldsymbol{x}\|_2] + E_{\boldsymbol{y}\sim p_{data}(\boldsymbol{y})}[\|G(F(\boldsymbol{y})) - \boldsymbol{y}\|_2]. \tag{5.79}$$

The total loss is given by

$$\mathfrak{L}_{cycleGAN}(G, F, D_X, D_Y) = \mathfrak{L}_{\text{GAN}}(G, D_Y, \boldsymbol{X}, \boldsymbol{Y}) + \mathfrak{L}_{\text{GAN}}(F, D_X, \boldsymbol{Y}, \boldsymbol{X}) + \lambda\mathfrak{L}_{cyc}(G, F), \tag{5.80}$$

where λ is a hyperparameter indicating the importance of cycle consistency loss. Then, the generator G and the backward generator F are obtained by

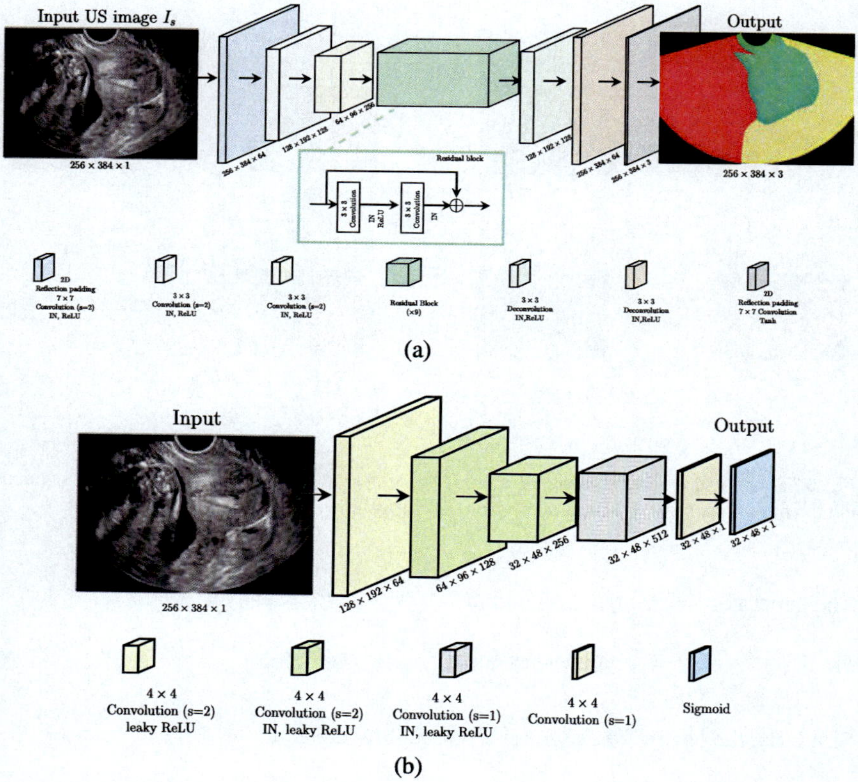

Fig. 5.36 Generator (**a**) and discriminator (**b**) architectures of cycleGAN. Here, "IN" stands for instance normalization [41] and "s" stands for stride

$$(G^*, F^*) = \arg\min_{G,F} \max_{D_Y, D_X} \mathfrak{L}_{cycleGAN}(G, F, D_X, D_Y). \tag{5.81}$$

For our generator networks G and F in Fig. 5.36, we adopt the architecture in [41]. For discriminator networks, the PatchGAN [33] is used. These two networks are shown in Fig. 5.36. For more details, refer to [33, 41, 106].

Mask R-CNN Next, we explore the semantic decomposition $f_{cl\text{-}loc_1}$ using Mask R-CNN [27]. Mask R-CNN, as shown in Fig. 5.37, is a framework proposed for instance segmentation, which detects objects in the input image while producing the semantic segmentation masks simultaneously.

The Mask R-CNN network for $f_{cl\text{-}loc_1}$ involves two stages: The first stage is to learn a function $f_{1,1} : I_s \mapsto \{(p_m, rx_m, ry_m, rw_m, rh_m)\}_{m=1}^{N^p}$, where I_s is an input image, p_m represents the probability of the m-th generated proposals being an object, and (rx_m, ry_m, rw_m, rh_m) denotes the upper left coordinates, width, and height of the m-th predicted bounding box, for $m = 1, \ldots, N^p$. The second stage is to learn a function $f_{1,2} : (I_s, \{(p_m, rx_m, ry_m, rw_m, rh_m)\}_{m=1}^{N^p}) \mapsto (\boldsymbol{c}, \boldsymbol{u}, \boldsymbol{S})$ to output the predicted class

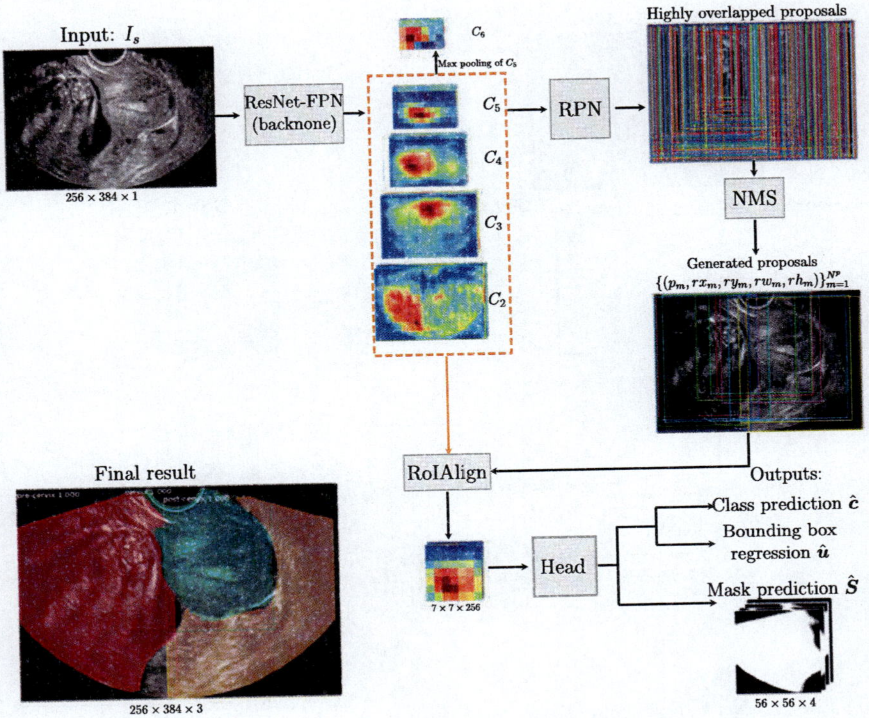

Fig. 5.37 Mask R-CNN architecture consisting of two stages. The first stage generates candidate proposals using RPN, and the second stage predicts the class, bounding box parameters, and a binary mask for each RoI using RoIAlign and Head network. ResNet-FPN is the backbone sharing features for both stages

label $\boldsymbol{c}$, bounding box coordinates $\boldsymbol{u}$, and a binary mask $\boldsymbol{S}$ for each RoI. We use ResNet-FPN [27] to share features between these stages, instead of learning two stages separately.

First stage—Backbone: The backbone of $f_{cl\text{-}loc_1}$ is ResNet-FPN [27], which is inspired by ResNet [28] and Feature Pyramid Network (FPN) [55]. ResNet is a substantially deep neural network that achieves good performances in image detection, localization, and segmentation. FPN is a top-down structure that outputs a feature pyramid from a single-scale input image.

As shown in Fig. 5.38, it outputs a feature pyramid, denoted as $[C_2, C_3, C_4, C_5]$, that will be used to extract multiple levels of RoI features according to their scales.

First stage—Region proposal network: Next, a region proposal network (RPN) uses features $[C_2, C_3, C_4, C_5]$ to generate region proposals with multiple scales and aspect ratios, where the proposals with high objectness score are passed to the second stage for detection and segmentation. The generated proposals tell the next stage where to look.

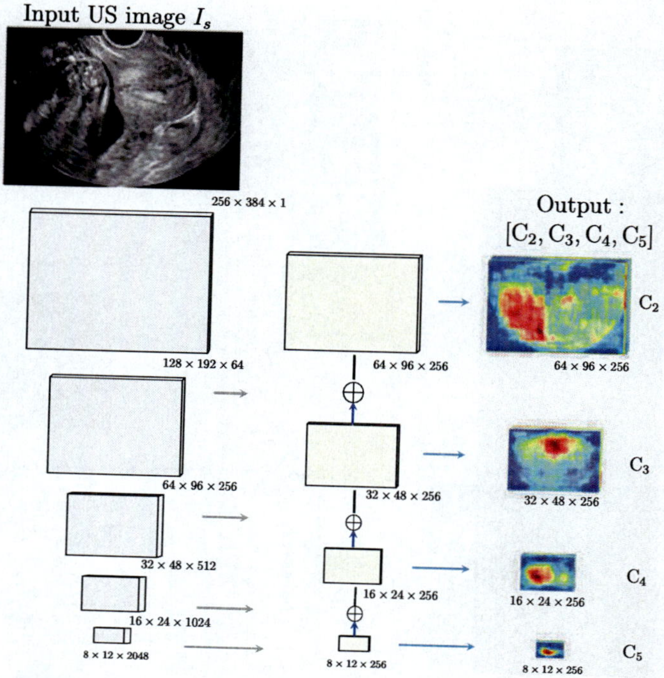

Fig. 5.38 ResNet-FPN structure. From up to bottom, the gray boxes denote the extracted features of the first to fifth blocks of ResNet. The gray, cyan, and blue arrows indicate 1×1 convolution with stride 1, 3×3 convolution with stride 1, and upsampling, respectively

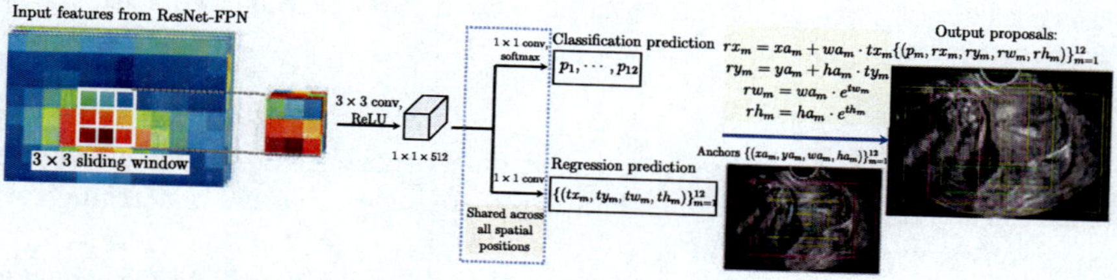

Fig. 5.39 RPN generates 12 proposals at each sliding window. If the input is C_5 that has the size $8 \times 12 \times 256$, then the output is $8 \times 12 \times 12$ proposals

As shown in Fig. 5.39, a 3×3 spatial sliding window is slid over the features $[C_2, C_3, C_4, C_5]$ and simultaneously predicts objectness scores $\{p_m\}_{m=1}^{12}$ and parameterized coordinates $\{t_m\}_{m=1}^{12}$, where m is index of reference boxes named anchors at each pixel position and $t_m = (tx_m, ty_m, tw_m, th_m) \in \mathbb{R}^4$ provides the information of the m-th anchor about the center coordinates, width, and height. Here, the number of anchors is 12 because we use four scales (32, 64, 128, 256) and three aspect ratios $(1:1, 1:2, 2:1)$.

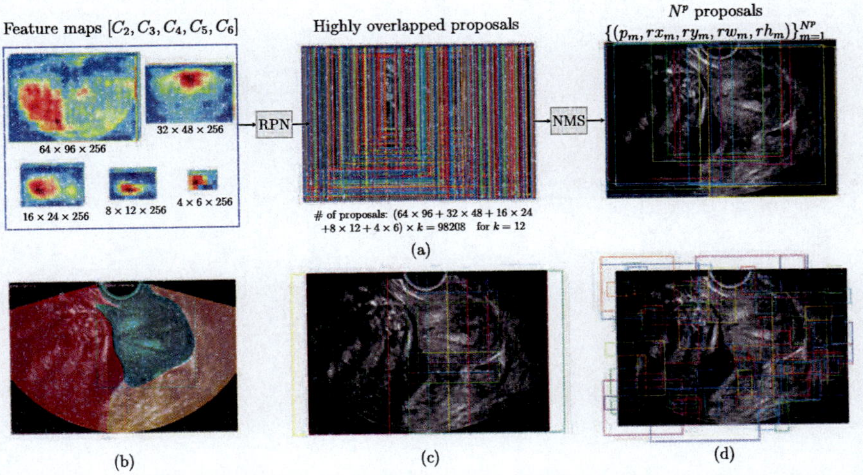

Fig. 5.40 Non-maximum suppression (NMS): **a** Redundancy reduction of proposals, **b** ground truths of pre-cervix, cervix, and post-cervix, **c** positive anchors, and **d** negative anchors

The RPN generates hundreds of thousands of proposals. However, most of them are highly overlapping, as shown in Fig. 5.40a. To reduce the redundancy, non-maximum suppression (NMS) [68] with an Intersection-over-Union (IoU) threshold of 0.7 is applied to the proposals based on their classification scores. Afterwards, the top-N^p ranked proposals are used for the second stage.

The corresponding loss function is

$$\begin{aligned} \mathfrak{L}_1 &= \frac{1}{N}\frac{1}{N_j^a}\sum_{j=1}^{N}\sum_{m=1}^{N_j^a}\mathfrak{L}_1(I_s^{(n)}, (p_m^{(j)}, t_m^{(j)})) \\ &= \frac{1}{N}\frac{1}{N_j^a}\sum_{j=1}^{N}\sum_{m=1}^{N_j^a}[\mathfrak{L}_{cross}(\hat{p}_m^{(j)}, p_m^{(j)}) + p_m^{(j)}\mathfrak{L}_r(\hat{t}_m^{(j)}, t_m^{(j)})], \end{aligned} \tag{5.82}$$

where $\{(I_s^{(j)}, (p_m^{(j)}, t_m^{(j)})) : j = 1, \ldots, N; m = 1, \ldots, N_j^a\}$ is a labeled training data. Here, the ground truth label $p_m^{(j)}$ is 1 if the m-th anchor of the j-th input is positive, and 0 if negative. We define positive anchors as those that have an IoU ≥ 0.7 with any ground truth box, and negative anchors as those that do not cover any object by more than 0.3 IoU. See examples in Fig. 5.40c and d. The $\mathfrak{L}_{cross}$ is a classification loss given by $\mathfrak{L}_{cross}(\hat{p}_m^{(j)}, p_m^{(j)}) = -p_m^{(j)}\log(\hat{p}_m^{(j)})$. The $\mathfrak{L}_r$ is the regression loss defined in [24].

The second stage—RoI extraction and RoIAlign: To deal with the unpredictable size of targets, multi-scale RoI features are extracted from the feature pyramid of the backbone according to their scales, as shown in Fig. 5.41. The level of the feature pyramid to extract RoI, denoted as k, can be computed as

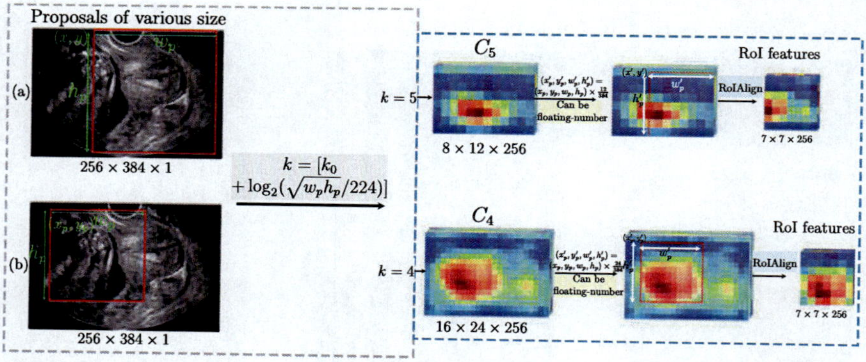

Fig. 5.41 RoI extraction and RoIAlign process

$$k = [k_0 + \log_2(\sqrt{w_p h_p}/224)], \tag{5.83}$$

where k_0 is set 4, $[\cdot]$ denotes a ceiling operation, h_p and w_p are the height and width of a proposal. It can be seen that feature maps of high resolution which preserves rich spatial infromation can be used to segment small RoI, while more semantic features are used to segment big RoI. This allows accurate localization and segmentation.

The extracted RoI can be floating-number, as shown in Fig. 5.41. To avoid the misalignment between the RoI and extracted features, a layer called RoIAlign [27] is proposed to maintain exact spatial location. This pixel-to-pixel alignment ensures accurate segmentation of the binary mask that requires fine spatial features. To be precise, the extracted floating-number RoI is first divided into bins of fixed size (e.g., 7×7), then bilinear interpolation is performed to calculate the values of four sampled locations in each bin, and finally max pooling is applied to achieve a small feature map. For further details, refer to [27].

The second stage—Head network: We apply a network Head [27], as shown in Fig. 5.42, to each small RoI feature map, in order to predict the class; we regress the bounding box, and segment the mask in RoIs.

Performance Comparison Fig. 5.43 shows performance comparison of four different methods for the decomposition $f_{cl\text{-}loc_1}$. The three models of FCdDN, cycle-GAN, and Mask R-CNN show poor performance in the accuracy and robustness of segmentation. On the other hand, U-Net produces reliable segmentation. Therefore, we adopt U-Net for $f_{cl\text{-}loc_1}$.

5.6.2.2 CL-Related Feature Extraction

This section is based on the paper [50]. Recall that the $f_{cl\text{-}loc_2}$ is a map:

$$f_{cl\text{-}loc_2} : \underbrace{I_s \odot \mathscr{S}_2}_{I_c} \mapsto Y, \tag{5.84}$$

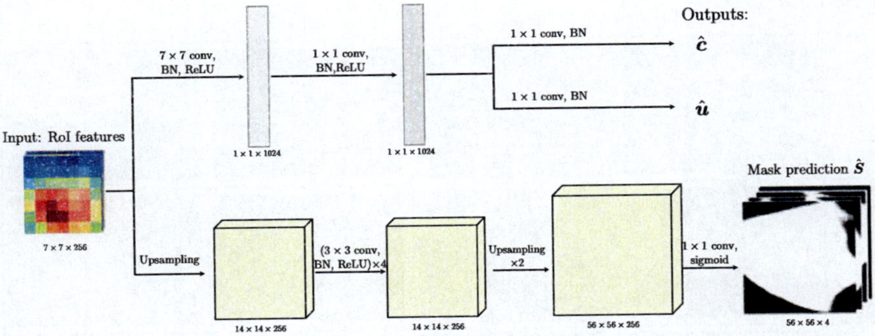

Fig. 5.42 Head structure

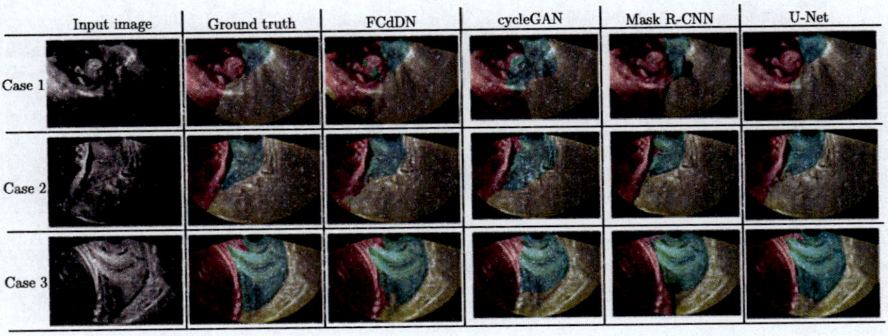

Fig. 5.43 Performance comparison of four different methods for the decomposition $f_{cl\text{-}loc_1}$ on test data

where Y is the binary image of CC and I_c is the image. It focuses on the cervix region to identify the CC, anterior and posterior lips, and posterior vaginal wall, taking into account both local and global information in the ROI that mimics the clinician's cervix finding process. The key is to predict the tracing of CC, where the output is expressed by a confidence map (denoted as C) representing the beliefs about the canal region at each pixel location in the image. Let $\boldsymbol{l}$ denote the pixel area occupying the CC. The confidence map is defined as a normalized distance map [96] of the CC $\boldsymbol{l}$:

$$C(\mathbf{x}) = e^{-\lambda \text{dist}(\mathbf{x}, \boldsymbol{l})} \quad \text{for each pixel position } \mathbf{x}, \tag{5.85}$$

where λ is a hyperparameter for normalization.

To obtain an accurate identification of the CC including internal os and external os, we auxiliary segment the cervix region into three anatomical structures: anterior lip $\mathscr{T}_1$, posterior lip $\mathscr{T}_2$, and posterior vaginal wall $\mathscr{T}_3$. We divide $f_{cl\text{-}loc_2}$ into two parts:

$$f_{cl\text{-}loc_2} = f^{post}_{cl\text{-}loc_2} \circ f^{main}_{cl\text{-}loc_2} : I_c \mapsto (C, \boldsymbol{T}) \mapsto Y, \tag{5.86}$$

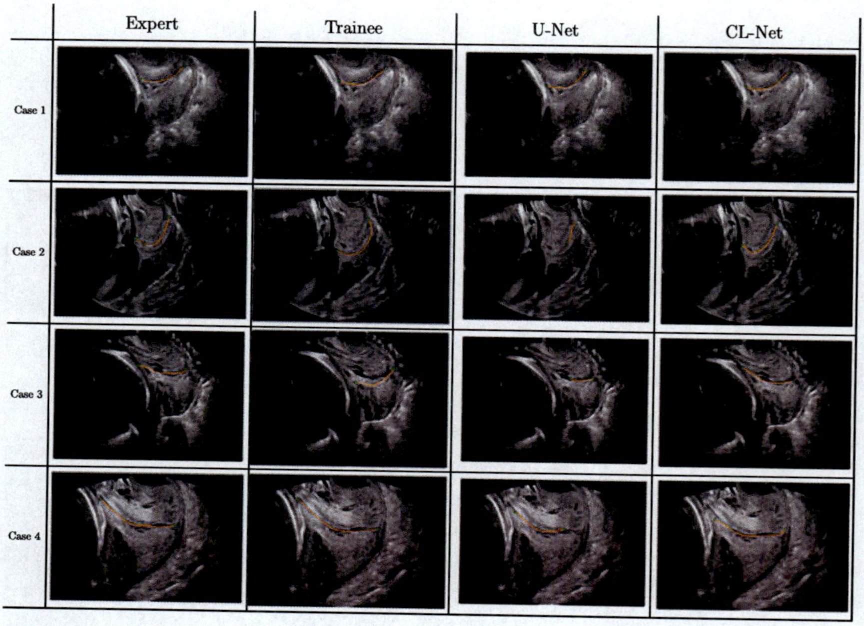

Fig. 5.44 Qualitative comparison among expert, trainee, U-Net w/o SL, and CL-Net: Identification of CC, internal os, and external os. "SL" stands for the supplementary learning of CL-related features

where f^{post} is the post-processing and $\boldsymbol{T} = (\mathcal{T}_1, \mathcal{T}_2, \mathcal{T}_3)$. The training of $f^{main}_{cl\text{-}loc_2}$ aims to segment the anatomical structures of the cervix region and find the CC region simultaneously with high confidence. We take a multi-task loss, which considers both segmentation loss for three structures ($\mathcal{T}_1$,$\mathcal{T}_2$ and $\mathcal{T}_3$) and mean squared error (MSE) of the confidence map.

The supplementary segmentation $\boldsymbol{T}$ is designed to help the network identify the CC by learning not only the surrounding anatomical features but also their spatial relationships with the CC. Therefore, to extract CL-related functions, we adopt a U-Net architecture that outputs two different images at the same time. The input is the ROI image I_c that is resized into 256×256 pixels. The confidence map learning and complementary segmentation tasks share the same learning parameters, except for those at the end of the network. This strategy ensures a robust CC identification by adequately learning the anatomical features.

The results in Fig. 5.44 show that CL-Net outperforms the single-output U-Net without the supplementary learning of CL-related features. The superiority of CL-Net is due to the auxiliary force of capturing CL-related anatomical structures, and this assisting force helps to find the CC stably. Our experiments show that, without using the auxiliary force, various U-Nets including U-Net, attention U-Net, and UNet++ have difficulty in selectively sensitising the thread-like thin CC, due to its low image contrast and heterogeneous echogenicity.

5.7 Discussion

AI algorithm-based obstetric ultrasound automation is being used as a turning point for next-generation ultrasound diagnostic devices by ultrasound companies (GE, Phillips, Siemens, Samsung Medison, etc.). Recently, automation of trans-abdominal ultrasound has achieved considerable success, whereas automation of transvaginal ultrasound is rarely studied. Cervical length (CL) measurement through vaginal ultrasonography is essential for the prevention and diagnosis of premature birth. Since CL measurement in mothers is performed by the vaginal test, the examination process is very inconvenient and takes a long time. In addition, the reliability of the examination is greatly affected by the knowledge and skill of the ultrasound operator (accurate cervical cross-sectional image acquisition and caliper placement), so the demand for automation technology in the clinic is very high. Through the automated technology of CL measurement using this vaginal ultrasound, the examination time is greatly shortened, the discomfort of the mother is resolved, and the diagnosis error is prevented, thereby promoting healthy pregnancy and childbirth.

The current success of deep learning is attributed to its superior ability to extract local and global interconnections by capturing spatial relationships between pixels in training data. However, while convolutional networks are effective at unifying spatial contexts, their ability to distinguish subtle differences between two very similar images is weak. These shortcomings are a major obstacle to automatic search of standard planes, which is essential for fetal ultrasound. Recently, there have been numerous studies on the automatic standard plane detection in abdominal US focusing on various groups, but they have not been successful in reaching the level of clinical use. According to our experience of this research, existing methods (e.g., CNN with Attention Gate, RNN, U-NET, YOLO, Faster R-CNN) have limitations in learning minute differences between US image frames. In addition, transvaginal US, which is in desperate need of automation, has hardly been studied compared to abdominal US. In transvaginal US, it is difficult to identify the cervix even if the image is enlarged. There have been several attempts to automate CL measurement using deep learning, but it seems difficult to achieve good performance without the aid of adjacent anatomical structures.

Fetal US images contain a wide variety of artifacts and noise, making them much more difficult to handle than other imaging modalities (e.g., CT and MRI). US image artifacts and noise are intricately related to the maternal, fetal position, scan direction, and transducer position. Therefore, the expertise of experts (anatomical characteristics, decision-making process) and understanding of characteristics of US imaging should be carefully reflected in machine learning. When using deep learning, it is important to note that if the loss function is limited to the thin cervical canal (surrounding structures and signal contrast are unclear and echo variability is very diverse). Therefore, rather than focusing excessively on the cervical canal, a strategy to comprehensively judge within the entire surrounding structure of the cervical canal is needed.

The hardest part is selection of a standard plane from a US video frame. A lesson from recent failures is that deep learning architecture for automating standard plane selection should include anatomical context information on minute anatomical changes between video frames. Deep learning should be designed to reflect the doctor's ability to make decisions in the context between frames of US video based on local/global knowledge of anatomical structures.

Acknowledgements This research was supported by Samsung Science & Technology Foundation (No. SRFC-IT1902-09). Cho and Seo were supported by a grant of the Korea Health Technology R&D Project through the Korea Health Industry Development Institute (KHIDI), funded by the Ministry of Health & Welfare, Republic of Korea (grant number : HI20C0127).

References

1. Alexe, B., Deselaers, T., Ferrari, V.: Measuring the objectness of image windows. IEEE Trans. Pattern Anal. Mach. Intell. **34**(11), 2189–2202 (2012)
2. Arisoy, R., Yayla, M.: Transvaginal sonographic evaluation of the cervix in asymptomatic singleton pregnancy and management options in short cervix. J. Pregnancy (2012)
3. Baumgartner, C.F., Kamnitsas, K., Matthew, J., Fletcher, T.P., Smith, S., Koch, L.M., Kainz, B., Rueckert, D.: Sononet: real-time detection and localisation of fetal standard scan planes in freehand ultrasound. IEEE Trans. Med. Imaging **36**(11), 2204–2215 (2017)
4. Beck, S., Wojdyla, D., Say, L., Betran, A.P., Merialdi, M., Requejo, J.H., Rubens, C., Menon, R., Van Look, P.F.: The worldwide incidence of preterm birth: a systematic review of maternal mortality and morbidity. Bull. World Health Organ. **88**, 31–38 (2010)
5. Bennett, N., Burridge, R., Saito, N.: A method to detect and characterize ellipses using the Hough transform. IEEE Trans. Pattern Anal. Mach. Intell. **21**(7), 652–657 (1999)
6. Berghella, V., Palacio, M., Ness, A., Alfirevic, Z., Nicolaides, K.H., Saccone, G.: Cervical length screening for prevention of preterm birth in singleton pregnancy with threatened preterm labor: systematic review and meta-analysis of randomized controlled trials using individual patient-level data. Ultrasound Obstet. Gynecol. **49**(3), 322–329 (2017)
7. Blencowe, H., Cousens, S., Oestergaard, M.Z., Chou, D., Moller, A.B., Narwal, R., Adler, A., Garcia, C.V., Rohde, S., Say, L., et al.: National, regional, and worldwide estimates of preterm birth rates in the year 2010 with time trends since 1990 for selected countries: a systematic analysis and implications. Lancet **379**(9832), 2162–2172 (2012)
8. Cai, Y., Droste, R., Sharma, H., Chatelain, P., Drukker, L., Papageorghiou, A.T. and Noble, J.A.: Spatio-temporal visual attention modelling of standard biometry plane-finding navigation. Med. Image Anal. **65**, 101762 (2020)
9. Campbell, S., Wilkin, D.: Ultrasonic measurement of fetal abdomen circumference in the estimation of fetal weight. BJOG: Int. J. Obstet. Gynaecol. **82**(9), 689–697 (1975)
10. Carvalho, M.H.B., Bittar, R.E., Brizot, M.L., Maganha, P.P.S., Borges da Fonseca, E.S.V., Zugaib, M.: Cervical length at 11–14 weeks' and 22–24 weeks' gestation evaluated by transvaginal sonography, and gestational age at delivery. Ultrasound Obstet. Gynecol.: Off. J. Int. Soc. Ultrasound Obstet. Gynecol. **21**(2), 135–139 (2003)
11. Chalana, V., Winter III, T.C., Cyr, D.R., Haynor, D.R., Kim, Y.: Automatic fetal head measurements from sonographic images. Acad. Radiol. **3**(8), 628–635 (1996)
12. Chen, H., Ni, D., Qin, J., Li, S., Yang, X., Wang, T., Heng, P.A.: Standard plane localization in fetal ultrasound via domain transferred deep neural networks. IEEE J. Biomed. Health Inform. **19**(5), 1627–1636 (2015)
13. Chen, J., Lu, Y., Yu, Q., Luo, X., Adeli, E., Wang, Y., Lu, L., Yuille, A.L., Zhou, Y.: Transunet: transformers make strong encoders for medical image segmentation (2021). arXiv:2102.04306

14. Chen, L.C., Papandreou, G., Schroff, F., Adam, H.: Rethinking Atrous convolution for semantic image segmentation (2017). arXiv:1706.05587
15. Cho, H.C., Sun, S., Hyun, C.M., Kwon, J.Y., Kim, B., Park, Y., Seo, J.K.: Automated ultrasound assessment of amniotic fluid index using deep learning. Med. Image Anal. **69**, 101951 (2021)
16. Coombe-Patterson, J.: Amniotic fluid assessment: amniotic fluid index versus maximum vertical pocket. J. Diagn. Med. Sonogr. **33**(4), 280–283 (2017)
17. Drozdzal, M., Vorontsov, E., Chartrand, G., Kadoury, S., Pal, C.: The importance of skip connections in biomedical image segmentation. In: Deep Learning and Data Labeling for Medical Applications, pp. 179–187. Springer, Berlin (2016)
18. Dubil, E.A., Magann, E.F.: Amniotic fluid as a vital sign for fetal wellbeing. Australas. J. Ultrasound Med. **16**(2), 62–70 (2013)
19. Espinoza, J., Good, S., Russell, E., Lee, W.: Does the use of automated fetal biometry improve clinical work flow efficiency? J. Ultrasound Med. **32**(5), 847–850 (2013)
20. Feldman, M.K., Katyal, S., Blackwood, M.S.: US artifacts. Radiographics **29**(4), 1179–1189 (2009)
21. Foi, A., Maggioni, M., Pepe, A., Rueda, S., Noble, J.A., Papageorghiou, A.T., Tohka, J.: Difference of gaussians revolved along elliptical paths for ultrasound fetal head segmentation. Comput. Med. Imaging Graph. **38**(8), 774–784 (2014)
22. Huazhu, F., Cheng, J., Yanwu, X., Wong, D.W.K., Liu, J., Cao, X.: Joint optic disc and cup segmentation based on multi-label deep network and polar transformation. IEEE Trans. Med. Imaging **37**(7), 1597–1605 (2018)
23. Kunihiko, F., Sei, M.: Neocognitron: a self-organizing neural network model for a mechanism of visual pattern recognition. In: Competition and Cooperation in Neural Nets, pp. 267–285. Springer, Berlin (1982)
24. Girshick, R.: Fast r-cnn. In: Proceedings of the IEEE International Conference on Computer Vision, pp. 1440–1448 (2015)
25. Hadlock, F.P., Harrist, R.B., Sharman, R.S., Deter, R.L., Park, S.K.: Estimation of fetal weight with the use of head, body, and femur measurements-a prospective study. Am. J. Obstet. Gynecol. **151**(3), 333–337 (1985)
26. Harrington, T.: Is the current measurement criteria appropriate for selecting women who require transvaginal assessment of cervical length in a low-risk population? Sonography **1**(2), 39–43 (2014)
27. He, K., Gkioxari, G., Dollar, P. and Girshick, R.: Mask r-cnn. In: Proceedings of the IEEE International Conference on Computer Vision, pp. 2961–2969 (2017)
28. He, K., Zhang, X., Ren, S., Sun, J.: Deep residual learning for image recognition. In: Proceedings of the IEEE Conference on Computer Vision and Pattern Recognition, pp. 770–778 (2016)
29. Hoskins, P.R., Martin, K., Thrush, A.: Diagnostic ultrasound: physics and equipment. CRC Press (2019)
30. Huang, G., Liu, Z., Van Der Maaten, L., Weinberger, K.Q.: Densely connected convolutional networks. In: Proceedings of the IEEE Conference on Computer Vision and Pattern Recognition, pp. 4700–4708 (2017)
31. Ibtehaz, N., Rahman, M.S.: Multiresunet: rethinking the u-net architecture for multimodal biomedical image segmentation. Neural Netw. **121**, 74–87 (2020)
32. Sergey Ioffe and Christian Szegedy. Batch normalization: Accelerating deep network training by reducing internal covariate shift. In International Conference on Machine Learning, pages 448-456. PMLR, 2015
33. Isola, P., Zhu, J.Y., Zhou, T., Efros, A.A.: Image-to-image translation with conditional adversarial networks. In: Proceedings of the IEEE Conference on Computer Vision and Pattern Recognition, pp. 1125–1134 (2017)
34. Goldenberg, R.J., Meis, P., Mercer, B., Moawad, A., Das, A.: The length of the cervix and the risk of spontaneous premature delivery. N. Engl. J. Med. **334**, 567–572 (1996)
35. Jaesung Jang and Chi Young Ahn: Industrial mathematics in ultrasound imaging. J. Korean Soc. Ind. Appl. Math. **20**(3), 175–202 (2016)

36. Jang, J., Park, Y., Kim, B., Lee, S.M., Kwon, J.Y., Seo, J.K.: Automatic estimation of fetal abdominal circumference from ultrasound images. IEEE J. Biomed. Health Inform. **22**(5), 1512–1520 (2017)
37. Jardim, S.M.G.V.B., Figueiredo, M.A.T.: Segmentation of fetal ultrasound images. Ultrasound Med. Biol. **31**(2), 243–250 (2005)
38. Jegou, S., Drozdzal, M., Vazquez, D., Romero, A., Bengio, Y.: The one hundred layers tiramisu: fully convolutional densenets for semantic segmentation. In: Proceedings of the IEEE Conference on Computer Vision and Pattern Recognition Workshops, pp. 11–19 (2017)
39. Jha, D., Smedsrud, P.H., Riegler, M.A., Johansen, D., De Lange, T., Halvorsen, P., Johansen, H.D.: Resunet++: an advanced architecture for medical image segmentation. In: 2019 IEEE International Symposium on Multimedia (ISM), pp. 225–2255. IEEE (2019)
40. Jin, J., Dundar, A., Culurciello, E.: Flattened convolutional neural networks for feedforward acceleration (2014). arXiv:1412.5474
41. Johnson, J., Alahi, A., Fei-Fei, L.: Perceptual losses for real-time style transfer and super-resolution. In: European Conference on Computer Vision, pp. 694–711. Springer, Berlin (2016)
42. Kagan, K.O., Sonek, J.: How to measure cervical length. Ultrasound Obstet. Gynecol. **45**(3), 358–362 (2015)
43. Kehl, S., Schelkle, A., Thomas, A., Puhl, A., Meqdad, K., Tuschy, B., Berlit, S., Weiss, C., Bayer, C., Heimrich, J., et al.: Single deepest vertical pocket or amniotic fluid index as evaluation test for predicting adverse pregnancy outcome (safe trial): a multicenter, open-label, randomized controlled trial. Ultrasound Obstet. Gynecol. **47**(6), 674–679 (2016)
44. Kim, B., Kim, K.C., Park, Y., Kwon, J.Y., Jang, J., Seo, J.K.: Machine-learning-based automatic identification of fetal abdominal circumference from ultrasound images. Physiol. Meas. **39**(10), 105007 (2018)
45. Kim, H.P., Lee, S.M., Kwon, J.Y., Park, Y., Kim, K.C., Seo, J.K.: Automatic evaluation of fetal head biometry from ultrasound images using machine learning. Physiol. Meas. **40**(6), 065009 (2019)
46. Krizhevsky, A., Sutskever, I., Hinton, G.E.: Imagenet classification with deep convolutional neural networks. Adv. Neural Inf. Process. Syst. **25**, 1097–1105 (2012)
47. Kurjak, A.: Donald School Textbook of Ultrasound in Obstetrics & Gynaecology. JP Medical Ltd (2017)
48. Kuusela, P., Wennerholm, U.-B., Fadl, H., Wesstrom, J., Lindgren, P., Hagberg, H., Jacobsson, B., Valentin, L.: Second trimester cervical length measurements with transvaginal ultrasound: a prospective observational agreement and reliability study. Acta Obstet. Gynecol. Scand. **99**(11), 1476–1485 (2020)
49. Kwan, A., Dudley, J., Lantz, E.: Who really discovered snell's law? Phys. World **15**(4), 64 (2002)
50. Sun, S., Kwon, H.Y., Kwon, J.Y., Yun, H.S., Park, S., Cho, H.C., Seo, J.K.: Deep learning-based automatic measurement of cervical length in transvaginal sonography, preprint (2022)
51. LeCun, Y., Bengio, Y., Hinton, G.: Deep learning. Nature **521**(7553), 436–444 (2015)
52. LeCun, Y., Boser, B., Denker, J.S., Henderson, D., Howard, R.E., Hubbard, W., Jackel, L.D.: Backpropagation applied to handwritten zip code recognition. Neural Comput. **1**(4), 541–551 (1989)
53. LeCun, Y., Bottou, L., Bengio, Y., Haffner, P.: Gradient-based learning applied to document recognition. Proc. IEEE **86**(11), 2278–2324 (1998)
54. Li, X.H.: Ultrasound scan conversion on TI's C64x+ DSPs. Application Report SPRAB32, Texas Instruments (2009)
55. Lin, T.Y., Dollar, P., Girshick, R., He, K., Hariharan, B., Belongie, S.: Feature pyramid networks for object detection. In: Proceedings of the IEEE Conference on Computer Vision and Pattern Recognition, pp. 2117–2125 (2017)
56. Lin, X., Wang, F., Guo, L., Zhang, W.: An automatic key-frame selection method for monocular visual odometry of ground vehicle. IEEE Access **7**, 70742–70754 (2019)
57. Litjens, G., Kooi, T., Bejnordi, B.E., Setio, A.A.A., Ciompi, F., Ghafoorian, M., Van Der Laak, J.A., Van Ginneken, B. and Sanchez, C.I.: A survey on deep learning in medical image analysis. Med. Image Anal. **42**, 60–88 (2017)

58. Livne, M., Rieger, J., Aydin, O.U., Taha, A.A., Akay, E.M., Kossen, T., Sobesky, J., Kelleher, J.D., Hildebrand, K., Frey, D., et al.: A u-net deep learning framework for high performance vessel segmentation in patients with cerebrovascular disease. Front. Neurosci. **13**, 97 (2019)
59. Long, J., Shelhamer, E., Darrell, T.: Fully convolutional networks for semantic segmentation. In: Proceedings of the IEEE Conference on Computer Vision and Pattern Recognition, pp. 3431–3440 (2015)
60. Looney, P., Stevenson, G.N., Nicolaides, K.H., Plasencia, W., Molloholli, M., Natsis, S., Collins, S.L.: Fully automated, real-time 3D ultrasound segmentation to estimate first trimester placental volume using deep learning. JCI Insight **3**(11) (2018)
61. Lu, R., Shen, Y.: Image segmentation based on random neural network model and gabor filters. In: 2005 IEEE Engineering in Medicine and Biology 27th Annual Conference, pp. 6464–6467. IEEE (2006)
62. Luntsi, G., Burabe, F.A., Ogenyi, P.A., Zira, J.D., Chigozie, N.I., Nkubli, F.B. and Dauda, M.: Sonographic estimation of amniotic fluid volume using the amniotic fluid index and the single deepest pocket in a resourcelimited setting. J. Med. Ultrasound **27**(2), 63 (2019)
63. Malinger, G., Paladini, D., Haratz, K.K., Monteagudo, A., Pilu, G.L., Timor-Tritsch, I.E.: Isuog practice guidelines (updated): sonographic examination of the fetal central nervous system. part 1: performance of screening examination and indications for targeted neurosonography. Ultrasound Obstet. Gynecol. **56**(3), 476–484 (2020)
64. Manning, F.A., Platt, L.D., Sipos, L.: Antepartum fetal evaluation: development of a fetal biophysical profile. Am. J. Obstet. Gynecol. **136**(6), 787–795 (1980)
65. Martin, D.J., Wells, I.T., Goodwin, C.R.: Physics of ultrasound. Anaesth. Intensiv. Care Med. **16**(3), 132–135 (2015)
66. Martin, J.A., Hamilton, B.E., Osterman, M.J., Driscoll, A.K.: Births: final data for 2019. National Vital Statistics Reports: From the Centers for Disease Control and Prevention, National Center for Health Statistics, National Vital Statistics System **70**(2), 1–51 (2021)
67. McLaughlin, R.A.: Randomized Hough transform: improved ellipse detection with comparison. Pattern Recognit. Lett. **19**(3–4), 299–305 (1998)
68. Neubeck, A., Van Gool, L.: Efficient non-maximum suppression. In: 18th International Conference on Pattern Recognition (ICPR'06), vol. 3, pp. 850–855. IEEE (2006)
69. Ng, A., Swanevelder, J.: Resolution in ultrasound imaging. Contin. Educ. Anaesth. Critical Care Pain **11**(5), 186–192 (2011)
70. Ni, D., Yang, X., Chen, X., Chin, C.T., Chen, S., Heng, P.A., Li, S., Qin, J., Wang, T.: Standard plane localization in ultrasound by radial component model and selective search. Ultrasound Med. Biol. **40**(11), 2728–2742 (2014)
71. Noble, W.S.: What is a support vector machine? Nat. Biotechnol. **24**(12), 1565–1567 (2006)
72. Ouahabi, A., Taleb-Ahmed, A.: Deep learning for real-time semantic segmentation: application in ultrasound imaging. Pattern Recogn. Lett. **144**, 27–34 (2021)
73. Paladini, D., Malinger, G., Birnbaum, R., Monteagudo, A., Pilu, G., Salomon, L.J., Timor-Trirtsch, I.E.: Isuog practice guidelines (updated): sonographic examination of the fetal central nervous system. Part 2: performance of targeted neurosonography. Ultrasound Obstet. Gynecol. **57**(4), 661–671 (2021)
74. Pathak, S.D., Haynor, D.R., Kim, Y.: Edge-guided boundary delineation in prostate ultrasound images. IEEE Trans. Med. Imaging **19**(12), 1211–1219 (2000)
75. Phelan, J.P., Smith, C.V., Broussard, P., Small, M.: Amniotic fluid volume assessment with the four-quadrant technique at 36-42 weeks' gestation. J. Reprod. Med. Obstet. Gynecol. **32**(7), 540–542 (1987)
76. Ponomarev, G.V., Gelfand, M.S., Kazanov, M.D.: A multilevel thresholding combined with edge detection and shape-based recognition for segmentation of fetal ultrasound images. In: Proceedings of challenge US: Biometric Measurements from Fetal Ultrasound Images, ISBI, pp. 17–19 (2012)
77. Prasad, D.K., Leung, M.K., Quek, C.: Ellifit: an unconstrained, non-iterative, least squares based geometric ellipse fitting method. Pattern Recognit. **46**(5), 1449–1465 (2013)

78. Pu, B., Li, K., Li, S., Zhu, N.: Automatic fetal ultrasound standard plane recognition based on deep learning and IIoT. IEEE Trans. Ind. Inform. (2021)
79. Purisch, S.E., Gyamfi-Bannerman, C.: Epidemiology of preterm birth. In: Seminars in Perinatology, vol. 41, pp. 387–391. Elsevier (2017)
80. Redmon, J., Divvala, S., Girshick, R., Farhadi, A.: You only look once: unified, real-time object detection. In: Proceedings of the IEEE Conference on Computer Vision and Pattern Recognition, pp. 779–788 (2016)
81. Ren, S., He, K., Girshick, R., Sun, J.: Faster R-CNN: towards real-time object detection with region proposal networks. Adv. Neural. Inf. Process. Syst. **28**, 91–99 (2015)
82. Ronneberger, O., Fischer, P., Brox, T.: U-net: convolutional networks for biomedical image segmentation. In: International Conference on Medical Image Computing and Computer-Assisted Intervention, pp. 234–241. Springer, Berlin (2015)
83. Rueda, S., Fathima, S., Knight, C.L., Yaqub, M., Papageorghiou, A.T., Rahmatullah, B., Foi, A., Maggioni, M., Pepe, A., Tohka, J., et al.: Evaluation and comparison of current fetal ultrasound image segmentation methods for biometric measurements: a grand challenge. IEEE Trans. Med. Imaging **33**(4), 797–813 (2013)
84. Russakovsky, O., Deng, J., Hao, S., Krause, J., Satheesh, S., Ma, S., Huang, Z., Karpathy, A., Khosla, A., Bernstein, M., et al.: Imagenet large scale visual recognition challenge. Int. J. Comput. Vision **115**(3), 211–252 (2015)
85. Rutherford, S.E., Smith, C.V., Phelan, J.P., Kawakami, K., Ahn, M.O.: Four-quadrant assessment of amniotic fluid volume. Interobserver and intraobserver variation. J. Reprod. Med. **32**(8), 587–589 (1987)
86. Salomon, L.J., Alfirevic, Z., Da Silva Costa, F., Deter, R.L., Figueras, F., Ghi, T.A., Glanc, P., Khalil, A., Lee, W., Napolitano, R., et al.: Isuog practice guidelines: ultrasound assessment of fetal biometry and growth. Ultrasound Obstet. Gynecol. **53**(6), 715–723 (2019)
87. Schlemper, J., Oktay, O., Schaap, M., Heinrich, M., Kainz, B., Glocker, B., Rueckert, D.: Attention gated networks: learning to leverage salient regions in medical images. Med. Image Anal. **53**, 197–207 (2019)
88. Sotiriadis, A., Papatheodorou, S., Kavvadias, A., Makrydimas, G.: Transvaginal cervical length measurement for prediction of preterm birth in women with threatened preterm labor: a meta-analysis. Ultrasound Obstet. Gynecol.: Off. J. Int. Soc. Ultrasound Obstet. Gynecol. **35**(1), 54–64 (2010)
89. Srivastava, N., Hinton, G., Krizhevsky, A., Sutskever, I., Salakhutdinov, R.: Dropout: a simple way to prevent neural networks from overfitting. J. Mach. Learn. Res. **15**(1), 1929–1958 (2014)
90. Stebbing, R.V., McManigle, J.E.: A boundary fragment model for head segmentation in fetal ultrasound. In: Proceedings of Challenge US: Biometric Measurements from Fetal Ultrasound Images, ISBI, pp. 9–11 (2012)
91. Sun, S., Kwon, J.Y., Park, Y., Cho, H.C., Hyun, C.M., Seo, J.K.: Complementary network for accurate amniotic fluid segmentation from ultrasound images. IEEE Access **9**, 108223–108235 (2021)
92. To, M.S., Skentou, C., Chan, C., Zagaliki, A., Nicolaides, K.H.: Cervical assessment at the routine 23-week scan: standardizing techniques. Ultrasound Obstet. Gynecol.: Off. J. Int. Soc. Ultrasound Obstet. Gynecol. **17**(3), 217–219 (2001)
93. Lingyun, W., Cheng, J.-Z., Li, S., Lei, B., Wang, T., Ni, D.: Fuiqa: fetal ultrasound image quality assessment with deep convolutional networks. IEEE Trans. Cybern. **47**(5), 1336–1349 (2017)
94. Shi, X., Chen, Z., Wang, H., Yeung, D.Y., Wong, W.K., Woo, W.C.: Convolutional LSTM network: a machine learning approach for precipitation nowcasting. In: Advances in Neural Information Processing Systems, pp. 802–810 (2015)
95. Lei, X., Oja, E., Kultanen, P.: A new curve detection method: randomized Hough transform (RHT). Pattern Recogn. Lett. **11**(5), 331–338 (1990)
96. Yin, S., Peng, Q., Li, H., Zhang, Z., You, X., Fischer, K., Furth, S.L., Tasian, G.E., Fan, Y.: Automatic kidney segmentation in ultrasound images using subsequent boundary distance regression and pixelwise classification networks. Med. Image Anal. **60**, 101602 (2020)

97. Yost, N.P., Bloom, S.L., Twickler, D.M., Leveno, K.J.: Pitfalls in ultrasonic cervical length measurement for predicting preterm birth. Obstet. Gynecol. **93**(4), 510–516 (1999)
98. Yu, F., Koltun, V.: Multi-scale context aggregation by dilated convolutions (2015). arXiv:1511.07122
99. Jinhua, Yu., Wang, Y., Chen, P.: Fetal ultrasound image segmentation system and its use in fetal weight estimation. Med. Biol. Eng. Comput. **46**(12), 1227–1237 (2008)
100. Yu, Y., Acton, S.T.: Speckle reducing anisotropic diffusion. IEEE Trans. Image Process. **11**(11), 1260–1270 (2002)
101. Zagzebski, J.A.: Essentials of ultrasound physics. Mosby (1996)
102. Zahedi-Spung, L.D., Raghuraman, N., Macones, G.A., Cahill, A.G., Rosenbloom, J.I.: Neonatal morbidity and mortality by mode of delivery in very preterm neonates. Am. J. Obstet. Gynecol. (2021)
103. Zhang, C., Bengio, S., Hardt, M., Recht, B., Vinyals, O.: Understanding deep learning requires rethinking generalization. In: 5th International Conference on Learning Representations, ICLR 2017, Toulon, France, April 24–26, 2017, Conference Track Proceedings. OpenReview.net (2017)
104. Zhou, Z., Rahman Siddiquee, M.M., Tajbakhsh, N., Liang, J.: Unet++: a nested u-net architecture for medical image segmentation. In: Deep Learning in Medical Image Analysis and Multimodal Learning for Clinical Decision Support, pp. 3–11. Springer, Berlin (2018)
105. Zongwei Zhou, Md., Siddiquee, M.R., Tajbakhsh, N., Liang, J.: Unet++: redesigning skip connections to exploit multiscale features in image segmentation. IEEE Trans. Med. Imaging **39**(6), 1856–1867 (2019)
106. Zhu, J.Y., Park, T., Isola, P., Efros, A.A.: Unpaired image-to-image translation using cycle-consistent adversarial networks. In: Proceedings of the IEEE International Conference on Computer Vision, pp. 2223–2232 (2017)

Chapter 6
Electrical Impedance Imaging

Hyeuknam Kwon, Ariungerel Jargal, and Jin Keun Seo

Abstract Recently, there has been marked progress in electrical impedance imaging in which cross-sectional image reconstructions inside the human body are pursued. These techniques also have wider applications as imaging methods in medicine, biotechnology, non-destructive testing, the monitoring of industrial processes, and in other areas. Their imaging techniques allow to visualize new contrast information of biological tissues and organs exhibiting distinct electrical properties depending on their physiological functions and pathological states. The mathematical models for bioimpedance imaging are expressed as nonlinear inverse problems involving time-harmonic Maxwell's equations with electrical tissue properties being described by frequency-dependent conductivity and permittivity. This chapter reviews electrical tissue property imaging modalities.

6.1 Introduction

Biological tissues and organs exhibit distinct electrical properties depending on their physiological functions and pathological states [21–23, 45, 56, 57]. Biological tissues can be considered as a three-dimensional arrangement of cells embedded in an extracellular matrix, where cells are composed of intracellular fluid and organelles surrounded by cell membranes. The complex conductivity distribution of the body varies with physiological functions such as respiration, blood flow and perfusion, and

H. Kwon
Division of Software, Yonsei University, Wonju, Korea
e-mail: hyeuknamkwon@yonsei.ac.kr

A. Jargal
Department of Mathematics, School of Applied Sciences, Mongolian University of Science and Technology, Ulaanbaatar, Mongolia

J. K. Seo (✉)
School of Mathematics and Computing (Computational Science and Engineering), Yonsei University, Seoul, Korea
e-mail: seoj@yonsei.ac.kr

J. K. Seo (ed.), *Deep Learning and Medical Applications*, Mathematics in Industry 40,
https://doi.org/10.1007/978-981-99-1839-3_6

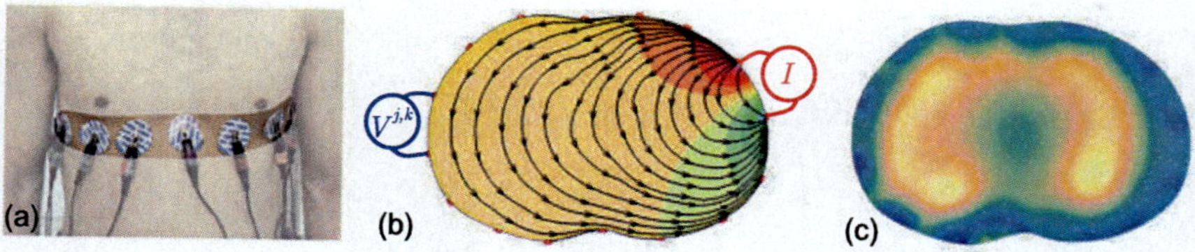

Fig. 6.1 **a** Electrodes attached to human chest, **b** electric current streamlines and voltage difference $V^{j,k}$ subject to the j-th pair of electrodes, (c) reconstructed image

nerve activity, and also with pathological states such as ischemia, bleeding, inflammation, and tumor. Hence, there have been numerous studies to measure or image bio-impedance.

Electrical Impedance Tomography (EIT) was introduced by Henderson and Webster [32] in 1978. It aims to provide tomographic imaging of tissue's electrical properties such as the conductivity (σ) and permittivity (ε), using multiple surface electrodes as shown in Fig. 6.1. Here, electrodes are used to measure the boundary current–voltage relations, partial information of a Neumann-to-Dirichlet map. In 1980, Calderón [11] presented the mathematical inverse problem of EIT under ideal assumptions; the inverse problem is to identify the coefficient σ entering the elliptic partial differential equation $\nabla \cdot (\sigma \nabla u) = 0$ in a domain Ω from the knowledge of the Neumann-to-Dirichlet map. Roughly speaking, the Neumann-to-Dirichlet map is equivalent to the Neumann function $N(\cdot, \cdot)|_{\partial\Omega\times\partial\Omega}$ corresponding to $\nabla \cdot (\sigma \nabla u) = 0$ in Ω. Calderón's idea of a complex geometric optics solution in the paper [11] initiated the uniqueness issue of the mathematical EIT problem, and it has become a key driving force in theoretical growth in inverse problems over the last 30 years [6, 41, 49, 72]. In 1984, Barber and Brown [8] developed an EIT version of the CT back-projection algorithm with a deep understanding of the conductivity equation in the context of EIT. They developed the first EIT device (Sheffield Mark 1) with one active current source to perform phantom and human experiments [46]. In 1986, Isaacson [35] suggested the concept of distinguishability between two different conductivity distributions, and the RPI group developed EIT systems with multiple active current sources to maximize the distinguishability [16].

Let us discuss what information can be obtained from EIT. For ease of explanation, let us neglect interfacial phenomena (such as contact impedance) between the electrodes and the body Ω. When a sinusoidal current of I mA at angular frequency ω is injected through a pair of electrodes, it produces the time-harmonic electrical field $\mathbf{E}$ and the current density $\mathbf{J}$ at angular frequency ω inside Ω. Denoting the resulting voltage difference between the same pair of electrodes by V, the impedance Z is given by

$$Z = \frac{V}{I} \approx \frac{1}{I^2} \int_{\Omega} \mathbf{J} \cdot \mathbf{E}\, d\mathbf{x} = \frac{1}{I^2} \int_{\Omega} \frac{1}{\sigma + i\omega\varepsilon} \mathbf{J} \cdot \mathbf{J}\, d\mathbf{x}.$$

Hence, the impedance Z is determined by the geometry of Ω, electrode position, and the distribution of the effective complex conductivity distribution $\gamma := \sigma + i\omega\varepsilon$.

In the case of an N-channel EIT system having N electrodes, we obtain the set $\{Z^{j,k} : j, k = 1, \cdots, N\}$, where $Z^{j,k}$ is the trans-impedance corresponding to the measured voltage difference $V^{j,k}$ between the k-th pair of electrodes subject to the injection current using the j-th pair of electrodes. Hence, the inverse problem can be viewed as recovering the impedivity $1/\gamma$ from the knowledge of the set $\{Z^{j,k} : j, k = 1, \cdots, N\}$.

Static EIT for recovering $\gamma = \sigma + i\omega\varepsilon$ has fundamental drawbacks because EIT data depends strongly on the boundary geometry and electrode positions, whereas it is much less sensitive to a local perturbation of σ away from the measuring electrodes. Taking account of these fundamental ill-posed structures, static EIT will continue to have problems with providing useful images for clinical applications [33, 57].

Time difference EIT (tdEIT) for recovering the time change $\frac{\partial}{\partial t}\gamma$ has much better posed structures, because a background data subtraction method effectively eliminates the technical difficulties of static EIT including the boundary geometry errors and electrode position uncertainties. Indeed, the time change of potential u^j subject to j-th current is approximately dictated by $\nabla \cdot (\frac{\partial}{\partial t}\gamma \nabla u^j) = -\nabla \cdot (\gamma \nabla \frac{\partial}{\partial t} u^j)$ in Ω, and therefore the data $\{\frac{d}{dt} Z^{j,k} : j, k = 1, \cdots, N\}$ is less affected by the boundary geometry error and electrode position uncertainties.

To overcome the fundamental limitations of static EIT using boundary voltage data only, additional measurements should be incorporated to measure γ. Due to the relation $\gamma \mathbf{E} = \mathbf{J} = \nabla \times \mathbf{B}$ with $\mathbf{B}$ being the magnetic flux density, MRI would be a top candidate for recovering γ. In 1989, Joy et al. [36] developed a technique to measure B_z where $\mathbf{B} = (B_x, B_y, B_z)$ is the internal magnetic flux density induced by an externally injected current through surface electrodes and z-axis is the direction of the main magnetic field of an MRI scanner. In 2000, Kwon et al. [43] proposed the $\mathbf{J}$-substitution algorithm based on the non-linear PDE$\nabla \cdot \left(\frac{|\mathbf{J}|}{|\nabla u|} \nabla u \right) = 0$ in Ω. This new approach using the non-linear PDE could produce high-resolution conductivity images by displaying $\sigma = |\mathbf{J}|/|\nabla u|$ [39]. This EIT method using MRI is called magnetic resonance electrical impedance tomography (MREIT). Recently, the Toronto group [50, 51] studied a method of recovering σ from knowledge of a single current density $|\mathbf{J}|$, called current density impedance imaging (CDII). But the major drawback of all these methods is the requirement of measuring the full components of $\mathbf{B}$, which requires rotating the imaging object inside the MRI scanner. Unfortunately, it is very difficult to use these approaches for human and animal experiments. Therefore, the most challenging issue of MREIT was to remove the practical technical difficulties of rotating the imaging object inside the MRI scanner. We should recover σ using only B_z, instead of the full component of $\mathbf{B}$.

In 2003, Seo et al. [60] invented the first constructive B_z-based MREIT algorithm, called the harmonic B_z algorithm, which removed the rotation process. Since its invention, imaging techniques in MREIT have advanced rapidly and now can perform state-of-the-art conductivity imaging of animal and human subjects [58, 78].

MREIT relies on measured magnetic field data at low frequencies which are influenced by the low-frequency conductivity distribution. In contrast to MREIT, magnetic resonance electrical properties tomography (MREPT) is based on standard RF field mapping techniques to measure the active magnetic RF field component (see

Table 6.1 Comparison of three electrical impedance imaging methods; electrical impedance tomography (EIT), magnetic resonance electrical impedance tomography (MREIT), magnetic resonance electrical properties tomography (MREPT)

	EIT	MREIT	MREPT
Data type	External	Internal	Internal
Measurement	Surface electrodes	Surface electrodes and MR machine	MR machine
Electric field frequency	Less than 1 MHz	few kHz	128 MHz at 3 T MRI

Table 6.1). A time-varying magnetic field inside the human body is affected by σ and ε via Maxwell's equations. Noting that the MRI scanner can acquire the field map at the Larmor frequency by relying on its RF subsystem, Haacke et al. [25] suggested the MREPT technique in 1991 to image $\gamma = \sigma + i\omega\varepsilon$ at the Larmor frequency.

Future studies of EIT, MREIT, and MREPT should overcome several technical barriers to advance the methods to the stage of routine clinical uses. We expect EIT, MREIT, MREPT to be a new clinically useful bio-imaging modality which manifests structural, functional, and pathological conditions of biological tissues and organs providing valuable diagnostic information.

6.2 Electrical Impedance Tomography (EIT)

6.2.1 *Conductivity and Permittivity*

Effective electrical conductivity (σ) and permittivity (ε) are determined by Ohm's law dealing with the relationship between **E** and **J**, under the assumption that there exist a time-harmonic electric field **E** and current density **J** inside the body Ω subject to a certain kind of external excitation.

The effective $\gamma = \sigma + i\omega\varepsilon$ is defined by Ohm's law in the sense of the ensemble average of the fields. The effective γ at point **x** describes the linear relationship between the ensemble mean current density and the ensemble mean electrical field over the voxel $\square_{\mathbf{x}}$ containing point **x**:

$$\int_{\square_{\mathbf{x}}} \mathbf{J}(\mathbf{x}')dx' \approx \gamma(\mathbf{x}) \int_{\square_{\mathbf{x}}} \mathbf{E}(\mathbf{x}')dx' \quad \text{for pairs of time-harmonic fields } (\mathbf{E}, \mathbf{J}).$$

Hence, the effective γ depends on the size of the voxel. We may define the effective γ at a macroscopic scale as an ensemble average of the pointwise value of γ via homogenization [47]. Noting that γ is a passive property, the pointwise γ does not exist in the real world. Hence, the pointwise γ should be understood as a microscopic γ. The pointwise conductivity and permittivity are assumed to be isotropic and independent of ω, whereas the effective conductivity and permittivity depend on the frequency ω, and it can be approximately represented by a symmetric matrix. Obtaining most of the frequency-dependent behavior of the effective admittivity, we may increase the amount of the measurable information and, therefore, the distinguishability among different functions and states. We may measure the effective admittivity spectra of biological tissues or organs for their tissue characterizations [20, 45, 56]. Recently, Ammari et al. [3] provided a rigorous mathematical analysis on the frequency-dependent effective admittivity.

Symbol	Name	Unit	Relation
ε	Permittivity	F/m	
σ	Conductivity	S/m	
$\gamma = \sigma + i\omega\varepsilon$	Admittivity	S/m	$\mathbf{J} = (\sigma + i\omega\varepsilon)\mathbf{E}$
μ	Permeability	H/m	$\mathbf{J} = \frac{1}{\mu}\nabla \times \mathbf{B}$

Bioimpedance imaging requires us to generate $\mathbf{J}$ inside the body Ω. The current density $\mathbf{J}$ can be produced either by injecting current into Ω through a pair of electrodes on the boundary $\partial\Omega$ or by feeding alternating current into a coil outside Ω. The relationship between $\mathbf{E}$ and $\mathbf{J}$ in a biological subject changes with the angular frequency ω due to the tissue structure involving its ion concentrations in extra- and intracellular fluids, cellular structure and density, molecular compositions, membrane characteristics and other factors. See Fig. 6.2. As a result, the effective σ and ε show a variable response over the frequency range from a few Hz to MHz [18, 30, 54, 74].

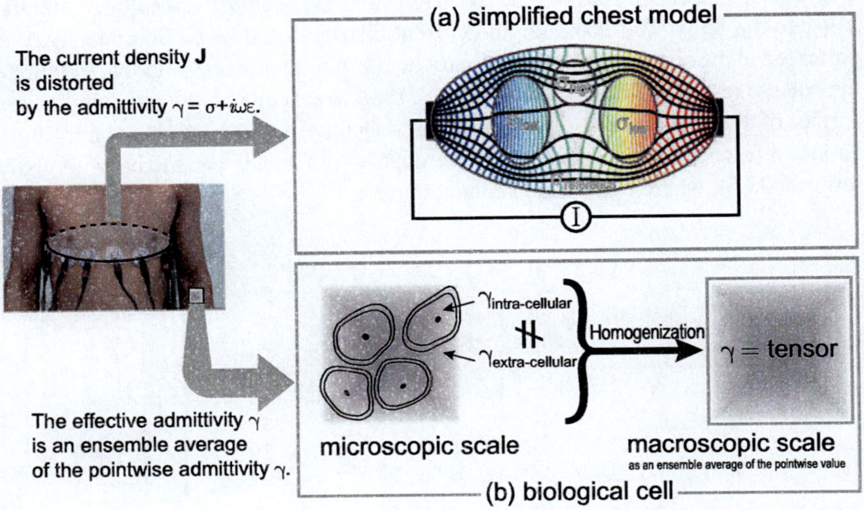

Fig. 6.2 The biological tissues that make up the human body can be considered conductive materials. **a** The admittivity distribution $\gamma = \sigma + i\omega\varepsilon$ deforms the current density $\mathbf{J}$, which is generated by the alternating current through boundary electrodes. **b** An ensemble average of the pointwise γ via homogenization [47] at a microscopic scale gives the effective γ at a macroscopic scale

6.2.2 *Forward Problem in EIT*

6.2.2.1 Ohm's Law

Before giving the mathematical model, let us consider physical phenomena of EIT. When sinusoidal current is injected to a physical domain (Ω), the corresponding potential (u) is measurable on the boundary of the region ($\partial\Omega$). The restriction of u to $\partial\Omega$ will be denoted by $u|_{\partial\Omega}$. We consider an alternating current (AC) which can be changed depending on time.

In the simplest case, we consider a circuit containing a resistor and sinusoidal current source in 1D as shown in Fig. 6.3. The current is flowing from the top to bottom surface electrodes, then the measured voltage is the difference between the top and the bottom surface potential

$$V = u|_{\text{top}} - u|_{\text{bottom}}. \tag{6.1}$$

The injected current and its corresponding measured voltage are proportional:

$$Z = \frac{V}{I} \text{ (Ohm's law)}. \tag{6.2}$$

The impedance (Z) is a material's capability to impede the flow of current.

The inverse value of impedivity is an admittivity ($\gamma = \sigma + i\omega\epsilon$) which depends on the length of the object and the surface area of electrodes. Permittivity (ε) is a measure of a material's electric polarizability.

We inject I through surface electrodes ($\mathcal{E}_+, \mathcal{E}_-$) to generate the resulting potential u inside the body. The potential and current distribution flow with conductivity is perturbed at the circle and ellipse anomaly in $2D$. As EIT is a non-invasive technique the voltage resulting from the application of the current can only be measured on the surface of the object (Fig. 6.4). The current lines gathering from a higher conductivity anomaly (circle, 10), on the contrary, escape into a lower conductivity anomaly (ellipse, 0.15), where background conductivity is 1.

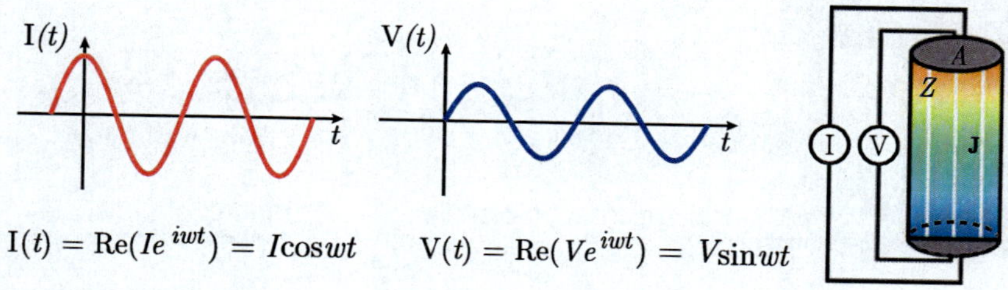

Fig. 6.3 Representation of a one-dimensional example. When alternating current is injected, the voltage (the magnitude and phase) is determined according to the length (L), area (A), and electrical properties (γ) of the object

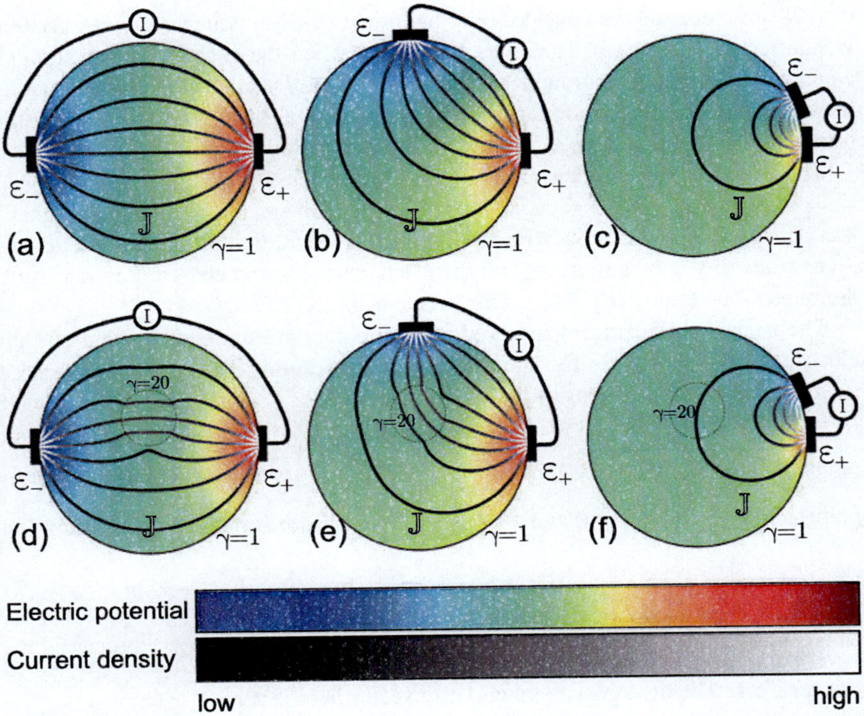

Fig. 6.4 Representation of two-dimensional examples. When alternating current is injected, the potential is determined according to the electrode position, electrode area, and electrical properties of the object. The objects in the first row (**a**, **b**, **c**) have homogeneous admittivity $\gamma = 1$ while the objects in the second row (**d**, **e**, **f**) ha high conductive anomaly of admitivity $\gamma = 20$. The red–green–blue color map shows electric potential distribution and the white–black curve represents current density J

6.2.2.2 Derivation of Elliptic PDE

When we inject current through an object, there are arise an electric field and a magnetic field. The relation between the electric field and magnetic field is shown in Maxwell's equations (Table 6.2). We use the Faraday's and Ampere's law for deriving

Table 6.2 Maxwell's equations for time-harmonic fields

Name	Equation
Gauss's law	$\nabla \cdot \mathbf{E} = \rho/\epsilon$
Gauss's law of magnetism	$\nabla \cdot \mathbf{H} = 0$
Faraday's law of induction	$\nabla \times \mathbf{E} = -\mathbf{i}\omega\mathbf{B}$
Ampere's circuital law	$\nabla \times \mathbf{H} = \mathbf{J} + \mathbf{i}\omega\mathbf{D}$

the EIT problem and we consider time-harmonic fields in which the time variation is sinusoidal. The idea of Faraday's law is that when the current flowing through a conductor is changed, an inductive electromotive force is created. An alternating magnetic field is created around it. Ampere's circuital law states the relationship between the current and the magnetic field created by this current.

There are relationships such as $\mathbf{B} = \mu\mathbf{H}$, $\mathbf{J} = \sigma\mathbf{E}$, $\mathbf{D} = \varepsilon\mathbf{E}$. In free space, $\varepsilon = \varepsilon_0 = 8.85 \times 10^{-12}$ and $\mu = \mu_0 = 4\pi \times 10^{-7}$. The magnetic field $\mathbf{H}$ and magnetic flux density $\mathbf{B}$ are related by permeability μ, the electric field $\mathbf{E}$ and current density $\mathbf{J}$ are related by conductivity σ, and the electric field $\mathbf{E}$ and electric flux density $\mathbf{D}$ are related by permittivity ε for linear, isotropic objects.

The basic field quantities involved in the domain are the electric field and the electric current density. In Faraday's law, after substituting the relation of magnetic field and magnetic flux density $\mathbf{B} = \mu\mathbf{H}$, we have $\nabla \times \mathbf{E} = -i\mu\omega\mathbf{H}$. If we assume $\mu = \mu_0$, we have $\nabla \times \mathbf{E} \approx 0$. It follows from Stoke's theorem that there exists a potential u satisfying $u(\mathbf{r}_2) - u(\mathbf{r}_1) = -\int_{C_{\mathbf{r}_1 - \mathbf{r}_2}} \mathbf{E} \cdot dl$ for any two points $\mathbf{r}_1$ (starting point) and $\mathbf{r}_2$ (ending point) and $-\nabla u \approx \mathbf{E}$ in Ω. From Ampere's circuital law, we have

$$\nabla \times \mathbf{H} = \mathbf{J} + i\omega\mathbf{D} = \sigma\mathbf{E} + i\omega\varepsilon\mathbf{E} = (\sigma + i\omega\epsilon)\mathbf{E} = -\gamma\nabla u,$$

where admittivity $\gamma = \sigma + i\omega\varepsilon$.

Remark 6.1 The divergence of a curl of a vector field is zero.

$$\begin{aligned}
\nabla \cdot (\nabla \times \mathbf{H}) &= \frac{\partial}{\partial x}(\nabla \times \mathbf{H})_x + \frac{\partial}{\partial y}(\nabla \times \mathbf{H})_y + \frac{\partial}{\partial z}(\nabla \times \mathbf{H})_z \\
&= \frac{\partial}{\partial x}\left(\frac{\partial H_z}{\partial y} - \frac{\partial H_y}{\partial z}\right) + \frac{\partial}{\partial y}\left(\frac{\partial H_x}{\partial z} - \frac{\partial H_z}{\partial x}\right) + \frac{\partial}{\partial z}\left(\frac{\partial H_y}{\partial x} - \frac{\partial H_x}{\partial y}\right) \\
&= \left(\frac{\partial}{\partial x}\frac{\partial H_z}{\partial y} - \frac{\partial}{\partial y}\frac{\partial H_z}{\partial x}\right) + \left(\frac{\partial}{\partial y}\frac{\partial H_x}{\partial z} - \frac{\partial}{\partial z}\frac{\partial H_x}{\partial y}\right) + \left(\frac{\partial}{\partial z}\frac{\partial H_y}{\partial x} - \frac{\partial}{\partial x}\frac{\partial H_y}{\partial z}\right) \\
&= 0.
\end{aligned}$$

Hence, $\nabla \cdot \nabla \times \mathbf{H} = 0$ and elliptic PDE can be governed by

$$-\nabla \cdot (\gamma\nabla u) = 0 \ \text{in} \ \Omega. \tag{6.3}$$

Remark 6.2 In the case of EIT for monitoring lung ventilation, the domain Ω in (6.3) would be the human thorax. Time-difference EIT has the unique ability to allow long-term, continuous monitoring of lung ventilation at the bedside [73]. The reconstruction algorithm involves the sensitivity matrix which depends mainly on the geometry of an imaging object and electrode positions. Hence, it would be desirable to segment the geometry of the body and to identify electrode positions in a simple way to improve the quality of the reconstructed image (Fig. 6.5).

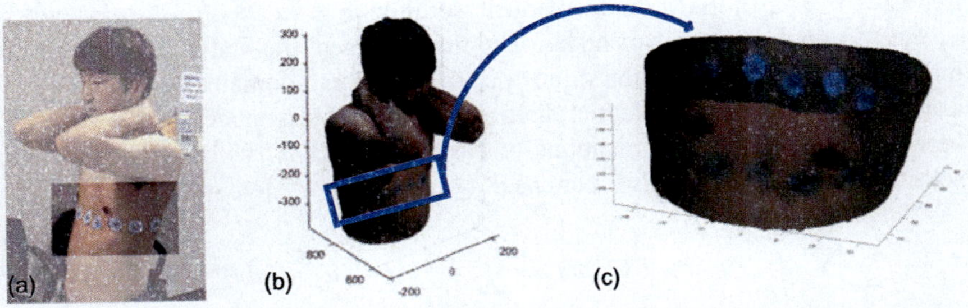

Fig. 6.5 **a** Human thorax with surface electrodes. **b** 3D scanned image of human thorax. **c** Definition of Ω from 3D scanned image

6.2.2.3 Finite Element Method

In the forward model, we want to find potential u with given γ. In practice, the material property γ may change abruptly. For example, a conductivity distribution inside the human body may have a jump along the boundary of two different organs. In this case, there exists no solution $u \in \mathbb{C}^2(\Omega)$ in the classical sense. We can solve the minimization and variational problem within the Sobolev space $\mathbf{H}^1(\Omega)$:

$$\begin{cases} -\nabla \cdot (\gamma \nabla u) = 0 & \text{in } \Omega, \\ (\gamma \nabla u) \cdot \mathbf{n} = f & \text{on } \partial\Omega. \end{cases} \tag{6.4}$$

We construct FEM space $\mathbf{V} \subset \mathbf{H}_0^1(\Omega_h)$:

$$\mathbf{V} = \{v \in \mathbb{C}(\Omega_h) : v|_{T_j} \text{ is linear for } j = 1, \cdots, M\} \cap \mathbf{H}_0^1(\Omega_h) \tag{6.5}$$

We multiply by the weight function $\phi \in \mathbf{V}$ and integrate by parts using Green's first identity to get

$$\int_\Omega -\nabla \cdot (\gamma \nabla u)\phi d\mathbf{r} = 0 \Rightarrow -\int_{\partial\Omega} \gamma\phi(\nabla u \cdot \mathbf{n})ds + \int_\Omega \gamma \nabla u \cdot \nabla\phi d\mathbf{r} = 0. \tag{6.6}$$

After rearranging and substituting boundary condition, we get

$$\int_\Omega \gamma \nabla u \cdot \nabla \phi d\mathbf{r} = \int_{\partial\Omega} \gamma\phi(\nabla u \cdot \mathbf{n})ds = \int_{\partial\Omega} \phi(\gamma\nabla u \cdot \mathbf{n})ds = \int_{\partial\Omega} f\phi ds.$$

According to the Lax–Milgram theorem [17], there exists a unique solution satisfying

$$\int_\Omega \gamma \nabla u \cdot \nabla \phi d\mathbf{r} = \int_{\partial\Omega} f\phi ds, \quad \forall \phi \in \mathbf{V}. \tag{6.7}$$

Let $u = \sum_{i=1}^{N} u_i \phi_i$ be a finite element solution $\mathbf{u} = [u_1, u_2, \cdots, u_N]^T$, where $u_i = u(\mathbf{r}_i)$ and N-number of nodes. The finite element method is a technique for numerically approximating the value of a function over a domain. Discretizing over an approximation of the potential distribution u, defined at nodes of the mesh and between by the linear combinations of basis functions ϕ_i, with weight functions selected to be the same basis functions ϕ_j gives a linear system of equations

$$\sum_i \sum_j \int_\Omega \gamma u_j \nabla\phi_i \cdot \nabla\phi_j d\mathbf{r} = \sum_j \int_{\partial\Omega} \phi_i (\gamma \nabla u_j \cdot \mathbf{n}) ds \quad \text{for rows } i.$$

Then $\mathbf{u}$ satisfies

$$\underbrace{\begin{pmatrix} \int_\Omega \gamma \nabla\phi_1 \cdot \nabla\phi_1 d\mathbf{r} & \cdots & \int_\Omega \gamma \nabla\phi_1 \cdot \nabla\phi_N d\mathbf{r} \\ \vdots & \int_\Omega \gamma \nabla\phi_i \cdot \nabla\phi_j d\mathbf{r} & \vdots \\ \int_\Omega \gamma \nabla\phi_N \cdot \nabla\phi_1 d\mathbf{r} & \cdots & \int_\Omega \gamma \nabla\phi_N \cdot \nabla\phi_N d\mathbf{r} \end{pmatrix}}_{A} \underbrace{\begin{pmatrix} u_1 \\ \vdots \\ u_j \\ \vdots \\ u_N \end{pmatrix}}_{\mathbf{u}} = \underbrace{\begin{pmatrix} \int_{\partial\Omega} f\phi_1 ds \\ \vdots \\ \int_{\partial\Omega} f\phi_i ds \\ \vdots \\ \int_{\partial\Omega} f\phi_N ds \end{pmatrix}}_{b} \tag{6.8}$$

Let $\tau = [T_1, \cdots, T_M]$ be the set f the triangular elements. The ij–th element of the matrix A can be decomposed into $a_{ij} = \sum_{T\in\tau} \int_T \gamma \nabla\phi_i \cdot \nabla\phi_j d\mathbf{r}$ and the matrix A is symmetric and sparse.

The forward problem (6.4) is solved by the finite element method. The finite element model (node points, FEM triangulation and basis function) is applied to the forward problem when calculating the potential u (see Fig. 6.6). The potential field of a disk with current injected between two adjacent electrodes and the potential u is monotonically decreasing from the source to sink electrodes.

The solution of the linear equation $A\mathbf{u} = b$ is shown in Fig. 6.6. If we assume $\mathbf{u}$ is a solution of (6.4), then $\mathbf{u} + c$ is also a solution of (6.4) for any constant c. It means that A is not invertible and the rank of A is $N - 1$. By dropping the first row and column of A, the linear equation can be solvable and $\hat{\mathbf{u}} = [A_{N-1}]^{-1}\hat{b}$. Using the knowledge of $\hat{\mathbf{u}}$, we obtain the solution $u = \sum_{j=2}^{N} \hat{u}_j \phi_j - c$, where c is a constant chosen so that $\int_{\partial\Omega_h} u ds = 0$.

6.2.3 *Inverse Problem in EIT*

6.2.3.1 Computation of Admittivity in Homogeneous Material

Assume that there are two pairs of electrodes that inject current ($\mathcal{E}_+^1$, $\mathcal{E}_-^1$) and measure voltage ($\mathcal{E}_+^2$, $\mathcal{E}_-^2$), as shown in Fig. 6.7. The electric potential u^i is determined by the

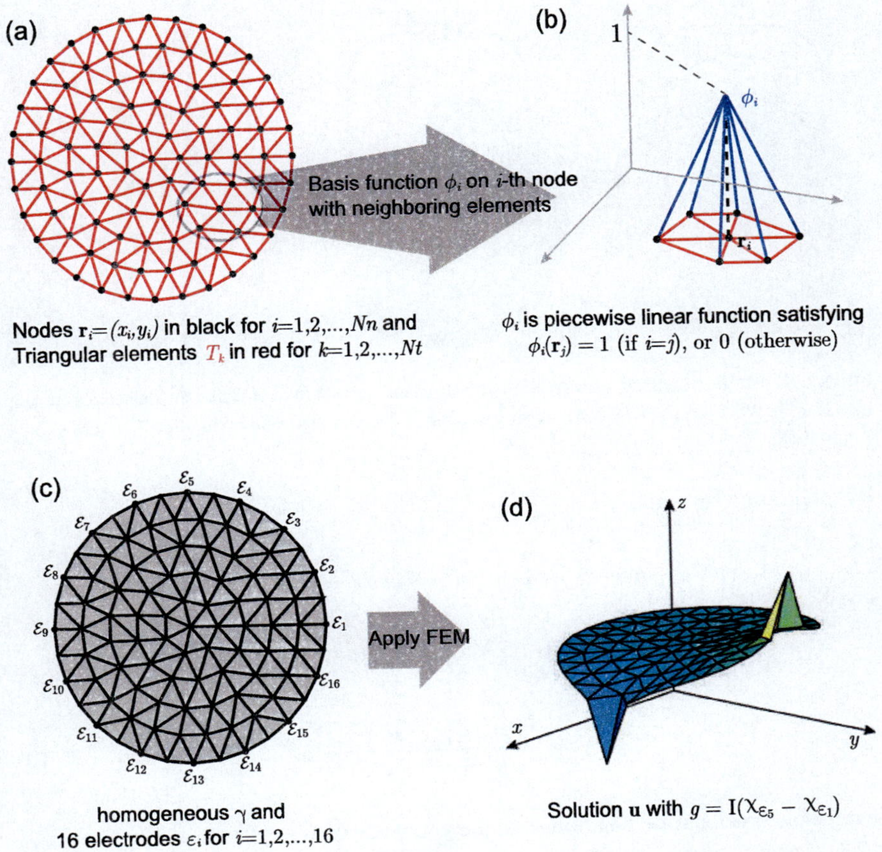

Fig. 6.6 **a** Nodes and triangular elements for the finite element method (FEM). **b** A linear basis function ϕ_i which is used in FEM. **c** Triangular mesh with homogeneous admittivity γ with 16 electrodes $\mathcal{E}_i$. **d** Numerical solution $\mathbf{u}$ in (6.8) using FEM

geometry Ω and the current driving electrodes $(\mathcal{E}_+^i, \mathcal{E}_-^i)$ for $i = 1, 2$:

$$\begin{cases} -\nabla\cdot(\gamma\nabla u^i) = 0 & \text{in } \Omega, \\ (\gamma\nabla u^i)\cdot\mathbf{n} = I(\delta_{\mathcal{E}_+^i} - \delta_{\mathcal{E}_-^i}) & \text{on } \partial\Omega, \end{cases} \tag{6.9}$$

where $\delta_{\mathcal{E}_+^i}$ is the Dirac delta function having peak at $\mathcal{E}_+^i$.

$$\int_\Omega -\nabla\cdot(\gamma\nabla u^1)u^2 d\mathbf{r} = 0,$$

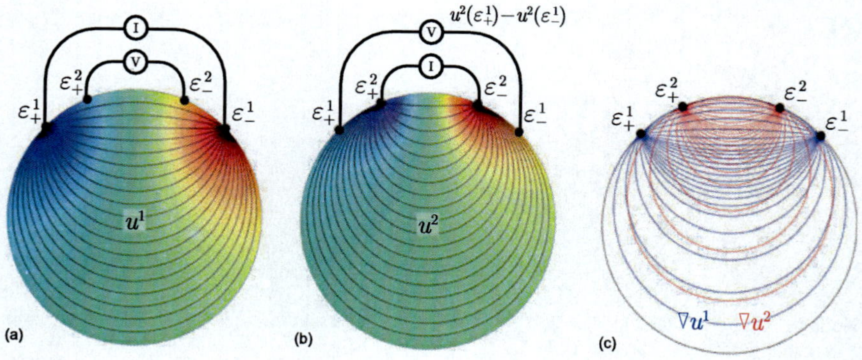

Fig. 6.7 Illustration of electric potential distribution and current map: **a** When $\mathscr{E}_\pm^1$ is used for current injection. **b** When $\mathscr{E}_\pm^2$ is used for current injection. **c** Two current maps ∇u^1 and ∇u^2 are shown simultaneously

By integration by parts,

$$-\int_{\partial\Omega} (\gamma \nabla u^1 \cdot \mathbf{n}) u^2 ds + \int_{\Omega} \gamma \nabla u^1 \cdot \nabla u^2 d\mathbf{r} = 0$$

$$\int_{\Omega} \gamma \nabla u^1 \cdot \nabla u^2 d\mathbf{r} = \int_{\partial\Omega} (\gamma \nabla u^1 \cdot \mathbf{n}) u^2 ds = I[(u^2(\mathscr{E}_+^1) - u^2(\mathscr{E}_-^1)],$$

$$\gamma = \frac{I[(u^2(\mathscr{E}_+^1) - u^2(\mathscr{E}_-^1)]}{\int_\Omega \nabla u^1 \cdot \nabla u^2 d\mathbf{r}}. \tag{6.10}$$

γ depends on electrode positions and the geometry of Ω.

6.2.3.2 Inbody's Impedance Measurement

Inbody's multi-frequency measurements method accurately measures intracellular water and extracellular body water by using multiple broadband frequencies in the range of 1–1000 kHz. The ability of the current to pass through cell membranes varies according to its frequency.

By taking the structural characteristics of the human body into consideration, the Inbody uses two current and voltage electrodes for each hand and foot, requiring the user to grasp and step on to a total of eight electrodes. The design significantly improves ease of use for repeated tests, even when the measuring posture is changed or when measurements are taken multiple times. The measurement always starts and ends at the same points–the wrists and the ankles–ensuring accurate results.

The human body is inhomogeneous. A human body Ω can be decomposed into five parts as Z_{RA}, Z_{LA}, Z_{RL}, Z_{LL}, and Z_B.

The sinusoidal current is injected into the six combinations of these five body parts and the six voltages V^1, V^2, V^3, V^4, V^5 and V^6, as shown in Fig. 6.8. The data is used to calculate the body compositions $V^{jk} = \frac{1}{I} \int_\Omega \gamma \nabla u^j \cdot \nabla u^k d\mathbf{r}$ and

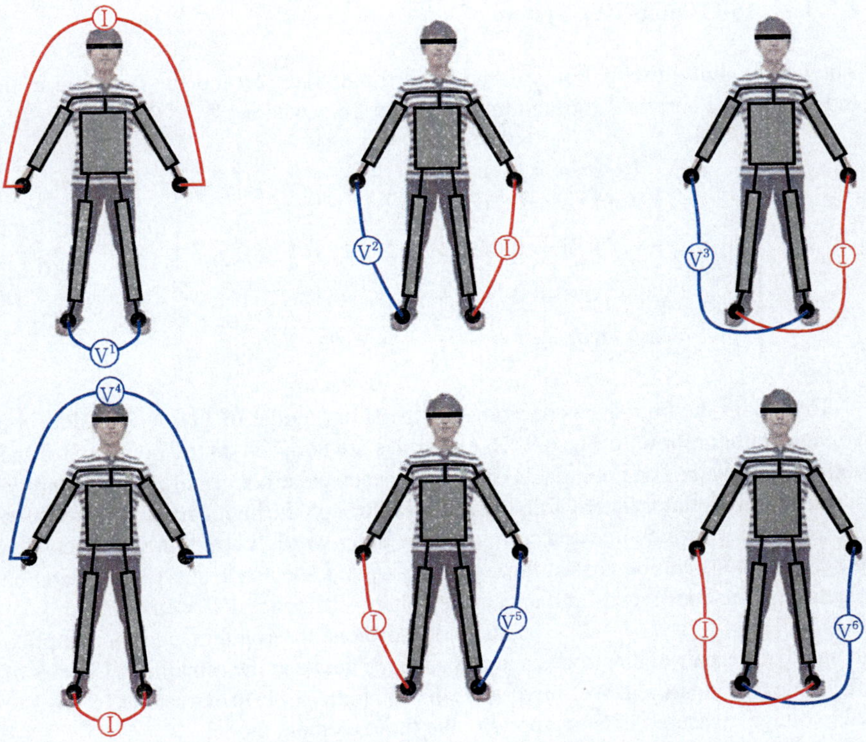

Fig. 6.8 Inbody measurement: six different ways of injecting current and measuring voltage are shown when the human body is considered as five parts with homogeneous electrical conductivity

$$Z^j = \frac{V^j}{I} = \frac{1}{I^2}\int_{\Omega} \frac{1}{\gamma}|\gamma \nabla u^j|^2 dx \text{ : measurable}$$

$$\begin{aligned} Z^1 = & \underbrace{\frac{1}{I^2}\int_{\Omega_{\text{RightArm}}} \frac{1}{\gamma}|\gamma \nabla u^1|^2 dx}_{Z_{\text{RightArm}}} + \underbrace{\frac{1}{I^2}\int_{\Omega_{\text{LeftArm}}} \frac{1}{\gamma}|\gamma \nabla u^1|^2 dx}_{Z_{\text{LeftArm}}} \\ & + \underbrace{\frac{1}{I^2}\int_{\Omega_{\text{Body}}} \frac{1}{\gamma}|\gamma \nabla u^1|^2 dx}_{Z_{\text{Body}}} + \\ & + \underbrace{\frac{1}{I^2}\int_{\Omega_{\text{RightLeg}}} \frac{1}{\gamma}|\gamma \nabla u^1|^2 dx}_{\approx 0} + \underbrace{\frac{1}{I^2}\int_{\Omega_{\text{LeftLet}}} \frac{1}{\gamma}|\gamma \nabla u^1|^2 dx}_{\approx 0}. \end{aligned}$$

6.2.3.3 16-Channel EIT System

The conductivity distribution γ is given in the domain Ω and the potential u^j is induced by the current I through the j-th pair of electrodes $(\mathcal{E}^j, \mathcal{E}^{j+1})$:

$$\begin{cases} -\nabla \cdot (\gamma \nabla u^j) = 0 & \text{in } \Omega, \\ (\gamma \nabla u^j) \cdot \mathbf{n} = 0 & \text{on } \partial\Omega \setminus \bigcup_i^{16} \mathcal{E}^i, \\ \int_{\mathcal{E}^i} (\gamma \nabla u^j) \cdot \mathbf{n} = 0 & \text{for } i \in \{1, \ldots, 16\} \setminus \{j, j+1\}, \\ \mathbf{n} \times \nabla u^j = 0 & \text{on } \mathcal{E}^i \text{ for } i = 1, \ldots, 16, \\ \int_{\mathcal{E}^j} (\gamma \nabla u^j) \cdot \mathbf{n}\, ds = I = -\int_{\mathcal{E}^{j+1}} (\gamma \nabla u^j) \cdot \mathbf{n}\, ds. \end{cases} \tag{6.11}$$

To explain the boundary conditions of the shunt model (6.11), we consider the voltage configuration. In Fig. 6.9, 16 electrodes are attached on the boundary of the abdomen domain. The potential is constant on each electrode. Also on the boundary surface is an insulator. There is no current flow through the boundary except between the 2 electrodes which injected current. In the other words, current injected between the 2 electrodes can be passed through the j–th sink electrode and (j+1)–th source electrode. The above measured data is the voltage difference $V^{jk} = u^j|_{\mathcal{E}^k} - u^j|_{\mathcal{E}^{k+1}}$ for $k = 1, \cdots, j-2, j+2, \cdots, 16$. In the adjacent drive pattern, current is applied to an adjacent pair of electrodes and the voltages between the remaining 13 pairs of electrodes are measured. We aim to solve the equation (6.11) to extract the $16 \times 13 = 208$ voltage measurements by applying the finite element model.

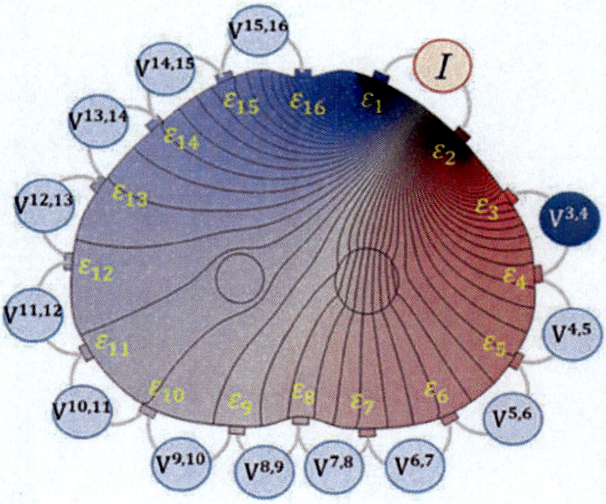

Fig. 6.9 16–channel EIT system. A pair of adjacent electrodes is used to apply a current, and voltage is measured using the adjacent electrodes on all remaining electrodes. The applied current induces a voltage distribution inside, and the distribution of electrical conductivity distorts the voltage distribution

In the inverse problem, the potential $\mathbf{V}$ is given and γ is unknown. Multiplying u^k by the first equation in (6.11) gives

$$
\begin{aligned}
-\int_{\Omega} \nabla \cdot (\gamma \nabla u^j) u^k d\mathbf{r} &= 0 \\
-\int_{\partial\Omega} \gamma \nabla u^j \cdot \mathbf{n} u^k ds + \int_{\Omega} \gamma \nabla u^j \cdot \nabla u^k d\mathbf{r} &= 0 \\
\int_{\Omega} \gamma \nabla u^j \cdot \nabla u^k d\mathbf{r} &= \int_{\partial\Omega} u^k \gamma \nabla u^j \cdot \mathbf{n} ds \\
\frac{1}{I}\int_{\Omega} \gamma \nabla u^j \cdot \nabla u^k d\mathbf{r} &= \frac{1}{I}\int_{\partial\Omega} u^k (\gamma \nabla u^j) \cdot \mathbf{n} ds \\
&= \frac{1}{I}\int_{\mathcal{E}^j \cup \mathcal{E}^{j+1}} u^k (\gamma \nabla u^j) \cdot \mathbf{n} ds \\
&= u^k|_{\mathcal{E}^j} - u^k|_{\mathcal{E}^{j+1}} \\
&= V^{kj}.
\end{aligned} \tag{6.12}
$$

For a given reference admittivity γ_0, the corresponding potential u_0 satisfies

$$
\frac{1}{I}\int_{\Omega} \gamma_0 \nabla u_0^j \cdot \nabla u_0^k d\mathbf{r} = V_0^{jk}. \tag{6.13}
$$

The difference $V^{jk} - V_0^{jk}$ satisfies

$$
\begin{aligned}
V^{jk} - V_0^{jk} &= \frac{1}{I}\int_{\Omega} \gamma \nabla u^j \cdot \nabla u^k d\mathbf{r} - \frac{1}{I}\int_{\Omega} \gamma_0 \nabla u_0^j \cdot \nabla u_0^k d\mathbf{r} && (6.14) \\
&= -\frac{1}{I}\int_{\Omega} (\gamma - \gamma_0) \nabla u^j \cdot \nabla u_0^k d\mathbf{r} && (6.15) \\
&\approx -\frac{1}{I}\int_{\Omega} (\gamma - \gamma_0) \nabla u_0^j \cdot \nabla u_0^k d\mathbf{r}. && (6.16)
\end{aligned}
$$

Discretizing Ω into M elements as $\Omega = \bigcup_m^M T_m$,

$$
-\sum_m^M \frac{1}{I}\int_{T_m} \left(\gamma\Big|_{T_m} - \gamma_0\Big|_{T_m}\right) \nabla u^j \cdot \nabla u_0^k d\mathbf{r} = V^{jk} - V_0^{jk}. \tag{6.17}
$$

A matrix form of the inverse problem is $\mathbb{S}\dot{\gamma} = \dot{\mathbf{V}}$, where $\dot{\gamma} = \gamma - \gamma_0$, $\dot{\mathbf{V}} = \mathbf{V} - \mathbf{V}_0$ and $\mathbf{V} = [V^{1,2}, ..., V^{15,16}, ..., V^{j,k}, ...]^T$.

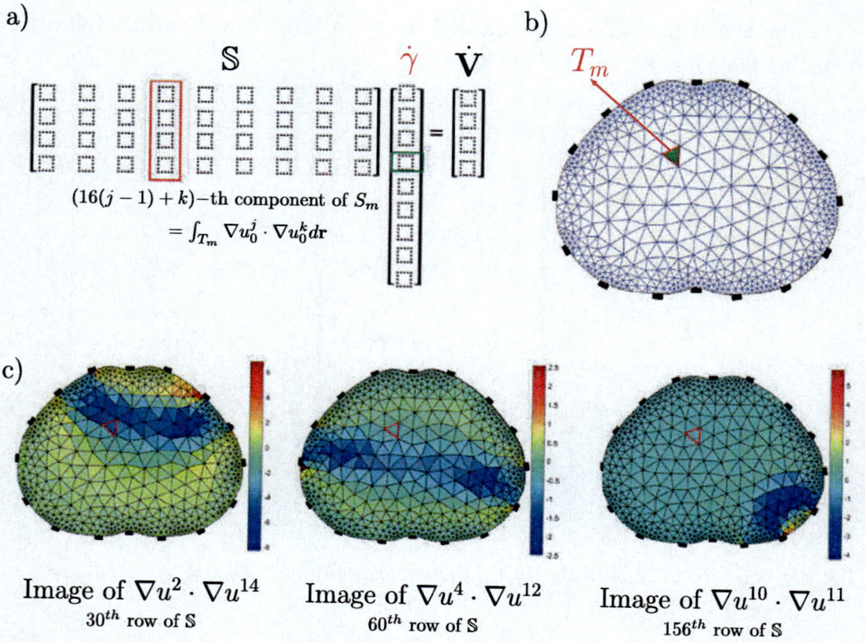

Fig. 6.10 FEM element (pixel). **a** Sensitivity matrix; **b** the position of pixel T_m; **c** potential u from pixel dipole

$$\mathbb{S} = \begin{pmatrix} | & \cdots & | & \cdots & | \\ -\frac{1}{I}\int_{T_1} \nabla u_0^j \cdot \nabla u_0^k d\mathbf{r} & \cdots & -\frac{1}{I}\int_{T_m} \nabla u_0^j \cdot \nabla u_0^k d\mathbf{r} & \cdots & -\frac{1}{I}\int_{T_M} \nabla u_0^j \cdot \nabla u_0^k d\mathbf{r} \\ | & \cdots & | & \cdots & | \end{pmatrix}. \tag{6.18}$$

Noting that the reciprocity principle

$$V^{jk} = u^j|_{\mathcal{E}^k} - u^j|_{\mathcal{E}^{k+1}} = \frac{1}{I}\int_{\Omega} \gamma \nabla u^j \nabla u^k d\mathbf{r} = u^k|_{\mathcal{E}^j} - u^k|_{\mathcal{E}^{j+1}} = V^{kj},$$

we have $(N-3)N/2$ independent data. See Fig. 6.10 for the sensitivity matrix.

Time-difference EIT is to provide images of changes (i.e. $\delta\gamma = \gamma - \gamma_0$) that satisfies

$$\begin{aligned}
0 &= \nabla\cdot\Big((\gamma_0+\delta\gamma)\nabla(u_0^j+\delta u^j)\Big)\\
&= \nabla\cdot(\gamma_0\nabla\delta u^j)+\nabla(\delta\gamma_0\nabla u_0^j)\\
\int_\Omega -\nabla\cdot(\gamma_0\nabla\delta u^j)u_0^k &= \int_\Omega \nabla\cdot(\delta\gamma\nabla u_0^j)u_0^k\\
\int_\Omega \gamma_0\nabla\delta u^j\cdot u_0^k &= \int_{\partial\Omega}(\gamma_0\nabla\delta u^j)\cdot\mathbf{n}\cdot u_0^k\\
&= \int_\Omega \delta\gamma\nabla u_0^j\cdot\nabla u_0^k\\
\int_\Omega \delta\gamma\nabla u_0^j\cdot\nabla u_0^k &= \int_{\partial\Omega}(\gamma_0\nabla\delta u^j)\cdot\mathbf{n}\cdot u_0^k\\
&= \delta u^j(\mathcal{E}_{k+1})-\delta u^j(\mathcal{E}_k) \;=\; \delta V^{jk}
\end{aligned} \tag{6.19}$$

6.2.4 Sensitivity Analysis

The inverse problem satisfies the matrix equation $\mathbb{S}\dot{\gamma}=\mathbf{V}$. The Jacobian matrix $\mathbb{S}$ is called the sensitivity matrix.

Let's consider singular value decomposition of $\mathbb{S}^T\mathbb{S}$:

$$\mathbb{S}^T\mathbb{S} = W\cdot\Sigma\cdot W^T = \begin{pmatrix} w_1 & \cdots & w_M\end{pmatrix}\begin{pmatrix}\lambda_1^2 & \cdots & 0\\ \vdots & \ddots & \vdots\\ 0 & \cdots & \lambda_M^2\end{pmatrix}\begin{pmatrix} w_1^T\\ \vdots\\ w_M^T\end{pmatrix}, \tag{6.20}$$

where v_i are eigenvectors and λ_i are singular values. Since the conditional number is very high (for example in Fig. 6.11, $\lambda_1/\lambda_{35}=2.7\times10^3$), the Jacobian matrix $\mathbb{S}$ is ill-conditioned. Imagine that $\dot{\mathbf{g}}$ is expressed as a linear combination of the eigenvectors $\dot{\mathbf{g}}=\sum_{i=1}^M a_i w_i$. The inverse problem is reduced to finding $(a_1, a_2, ..., a_M)$ such that

$$\mathbb{S}^T\mathbb{S}\left(\sum_{i=1}^M a_i w_i\right)=\mathbb{S}^T\dot{\mathbf{V}} \;\Leftrightarrow\; \sum_{i=1}^M a_i\lambda_i^2 w_i=\mathbb{S}^T\dot{\mathbf{V}}. \tag{6.21}$$

If $\lambda_j\approx 0$, then

$$a_i=\frac{1}{\lambda_i^2}\frac{\langle\mathbb{S}^T\dot{\mathbf{V}},w_i\rangle}{\langle w_i,w_i\rangle}\approx\infty\cdot\frac{\langle\mathbb{S}^T\dot{\mathbf{V}},w_i\rangle}{\langle w_i,w_i\rangle}. \tag{6.22}$$

Hence, a small error in $\dot{\mathbf{V}}$ produces a large error in a_i.

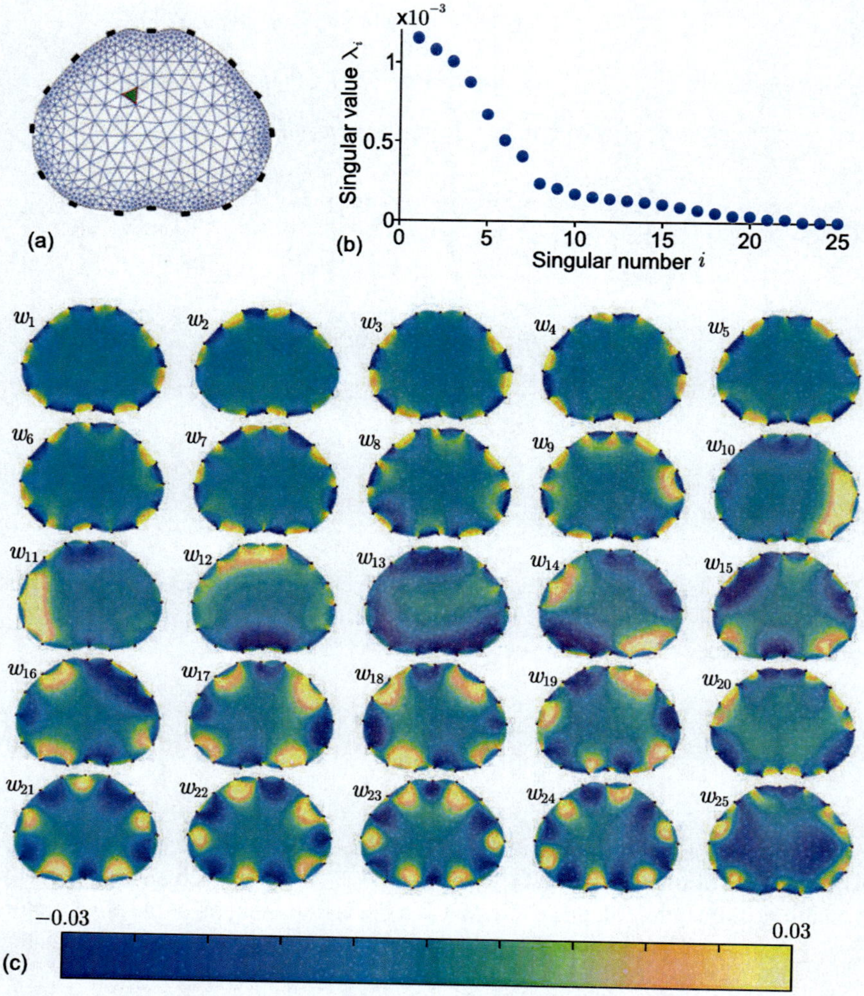

Fig. 6.11 The singular values and singular vector of $\mathbb{S}^T\mathbb{S}$ are computed when we use 16 electrodes with adjacent pairs. **a** Abdominal model with electrodes (black) and mesh (blue line). **b** Singular values of $\mathbb{S}^T\mathbb{S}$. Of the over 500 singular values, only the first 25 are displayed because after that they are almost zero. **c** Visualization of the first 25 eigenvectors w_i $(i = 1, 2, \ldots, 25)$ of $\mathbb{S}^T\mathbb{S}$

6.2.4.1 Ill-Posed Problem and Regularization

Ill-posed inverse problems are defined as opposites of well-posed direct problems [26]. A problem is well-posed if the following conditions hold: A solution exists, the solution is unique, and the solution depends continuously on the input. Since the matrix $\mathbb{S}$ is ill-conditioned, $\mathbb{S}\dot{\gamma} = \dot{\mathbf{V}}$ is ill-posed. Moreover, the number of unknowns

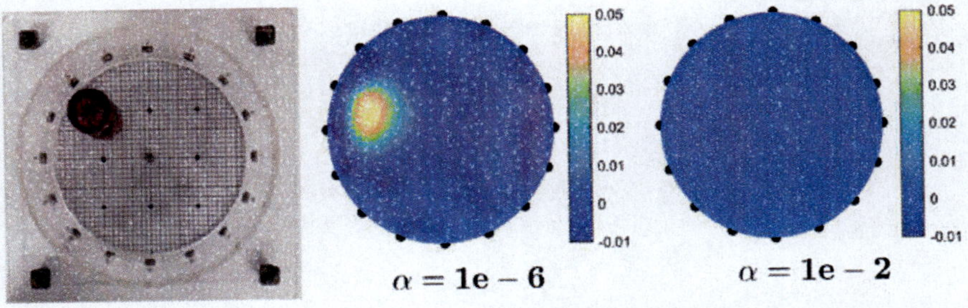

Fig. 6.12 Reconstructed images using Tikhonov regularization with different regularization parameters

(i.e. pixel numbers, e.g. 5000) in $\dot{\gamma}$ is bigger than the number of equations (i.e. number of current–voltage patterns, e.g. 208). Hence, there are infinitely many solutions.

In order to deal with this ill-posedness, we usually use a regularization technique: From (6.21), we have

$$(\alpha I + \mathbb{S}^T\mathbb{S})\left(\sum_{i=1}^{M} a_i w_i\right) = \mathbb{S}^T\dot{\mathbf{V}} \Rightarrow \sum_{i=1}^{M} a_i\left(\alpha + \lambda_i^2\right) w_i = \mathbb{S}^T\dot{\mathbf{V}}. \tag{6.23}$$

The standard reconstruction algorithm is based on the conventional regularized model-fitting method:

$$\dot{\gamma} = \text{argmin}_{\dot{\gamma}} \|\dot{\mathbf{V}} - \mathbb{S}\dot{\gamma}\|^2 + \lambda Reg(\dot{\gamma}), \tag{6.24}$$

where $\|\cdot\|$ is the standard Euclidean norm, Reg is a regularization operator, and $\lambda > 0$ is the regularization parameter.

We will deal with two regularization parameters; Tikhonov regularization, and total variation. First, Tikhonov regularization is when $Reg(\cdot) := \|\cdot\|_2^2$. Thus, the corresponding solution is $\dot{\gamma}^*$:

$$\dot{\gamma}^* = \text{argmin}_{\dot{\gamma}} \|\mathbb{S}\dot{\gamma} - \dot{\mathbf{V}}\|^2 + \alpha\|\dot{\gamma}\|_2^2, \tag{6.25}$$

where α is regularization parameter [75]. Here the objective function is $\Phi(\dot{\gamma}) = \|\mathbb{S}\dot{\gamma} - \dot{\mathbf{V}}\|^2 + \alpha\|\dot{\gamma}\|_2^2$. The minimizer $\dot{\gamma}^*$ satisfies $\nabla\Phi(\dot{\gamma}^*) = 2\mathbb{S}^T(\mathbb{S}\dot{\gamma} - \dot{\mathbf{V}}) + 2\alpha\dot{\gamma} = 0$ and its solution can be obtained as

$$\dot{\gamma}^* = (\mathbb{S}^T\mathbb{S} + \alpha I)^{-1}\mathbb{S}^T\dot{\mathbf{V}}, \tag{6.26}$$

Figure 6.12 shows $\dot{\gamma}^*$ using (6.25) with different regularization parameters.

Total variation regularization uses $Reg(\cdot) := \|D\dot{\gamma}\|_1$. Thus, the corresponding solution is $\dot{\gamma}^*$:

$$\dot{\gamma}^* = \text{argmin}_{\dot{\gamma}} \frac{1}{2}\|\mathbb{S}\dot{\gamma} - \dot{\mathbf{V}}\|^2 + \lambda\|D\dot{\gamma}\|_1.$$

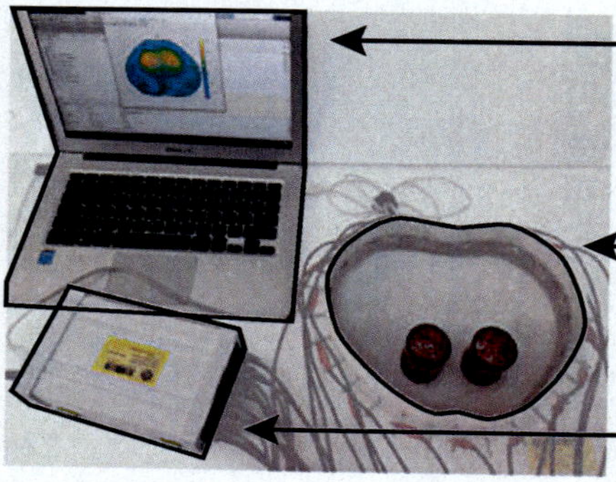

Fig. 6.13 Picture of phantom experiment. Abdomen-shaped tank is filled with saline water and 2 glass cups. In order to measure voltage while injecting current, a multi-channel electrical impedance device is connected to the tank. A laptop controls the device and reconstructs conductivity distribution

The corresponding dual problem is $\text{argmin}_{\dot{\gamma}} \frac{1}{2}\|\mathbb{S}\dot{\gamma} - \dot{\mathbf{V}}\|^2 + \lambda\|z\|_1$ subject to $z = D\dot{\gamma}$ and the augmented Lagrangian is

$$L(\dot{\gamma}, y, z) = \frac{1}{2}\|\mathbb{S}\dot{\gamma} - \dot{\mathbf{V}}\|^2 + \lambda\|z\|_1 + y^T(D\dot{\gamma} - z) + \frac{\rho}{2}\|D\dot{\gamma} - z\|^2.$$

So, the solution can be obtained using the Alternating Direction Method of Multipliers (ADMM):

1. Initial guess for $\dot{\gamma}^{(0)}$, $y^{(0)}$ and $z^{(0)}$.
2. Solve problems

$$\begin{cases} \dot{\gamma}^{(k+1)} := (\mathbb{S}'\mathbb{S} + \rho D^T D)^{-1}(\mathbb{S}\dot{\mathbf{V}} - D^T(y^{(k)} - \rho z^{(k)})). \\ z^{(k+1)} := \text{Shrinkage function}_{\lambda/\rho}(D\dot{\gamma}^{(k+1)} + \frac{1}{\rho}y^{(k)}), \\ y^{(k+1)} := y^{(k)} + \rho(D\dot{\gamma}^{(k+1)} - z^{(k+1)}). \end{cases}$$

Figure 6.13 shows a EIT system phantom. Figure 6.14 shows a comparison of two regularization models.

Remark 6.3 The inverse problem in EIT has major issues, e.g. it is nonlinear and ill-posed. There are various other issues such as geometry error, electrode position error, measurement noise, domain truncation error and current-voltage pattern effect. We should note that $\mathbf{S}^\gamma$ can be viewed as a highly nonlinear function of γ. Moreover, $\mathbf{V}$ depends mainly on the boundary geometry and the electrode positions, whereas its dependence on a local perturbation of $\boldsymbol{\gamma}$ is relatively small. Barber and Brown [7] made the following observation: *If electrodes are spaced 10 cm apart around the thorax, variation in positioning of 1 mm will produce errors of 1% in the data* $\mathbf{V}$. Such a 1% error is too large for most clinical applications. Hence, the static EIT problem is highly nonlinear and severely ill-posed. Even if infinitely many trans-impedance

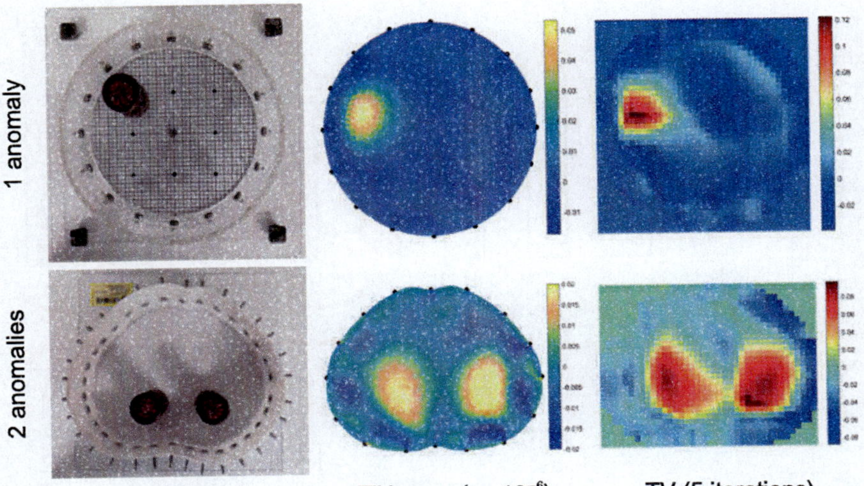

Fig. 6.14 Reconstructed images of time-difference EIT

data $Z^{j,k}$ are available, it seems to be very difficult to recover the distribution of the impedivity $1/\gamma$. Instead of conductivity imaging, some researchers are focusing on detection of anomalies such as cancer instead of imaging [4, 42].

6.2.5 *Deep Learning-Based EIT*

Whether an inverse problem is well-posed may be dependent on how the solution is expressed. Many problems are ill-posed because we are overly ambitious or lacking in expressiveness. For clinical EIT to succeed in commercialization, it is necessary to acknowledge its limitations and make an effort to provide an impedance image that fits a special purpose for tracking projected imaging features for medical diagnosis. For example, for a 16-channel EIT system, we have to deal with the uncertainty of a number of free parameters (pixel dimension−data dimension = 16384 − 208).

In order to solve nonlinear and ill-posed inverse problems in EIT, deep learning techniques have been recently adopted [13, 14, 27, 66]. Deep learning methods appear to have a strong capability to explore the prior information of the expected images via training data, which allows one to deal with the uncertainty of solutions to ill-posed inverse problems. Deep learning frameworks may provide a nonlinear regression on training data which acts as learning complex prior knowledge on the output. In [66], an image reconstruction algorithm is created based on a variational auto-encoder (VAE) and multi-layer perceptron (MLP) (Fig. 6.15). The first network, Fig. 6.15a, is a VAE network that allows one to achieve compact representation (or low–dimensional manifold learning) for prior information of lung EIT images.

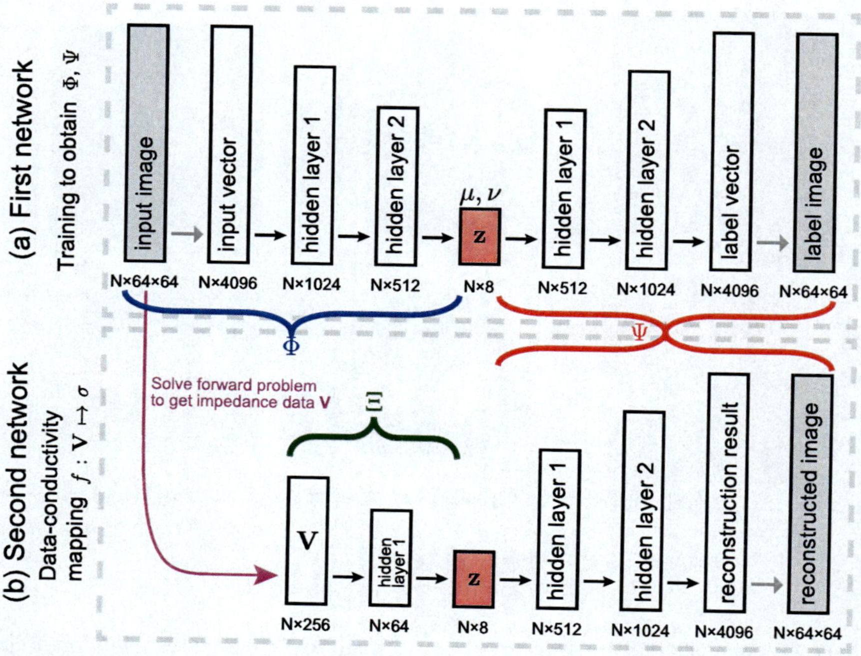

Fig. 6.15 Architecture of the learning based image reconstruction method in [66]. **a** First network learns an 8-dimensional manifold representation **z** for getting prior knowledge of EIT images using variational autoencoder (VAE). **b** Second network trains a map $f : \mathbf{V} \to \sigma$ from impedance data **V** to conductivity distribution σ. Here **z** and Ψ were given by the first network

For the second step, Fig. 6.15b, only the decoder part of this network is used. This decoder part takes only very few latent variables and transforms them back to produce an image on the learned manifold of meaningful reconstructions. The second network now takes a data vector and maps it onto the latent variables, which are then input to the decoder. This approach exploits the potential of neural networks for constructing low–dimensional nonlinear representations of approximate solution maps.

6.2.6 Applications

There are many areas where EIT can be applied. One of the most active fields is cardiopulmonary function monitoring. The high air pressure in mechanical ventilators often causes parts of the lungs to over-inflate, resulting in ventilator-associated lung injury (VALI). To address these unmet clinical needs, dynamic conductivity images from EIT can be used for lung protective ventilation (LPV) to find an optimal PEEP (positive end-expiratory pressure) value for each patient and for continuous bedside

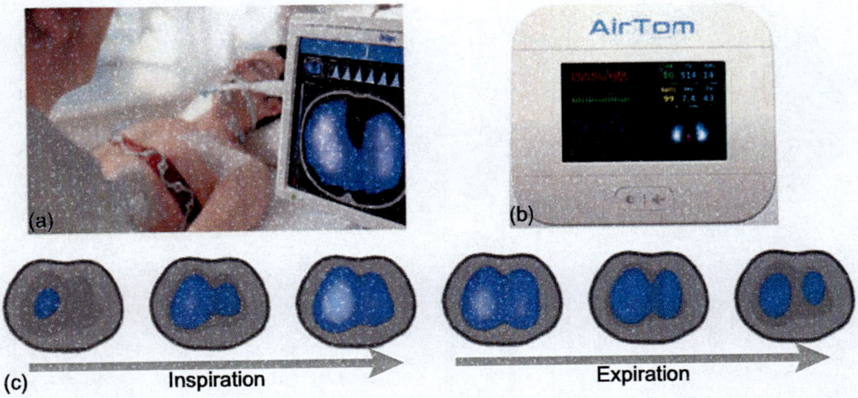

Fig. 6.16 EIT ventilation monitoring. **a** Electrode belt with device screen from Dräger. **b** Device display of showing lung image and other clinical information from BiLab. **c** Lung ventilation image while inspiration and expiration from Timpel

monitoring of regional air distributions in the lungs. There are many related companies, e.g. Dräger [19] in Germany, BiLab [9] in South Korea, Timpel [76] in Brazil, Swisstom [71] which was acquired by Sentec [55] in Switzerland. Figure 6.16 shows the use of EIT technology for lung monitoring, including electrode belts, devices, display screens, and lung conductivity distributions.

Another possible application is to estimate abdominal obesity which requires static images. Abdominal obesity is closely linked to metabolic syndrome as well as a risk factor for various other health conditions. Static conductivity images from EIT can provide regional distribution of abdominal fats, such as subcutaneous and visceral fats for continuous self-monitoring to track body fat status as part of a daily routine. In 2017, Ammari et al. [5] provides a reconstruction method for imaging the abdominal conductivity distribution using some electrodes surrounding the abdomen. The static EIT image reconstruction technologies in [5, 44] have been validated through several numerical experiments (e.g. Fig. 6.17a, b), suggesting that EIT can provide human abdominal conductivity images for estimating abdominal obesity. An example of using human bioimpedance is the device of InBody Co., Ltd. [34]. InBody measured abdominal obesity (without abdominal image) using bioimpedance obtained using four electrodes in the human abdomen (Fig. 6.17c, d).

Sleep apnea, which affects more than 100 million people, is a potentially serious sleep disorder in which breathing starts and stops repeatedly. The most common type of sleep apnea occurs during sleep when breathing is blocked due to the collapse of the throat's relaxed tongue and/or fatty tissues. The EIT technique can be used for patient-friendly home sleep tests (HSTs) of patients with apnea by providing images of the opening and closing of the airway and/or the air filling the lungs at every breath. Currently, changes in the upper airway size can be estimated with good

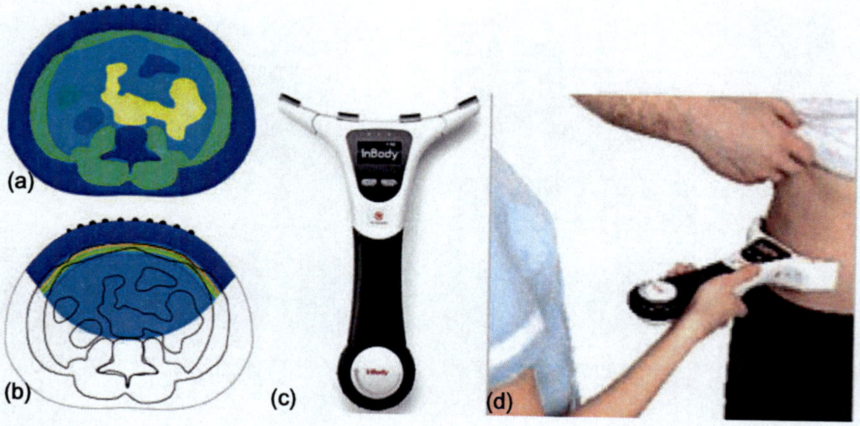

Fig. 6.17 Abdominal fat measurement using bio-impedance. **a** Abdominal electrical conductivity distribution used in numerical experiments. **b** Reconstructed anterior abdominal EIT image with 10 surface electrodes using bio-impedance from (**a**) in numerical experiments [44]. **c** A commercial device of InBody Co., Ltd. [34] that provides indicators of abdominal obesity such as subcutaneous fat mass, visceral fat mass, abdominal fat ratio, visceral/subcutaneous fat cross-sectional area ratio, and visceral fat cross-sectional area. **d** Demonstration of use of product (**c**)

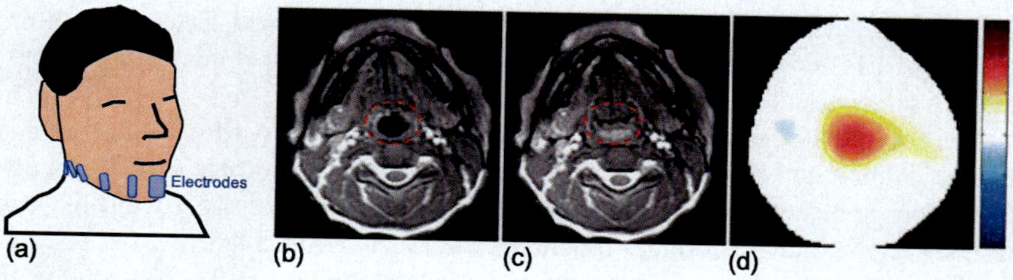

Fig. 6.18 Images for upper airway narrowing or collapse. **a** Surface electrodes attached to the lower face. **b** Magnetic resonance (MR) image when holding the breath at the end of normal inhalation (open airway). **c** MR image in the middle of the swallowing maneuver (closed airway). **d** EIT image reconstructed by subtracting bio-impedance for airway open status from airway closed status

accuracy, but shape estimation needs future improvements in the EIT image quality. The research in [40] shows the EIT applied to the upper airway image. By attaching surface electrodes to the lower jaw, the electrical conductivity distribution according to the opening and closing of the upper airway is visualized and verified using MR images (Fig. 6.18).

6.3 Magnetic Resonance Electrical Impedance Tomography (MREIT)

Since numerous endeavors in EIT research during the last three decades have shown us that the boundary data in EIT may not be sufficient to achieve robust reconstructions of γ, we should find a different method to probe $\mathbf{J}$ and $\mathbf{E}$. Due to the relation $\gamma\mathbf{E} = \mathbf{J} = \nabla \times \mathbf{B}$ with $\mathbf{B}$ being the magnetic flux density, MRI would be a top candidate to estimate $\mathbf{J}$ for conductivity imaging. In 1989, a research group at Toronto University developed a current density imaging (CDI) technique [36] using MRI to visualize $\mathbf{J}$ in Ω due to an injection current through surface electrodes. The major drawback of CDI is the requirement of object rotation inside the MRI scanner to acquire all three components of the induced magnetic flux density, as MRI can measure only the z-component of $\mathbf{B} = (B_x, B_y, B_z)$, where the z-axis is the axial magnetization direction of the MRI scanner. Hence, we must rotate the object in order to acquire three components of $\mathbf{B}$, which leads to serious technical difficulties.

Although there have been numerous attempts to deal with the requirement of the object rotation since the 1990s, there remain serious technical difficulties in handling this drawback, which seriously limits the clinical applicability of this tool. In order to reach the stage of animal and human imaging, we should recover σ using only B_z data to avoid object rotation. Until 2000, conductivity imaging using only B_z data seemed impossible. According to Maxwell's equations, the current density is directly related to the three components of $\mathbf{B}$, and σ must be computed from the relationship $\mathbf{J} = \sigma\mathbf{E}$. Hence, most researchers considered that only B_z data is insufficient for recovering σ.

In 2003, Seo et al. [60] carefully investigated the nonlinear relationship between the conductivity and the measured data via the Biot–Savart law, making a key observation that $\nabla^2 B_z$ probes changes in the logarithm of σ along any equipotential curve in each imaging slice. In this method, two different currents are injected into the body to generate two linearly independent current densities $\mathbf{J}^1$ and $\mathbf{J}^2$. They showed that if $|(\mathbf{J}^1 \times \mathbf{J}^2) \times (0, 0, 1)| \neq 0$ at every position in the imaging slice, $\nabla\sigma \times (0, 0, 1)$ could be precisely reconstructed. They used a geometric index theory in mathematics to prove rigorously that the area of the parallelogram is non-zero when the two pairs of surface electrodes are appropriately attached. Taking advantage of these mathematical observations, they found a representation formula for the conductivity which leads to the development of a constructive irrotational MREIT algorithm termed the harmonic B_z algorithm. This representation formula exists in an implicit form due to the non-linear relationship between the conductivity and the measured data, but it was designed to use the fixed point theory. This means that the formula has a contraction mapping property such that an iterative method can be used. An interesting aspect is that this method takes advantage of the major drawback of EIT, ill-posedness, in that the overall flow of the current density is insensitive to local perturbations in the conductivity distribution. Indeed, the harmonic B_z method takes advantage of this fact to make the algorithm work. After the invention of the harmonic B_z algorithm,

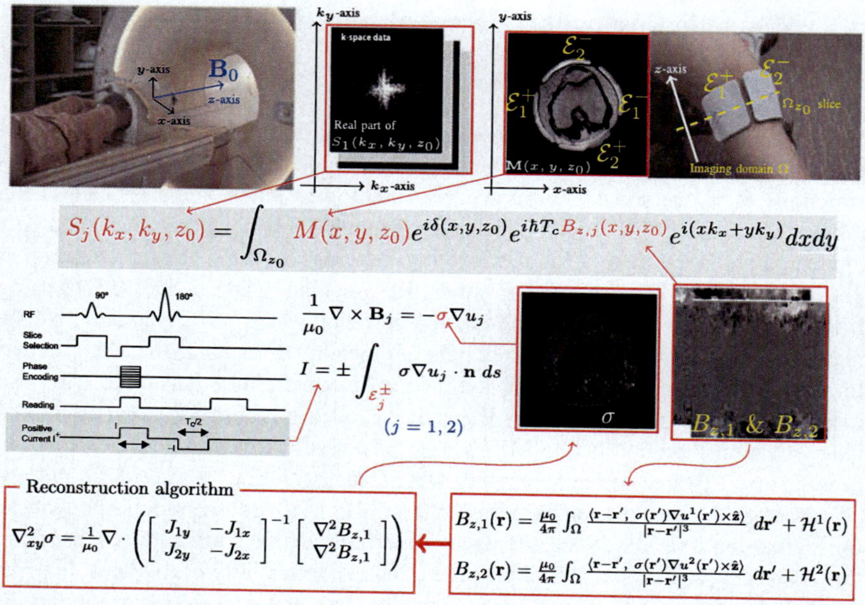

Fig. 6.19 Mathematics-oriented research overcomes technical barriers in electrical tissue property imaging [59]. The harmonic B_z-algorithm is the first imaging technique for static conductivity and current density distributions not to require rotation of the object inside the MRI [56, 58, 59]

imaging techniques in MREIT have advanced rapidly and now can offer state-of-the-art conductivity imaging using MRI animal and human experiments [58, 78]. See Fig. 6.20.

The MREIT image reconstruction procedure is as follows [56–58]:

Step 1. As shown in Fig. 6.19, we attach two pairs of electrodes $\mathcal{E}_1^{\pm}$ and $\mathcal{E}_2^{\pm}$ on the surface of Ω.

Step 2. Inject two linearly independent currents through the pairs of electrodes $\mathcal{E}_1^{\pm}$ and $\mathcal{E}_2^{\pm}$ to produce two linearly independent current densities $\mathbf{J}^1$ and $\mathbf{J}^2$ inside Ω, respectively.

Step 3. Acquire k-space data from an MRI scanner. The MR spectrometer provides the complex k-space data $\mathcal{S}_j$:

$$\mathcal{S}_j(k_x, k_y, z_0) = \int_{\Omega_{z_0}} M(x, y, z_0) e^{i\delta(x,y,z_0)} e^{i\hbar T_c B_{z,j}(x,y,z_0)} e^{i(xk_x + yk_y)}\, dxdy. \tag{6.27}$$

where M is a conventional MR magnitude image, δ is any systematic phase artifact, $\hbar = 26.75 \times 10^7$rad/T $\cdot$ s is the gyromagnetic ratio of hydrogen and T_c the current pulse width in seconds. Taking two-dimensional discrete Fourier transformations to the k-space MR signal and applying a two-dimensional phase-unwrapping operator, we obtain the data obtain the complex MR images $B_{z,1}$ and $B_{z,2}$ as shown in Fig. 6.19. For details, please refer to [56, 58].

Step 4. Get simulated $\mathbf{J}_j = -\sigma \nabla u_j[\sigma]$.

- Use the MR magnitude image M to segment $\partial\Omega$ and $\mathcal{E}_j^{\pm}$.
- Computation of $\mathbf{E}_j = -\nabla u_j[\sigma]$:

$$\nabla u_j[\sigma] = \beta_j \nabla \tilde{u}_j \quad .$$

where $\tilde{u}_j$ is the solution of

$$\begin{cases} \nabla \cdot (\sigma \nabla \tilde{u}_j) = 0 \text{ in } \Omega \\ \tilde{u}_j|_{\mathcal{E}_j^{\pm}} = \pm 1, \quad -\sigma \frac{\partial \tilde{u}_j}{\partial \mathbf{n}}|_{\partial\Omega\setminus(\mathcal{E}_j^{+} \cup \mathcal{E}_j^{-})} = 0 \end{cases}$$

and $\beta_j = I \left(\int_{\mathcal{E}_j^{+}} \sigma \frac{\partial \tilde{u}_j}{\partial \mathbf{n}} \right)^{-1}$.

Step 5. For $\sigma^0 = 1$, compute $\mathbb{A}[\sigma^0]$ where $\mathbb{A}[\sigma]$ is defined by

$$\mathbb{A}[\sigma] := \begin{bmatrix} \sigma \frac{\partial u_1[\sigma]}{\partial x} & -\sigma \frac{\partial u_1[\sigma]}{\partial y} \\ \sigma \frac{\partial u_2[\sigma]}{\partial x} & -\sigma \frac{\partial u_2[\sigma]}{\partial y} \end{bmatrix}.$$

Step 6. Compute

$$F_{\sigma^0} := \mathbb{A}[\sigma^0]^{-1} \begin{bmatrix} \nabla^2 B_{z,1} \\ \nabla^2 B_{z,2} \end{bmatrix} (1 - \chi_{\Omega^\delta}),$$

where Ω^δ is a subdomain of Ω:

$$\Omega^\delta := \{(x, y, z) \in \Omega \ : \ |M| < \delta_M, \ \left|\det \mathbb{A}[\sigma^0]\right| < \delta_A\}.$$

Here, δ_M and δ_A are small positive numbers depending on the signal-tonoise ratio (SNR) of the measured B_z data.

Step 7. For each slice Ω_{z_0}, solve the two-dimensional Poisson equation

$$\begin{cases} \nabla^2_{xy} \ln \sigma^1(x, y, z_0) = \nabla_{xy} \cdot F_{\sigma^0}(x, y, z_0) & \text{for } (x, y, z_0) \in \Omega_{z_0} \\ \ln \sigma^1 = 0 & \text{on } \partial\Omega_{z_0}. \end{cases} \tag{6.28}$$

Step 8. Scale σ^1 as

$$\sigma^1 \quad \leftarrow \quad \frac{V_1^+ - V_1^-}{u_1[\sigma^1]|_{\mathcal{E}_2^+} - u_1[\sigma^1]|_{\mathcal{E}_2^-}} \sigma^1,$$

where $V_1^+ - V_1^-$ is the measured voltage difference between electrodes $\mathcal{E}_2^+$ and $\mathcal{E}_2^-$ due to the injection current.

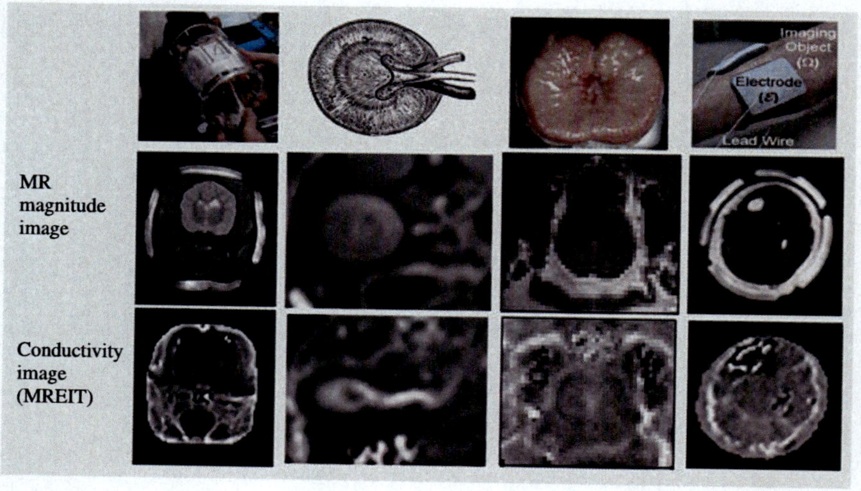

Fig. 6.20 MREIT animal and human experiments [59]. After the invention of the harmonic B_z algorithm, imaging techniques in MREIT have been advanced rapidly and now can offer state-of-the-art conductivity imaging using MRI animal and human experiments [58]

Step 9. If necessary, repeat steps 4 to 7, replacing σ^0 by σ^1 and denoting the updated conductivity as σ^2. Repeat the process to improve the quality of the reconstructed conductivity image.

We should emphasize the amount of current passed through a human body to follow FDA guidelines for the clinical use of MREIT. Currently, the major technical barrier to be overcome for the clinical use of MREIT is to reduce the amount of current to a level that does not produce undesirable nerve or muscle stimulation. This is because the quality of the reconstructed images is approximately proportional to the signal-to-noise ratio (SNR) of the measured B_z data, which is proportional to the amount of injection current. For FDA approval, it would be desirable to reduce the current amplitude to 1 mA. In theory, the SNR of B_z data can be maintained with the reduced current amplitude by increasing the current injection time T_c within one pulse repetition time (TR). In practice, without optimizing pulse sequence, this causes deterioration of SNR in both the MR phase image and the MR magnitude image. We need an innovative data processing methods as well as improved measurement techniques to reduce the amount of current.

6.4 Electrical Properties Tomography

MREIT can provide σ at low frequencies below a few kHz, whereas EPT produces $\gamma = \sigma + i\omega\varepsilon$ at the Larmor frequency of 128 MHz at 3 T MRI. Since σ and ε change with frequency, MREIT and EPT may provide different images of the same biological tissue. EPT uses the standard RF field mapping techniques to measure the active magnetic RF field component, which is influenced by γ.

REPT uses an RF coil in an MR scanner to feed a sinusoidal current at the Larmor frequency ω. It produces a time harmonic magnetic field $\mathbf{H} = (H_x, H_y, H_z)$ in the imaging subject Ω inside the MR scanner. Then, B_1 mappings [2, 56, 70] allow one to measure the positive rotating magnetic field $H^+ := \frac{1}{2}(H_x + iH_y)$. This H^+ is influenced by γ in the following way:

$$-\nabla^2\mathbf{H} = \nabla \ln \gamma \times [\nabla \times \mathbf{H}] - i\omega\mu_0\gamma\mathbf{H} \quad \text{in } \Omega, \tag{6.29}$$

where $\mu_0 = 4\pi \times 10^{-7}$ H/m is the magnetic permeability of free space. The inverse problem of EPT is to identify $\gamma =$ from the measured data $H^+ = \frac{1}{2}(H_x + iH_y)$. If $\nabla\gamma = \mathbf{0}$, then (6.29) becomes

$$-\nabla^2\mathbf{H} = i\omega\mu_0\gamma\mathbf{H} \quad \text{in } \Omega. \tag{6.30}$$

Hence, under the assumption of local homogeneity [38, 77], γ can be obtained directly by

$$\gamma = \frac{1}{i\omega\mu_0}\frac{\nabla^2 H^+}{H^+} \quad \text{in } \Omega. \tag{6.31}$$

However, this direct formula neglecting the contribution of $\nabla \ln \gamma \times (\nabla \times \mathbf{H})$ produces serious artifacts along the complex conductivity interfaces. Seo et al. [64] showed, both theoretically and experimentally, that the reconstruction error is fundamental and comes from ignoring the contribution of $\nabla \ln \gamma \times (\nabla \times \mathbf{H})$. Figure 6.21 shows an EPT experiment applied for breast cancer diagnosis [37].

It should deal with the contribution of $\nabla \ln \gamma \times (\nabla \times \mathbf{H})$. Recently, Song and Seo [69] developed a reconstruction method removing the assumption of $(\frac{\partial}{\partial x}, \frac{\partial}{\partial y})\gamma = \mathbf{0}$. This method still requires the assumption of $\frac{\partial \gamma}{\partial z} = 0$. In this method, the reconstruction problem is converted to solve a semilinear elliptic PDE with coefficients that only depend on H^+. We refer to [56, 58] for a review of EPT.

6.5 Discussion and Conclusion

Recently, there has been marked progress in electromagnetic property and mechanical property imaging techniques in which cross-sectional image reconstructions inside the human body are pursued. These techniques also have wider applications as imaging methods in medicine, biotechnology, non-destructive testing, the monitoring of industrial processes, and in other areas (Fig. 6.22).

Lessons from three decades of mathematical techniques in medical imaging have shown that the symbiotic interplay among theoretical mathematics, computational mathematics, and experiments is crucial for understanding and solving these

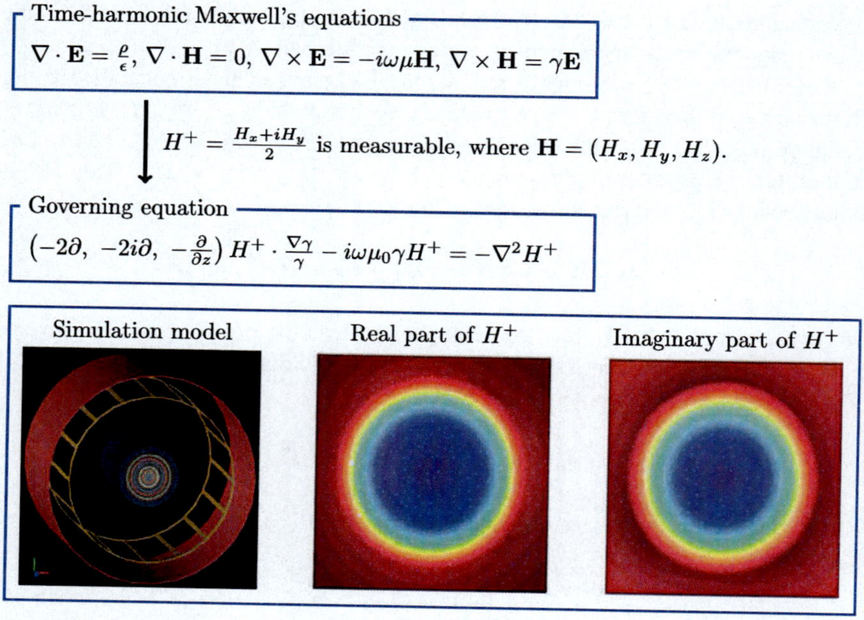

Fig. 6.21 EPT produces both σ and ε images at the Larmor frequency of an MRI scanner based on B_1-mapping techniques, whereas MREIT produces σ at low frequency below 1 KHz. Since σ varies with frequency, MREIT and EPT provide different σ images

nonlinear problems in practice. It is necessary to understand practical limitations imposed by measurement methods. In order to effectively deal with nonlinear inverse problems in medical imaging, we suggest to follow the steps below [57].

1. Understand underlying physical phenomena and the constraints imposed on the problem, which may enable us to improve solutions of nonlinear inverse problems. Physics, chemistry and also biology play crucial roles here. No attempt is made to be comprehensive in terms of physics, chemistry and biology.
2. Understand forward problems which usually are the processes of information loss. They provide strategic insights in seeking solutions of nonlinear inverse problems. We describe underlying principles so that readers can understand their mathematical formulations.
3. Formulate forward problems in such a way that we can deal with them systematically and quantitatively.
4. Understand how to probe the imaging object and what is measurable using available engineering techniques. Practical limitations associated with the measurement sensitivity and specificity, noise, artifacts, interface between target object and instrument, data acquisition time and so on must be properly understood and analyzed.
5. Understand what is feasible in a specific nonlinear inverse problem.

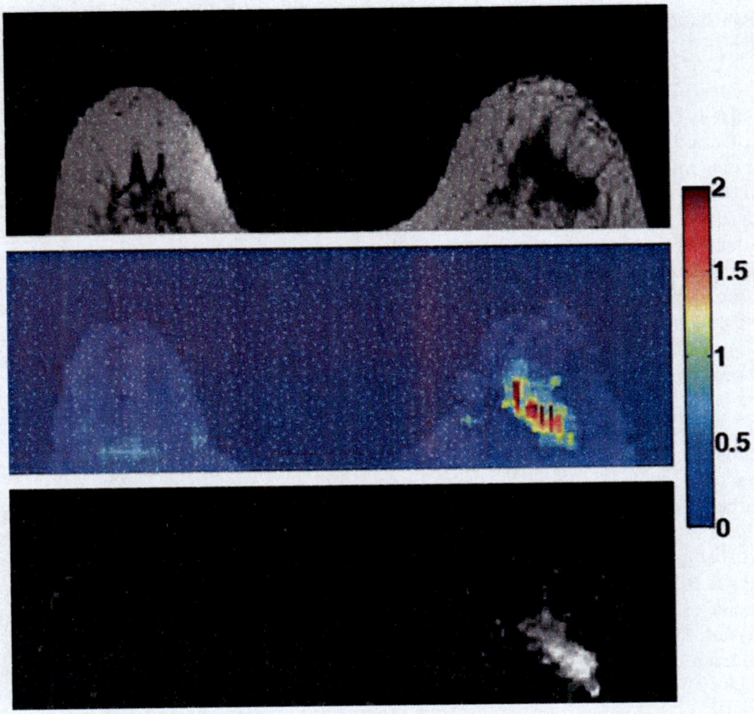

Fig. 6.22 Breast tumor patient study using EPT. Top: Turbo spin echo (magnitude), center: reconstructed conductivity image, bottom: contrast enhanced dynamic image. The dynamic image shows the region of the tumor [37]

6. Formulate proper nonlinear inverse problems by defining image contrast associated with physical quantities. Mathematical formulations should include any interrelation between those qualities and measurable data.
7. Construct inversion methods to produce images of contrast information.
8. Develop computer programs and properly address critical issues of numerical analysis.
9. Customize the inversion process by including a priori information.
10. Validate results by simulations and experiments.

Acknowledgements This research was supported by Samsung Science & Technology Foundation (No. SRFC-IT1902-09). Seo was supported by a grant of the Korea Health Technology R&D Project through the Korea Health Industry Development Institute (KHIDI), funded by the Ministry of Health & Welfare, Republic of Korea (grant number: HI20C0127).

References

1. Adler, A., Arnold, J., Bayford, R., Borsic, A., Brown, B., Dixon, P., Faes, T., Frerichs, I., Gagnon, H., Garber, Y., Grychtol, B., Hahn, G., Lionheart, W., Malik, A., Patterson, R., Stocks, J., Tizzard, A., Weiler, N., Wolf, G.: GREIT: A unified approach to 2D linear EIT reconstruction of lung images. Physiol. Meas. **30**, S35–S55 (2009)
2. Akoka, S., Franconi, F., Seguin, F., le Pape, A.: Radiofrequency map of an NMR coil by imaging. Magn. Reson. Imag. **11**, 437–441 (2009)
3. Ammari, H., Garnier, J., Giovangigli, L., Jing, W., Seo, J.K.: Spectroscopic imaging of a dilute cell suspension (2013). arXiv:1310.1292
4. Ammari, H., Kang, H.: Polarization and moment tensors with applications to inverse problems and effective medium theory. In: Applied Mathematical Science, vol. 162. Springer (2007)
5. Ammari, H., Kwon, H., Lee, S., Seo, J.K.: Mathematical framework for abdominal electrical impedance tomography to assess fatness. SIAM J. Imag. Sci. **10**(2), 900–919 (2017)
6. Astala, K., Päivärinta, L.: Calderon's inverse conductivity problem in the plane. Ann. Math. **163**, 265–299 (2006)
7. Barber, D.C., Brown, B.H.: Errors in reconstruction of resistivity images using a linear reconstruction technique. Clin. Phys. Physiol. Meas. **9**(Suppl A), 101–104 (1988)
8. Barber, D.C., Brown, B.H.: Applied potential tomography. J. Phys. E: Sci. Instrum. **17**, 723–733 (1984)
9. BiLab. http://bilabhealthcare.com/
10. Brown, B.H., Leathard, A.D., Lu, L., Wang, W., Hampshire, A.: Measured and expected Cole parameters from electrical impedance tomographic spectroscopy images of the human thorax. Physiol. Meas. **16**, A57–67 (1995)
11. Calderón, A.P.: On an inverse boundary value problem. In: Seminar on Numerical Analysis and Its Applications to Continuum Physics, Soc. Brasileira de Matemàtica, pp. 65–73 (1980)
12. Candès, E.J., Romberg, J.K., Tao, T.: Stable signal recovery from incomplete and inaccurate measurements. Commun. Pure Appl. Math. **59**, 1207–1223 (2006)
13. Capps, M., Mueller, J.L.: Reconstruction of organ boundaries with deep learning in the D-bar method for electrical impedance tomography. IEEE Trans. Biomed. Eng. **68**, 826–833 (2021)
14. Chen, Z., Yang, Y., Jia, J., Bagnaninchi, P.: Deep learning based cell imaging with electrical impedance tomography. In: IEEE International Instrumentation and Measurement Technology Conference, pp. 1–6 (2020)
15. Cheney, M., Isaacson, D., Newell, J., Goble, J., Simske, S.: NOSER: An algorithm for solving the inverse conductivity problem. Int. J. Imag. Syst. Technol. **2**, 66–75 (1990)
16. Cheney, M., Isaacson, D., Newell, J.C.: Electrical impedance tomography. SIAM Rev. **41**, 85–101 (1999)
17. Chipot, M.: Elliptic equations: an introductory course. Springer Science & Business Media (2009)
18. Cole, K.S.: Membranes, Ions, and Impulses. University of California Press, Berkeley (1972)
19. Drägerwerk AG & Co. KGaA. https://www.draeger.com/
20. Fricke, H.: A mathematical treatment of the electrical conductivity of colloids and cell suspensions. J. Gen. Physiol. **6**, 375–384 (1924)
21. Gabriel, C., Gabriel, S., Corthout, E.: The dielectric properties of biological tissues: I. literature survey. Phys. Med. Biol. **41**, 2231–2249 (1996)
22. Gabriel, S., Lau, R.W., Gabriel, C.: The dielectric properties of biological tissues: II. measurements in the frequency range 10Hz to 20GHz. Phys. Med. Biol. **41**, 2251–2269 (1996)
23. Geddes, L.A., Baker, L.E.: The specific resistance of biological material: a compendium of data for the biomedical engineer and physiologist. Med. Biol. Eng. **5**, 271–293 (1967)
24. Grimnes, S., Martinsen, O.G.: Bioimpedance and Bioelectricity Basics. Academic Press, London, UK (2000)
25. Haacke, E.M., Petropoulos, L.S., Nilges, E.W., Wu, D.H.: Extraction of conductivity and permittivity using magnetic resonance imaging. Phys. Med. Biol. **36**, 723–733 (1991)

26. Hadamard, J.: Sur les problémes aux dérivées partielles et leur signification physique. Princeton University Bulletin, pp. 49–52 (1902)
27. Hamilton, S.J., Hauptmann, A.: Deep D-bar: real-time electrical impedance tomography imaging with deep neural networks. IEEE Trans. Med. Imag. **37**, 2367–2377 (2018)
28. Hanke, M., Harrach, B., Hyvӧnen, M.: Justification of point electrode models in electrical impedance tomography. Math. Models Methods Appl. Sci. **21**(06), 1395–1413 (2011)
29. Hanke, M., Brühl, M.: Recent progress in electrical impedance tomography. Inverse Prob. **19**, S65–S90 (2003)
30. Hao, T.: Electrorheological Fluids: the Non-aqueous Suspensions. Elsevier (2011)
31. Harrach, B., Ullrich, M.: Monotonicity-based shape reconstruction in electrical impedance tomography. SIAM J. Math. Anal. **45**(6), 3382–3403 (2013)
32. Henderson, R.P., Webster, J.G.: An impedance camera for spatially specific measurements of the Thorax. IEEE Trans. Biomed. Eng. **25**, 250–254 (1978)
33. Holder, D.: Electrical Impedance Tomography: Methods, History and Applications. IOP Publishing, Bristol, UK (2005)
34. Inbody technology. http://www.inbody.com/
35. Isaacson, D.: Distinguishability of conductivities by electric current computed tomography. IEEE Trans. Med. Imag. MI-5, 91–95 (1986)
36. Joy, M.L., Scott, G.C., Henkelman, R.M.: In-vivo detection of applied electric currents by magnetic resonance imaging. Mag. Res. Imag. **7**, 89–94 (1989)
37. Katscher, U., Kim, D., Seo, J.K.: Recent progress and future challenges in MR electric properties tomography. In: Computational and Mathematical Methods in Medicine, 2013, Article ID 546562 (2013)
38. Katscher, U., Voigt, T., Findeklee, C., Vernickel, P., Nehrke, K., Dossel, O.: Determination of electrical conductivity and local SAR via B1 mapping. IEEE Trans. Med. Imag. **28**, 1365–1374 (2009)
39. Khang, H.S., Lee, B.I., Oh, S.H., Woo, E.J., Lee, S.Y., Cho, M.H., Kwon, O.I., Yoon, J.R., Seo, J.K.: J-substitution algorithm in magnetic resonance electrical impedance tomography (MREIT): Phantom experiments for static resistivity images. IEEE Trans. Med. Imag. **21**(6), 695–702 (2002)
40. Kim, Y.E., Woo, E.J., Oh, T.I., Kim, S.W.: Real-time identification of upper airway occlusion using electrical impedance tomography. J. Clin. Sleep Med. **15**, 563–571 (2019)
41. Kohn, R., Vogelius, M.: Determining conductivity by boundary measurements. Commun. Pure Appl. Math. **37**, 113–123 (2006)
42. Kwon, O., Seo, J.K., Yoon, J.R.: A real-time algorithm for the location search of discontinuous conductivites with one measurement, pp. 1–29. LV (2002). (Comm. Pure Appl. Math.)
43. Kwon, O., Woo, E.J., Yoon, J.R., Seo, J.K.: Magnetic resonance electrical impedance tomography (MREIT): simulation study of J-substitution algorithm. IEEE Trans. Biomed. Eng. **49**(2), 160–167 (2002)
44. Lee, K., Yoo, M., Jargal, A., Kwon, H.: Electrical impedance tomography-based abdominal subcutaneous fat estimation method using deep learning. In: Computational and Mathematical Methods in Medicine (2020)
45. Martinsen, O.G., Grimnes, S.: Bioimpedance and Bioelectricity Basics, 2nd edn. Academic Press (2011)
46. Metherall, P., Barber, D.C., Smallwood, R.H., Brown, B.H.: Three-dimensional electrical impedance tomography. Nature **380**, 509–512 (1996)
47. Milton, G.W.: The Theory of Composites. Cambridge University Press (2002)
48. Mueller, J., Siltanen, S., Isaacson, D.: A direct reconstruction algorithm for electrical impedance tomography. IEEE Trans. Med. Imag. **21**(6), 555–559 (2002)
49. Nachman, A.: Global uniqueness for a two-dimensional inverse boundary problem. Ann. Math. **143**, 71–96 (1996)
50. Nachman, A., Tamasan, A., Timonov, A.: Conductivity imaging with a single measurement of boundary and interior data. Inverse Prob. **23**, 2551–2563 (2007)

51. Nachman, A., Tamasan, A., Timonov, A.: Reconstruction of planar conductivities in subdomains from incomplete data. SIAM J. Appl. Math. **70**(8), 3342–3362 (2010)
52. Romsauerova, A., McEwan, A., Horesh, L., Yerworth, R., Bayford, R.H., Holder, D.S.: Multi-frequency electrical impedance tomography (EIT) of the adult human head: initial findings in brain tumours, arteriovenous malformations and chronic stroke, development of an analysis method and calibration. Physiol. Meas. **27**, S147-61 (2006)
53. Santosa, F., Vogelius, M.: A back-projection algorithm for electrical impedance imaging. SIAM J. Appl. Math. **50**, 216–243 (1990)
54. Schwan, H.P.: Electrical properties of tissue and cell suspensions. Adv. Biol. Med. Phys. **5**, 147–209 (1957)
55. Sentec AG. https://www.sentec.com/
56. Seo, J.K., Woo, E.J., Katcher U., Wang, Y.: Electro-Magnetic Tissue Properties MRI. Imperial College Press (2013)
57. Seo, J.K., Woo, E.J.: Nonlinear Inverse Problems in Imaging. Wiley (2013)
58. Seo, J.K., Woo, E.J.: Magnetic resonance electrical impedance tomography (MREIT). SIAM Rev. **53**, 40–68 (2011)
59. Seo, J.K., Woo, E.J.: Electrical tissue property imaging at low frequency using MREIT. IEEE Trans. Biomed. Eng. **61**, 1390–1399 (2013)
60. Seo, J.K., Yoon, J.R., Woo, E.J., Kwon, O.: Reconstruction of conductivity and current density images using only one component of magnetic field measurements. IEEE Trans. Biomed. Eng. **50**, 1121–1124 (2003)
61. Seo, J.K., Pyo, H.C., Park, C., Kwon, O., Woo, E.J.: Image reconstruction of anisotropic conductivity tensor distribution in MREIT: computer simulation study. Phys. Med. Biol. **49**, 4371–4382 (2004)
62. Seo, J.K., Lee, J., Kim, S.W., Zribi, H., Woo, E.J.: Frequency-difference electrical impedance tomography (fdEIT): algorithm development and feasibility study. Physiol. Meas. **29**, 929–944 (2008)
63. Seo, J.K., Jeon, K., Lee, C.O., Woo, E.J.: Non-iterative harmonic Bz algorithm in MREIT. Inverse Prob. **27**, 085003 (2011)
64. Seo, J.K., Ghim, M., Lee, J., Choi, N., Woo, E.J., Kim, H.J., Kwon, O.I., Kim, D.: Error analysis for electrical property imaging using MREPT. IEEE Trans. Med. Imag. **31**(3), 430–437 (2012)
65. Seo, J.K., Kim, D.H., Lee, J., Kwon, O.I., Sajib, S.Z.K., Woo, E.J.: Electrical tissue property imaging using MRI at dc and Larmor frequency. Inverse Prob. **28**, 084002 (2012)
66. Seo, J.K., Kim, K.C., Jargal, A., Lee, K., Harrach, B.: A learning-based method for solving ill-posed nonlinear inverse problems: a simulation study of lung EIT. SIAM J. Imag. Sci. **12**, 1275–1295 (2019)
67. Somersalo, E., Cheney, M., Isaacson, D., Isaacson, E.: Layer stripping: a direct numerical method for impedance imaging. Inverse Prob. **7**, 899–926 (1991)
68. Somersalo, E., Cheney, M., Isaacson, D.: Existence and uniqueness for electrode models for electric current computed tomography. SIAM J. Appl. Math. **52**, 1023–40 (1992)
69. Song, Y., Seo, J.K.: Conductivity and permittivity image reconstruction at the Larmor frequency using MRI. SIAM J. Appl. Math. **73**, 2262–2280 (2013)
70. Stollberger, R., Wach, P.: Imaging of the active B1 field in vivo. Magn. Reson. Med. **35**, 246–251 (1996)
71. Swisstom AG. http://www.swisstom.com/
72. Sylvester, J., Uhlmann, G.: A global uniqueness theorem for an inverse boundary value problem. Ann. Math. **125**, 153–169 (1987)
73. Teschner, E., Imhoff, M.: Electrical impedance tomography: realization of regional ventilation monitoring. Dräger technology for life. http://www.draeger.com
74. Thuéry, J.: Microwaves: Industrial, Scientific and Medical Applications. Artech House Inc., London (1992)
75. Tikhonov, A.N., Arsenin, V.Y.: Solutions of Ill-Posed Problems. New York, vol. 1, 487 (1977)
76. Timpel Medical. https://www.timpelmedical.com/

77. Wen, H.: Noninvasive quantitative mapping of conductivity and dielectric distributions using RF wave propagation effects in high-field MRI. In: Yaffe, M.J., Antonuk, L.E. (eds.) Proceedings of SPIE, vol. 5030, pp. 471–477. San Diego, CA (2003)
78. Woo, E.J., Seo, J.K.: Magnetic resonance electrical impedance tomography (MREIT) for high-resolution conductivity imaging. Physiol. Meas. **29**, R1–R26 (2008)
79. Woo, E.J., Lee, S.Y., Mun, C.W.: Impedance tomography using internal current density distribution measured by nuclear magnetic resonance. SPIE **2299**, 377–385 (1994)
80. Zolgharni, M., Ledger, P.D., Armitage, D.W., Holder, D.S., Griffiths, H.: Imaging cerebral haemorrhage with magnetic induction tomography: numerical modeling. Physiol. Meas. **30**, S187–S200 (2009)

Chapter 7
Deep Learning for Ill Posed Inverse Problems in Medical Imaging

Chang Min Hyun and Jin Keun Seo

Abstract Recently, with the significant developments in deep learning (DL) techniques, solving underdetermined inverse problems has become one of the major concerns in the medical imaging domain, where underdetermined problems are motivated by the willingness to provide high-resolution medical images with as little data as possible, by optimizing data collection in terms of minimal acquisition time, cost-effectiveness, and low invasiveness. DL methods appear to have a strong capability to explore the prior information of the expected images via training data, which allows one to deal with the uncertainty of solutions to ill-posed inverse problems. This chapter aims to discuss some mathematical interpretations of DL-based nonlinear low-dimensional representations of expected solutions to ill-posed inverse problems.

7.1 Introduction

Inverse problems arise in finding physical quantities (e.g., attenuation in CT, nuclear spin density in MRI, and electrical tissue properties in EIT) that are observable or measurable and whose values change with position and time to form signals. Whether or not an inverse problem is well-posed may be dependent on how the solution is expressed. Many problems are ill-posed because we are overly ambitious or lacking in expressiveness. There are roughly two types of inverse problems. Type 1 is characterized by having data that is much smaller in dimension compared to the input (i.e., undersampled models that violate the Nyquist criteria in the sense that the number of equations is much smaller than the number of unknowns) [16, 18] (described in Fig. 7.1); Type 2 refers to inaccurate forward models with data contaminated by various noise and artifacts (e.g., inverse problems with forward

C. M. Hyun · J. K. Seo (✉)
School of Mathematics and Computing (Computational Science and Engineering), Yonsei University, Seoul, Korea
e-mail: seoj@yonsei.ac.kr

C. M. Hyun
e-mail: chammyhyun@yonsei.ac.kr

J. K. Seo (ed.), *Deep Learning and Medical Applications*, Mathematics in Industry 40,
https://doi.org/10.1007/978-981-99-1839-3_7

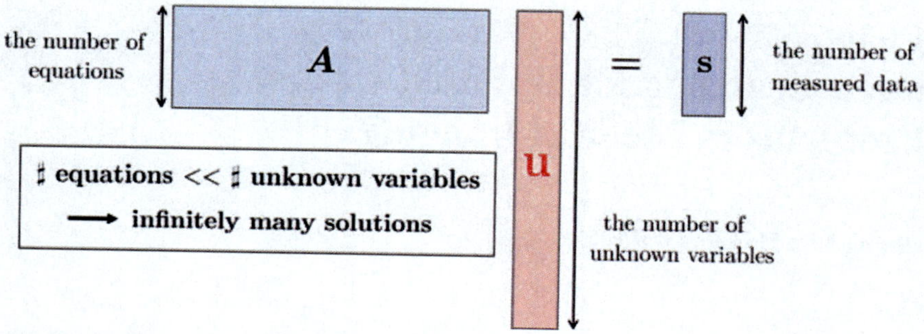

Fig. 7.1 Underdetermined linear inverse problem. In compressed sensing MRI and sparse-view CT, the dimension of measured data is much smaller than that of unknowns to be reconstructed. Without imposing a suitable constraint on the solution, we cannot solve this ill-posed problem

modeling errors associated with various uncertain factors and with the measured data being insensitive to local perturbation of the input) [26].

Why do we pay attention to underdetermined problems (fewer equations than unknowns) in CT and MRI? In medical imaging (e.g., CT and MRI), we want to provide high-resolution medical images, while optimizing data collection in terms of minimal acquisition time, cost-effectiveness and low invasiveness. The willingness to provide high-resolution images with as little data as possible leads to ill-posed inverse problems in the classical sense.

Solving the inverse problem requires setting up a mathematical model of the underlying physical phenomenon as a forward problem. Correct formulation of the forward problem is essential to obtain a meaningful solution of the related inverse problem. Classically, before deep learning, "well-posedness" was considered when setting up forward problems. For clear explanation, let $\mathbf{u} \in \mathbb{C}^{N \times N}$ represent a CT or MR image to be reconstructed, where N^2 is the number of pixels and $\mathbb{C}$ is the set of complex numbers. Let **s** denote a measured data (or signal) corresponding to the image **u**. In this chapter, **A** denotes a linear operator; **A** is a Fourier transform for MRI and a Radon transform for CT.

According to Hadamard [14], the linear system $\mathbf{Au} = \mathbf{s}$ is *"well-posed"* if the following three conditions hold:

- Existence: A solution exists. This means that **s** lies on the range space of **A**.
- Uniqueness: For each set of data **s**, $\mathbf{Au} = \mathbf{s}$ has a unique solution. This means that the null space of **A** is trivial.
- Stability: A solution depends continuously on the data. If the condition number of **A** is very large, **A** is ill-conditioned.

Conventional CT and MRI data collections are designed to ensure that the corresponding system $\mathbf{Au} = \mathbf{s}$ is well-posed. This means that the number of equations (data) is similar to the number of unknowns (pixels of image).

Let us briefly explain how sampling data in MRI relates to spatial resolution of the MR image. For convenience of explanation, consider a square image of width FOV.

Assume that the image is composed of N^2 square pixels, where the pixel width is $\Delta x = FOV/N$. Noting that MRI measures approximately an image's Fourier transform, k-space data are sampled through Fourier space. In order to reliably reconstruct an MR image by applying the inverse Fourier transform to the discrete k-space samples, the spacing (Δk) and the total sample range in Fourier space must satisfy $\Delta k \approx \frac{1}{FOV}$ and $\frac{1}{\Delta x} \approx k$-space FOV. Hence, in order for the MRI reconstruction problem to be well-posed from the classical mathematical point of view (i.e., the number of unknowns = the number of equations), the relationship between the spatial resolution in the image space and k-space sampling in the frequency space roughly satisfy the following condition:

$$\sharp \text{ pixels in image} \quad \approx \quad \sharp \text{ sampling in } k\text{-space}, \tag{7.1}$$

where $\sharp$ stands for "the number of". This rule is called Nyquist sampling in k-space, where Nyquist sampling rate means a sufficient condition for a sample rate that permits the sampled data to provide perfect reconstruction. Undersampled MRI violates this Nyquist sampling criteria (7.1), for the purpose of reducing data acquisition time. Compressed sensing MRI is a typical example of underdetermined problems [11].

Conventional CT measures approximately an image's Radon transform such that

$$\sqrt{\sharp \text{ pixel width}} \quad \approx \quad \sharp \text{ projection angles} \approx \quad \sharp \text{ detectors}. \tag{7.2}$$

In CT, the existence of a solution in the strict sense is not guaranteed, as most of the measured sinogram data do not match any image's Radon transformation. Sparse-view CT is a typical example of underdetermined problems that violate (7.2).

Solving a highly underdetermined problem requires dealing with the uncertainty caused by data with the dimension (the number of equations) much smaller than the pixel dimension of images (the number of unknowns). To convert an ill-posed problem into a well-posed one, we need a suitable data sampling strategy involving **A** and choose a highly reduced solution space (or manifold), denoted by $\mathcal{M}$, so that these choices allow us to satisfy the *restricted isometry property* (RIP) condition [5, 6, 18]:

$$\frac{1}{c}\|\mathbf{u} - \mathbf{u}'\| \leq \|\mathbf{A}\mathbf{u} - \mathbf{A}\mathbf{u}'\| \leq c\|\mathbf{u} - \mathbf{u}'\| \quad \text{for all } \mathbf{u}, \mathbf{u}' \in \mathcal{M}, \tag{7.3}$$

where c is a positive constant. Then, the underdetermined problem $\mathbf{A}\mathbf{u} = \mathbf{s}$ becomes the following constrained problem:

$$\left\| \begin{array}{l} \text{Solve } \mathbf{A}\mathbf{u} = \mathbf{s} \\ \text{subject to the constraint } \mathbf{u} \in \mathcal{M} \end{array} \right. \tag{7.4}$$

The challenge here is what is the solution space $\mathcal{M}$? What kind of prior information about the solution $\mathbf{u}$ constitutes $\mathcal{M}$? It is very difficult to compactly represent $\mathcal{M}$, using as few latent variables as possible, while making it computationally tractable.

This $\mathcal{M}$ can be a linear subspace, a union of linear subspaces (associated with sparse representation), or a nonlinear manifold. Traditionally, norm-based regularization techniques have been widely used to roughly impose prior knowledge of expected images:

$$\mathbf{u} = \underset{\mathbf{u}}{\operatorname{argmin}} \frac{1}{2} \|\mathbf{u} - \mathbf{A}^{\dagger}\mathbf{s}\|_{\ell_2}^2 + \lambda \text{Reg}(\mathbf{u}), \tag{7.5}$$

where Reg($\mathbf{u}$) is a regularization term enforcing some properties of expected solutions and λ is the regularization parameter controlling the trade-off between the data fidelity (i.e., $\mathbf{u} \approx \mathbf{A}^{\dagger}\mathbf{s}$) and regularity. For example, Reg($\mathbf{u}$) can be $\|\nabla\mathbf{u}\|_{\ell_1}$ (ℓ_1-norm enforcing the sparsity of $\nabla\mathbf{u}$) or $\|\mathbf{h}\|_{\ell_1}$ with $\mathbf{h}$ being $\mathbf{u} = \mathbf{D}\mathbf{h}$ (sparse representation of $\mathbf{u}$ over a dictionary $\mathbf{D}$). This approach is known to be effective in reducing noise in images. However, in medical imaging, these regularized data fitting methods may not selectively preserve small clinically useful features [18].

DL techniques have expanded our ability by sophisticated "disentangled representation learning" though training data, and appear to overcome limitations of existing mathematical methods in handling various ill-posed problems. The DL approach is a completely different paradigm from the classical regularized data-fitting approaches that use a "single" data-fidelity with regularization, and has excellent ability to learn complex prior knowledge of the output by effectively utilizing prior and additional information as a "group" data fidelity [18]. Deep learning methods can provide nonlinear regression on training data, which serves to learn complex prior knowledge about the output. See Fig. 7.2.

This chapter is based on the paper [18], which attempts to explain why deep learning methods have exhibited excellent performance in the above-mentioned underdetermined inverse problems. For a theoretical analysis, we introduce the M-RIP condition for deep learning-based solvability of ill-posed inverse problems (Type 1) in medicine. Assuming that medical data are on or near a low-dimensional manifold

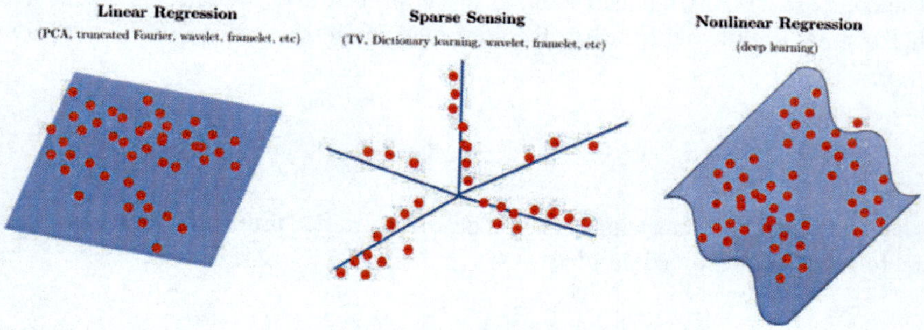

Fig. 7.2 Data-driven regression. Principal component analysis (PCA) is a linear regression (left). Sparse sensing (middle) can be considered a piecewise-linear regression because the solution is expected to be in a union of the low-dimensional linear subspaces. The deep learning approach is a nonlinear regression (right) over training data, which regresses an underlying data distribution into a low-dimensional manifold

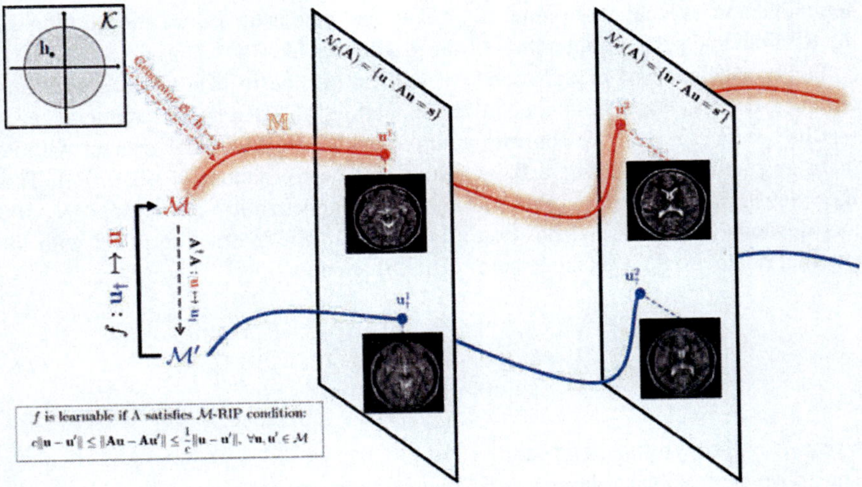

Fig. 7.3 Solvability of the underdetermined problem $\mathbf{Au} = \mathbf{s}$. Solving $\mathbf{Au} = \mathbf{s}$ can be achieved by learning $f : \mathbf{u}_\dagger = \mathbf{A}^\sharp \mathbf{s} \mapsto \mathbf{u} = \mathbf{A}_{\text{full}}^{-1} \mathbf{s}_{\text{full}}$ in (7.24) with probing the solution manifold $\mathcal{M}$. If $\mathbf{A}$ satisfies the $\mathcal{M}$-RIP condition, then $\mathbf{A}^\dagger \mathbf{A} : \mathcal{M} \mapsto \mathcal{M}'$ is one-to-one, i.e., $\mathcal{N}_\mathbf{s}(\mathbf{A}) \cap \mathcal{M} = \{\mathbf{u}\}$ is unique. In general, f is nonlinear and the degree of non-linearity depends on the sampling strategy of $\mathbf{s}$ and the degree of bending of the solution manifold. Here, the manifold $\mathcal{M}$ can be viewed as nonlinear regression over a set of all realistic MR images $\mathbb{M}$

embedded in high-dimensional ambient space, we need to fit a nonlinear solution manifold to training data. The schematic concept is described in Fig. 7.3. Manifold learning as a low-dimensional representation of high-resolution medical images would be an important future research topic.

7.2 Undersampled Magnetic Resonance Imaging (MRI)

7.2.1 MR Physics

MRI visualizes the amount of hydrogen atoms inside biological tissue by measuring magnetic moments inside the human body in the form of a cross-sectional image. MRI uses a variety of techniques to localize magnetic moments to provide cross-sectional images of the net magnetization vector density inside the human body. MRI uses several magnetic field sources, which can be broadly classified into external fields generated by the MRI scanner and internal fields that are emitted through nuclear resonance of the body. The external fields include the main field, RF-field, and gradient field that are designed to use the nuclear magnetic resonance (NMR) phenomenon determined by the interaction of a nuclear spin with the external magnetic field and its local environments. The internal field is measured after turning off the RF field and proper application of the gradient field.

For convenience of explanation, we will outline basic MR physics, ignoring complex details. For a fundamental understanding of MRI, we recommend reading the books [15, 31]. We consider a human body inside an MRI scanner with its main magnetic field $\mathbf{B}_0$, where $\mathbf{B}_0$ is assumed to be constant $\mathbf{B}_0 = (0, 0, B_0)$. This $\mathbf{B}_0 = (0, 0, B_0)$ produces a distribution of net magnetization $\mathbf{M} = (M_x, M_y, M_z)$ that depends on time t and position $\mathbf{r} = (x, y, z)$. The interaction of $\mathbf{M}$ with the external magnetic field $\mathbf{B}_0$ is dictated by the Bloch equation:

$$\frac{\partial}{\partial t}\begin{pmatrix} M_x \\ M_y \\ M_z \end{pmatrix} = \gamma \begin{vmatrix} \hat{\mathbf{x}} & \hat{\mathbf{y}} & \hat{\mathbf{z}} \\ M_x & M_y & M_z \\ 0 & 0 & B_0 \end{vmatrix} = \begin{pmatrix} \gamma B_0 M_y \\ -\gamma B_0 M_x \\ 0 \end{pmatrix}, \tag{7.6}$$

where γ is the gyromagnetic ratio, $\hat{\mathbf{x}} = (1, 0, 0)$, $\hat{\mathbf{y}} = (0, 1, 0)$, and $\hat{\mathbf{z}} = (0, 0, 1)$. The above expression explains how $\mathbf{B}_0$ causes $\mathbf{M}$ to precess around the $\hat{\mathbf{z}}$-axis at the angular frequency of γB_0.

To extract a signal of $\mathbf{M}$, we apply a second magnetic field $\mathbf{B}_1$ perpendicular to $\mathbf{B}_0 = (0, 0, B_0)$ to flip $\mathbf{M}$ toward the xy-direction to produce its xy-component. Here, $\mathbf{B}_1$ can be a radio-frequency (RF) magnetic field that is generated by RF-coils through which we inject sinusoidal current at the Larmor frequency γB_0. After terminating the RF pulse, we apply a phase encoding gradient in the y direction which makes the spin phase change linearly in the phase encoding direction. Then we apply the frequency encoding gradient field. Through multiple phase encodings, we can collect a set of k-space data.

In 2D Fourier imaging with Cartesian k-space sampling, the relation between the fully sampled MR data $\mathbf{s}_{\text{full}}$ and the image of $\mathbf{u}$ can be expressed as

$$\underbrace{\mathbf{u}(x, y) = \sum_{k_x=1-N/2}^{N/2} \sum_{k_y=1-N/2}^{N/2} \mathbf{s}_{\text{full}}(k_x, k_y)\, e^{2i\pi(k_x x + k_y y)/N}}_{\mathbf{u}=\mathbf{A}_{\text{full}}^{-1}\mathbf{s}_{\text{full}}} \tag{7.7}$$

for $x, y = 1 - N/2, \ldots, 0, \ldots, N/2$. Here, $\mathbf{s}_{\text{full}}(k_x, k_y)$ denote the MR-signal received at k-space position $(2\pi k_x/N, 2\pi k_y/N)$. The frequency encoding is along the k_x-axis and the phase encoding is along the k_y-axis.

The MRI scan time is roughly proportional to the number of time-consuming phase encoding steps in k-space. Because MRI scans entrap patients in uncomfortable and narrow space for long periods of time, shortening MRI scan times can help improve patient satisfaction, reduce motion artifacts from patient movement, and reduce medical costs. This is the motivation behind undersampled MRI, which aims to accelerate MRI scans by skipping phase encoding lines in k-space while eliminating aliasing.

There are many well-organized books/tutorials about basics of MRI. We refer the book [15] by Haacke et al. for detailed explanations on MRI.

7.2.2 *Towards Highly Undersampled MRI*

High-speed and high-resolution MRI is to significantly reduce MRI scan time while maintaining spatial resolution. The high-speed MRI can help increase patient satisfaction, reduce motion artifacts from patient movement, and reduce medical costs. In addition, this high-speed MRI is expected to be in high demand as a technology needed to examine the brains of children and fetuses who have difficulty controlling movement without anesthesia. Since the MRI scan time is roughly proportional to the number of phase encoding steps in k-space, the simplest way to reduce the MRI scan time is to increase the subsampling rate in the phase encoding direction.

The goal is to find an optimal reconstruction method using highly undersampled data that does not compromise reconstruction accuracy. The highly undersampled MRI reconstruction problem is a highly ill-posed problem that is to solve the following linear system with much fewer equations than unknowns:

$$\underbrace{\mathbf{A}}_{\mathscr{S}_{\text{sub}}(\mathbf{A}_{\text{full}})} \mathbf{u} = \underbrace{\mathbf{s}}_{\mathscr{S}_{\text{sub}}(\mathbf{s}_{\text{full}})} \tag{7.8}$$

where

- the relation between a 2D MR image $\mathbf{u}$ and the corresponding k-space data $\mathbf{s}_{\text{full}}$ is expressed as $\mathbf{A}_{\text{full}}\mathbf{u} = \mathbf{s}_{\text{full}}$ in (7.7);
- $\mathscr{S}_{\text{sub}}$ denotes a subsampling operator;
- $\mathbf{s} = \mathscr{S}_{\text{sub}}(\mathbf{s}_{\text{full}})$ is undersampled k-space data violating the Nyquist sampling criterion.

In standard (i.e., fully sampled) MRI, the Nyquist sampling makes the matrix $\mathbf{A}_{\text{full}}$ invertible, where $\mathbf{A}_{\text{full}}^{-1}$ corresponds to the discrete inverse Fourier transform. On the other hand, the system $\mathbf{Au} = \mathbf{s}$ in (7.8) has infinitely many solutions because $\mathbf{A}$ is a highly undersampled matrix that is not invertible. In this case, we can use the Moore–Penrose inverse, denoted by $\mathbf{A}^\dagger$, to get the minimum norm solution $\mathbf{A}^\dagger\mathbf{s}$, which differs significantly from the true solution $\mathbf{u} = \mathbf{A}_{\text{full}}^{-1}\mathbf{s}_{\text{full}}$. In this case, one may use the Moore–Penrose inverse, denoted by $\mathbf{A}^\dagger$, to get the minimum norm solution $\mathbf{A}^\dagger\mathbf{s}$, which is mostly different from the ground truth solution $\mathbf{u} = \mathbf{A}_{\text{full}}^{-1}\mathbf{s}$. Here, the Moore–Penrose inverse $\mathbf{A}^\dagger$ can be expressed as

$$\mathbf{A}^\dagger = \mathbf{A}_{\text{full}}^{-1}\mathscr{S}_{\text{sub}}^* = (\mathbf{A}^*\mathbf{A})^{-1}\mathbf{A}^*, \tag{7.9}$$

where $\mathbf{A}^*$ denotes the conjugate transpose of $\mathbf{A}$, $\mathscr{S}_{\text{sub}}^*$ is the dual operator of $\mathscr{S}_{\text{sub}}$, which can be understood as the zero padding operator corresponding to the subsampling $\mathscr{S}_{\text{sub}}$.

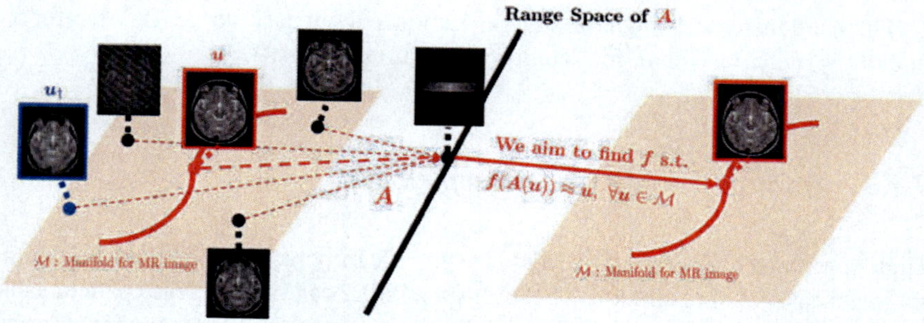

Fig. 7.4 Undersampled MRI reconstruction problem. Without a suitable constraint on the solution, the reconstruction problem of $\mathbf{s} = \mathbf{A}\tilde{\mathbf{u}}$ has infinitely many solutions (left). The goal is to extract $\mathbf{u} = \mathbf{A}_{\text{full}}^{-1}\mathbf{s}_{\text{full}}$ from infinitely many solutions. This is equivalent to finding a one-to-one map f such that $f(\mathbf{Au}) = \mathbf{u}$, $\forall \mathbf{u} \in \mathscr{M}$ (right)

Let's start with the simplest undersampled MRI problem, where $\mathbf{s}$ is data uniformly subsampled by a factor of 2. Then, the corresponding system can be expressed by

$$\begin{pmatrix} \cdots & \mathbf{s}_{\text{full}}(\frac{N}{2}-1, \frac{N}{2}) & \mathbf{s}_{\text{full}}(\frac{N}{2}, \frac{N}{2}) \\ \cdots & 0 & 0 \\ \cdots & \mathbf{s}_{\text{full}}(\frac{N}{2}-1, \frac{N}{2}-2) & \mathbf{s}_{\text{full}}(\frac{N}{2}, \frac{N}{2}-2) \\ \cdots & 0 & 0 \\ \cdots & \vdots & \vdots \end{pmatrix}, \tag{7.10}$$

The $\mathbf{A}^{\dagger}\mathbf{s}$ produces the following two-folded image by the Poisson summation formula [29]:

$$\mathbf{A}^{\dagger}\mathbf{s} = \mathbf{u} + \mathbf{u}^{\natural}, \quad \mathbf{u}^{\natural}(x, y) = \mathbf{u}(x, y + N/2), \tag{7.11}$$

Hence, $\mathbf{A}^{\dagger}\mathbf{s}$ is an unwanted image. Then, how can we extract $\mathbf{u} = \mathbf{A}^{-1}\mathbf{s}_{\text{full}}$ out of so many solutions?

Let us briefly explain about the solution space of $\mathbf{Au} = \mathbf{s}$ by reviewing undergraduate linear algebra. If m is the number of rows of $\mathbf{A}$, then the null space $\{\mathbf{u} : \mathbf{Au} = 0\}$ has $N - m$ bases $\{\phi_1, \phi_2, \ldots, \phi_{N-m}\}$. If $\mathbf{u}^*$ is a solution of $\mathbf{Au} = \mathbf{s}$, then all the linear combinations $\mathbf{u}^* + \sum_{j=1}^{N-m} c_j\phi_j$ satisfy $\mathbf{Au} = \mathbf{s}$. It is impossible to pick the right one out of these huge combinations, without having some knowledge on the true solution. See Fig. 7.4.

7.2.2.1 Compressed Sensing MRI

The compressed sensing technique uses image sparsity that is enforced to compensate for undersampled data [4, 6, 7, 9, 11, 25]. Random sampling is used for compressed sensing MRI, because the null space of $\mathbf{A}$ is a set of noise-like images, which can be

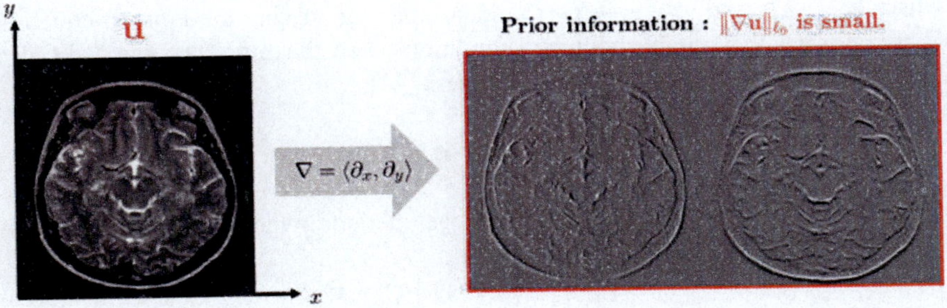

Fig. 7.5 Total variation-based compressed sensing method is based on the sparsity of the image gradient (i.e., $\|\nabla \mathbf{u}\|_{\ell_0}$). The image on the right shows that a human MR brain image has large sparsity on the gradient domain. Here, the gray-scale corresponds to the zero value

effectively handled by enforcing a sparsity-inducing prior. To be precise, CS-MRI can find a solution by the following regularized data-fitting method:

$$\mathbf{u} = \underset{\mathbf{u}}{\text{argmin}} \frac{1}{2} \|\mathbf{u} - \mathbf{A}^{\dagger}\mathbf{s}\|_{\ell_2}^2 + \lambda \|\Gamma(\mathbf{u})\|_{\ell_1}, \tag{7.12}$$

where $\Gamma(\mathbf{u})$ represents a transformation capturing the sparsity pattern of $\mathbf{u}$ and λ is the regularization parameter controlling the trade-off between the residual norm and regularity. Here, the term $\|\mathbf{u} - \mathbf{A}^{\dagger}\mathbf{s}\|_{\ell_2}^2$ forces the residual $\mathbf{u} - \mathbf{A}^{\dagger}\mathbf{s}$ to be small, whereas $\|\Gamma(\mathbf{u})\|_{\ell_1}$ enforces the sparsity of $\Gamma(\mathbf{u})$. In CS-MRI, a priori knowledge of MR images is converted to a sparsity of $\Gamma(\mathbf{u})$ with a suitable choice of Γ. The total variation (TV)-based CS method adopts $\Gamma(\mathbf{u}) = \nabla \mathbf{u}$ to impose a sparsity of the image gradient (i.e., $\|\nabla \mathbf{u}\|_{\ell_1}$), as shown in Fig. 7.5. This TV regularization is widely used for noise reduction and is based on the assumption that noise in an image has high TV. Reducing the total variation $\|\nabla \mathbf{u}\|_{\ell_1}$ subject to $\|\mathbf{u} - \mathbf{A}^{\dagger}\mathbf{s}\|_{\ell_2}^2 \approx 0$ removes unwanted noise while preserving image features such as edges. The minimizer $\mathbf{u}$ in (7.12) can be regarded as a solution of the corresponding Euler-Lagrange equation:

$$-\lambda \nabla \cdot \left(\frac{\nabla \mathbf{u}}{|\nabla \mathbf{u}| + \varepsilon} \right) + (\mathbf{u} - \mathbf{A}^{\dagger}\mathbf{s}) = 0, \quad \varepsilon \approx 0. \tag{7.13}$$

Here, for convenience, we abuse the notations of ∇ and $\mathbf{u}$ to use the discrete and continuous forms at the same time. For a computational point of view, the problem (7.12) can be solved by the iterative shrinkage-thresholding algorithm (ISTA) [2]. Writing $\mathbf{v} = \nabla \mathbf{u}$, the minimization problem (7.12) is equivalent to

$$\mathbf{v} = \underset{\mathbf{v}}{\text{argmin}} \ \frac{1}{2} \|\mathscr{U}\mathbf{v} - \mathbf{u}^{\dagger}\|_{\ell_2}^2 + \lambda \|\mathbf{v}\|_{\ell_1}, \tag{7.14}$$

where $\mathbf{u}^\dagger = \mathbf{A}^\dagger \mathbf{s}$ and $\mathscr{U} : \mathbf{v} \to \mathbf{u}$ is a linear operator solving the Poisson equation $\nabla^2 \mathbf{u} = \nabla \cdot \mathbf{v}$ with a suitable boundary condition. Then, the minimizer $\mathbf{v}$ in (7.14) can be achieved by the following iterative process [24]:

$$\mathbf{v}_{n+1} = \nabla \mathscr{U} \left(\mathscr{S}_{\lambda\alpha} \left\{ \mathbf{v}_n - \alpha \mathscr{U}^*(\mathscr{U} \mathbf{v}_n - \mathbf{u}^\dagger) \right\} \right), \tag{7.15}$$

where α is a step size and $\mathscr{S}_\tau$ is the shrinkage operator given by

$$\mathscr{S}_\tau(\mathbf{v})_i = \mathrm{sign}(v_i)(|v_i| - \tau)_+ \tag{7.16}$$

and $\mathscr{U}^*$ is the adjoint operator of $\mathscr{U}$ given by $\langle \mathscr{U} \mathbf{v}, \mathbf{u} \rangle = \langle \mathbf{v}, \mathscr{U}^* \mathbf{u} \rangle$ for all $\mathbf{u}$ and $\mathbf{v}$. Here, $(\cdot)_+$ is an operator that changes negative values to zero. Then, $\mathbf{u} = \mathscr{U}(\mathbf{v}_n)$ for a large n can be a good approximation of the minimizer in (7.12). Hence, the TV method removes some high oscillations through the shrinkage thresholding without exception. TV might not be able to selectively preserve clinically useful features that are small in terms of $\dfrac{\|\text{small feature}\|_{\ell_2}^2}{\|\nabla \text{small feature}\|_{\ell_1}}$.

Remark 7.1 Let's briefly explain the basic concept of ISTA [2] regarding (7.15). Consider

$$\mathbf{u} = \underset{\mathbf{u}}{\mathrm{argmin}} \underbrace{\frac{1}{2} \|\mathbf{A}\mathbf{u} - \mathbf{b}\|_{\ell_2}^2 + \lambda \|\mathbf{u}\|_{\ell_1}}_{\Phi_\lambda(\mathbf{u})}, \tag{7.17}$$

The following gradient decent method is used to solve (7.14):

- Choose an initial guess $\mathbf{u}_0$.
- For iterations $n = 1, \ldots$, update $\mathbf{u}_{n+1} = \mathbf{u}_n - \alpha_n \nabla \Phi(\mathbf{u}_n)$.

If we perform a quadratic approximation to Φ_0 while leaving the ℓ_1-norm term as is, the above iteration can be written as the following proximal gradient descent:

$$\mathbf{u}_{n+1} = \underset{\mathbf{u}}{\mathrm{argmin}} \left(\Phi_0(u_n) + (\mathbf{u} - \mathbf{u}_n) \cdot \nabla \Phi_0(\mathbf{u}_n) + \frac{1}{2\alpha} \|\mathbf{u} - \mathbf{u}_n\|_{\ell_2}^2 + \lambda \|\mathbf{u}\|_{\ell_1} \right), \tag{7.18}$$

By removing all constant terms, the above expression can be expressed as:

$$\mathbf{u}_{n+1} = \underset{\mathbf{u}}{\mathrm{argmin}} \left(\frac{1}{2} \|\mathbf{u} - (\mathbf{u}_n - \alpha \nabla \Phi_0(\mathbf{u}_n))\|_{\ell_2}^2 + \lambda\alpha \|\mathbf{u}\|_{\ell_1} \right), \tag{7.19}$$

The above expression means that $\mathbf{u}_{n+1} \approx \mathbf{u}_n - \alpha \nabla \Phi_0(\mathbf{u}_n)$ and $\|\mathbf{u}_{n+1}\|_{\ell_1} \approx 0$. Then, a simple computation gives

$$\mathbf{u}_{n+1} = \mathscr{S}_{\lambda\alpha} \left(\mathbf{u}_n - \alpha \nabla \Phi_0(\mathbf{u}_n) \right). \tag{7.20}$$

Compressed sensing is based on the assumption that human MR images can be sparsely represented by a basis $\{\mathbf{d}_k\}_{k=1}^N$; i.e., if $\mathbf{u}$ is a human MR image, then

$$\mathbf{u} = \mathbf{D}\mathbf{h} \quad \text{s.t.} \ \|\mathbf{h}\|_{\ell_0} \ll N, \tag{7.21}$$

where $\mathbf{D}$ is a matrix whose k-th column corresponds to $\mathbf{d}_k$ and $\|\mathbf{h}\|_{\ell_0}$ is the number of non-zero entries of $\mathbf{h}$. The $\{\mathbf{d}_k\}_{k=1}^N$ can be various wavelet bases [10, 23]. With $\Gamma(\mathbf{u}) = \mathbf{D}\mathbf{h}$ in (7.12), CS-MRI is solved by

$$\mathbf{u} = \mathbf{D}\mathbf{h}, \quad \mathbf{h} = \underset{\mathbf{h}}{\text{argmin}} \ \|\mathbf{D}\mathbf{h} - \mathbf{A}^{\dagger}\mathbf{s}\|_{\ell^2}^2 + \lambda\|\mathbf{h}\|_{\ell^1}. \tag{7.22}$$

The compressed sensing can be explained by the concept of *spark of* $\mathbf{A}$, which is the smallest number of linearly dependent columns of $\mathbf{A}$ [9]. Donoho and Elad [9] showed that if $\mathbf{u}$ and $\mathbf{u}'$ satisfy $\mathbf{A}\mathbf{u} = \mathbf{A}\mathbf{u}'$ and $\max(\|\mathbf{u}'\|_0, \|\mathbf{u}\|_0) < \frac{\text{spark}(\mathbf{A})}{2}$, then $\mathbf{u} = \mathbf{u}'$. Here, $\|\mathbf{u}\|_0$ indicates the number of non-zero entries of $\mathbf{u}$. Note that, for $k = 1, 2, 3, \ldots$, the set $\mathscr{A}_k := \{\mathbf{u} : \ \|\mathbf{x}\|_0 \leq k\}$ is the union of all possible k–dimensional linear subspace. The above uniqueness can be explained as follows: For $\mathbf{u}, \mathbf{u}' \in \mathscr{A}_k$ with $k < \frac{\text{spark}(\mathbf{A})}{2}$, $\mathbf{u} \neq \mathbf{u}'$ if and only if $\mathbf{A}\mathbf{u} \neq \mathbf{A}\mathbf{u}'$. This means that $\mathbf{A}\mathbf{u} = \mathbf{s}$ has at most one solution within the restricted set $\mathscr{A}_k$ for $k < \frac{\text{spark}(\mathbf{A})}{2}$. Hence, one may consider to solve the underdetermined problem by minimizing $\|\mathbf{u}\|_0$ subject to the constraint $\mathbf{A}\mathbf{u} = \mathbf{s}$. Unfortunately, ℓ_0-minimization is extremely difficult (NP-hard) due to lack of convexity; we cannot use Newton's iteration.

Donoho and Elad [9] observed that the ℓ_0-minimization problem (non-convex) can be relaxed by the ℓ_1-minimization problem (convex) under a condition related to the *restricted isometry property* (RIP) condition, that was introduced by Candes and Tao [5, 6]. If $\mathbf{A}$ has a RIP of order $2k$, then the underdetermined linear system has "*good-distinguishability*" within the k-sparse set $\mathscr{S}_k$.

However, in medical imaging, this sparse prior (or piece-wise linear regression) may not be appropriate for preserving small features that contain clinically useful information. See a simple denoising example in Fig. 7.6.

$$f_{\text{TV}}^{\lambda}(\mathbf{u}_{\dagger}) = \underset{\mathbf{u}}{\text{argmin}} \ \|\mathbf{u} - \mathbf{u}_{\dagger}\|_{\ell^2}^2 + \lambda\|\nabla\mathbf{u}\|_{\ell^1}$$

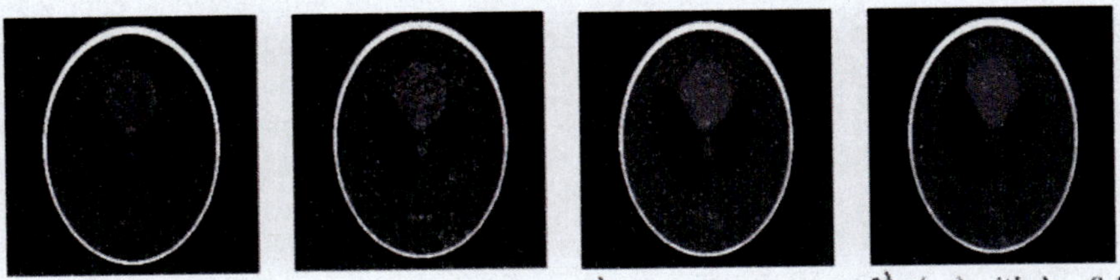

Reference | $\mathbf{u}_{\dagger}$ | $f_{\text{TV}}^{\lambda}(\mathbf{u}_{\dagger})$ with $\lambda = 0.2$ | $f_{\text{TV}}^{\lambda}(\mathbf{u}_{\dagger})$ with $\lambda = 0.4$

Fig. 7.6 TV implementation for a simple denoising problem. The image $\mathbf{u}_{\dagger}$ is generated by adding Gaussian noise to the reference Logan-phantom image. The TV method tends to remove small anomalies and its performance highly depends on choice of the regularization parameter λ

7.2.3 Deep Learning Approach

To solve $\mathbf{Au} = \mathbf{s}$ uniquely, the solution has to be restricted to a low-dimensional manifold $\mathcal{M}$, where all realistic MR images are lying near $\mathcal{M}$. With the use of the solution manifold $\mathcal{M}$, we expect to obtain $f(\mathbf{u}_\dagger)$ by solving the following constraint problem:

$$\begin{cases} \text{Solve } \mathbf{Au} = \mathbf{s} \\ \text{subject to the constraint } \mathbf{u} \in \mathcal{M} \end{cases}. \tag{7.23}$$

Our simplified goal of undersampled MRI reconstruction is to find a function f given by

$$f : \mathbf{u}_\dagger \in \mathcal{M}' \mapsto \mathbf{u} \in \mathcal{M}, \tag{7.24}$$

where

$$\mathcal{M}' := \{\mathbf{u}_\dagger = \mathbf{A}^\dagger \mathbf{Au} \ : \ \mathbf{u} \in \mathcal{M}\}. \tag{7.25}$$

However, $\mathcal{M}$ is unknown. The key issue is to find a generator $\mathfrak{G} : \mathbf{h} \in \mathcal{H} \mapsto \mathbf{u}$ satisfying

$$\mathcal{M} = \{\mathbf{u} : \mathbf{u} = \mathfrak{G}(\mathbf{h}) \text{ and } \mathbf{h} \in \mathcal{H}\} \tag{7.26}$$

and

$$c\|\mathbf{h} - \mathbf{h}'\| \leq \|\mathfrak{G}(\mathbf{h}) - \mathfrak{G}(\mathbf{h}')\| \leq \frac{1}{c}\|\mathbf{h} - \mathbf{h}'\| \quad \text{for some } c \in (0, 1], \tag{7.27}$$

where $\mathcal{H}$ denotes a subset of $\mathbb{R}^{\mathfrak{d}_{\text{mfd}}}$ and $\mathfrak{d}_{\text{mfd}} \ll N$ is the Hausdorff dimension of the manifold. With this manifold constraint, the underdetermined reconstruction problem can be well-posed in the sense that there exist a unique $\mathbf{h} \in \mathcal{H}$ such that $\mathbf{A}\mathfrak{G}(\mathbf{h}) = \mathbf{s}$. A necessary condition for the solvability of $\mathbf{A}\mathfrak{G}(\mathbf{h}) = \mathbf{s}$ is $\mathfrak{d}_{\text{mfd}} \leq N - m$.

Unfortunately, finding the generator $\mathfrak{G}$ can be a very difficult task. The variational autoencoder (VAE) [20] and GANs [12] can be used to find the generator $\mathfrak{G}$ via a training dataset $\{\mathbf{u}^{(k)}\}_{k=1}^{\mathfrak{n}_{\text{data}}}$. Although autoencoder-based approaches show remarkable performance in several applications [8, 19, 32, 33], for high-dimensional data, AEs do not seem to work well, producing blurring and loss of small details. Although GANs [1, 13, 22] have achieved remarkable success in generating various realistic images, there exist some limitations in synthesizing high-resolution medical data. Improving performance of AEs and GANs in high-dimensional medical image applications is still a challenging issue [18].

Since it is very difficult to know the manifold $\mathcal{M}$, one can achieve the reconstruction map f as follows:

$$f \ = \ \underset{f \in \mathbb{NN}}{\text{argmin}} \sum_{k=1}^{\mathfrak{n}_{\text{data}}} \|f(\mathbf{u}_\dagger^{(k)}) - \mathbf{u}^{(k)}\|_{\ell^2}^2, \tag{7.28}$$

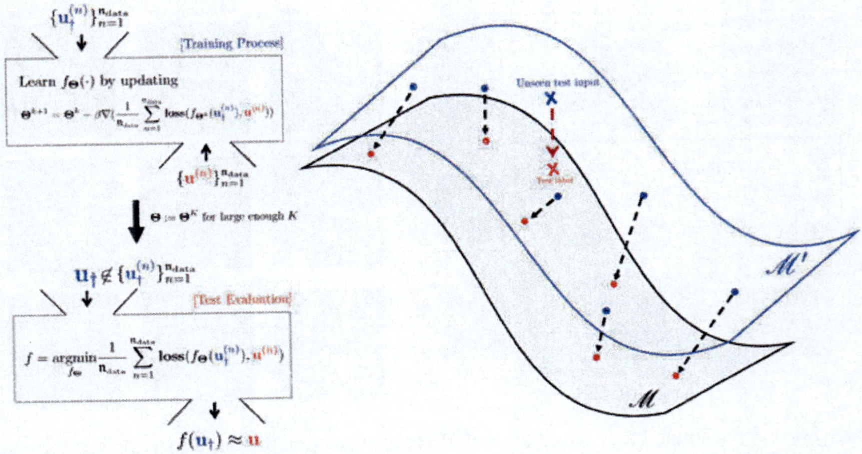

Fig. 7.7 Deep learning framework for learning the reconstruction map $f : \mathbf{u}_\dagger \in \mathcal{M}' \mapsto \mathbf{u} \in \mathcal{M}$. For a given deep learning network and a training dataset $\{\mathbf{u}_\dagger^{(k)}, \mathbf{u}^{(k)}\}_{k=1}^{\mathfrak{n}\text{ data}}$, a reconstruction map f is learned by minimizing the discrepancy between outputs $\{f(\mathbf{u}_\dagger^{(k)})\}_{k=1}^{\mathfrak{n}\text{ data}}$ and the corresponding labels $\{\mathbf{u}^{(k)}\}_{k=1}^{\mathfrak{n}\text{ data}}$

where $\mathbb{NN}$ denotes a set of functions described in a special form of neural network and $\{(\mathbf{u}^{(k)}, \mathbf{u}_\dagger^{(k)}) : k = 1, \dots, \mathfrak{n}_{\text{data}}\}$ is a training data. See Fig. 7.7.

Hyun et al. [18] observed that the subsampling strategy $\mathcal{S}_{\text{sub}}$ is important for the uniqueness of solution $\mathbf{u}$ on the manifold $\mathcal{M}_{\text{image}}$ among all the possible solutions. Precisely, a proper subsampling strategy $\mathcal{S}_{\text{sub}}$ is related to the following manifold restricted isometry property (RIP) condition: The matrix $\mathbf{A}$ associated with $\mathcal{S}_{\text{sub}}$ is said to satisfy the $\mathcal{M}$-RIP condition if there exists a constant $c \in (0, 1]$ such that

$$c\|\mathbf{u} - \mathbf{u}'\| \le \|\mathbf{A}\mathbf{u} - \mathbf{A}\mathbf{u}'\| \le \frac{1}{c}\|\mathbf{u} - \mathbf{u}'\| \text{ for all } \mathbf{u}, \mathbf{u}' \in \mathcal{M}. \tag{7.29}$$

Hyun et al. [18] observed the following:

- If $\mathbf{A}$ satisfies the $\mathcal{M}$-RIP condition in (7.29), then

$$\mathbf{A}^\dagger\mathbf{A} : \mathcal{M} \to \mathcal{M}' \text{ is one-to-one.} \tag{7.30}$$

- The reconstruction map $f : \mathbf{u}_\dagger \in \mathcal{M}' \mapsto \mathbf{u} \in \mathcal{M}$ is learnable if $\mathbf{A}$ satisfies the $\mathcal{M}$-RIP condition (7.29).

Given a highly undersampling operator $\mathcal{S}_{\text{sub}}$, the map f can be viewed as an image restoration function with filling-in missing data or unfolding image data; therefore, f depends on the image structure. The nonlinearity of f is affected by $\mathcal{S}_{\text{sub}}$ and the degree of bending of the manifold $\mathcal{M}$.

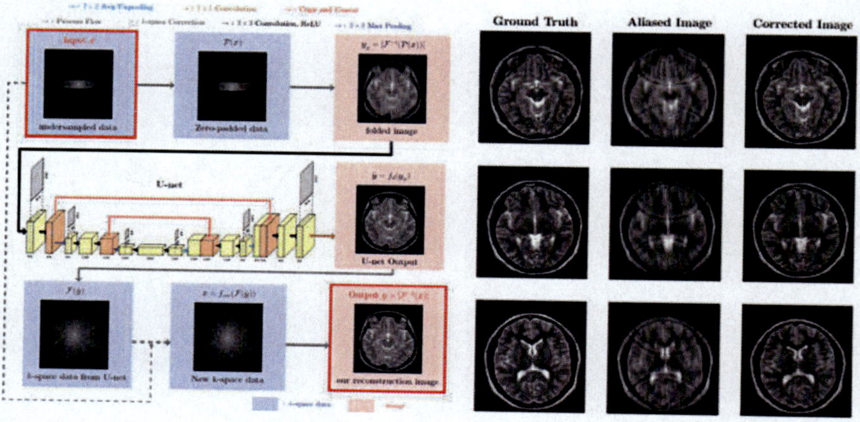

Fig. 7.8 Deep learning-based undersampled MRI reconstruction proposed by [16]. The method consists of two stages: deep learning using U-net and k-space correction. The U-net successfully recovers the zero-padded part of the k-space data. This figure was extracted from [16]

U-net [28] is a deep learning network that can effectively deals with the undersampled MRI reconstruction problem while implicitly learning the underlying data distributions $\mathscr{M}$ and $\mathscr{M}'$ through training data [16–18], as seen in Fig. 7.8. The network architecture comprises a contraction path $\phi : \mathbf{u}_\dagger \mapsto \mathbf{h}$ and an expansion path $\psi : \mathbf{h} \mapsto \mathbf{u}$, so that the reconstruction f is given by $f(\mathbf{u}_\dagger) = \psi \circ \phi(\mathbf{u}_\dagger)$. At the first layer in U-net, the input $\mathbf{u}_\dagger$ is convolved with the set of convolution filters θ_1 and added by bias $\mathbf{c}_1$ so that it generates a set of feature maps $\mathbf{h}_1$, given by

$$\mathbf{h}_1 = \mathrm{ReLU}(\theta_1 \circledast^1 \mathbf{u}_\dagger + \mathbf{c}_1), \tag{7.31}$$

where ReLU is the rectified linear unit defined pixelwisely by $\mathrm{ReLU}(p) = \max\{p, 0\}$ and $\circledast^1$ stands for the convolution with stride 1. We repeat this process to get $\mathbf{h}_2 = \mathrm{ReLU}(\theta_2 \circledast^1 \mathbf{h}_1 + \mathbf{c}_2)$ and apply max pooling to get $\mathbf{h}^{(3)}$. Through this contracting path, we can obtain low-dimensional feature maps by applying either convolution or max pooling. In the expansive path, we use the 2×2 average unpooling instead of max-pooling to restore the size of the output. To restore details in image, the upsampled output is concatenated with the correspondingly feature from the contracting path. At the last layer a 1×1 convolution without ReLU activation function is used to combine each feature with one integrated feature.

With U-net, the reconstruction map $f : \mathbf{u}_\dagger \mapsto \mathbf{u}$, as a function of $\Theta = \{\theta_1, \mathbf{c}_1, \theta_2, \mathbf{c}_2, \ldots\}$, is learned as follows:

$$f = \underset{f_\Theta}{\operatorname{argmin}} \frac{1}{\mathbf{n}_{\text{data}}} \sum_{k=1}^{\mathbf{n}_{\text{data}}} \| f_\Theta(\mathbf{x}^{(k)}) - \mathbf{y}^{(k)} \|_{\ell^2}^2. \tag{7.32}$$

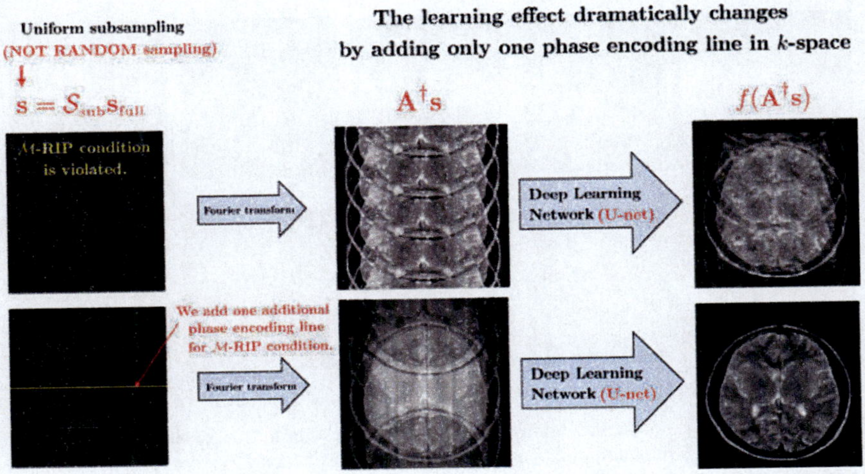

Fig. 7.9 (Top) Uniform subsampling fails to satisfy M-RIP condition. DL is *NOT* a magic. There exist location uncertainties under the uniform subsampling. This is the main reason why f is not learnable under the uniform subsampling. (Bottom) Learning effect dramatically changes by adding only one phase encoding line to a uniformly subsampled k-space data

The trained reconstruction map f shows remarkable performance in the undersampled MRI reconstruction problem, as shown in Fig. 7.8 [16–18].

An interesting observation in [16] is that the reconstruction map f with U-net is not learnable under uniform sampling, but adding only one low-frequency phase encoding line can lead the significant improvement on learning the map f as shown in Fig. 7.9. Hyun et al. [18] analyzed the reason by means of the $\mathscr{M}$-RIP condition in (7.29).

The basic idea of the analysis is as follows: Fig. 7.10b and d show $u_\dagger$ images obtained by strict uniform subsampling with factors 2 and 4, respectively. Using only these images, it is not possible to identify whether the anomaly is at the top or bottom. This location uncertainty can be understood as the violation of $\mathscr{M}$-RIP condition and, therefore, the reconstruction map f is not learnable. On the other hand, Fig. 7.10c and e show $u_\dagger$ images obtained by uniform subsampling with factors 2 and 4 + few low frequency phase encoding lines. The additional low-frequency lines allow us to deal with the location uncertainty, which can be understood as preventing the violation of $\mathscr{M}$-RIP condition. Detailed analysis would be provided in the following section.

7.2.3.1 Uniform Subsampling

For ease of explanation, the image size is assumed to be $N \times N$, where N is a multiple of 4. According to the Poisson summation formula, the discrete Fourier transform

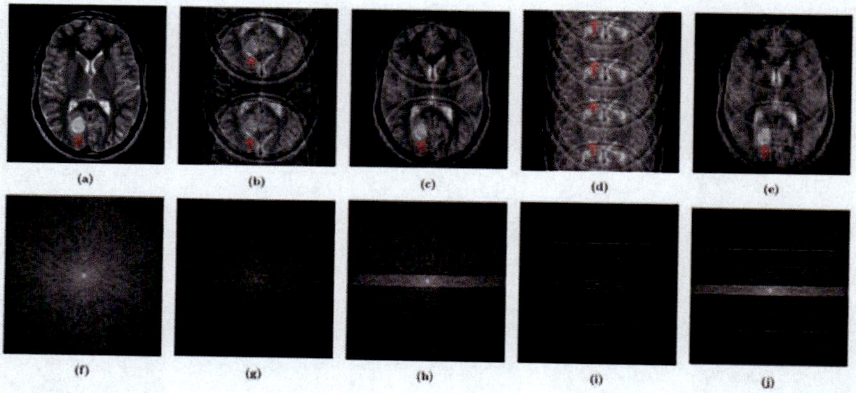

Fig. 7.10 MR images of human brain with a tumor at the bottom. Images **a**, **b**, **c**, **d**, and **e** are reconstructed from **f** full sampling, **g** uniform subsampling of factor 2, **h** uniform subsampling of factor 2 with added some low frequencies, *i* uniform subsampling of factor 4, and **j** uniform subsampling of factor 4 with added low frequencies, respectively. In **b** and **d**, tumor-like lesions are found at both the top and bottom; one is a copy of the other. Hence, there exists a location uncertainty in the uniform sampling. However, in the reconstructed image **c** and **e** using the uniform subsampling of factor 2 and 4 with added low frequencies, the tumors are clearly located at the bottom. The location uncertainty can hence be addressed by adding a few low frequencies in k-space. This figure was extracted from [16]

of the uniformly subsampled data with factor 4 provides an four-folded image $\mathbf{u}_\dagger$ as follows [29]:

$$\mathbf{u}_\dagger(x, y) = \mathbf{A}^\dagger \mathbf{A}\mathbf{u} = \frac{1}{4} \sum_{y' \equiv y \pmod{\frac{N}{4}}} \mathbf{u}(x, y'), \tag{7.33}$$

where $y' \equiv y \pmod{\frac{N}{4}}$ represent that both y and y' leave the same remainder when divided by $\frac{N}{4}$. Unfortunately, there exists an uncertainty that makes it impossible to reconstruct $\mathbf{u}$ from $\mathbf{u}_\dagger$, and therefore the reconstruction map f is not learnable. To investigate the reason, we define the following:

$$\Psi_{\text{ufm}} := \mathscr{N}_0(\mathbf{A}^\dagger \mathbf{A}) = \text{Span}\{\psi^{0,\eta}_{x_*,y_*} : x_*, y_* \in \mathbb{Z}_n, \eta = 1, 2, 3\}, \tag{7.34}$$

where $\mathbb{Z}_n := \{1, \ldots, N\}$ for any positive integer N and $\psi^{0,\eta}_{x_*,y_*}$ is given by

$$\psi^{0,\eta}_{x_*,y_*}(x, y) = \begin{cases} 1 & \text{if } (x, y) = (x_*, y_*), \\ -1 & \text{if } (x, y) = (x_*, y_*) + (0, \frac{n}{4}\eta), \\ 0 & \text{otherwise.} \end{cases} \tag{7.35}$$

Here, $y_* + \frac{\sqrt{l}}{4}\eta$ should be understood as modulo n.

Theorem 1 (Hyun et al. [18]) *There exists a non-zero* $\psi \in \Psi_{ufm}$ *and* $\mathbf{u} \in \mathscr{M}_{image}$ *such that* $\mathbf{u} + \psi \in \mathscr{M}_{image}$.

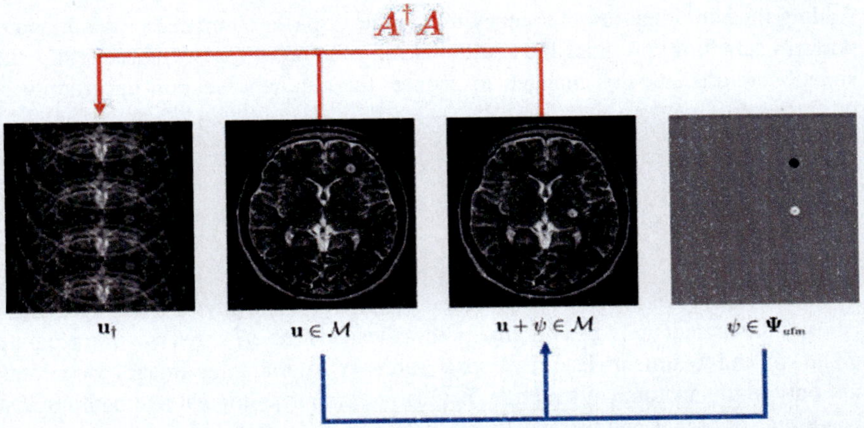

Fig. 7.11 Location uncertainty when using a uniform subsampling with factor 4. Let us consider two different MR images ($\mathbf{u}$ and $\mathbf{u} + \psi$), where only the small anomaly location is different. If we take $\mathbf{A}^\dagger\mathbf{A}$ to the images, the same outputs $\mathbf{u}_\dagger = \mathbf{A}^\sharp\mathbf{A}\mathbf{u} = \mathbf{A}^\sharp\mathbf{A}(\mathbf{u} + \psi)$ are produced

The observation implies that the $\mathcal{M}$-RIP condition does not hold, as f requires the following contradictory two conditions $f(\mathbf{u}_\dagger) = \mathbf{u}$ and $f(\mathbf{u}_\dagger) = \mathbf{u} + \psi$, where $\mathbf{u}_\dagger = \mathbf{A}^\dagger\mathbf{A}\mathbf{u} = \mathbf{A}^\dagger\mathbf{A}(\mathbf{u} + \psi)$. The location of a small anomaly cannot be determined, and, therefore, there are many location uncertainties under the uniform subsampling, as shown in Fig. 7.11. This is the main reason why f is not learnable under the uniform subsampling.

7.2.3.2 Uniform Sampling with Adding One Phase Encoding Line

Let $\mathscr{S}_{\text{sub}}$ be the uniform subsampling of factor 4 upon adding one phase encoding line. Then, $\mathbf{u}_\dagger = \mathbf{A}^\dagger \mathbf{s}$ can be decomposed into two parts:

$$\mathbf{u}_\dagger(x, y) = (\mathbf{u}_\dagger)_1(x, y) + (\mathbf{u}_\dagger)_2(x, y), \tag{7.36}$$

where $(\mathbf{u}_\dagger)_1$ is the uniform sampling part given by

$$(\mathbf{u}_\dagger)_1(x, y) := \frac{1}{4} \sum_{y' \equiv y \pmod{\frac{N}{4}}} \mathbf{y}(x, y') \tag{7.37}$$

and $(\mathbf{u}_\dagger)_2$ is the single phase encoding part given by

$$(\mathbf{u}_\dagger)_2(x, y) := \sum_{y' \in \mathbb{Z}_N} \mathbf{y}(x, y') e^{2\pi i (y-y')\Delta k}. \tag{7.38}$$

Adding the additional low-frequency line in the k-space (compared to the previous uniform sampling) provides the additional information of $(\mathbf{u}_{\dagger})_2$. Subsequently, the situation is dramatically changed to counter the anomaly-location uncertainty in uniform sampling.

7.3 Discussion

Medical imaging has evolved to visualize the anatomical and physiological characteristics of the human body in a minimally invasive and cost-effective way for accurate diagnosis and treatment. The pixel/voxel values of tomographic images are related not only to the anatomical structure, but also to the physiological and pathological conditions of organs and tissues.

When developing a new medical imaging technique that has not yet been commercialized, it is necessary to comprehensively consider usability, image quality limitations, data acquisition, operating costs, convenience, non-invasiveness, and others. As long as there are no concerns about the radiation dose exposure and CT equipment cost, it is not difficult to develop high-resolution CT imaging systems. Similarly, if there are no concerns about MR data acquisition time, motion artifacts, and MRI scanner cost, it is not difficult to develop high-resolution MRI imaging systems. If there were no concerns about the amount of electric current injected into the human body and the operating cost, MREIT, which provides high-resolution electrical conductivity distribution, would have already been commercialized and used in clinical practice. If we accept the fundamental imaging limitations of EIT systems (i.e., the very low sensitivity of data to local conductivity perturbations inside the human body) and can easily attach multiple electrodes to the human body surface in just a few seconds, then EIT can already be well used in clinical practice. In MR and ultrasonic elastography, robustness and repeatability may not be an issue if the uncertainties of stress and strain measurements are well addressed. Lessons from three decades in medical imaging show that mathematical methodologies built only on ideal situations may be inappropriate for clinical practice and, in the worst case, can be obstacles to achieving the ultimate goal in some practical aspects.

In the early stages of research, it would be desirable to conduct studies under ideal conditions for the development of basic mathematical theories. However, if researchers want to achieve the ultimate goal, they should try to undertake studies in real, rather than ideal, situations as soon as possible. As can be seen from the studies of the past several decades, if theory-centered research is conducted for a long period without experimental verification, a sense of scientific balance is lost, and technique-centered research is conducted rather than validity. A researcher who ignores the importance of practicality and focuses on the theory itself can be easily exposed to the danger of falling into self-satisfaction, boasting only genius and eloquence. The real combat-focused training of UFC fighters seems to apply to our scientific community as well. When we get hit in the face and feel what the pain is, we can effectively correct our own problems and actively learn new skills. This hands-on training will enable us to be creative and respond appropriately when attacked. On the other hand, like traditional martial arts, training in basic posture, attack, and defense every day in front of a tree doesn't seem appropriate in these rapidly changing times. Training and learning should be gained by interacting with others, not alone.

The recent development of deep learning in medical imaging means that research results do not stay in the laboratory, but contribute greatly to reaching the level of practical use. The advance of deep learning seems to be due to the rapid development of high-performance GPU computing and big data analysis/utilization technology and the active investment of big companies. In the medical imaging field, the reliability of deep learning technology will be improved as high-quality training data are accumulated. Many experiments have shown that well-trained neural networks work only in the immediate vicinity of the regression manifold generated from the training data. Even if two images are almost the same from the viewpoint of radiologists, deep neural networks may produce different results, because they are vulnerable to various noise-like perturbations. Hence, normalizing data is an important part of improving a network's generalization ability (by enhancing out-of-distribution robustness), but it can be very challenging. Data normalization and standardization can reduce diversity in images caused by variation among scanners or imaging protocols [18].

Despite numerous media reports and papers on deep learning, many doctors in the medical community believe that deep learning will only be used in very limited medical business environments. AI algorithms should be explainable and transparent in order that doctors can backtrack AI diagnosis. AI algorithms should be properly configured to reduce black box prediction as much as possible.

In ancient Egypt within a time period ranging from 3000 BC to 300 BC, the Aha problem $\frac{3}{2}x + 4 = 10$ was solved using a kind of iterative method, called the method of false position: Choose a guessed answer and iteratively adjust the answer to get the correct answer. This is because Egyptians at that time did not have a mathematical way to directly find the solution $x = \left(\frac{3}{2}\right)^{-1}(10-4)$. Similarly, many ill-posed nonlinear problems have been solved using complex iterative methods over decades, and the answers are very different from the true ones. This seems to be because we don't know how to handle the nonlinear solution space with low-dimensional manifolds. The famous debate between Riemann (1826 $\sim$ 1866) and Weierstrass (1815 $\sim$ 1897) regarding the convergence issue of the minimization problem in Dirichlet's principle was the absence of Sobolev space (as a proper solution space) and Rellich's compactness at that time. We should note that well-known well-posed problems (e.g., Dirichlet problem for the Laplace equation) can be ill-posed if suitable solution spaces are not defined. For example, it is well-known that $u = 0$ is the unique solution of the Laplace equation $\nabla \cdot \nabla u = 0$ in the 2D domain $\Omega = \{(r\cos\theta, r\sin\theta) \mid 0 < r < 1,\ 0 < \theta < \frac{3\pi}{2}\}$ with the homogeneous Dirichlet boundary condition $u|_{\partial\Omega} = 0$. However, without the constraint of the Sobolev space $H^1(\Omega) = \{u \mid \int_\Omega |u|^2 + |\nabla u|^2 < \infty\}$ in the solution space, there exist infinitely many solutions to the Dirichlet problem such as $u = (r^{\frac{2}{3}n} - r^{\frac{-2}{3}n})\sin(\frac{2}{3}\theta)$ for $n = 0, 1, 2, \ldots$ [29]. In other words, without the constraint of $u \in H^1(\Omega)$, we cannot define

$$u = \underset{u \text{ s.t. } u|_{\partial\Omega}=0}{\text{argmin}} \int_{\Omega} |\nabla u|^2. \tag{7.39}$$

Before the twentieth century, the Hilbert space $H^1(\Omega)$ and measure theory had not been introduced; thus, there was insufficient knowledge to validate Dirichlet's principle in a rigorous way. This issue of ascertaining the existence of a minimizer might have been a possible motivation for the development of the notion of compactness [31] including the Rellich-Kondrachov theorem [21, 27]. Significantly reducing the solution space is the key to solving the problem.

Similarly, many ill-posed problems $\mathbf{Au} = \mathbf{s}$ may remain unresolved because we do not have a reduced solution space (i.e., the proper low-dimensional solution manifold). The current mathematical system uses too large a solution space (Euclidean space), so there are too many solutions. A medical image (256 gray-scale level, 256×256 size) can be regarded as a point $\mathbf{x} = (x_1, \ldots, x_{256^2})$ in pixel-dimensional Euclidean space, where x_j (i.e., jth axis coordinate) corresponds to the gray-scale intensity at the jth. Hence, the number of all possible images is $256^{256\times256}$. Even if we collect one million tomographic images every day for 100 years, the total number is far less than $256^{4\times4}$, the number of all possible 256 gray-scale level images having 4×4 pixels. Hence, even this massive collection of tomography images occupies a much smaller area than 0.000000000000001% of the total images. Hence, more than 99.99999% of the images look like noise images, and likewise most solutions of $\mathbf{Au} = \mathbf{s}$ will look like noise. We cannot solve the ill-posed problem $\mathbf{Au} = \mathbf{s}$ without significantly reducing the solution space. Since DL methods have an excellent ability to learn data representations, they may have the ability to generate low-dimensional manifolds.

Acknowledgements This research was supported by Samsung Science & Technology Foundation (No. SRFC-IT1902-09). Seo was supported by a grant of the Korea Health Technology R&D Project through the Korea Health Industry Development Institute (KHIDI), funded by the Ministry of Health & Welfare, Republic of Korea (grant number : HI20C0127).

References

1. Arjovsky, M., Chintala, S., Bottou, L.: Wasserstein gan (2017). arXiv:1701.07875
2. Beck, A., Teboulle, M.: A fast iterative shrinkage-thresholding algorithm for linear inverse problems. SIAM J. Imag. Sci. **2**(1), 183–202 (2009)
3. Bayaraa, T., Hyun, C.M., Jang, T.J., Lee, S.M., Seo, J.K.: A two-stage approach for beam hardening artifact reduction in low-dose dental CBCT. IEEE Access (2020)
4. Bruckstein, A.M., Donoho, D.L., Elad, M.: From sparse solutions of systems of equations to sparse modeling of signals and images. SIAM Rev. **51**, 3481 (2008)
5. Candes, E.J., Tao, T.: Decoding by linear programming. IEEE Trans. Inf. Theory **51**, 4203–4215 (2005)
6. Candes, E.J., Romberg, J., Tao, T.: Robust Uncertainty Principles: exact signal reconstruction from highly incomplete frequency information. IEEE Trans. Inf. Theory **52**, 489–509 (2006)
7. Candes, E.J., Tao, T.: Reflections on compressed sensing. IEEE Inf. Theory Soc. Newsl. **58**, 20–23 (2008)

8. Chang, J.H.R., Li, C., Poczos, B., Kumar, B.V.K.V., Sankaranarayanan, A.C.: One network to solve them all solving linear inverse problems using deep projection models. In: 2017 IEEE International Conference on Computer Vision (ICCV) , pp. 5889–5898 (2017)
9. Donoho, D.L., Elad, M.: Optimally sparse representation in general (nonorthogonal) dictionaries via ‘ 1 minimization. Proc. Natl. Acad. Sci. U.S.A. **100**, 2197–2202 (2003)
10. Daubechies, I., Defrise, M., De Mol, C.: An iterative thresholding algorithm for linear inverse problems with a sparsity constraint. Commun. Pure Appl. Math. **57**, 1413–1457 (2004)
11. Donoho, D.L.: Compressed sensing. IEEE Trans. Inf. Theory **52**, 1289–1306 (2006)
12. Goodfellow, I., Pouget-Abadie, J., Mirza, M., Xu, B., Warde-Farley, D., Ozair, S., Courville, A., Bengio, Y.: Generative adversarial nets. Adv. Neural. Inf. Process. Syst. **27**, 2672–2680 (2014)
13. Gulrajani, I., Ahmed, F., Arjovsky, M., Dumoulin, V., Courville, A.: Improved training of Wasserstein gans (2017). arXiv:1704.00028
14. Hadamard, J.: Sur les problemes aux derivees partielles et leur signification physique. Bull. Univ. Princeton **13**, 49–52 (1902)
15. Haacke, E., Brown, R., Thompson, M., Venkatesan, R.: Magnetic Resonance Imaging Physical Principles and Sequence Design. Wiley, New York (1999)
16. Hyun, C.M., Kim, H.P., Lee, S.M., Lee, S., Seo, J.K.: Deep learning for undersampled MRI reconstruction. Phys. Med. Biol. **63**(13) (2018)
17. Hyun, C.M., Kim, K.C., Cho, H.C., Choi, J.K., Seo, J.K.: Framelet pooling aided deep learning network: the method to process high dimensional medical data. Mach Learn.: Sci. Technol. **1**, 015009 (2020)
18. Hyun, C.M., Baek, S.H., Lee, M., Lee, S.M., Seo, J.K.: Deep learning-based solvability of underdetermined inverse problems in medical imaging. Med, Image Anal (2021)
19. Jalali, S., Yuan, X.: Using auto-encoders for solving ill-posed linear inverse problems (2019). arXiv:1901.05045
20. Kingma, D.P., Welling, M.: Auto-encoding variational bayes (2013). arXiv:1312.6114
21. Kondrachov, W.: Certain properties of functions in the space L^p. Dokl. Akad. Nauk SSSR **48**, 535–538 (1945)
22. Karras, T., Aila, T., Laine, S., Lehtinen, J.: Progressive growing of GANs for improved quality stability, and variation. ICLR (2018)
23. Mallat, S.G.: A Wavelet Tour of Signal Processing. Academic (2009)
24. Michailovich, O.V.: An iterative shrinkage approach to total-variation image restoration. IEEE Trans. Image Process. **20**(5), 1281–1299 (2010)
25. Lustig, M., Donoho, D.L., Pauly, J.M.: Sparse MRI: the application of compressed sensing for rapid MR imaging. Magn. Reson. Med. **58**, 1182–1195 (2007)
26. Park, H.S., Baek, J., You, S.K., Choi, J.K., Seo, J.K.: Unpaired image denoising using a generative adversarial network in X-ray CT. IEEE Access (2019)
27. Rellich, F.: Ein Satz über mittlere Konvergenz, Göttingen Nachr. Acta Math. **141**, 165–186 (1930)
28. Ronneberger, O., Fischer, P., Brox, T.: U-net: Convolutional networks for biomedical image segmentation. In: Proceedings of the Medical Image Computing and Computer Assisted Intervention, pp. 234–241. Springer (2015)
29. Seo, J.K., Woo, E.J.: Nonlinear Inverse Problems in Imaging. Wiley, Chichester (2013)
30. Seo, J.K., Zorgati, H.: Compactness and Dirichlet’s principle. J Korean Soc. Indust. Appl. Math. **18**(2), 193–207 (2014)
31. Seo, J.K., Woo, E.J., Katscher, U., Wang, Y.: Electro-Magnetic Tissue Properties MRI. Imperial College Press (2014)
32. Seo, J.K., Kim, K.C., Jargal, A., Lee, K., Harrach, B.: A learning-based method for solving ill-posed nonlinear inverse problems: a simulation: a simulation study of Lung EIT. SIAM J, Imaging Sci (2019)
33. Tezcan, K.C., Baumgartner, C.F., Luechinger, R., Pruessmann, K.P., Konukoglu, E.: MR image reconstruction using deep density priors. IEEE Trans. Med, Imaging (2018)